NUCLEAR MAGNETIC RESONANCE SPECTROSCOPY IN MOLECULAR BIOLOGY

THE JERUSALEM SYMPOSIA ON QUANTUM CHEMISTRY AND BIOCHEMISTRY

Published by the Israel Academy of Sciences and Humanities, distributed by Academic Press (N.Y.)

1st JERUSALEM SYMPOSIUM: *The Physicochemical Aspects of Carcinogenesis* (October 1968)
2nd JERUSALEM SYMPOSIUM: *Quantum Aspects of Heterocyclic Compounds in Chemistry and Biochemistry* (April 1969)
3rd JERUSALEM SYMPOSIUM: *Aromaticity, Pseudo-Aromaticity, Antiaromaticity* (April 1970)
4th JERUSALEM SYMPOSIUM: *The Purines: Theory and Experiment* (April 1971)
5th JERUSALEM SYMPOSIUM: *The Conformation of Biological Molecules and Polymers* (April 1972)

Published by the Israel Academy of Sciences and Humanities, distributed by D. Reidel Publishing Company (Dordrecht, Boston and London)

6th JERUSALEM SYMPOSIUM: *Chemical and Biochemical Reactivity* (April 1973)

Published and distributed by D. Reidel Publishing Company (Dordrecht, Boston and London)

7th JERUSALEM SYMPOSIUM: *Molecular and Quantum Pharmacology* (March/April 1974)
8th JERUSALEM SYMPOSIUM: *Environmental Effects on Molecular Structure and Properties* (April 1975)
9th JERUSALEM SYMPOSIUM: *Metal-Ligand Interactions in Organic Chemistry and Biochemistry* (April 1976)
10th JERUSALEM SYMPOSIUM: *Excited States in Organic Chemistry and Biochemistry* (March 1977)

VOLUME 11

NUCLEAR MAGNETIC RESONANCE SPECTROSCOPY IN MOLECULAR BIOLOGY

PROCEEDINGS OF THE ELEVENTH JERUSALEM SYMPOSIUM ON QUANTUM CHEMISTRY AND BIOCHEMISTRY HELD IN JERUSALEM, ISRAËL, APRIL 3-7, 1978

Edited by

BERNARD PULLMAN

Université Pierre et Marie Curie (PARIS VI)
Institut de Biologie Physico-Chimique
(Fondation Edmond de Rothschild), Paris, France

D. REIDEL PUBLISHING COMPANY

DORDRECHT : HOLLAND / BOSTON : U.S.A.
LONDON : ENGLAND

Library of Congress Cataloging in Publication Data

Jerusalem Symposium on Quantum Chemistry and Biochemistry, 11th, 1978.
Nuclear magnetic resonance spectroscopy in molecular biology.

(The Jerusalem symposia on quantum chemistry and biochemistry ; v. 11)
Includes bibliographical references and index.
1. Nuclear magnetic resonance spectroscopy–Congresses. 2. Molecular biology–Technique–Congresses. I. Pullman, Bernard, 1919- II. Title. III. Series.
QH324.9.N8J47 1978 574.1'92'028 78-13426
ISBN 90-277-0932-7

Published by D. Reidel Publishing Company,
P.O. Box 17, Dordrecht, Holland

Sold and distributed in the U.S.A., Canada, and Mexico
by D. Reidel Publishing Company, Inc.
Lincoln Building, 160 Old Derby Street, Hingham,
Mass. 02043, U.S.A.

Printed in The Netherlands

PREFACE

The 11th Jerusalem Symposium continued the tradition of the pleasant and exciting meetings which once a year gather distinguished scientists, the world's most renowned experts in a specific field of quantum chemistry and biochemistry, in the impressive surroundings of the Israel Academy of Sciences and Humanities.

We wish to thank all those who made this meeting possible and contributed to its success: the Baron Edmond de Rothschild whose continuous generosity guarantees the perenniality of our venture, the Israel Academy of Sciences and Humanities and in particular its President Professor A. Dvoretsky, the Hebrew University of Jerusalem, and Professor David Ginsburg, for their devoted collaboration. Mrs. Abigail Hyam must be thanked for her decisive contribution to the efficiency and success of the local arrangements.

Bernard Pullman

TABLE OF CONTENTS

P-31 NMR OF NUCLEIC ACIDS. BOND ANGLE AND TORSIONAL EFFECTS

DAVID G. GORENSTEIN
Department of Chemistry, University of Illinois, Chicago Circle
Chicago, Illinois 60680, U.S.A.

1. INTRODUCTION

The P-31 Nucleus has proven to be an extremely useful n.m.r. probe of phosphates in biochemical systems. In order to maximize the amount of information obtainable from a study of P-31 n.m.r. spectra, we must be able to assess the extent to which structural and environmental changes perturb P-31 chemical shifts. In this paper we wish to first discuss those factors which alter P-31 chemical shifts and then to use this information as a probe of the structure of nucleic acids.

We have recently described an empirical correlation between P-31 chemical shifts and O-P-O bond angles in phosphates (1). As shown in Figure 1, for a wide variety of different alkyl phosphates (mono-, di-, and triesters, cyclic and acyclic, neutral, monoanionic, and dianionic structures) a decrease in the smallest O-P-O bond angle in the molecule results in a deshielding of the P-31 nucleus. The correlation possibly provides an explanation for the unusual downfield shift observed upon ionization of an acyclic monoanion. The ca 3° reduction in the O-P-O bond angle of the dianionic phosphate is consistent with the 4 ppm downfield shift. Charge alone is not responsible for the deshielding since the acyclic monoanion and free acid have similar chemical shifts (and significantly, similar O-P-O bond angles).

Pursuing the theoretical foundation for this bond angle effect we have indeed been able to calculate a bond angle sensitivity to the P-31 shifts. Most significantly, however, these calculations also revealed a torsional angle sensitivity to these shifts. These semi-empirical molecular orbital, chemical shift calculations (see details in reference 2) suggested that a phosphate diester in a gauche,gauche (g,g) conformation (structure I Figure 2) should have a P-31 chemical shift substantially upfield (by at least several ppm) from a phosphate diester in a gauche, trans conformation (g,t; structure II Figure 2)

B. Pullman (ed.), Nuclear Magnetic Resonance Spectroscopy in Molecular Biology, 1-15.

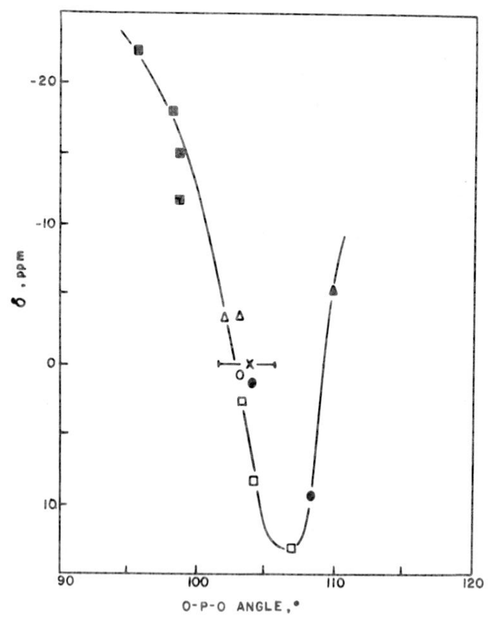

Figure 1. ^{31}P chemical shift of phosphate esters vs. O-P-O bond angle (■, five-membered cyclic esters; Δ, monoester dianions; x, acyclic monoester monoanions; O, acyclic diester monoanions; ●, diester free acids; □, six-membered cyclic esters; ▲, Li_3PO_4). Solid line has no theoretical significance.

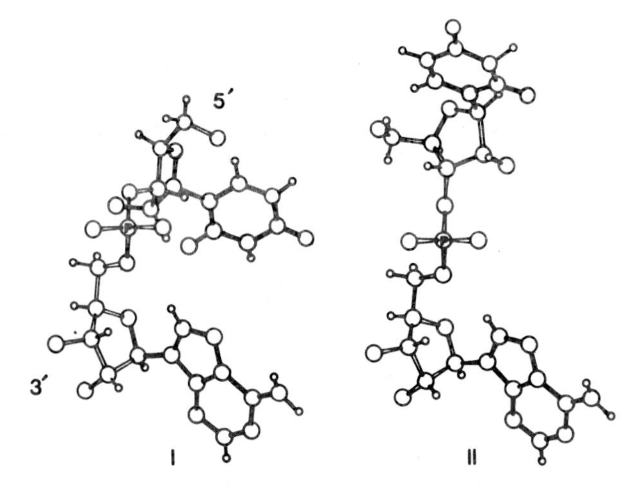

Figure 2. Structure of a dinucleoside monophosphate, UpA in the gauche, gauche, I, and gauche, trans, II, phosphate ester conformations. From Sundaralingam (21).

Actually the bond angle effect and torsional angle effect are not unrelated phenomena since Gorenstein, et al (3-5) and Perahia and Pullman (6) have established a coupling of phosphate ester bond angles with these torsional angles. As shown in Figure 3 a bond angle/ torsional angle energy surface for dimethyl phosphate was computed with the CNDO molecular orbital method (see computational details in

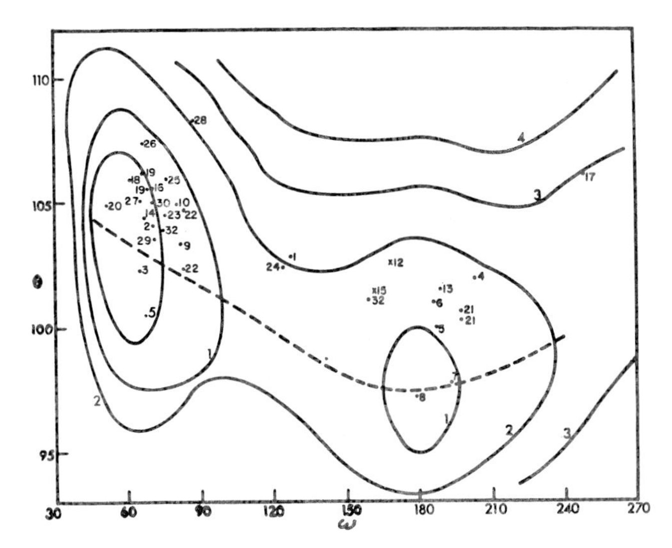

Figure 3. Bond angle (θ)-torsional angle (ω) countour map of dimethyl phosphate monoanion. The other dihedral angle was fixed at +60°. Isoenergy contours are in kcal/mol over the lowest energy geometry ($\omega = \omega' \sim 60°$, $\theta \sim 103.4°$). Points (●) and numbers refer to ester geometries.

refs 3-5) and demonstrates that rotation about the P-O ester bond produces significant bond angle distortion. Rotation about this bond from a gauche to a trans conformation results in ca a 5° reduction in the RO-P-OR bond angle θ. Thus the CNDO optimized angle in dimethyl phosphate is 103.4° for the g,g conformation, 97.5° for the g,t conformation, and 92.0° for the t,t conformation. Other molecules such as neutral phosphate triesters and acetals behave similarly and these distortions are confirmed by ab initio molecular orbital calculations and structural comparisons from X-Ray diffraction studies (3-6, and points in Figure 3). The coupling of these two geometric parameters is most significant for our P-31 studies since it means that a perturbation in the P-31 chemical shift of a phosphate ester will generally describe a single structural change. Thus, ambiguity over the origin of a given P-31 shift perturbation is often eliminated.* What experimental evidence can we provide to support this hypothesis?

In a number of six-membered ring model systems in which the torsional angles are rigidly defined by some molecular constraint, such as the two diastereomeric phosphate triesters 1 and 2

---

*Our proposal that P-31 chemical shifts serve as a probe of torsional geometries is made even more meaningful when one considers that other possible factors that could affect P-31 shifts in phosphate esters apparently are generally quite unimportant (7,8). Thus, we have recently noted that interaction of a mono- or dianionic phosphate monoester with strong hydrogen bonding donors (the active site of ribonuclease A) results in small P-31 chemical shift perturbations.

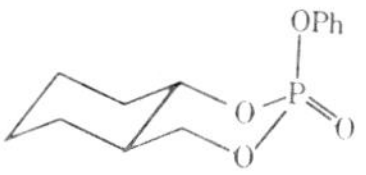

1 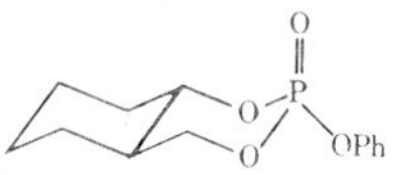 2

the esters with an axial P-OR bond have P-31 chemical shifts 2-6 ppm upfield from the esters with an equatorial P-OR bond (see also references in 8). In 1 and 2 a chair to chair ring flip is prevented by the trans decalin-type ring junction and the conformation about the endocyclic ester bonds relative to the exocyclic bond is gauche for the axial ester and trans for the equatorial ester. Thus substantial deshielding is observed in the trans relative to the gauche conformation, as predicted by the calculations.

2. P-31 CHEMICAL SHIFTS OF SINGLE-STRANDED NUCLEIC ACIDS

As a further test of our P-31 shift-torsional angle hypothesis we have studied the temperature dependence to the P-31 chemical shifts of a number of dinucleoside monophosphates, a dinucleotide (hereafter, both referred to generally as dimers) and homopoly-ribonucleic acids. It has been established mainly through ORD, CD, uv/vis absorption, and H-1 nmr spectroscopy that these single-stranded nucleic acids undergo a "helix-coil" transition with a characteristic melting temperature representing the temperature at which half the nucleic acid is in a base stacked, helix conformation and the other half is in a random coil, largely unstacked conformation (see Ts'o, reference 9, for example). Since the phosphate ester conformation in the helical state is g,g (10 and Figure 2), it was hoped that upon raising the temperature of these phosphate diesters, a shift to lower field would occur, indicating an increase in the proportion of non-gauche conformations.

As shown in Figures 4-6 and as contained in reference 8, the proposed shift in the P-31 signal of <u>all</u> dimers and polyribonucleic acids without exception is observed, with a .13-.64 ppm downfield shift, occurring with increasing temperature. (Experimental details are contained in 8.) These shifts are <u>not</u> an intrinsic property of all phosphate diesters or an experimental artifact since simple model systems fail to show this large downfield shift with increasing temperature. Thus, shown in Figure 7 is the temperature dependence to the P-31 chemical shift of dimethyl phosphate, diethyl phosphate, debenzyl phosphate and 2',3'-cyclic cytidine phosphate (cCMP) (pH 7, 10^{-3} $\underline{M}$ EDTA). In contrast to the dimeric and polymeric nucleic acids, the dimethyl and diethyl phosphates do now show any significant P-31 chemical shift variation with temperature (in fact, there may be a slight ca. .04 ppm upfield shift with increasing temperature for the dimethyl phosphate). It is especially significant that the P-31 shift with temperature for cCMP is so dramatically different from that of the acyclic diesters, shifting .5 ppm upfield with increasing temperature. The cyclic mononucleotide, of course, contains a rigidly constrained five-membered cyclic, monoanionic, phosphate diester. Since phosphate ester conformational effects cannot influence the P-31 chemical shift temperature dependence of cCMP, this molecule is perhaps a better standard than the conformationally flexible, acyclic dialkyl phosphates. The upfield shift with temperature for cCMP is due simply to our choice of C_6F_6 as the lock standard since the chemical shifts of the lock standards will themselves be temperature dependent. Using a D_2O lock, the cCMP signal actually shifts very little, but now, of course, relative to the D_2O lock, the nucleic acids would shift .7 to 1.3 ppm downfield with increasing temperature. The smaller but real downfield shift for the simple acylic diesters, dimethyl and diethyl phosphate, under D_2O lock conditions likely represents a real increase in the proportion of non-g,g conformations (the lowest energy structure is g,g for these molecules; see 3-6 and references therein).

The sigmoidal type melting curves shown in Figures 4-6 are consistent with melting curves obtained by other spectroscopic and physical probes. In addition P-31 chemical shift differences between the various nucleic acids are likely mostly attributable to differences in averaged phosphate ester conformational states which in turn result from base-stacking interaction differences among the nucleic acid bases. Thus the signal broadness and upfield shift for the neutral poly G (line width >25 Hz at 15° or lower) is indicative of the more rigid, helical conformation for this polymer (see also 14). Aggregation of the poly G involving hydrogen bonding interactions is also probably partially responsible for these differences. In the strongly alkaline solution the guanosine base is negatively charged, which will disrupt the base stacking and helical structure. The signal sharpness (line width <2 Hz) and the large downfield shift reflect the much greater degree of motional freedom in the more flexible random coil. The small change in P-31 shift with temperature for poly G in strongly alkaline solution is consistent with our interpretation that poly G has little helical structure at any temperature under these conditions.

4. DOUBLE-STRANDED NUCLEIC ACIDS

The single-stranded nucleic acids are not ideal models for locking a phosphate diester into a g,g conformational state since considerable conformational flexibility still exists in the single-stranded "helical" nucleic acids even at 0° (9).

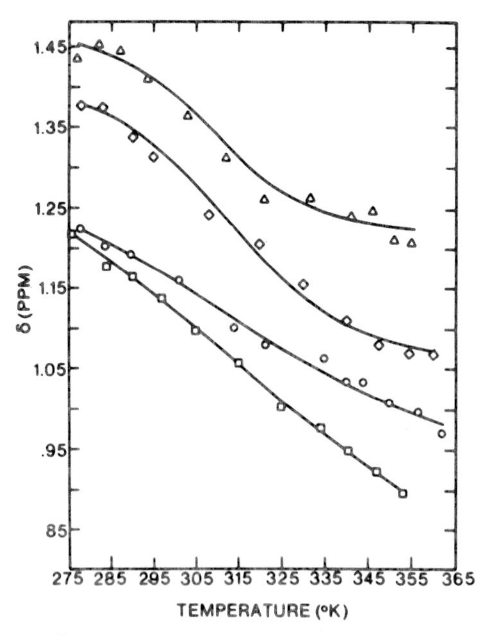

FIGURE 4: ^{31}P chemical shift vs. temperature for deoxyribodimers, d-pApT (diester signal only, △), d-ApT (◇), TpT (○), and d-ApA (□).

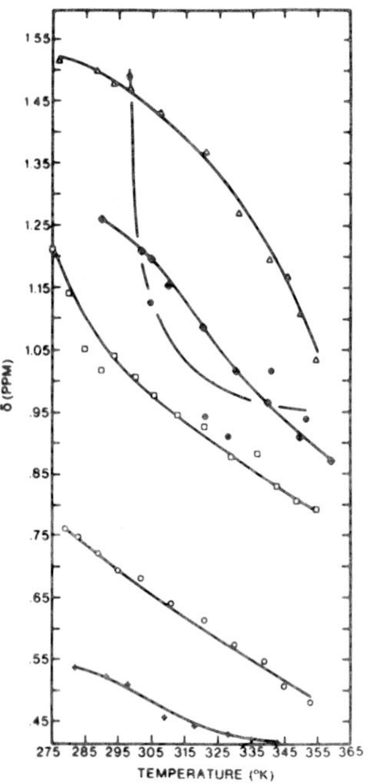

FIGURE 5: ^{31}P chemical shift vs. temperature for polynucleic acids, poly(C) (△), poly(A) (♦), poly(U) (0.5 M NaCl, □), poly(U) (○), poly(G) (pH 11.48, +), poly(G) (pH 10, ●).

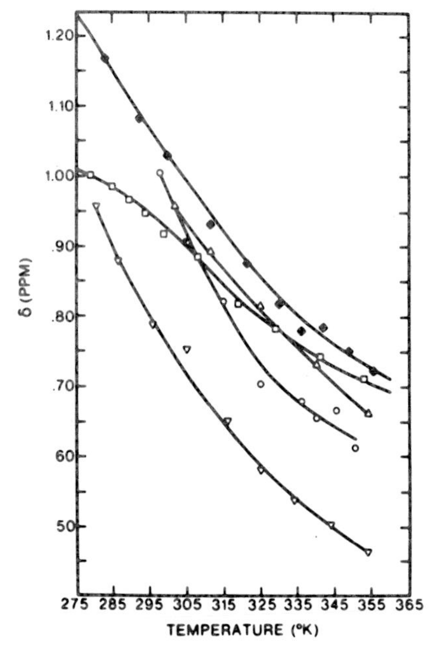

FIGURE 6: ^{31}P Chemical shift vs. temperature for ribodimers, UpU (□), ApA (△), ApU (♦), and CpC (▽) GpC (○),

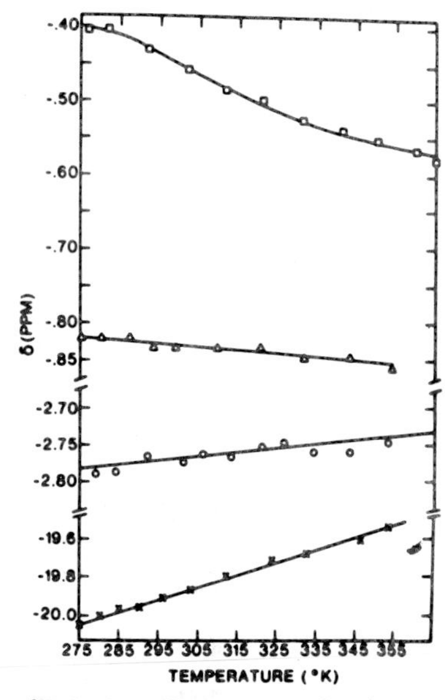

FIGURE 7: ^{31}P chemical shift vs. temperature for reference compounds, dibenzyl phosphate (□), diethyl phosphate (△), dimethyl phosphate (○), 2′,3′-cCMP (×). Note the ^{31}P chemical shift scale for cCMP has been reduced by one-fourth from other scales.

All chemical shifts were referenced to 85% phosphoric acid (0.0 ppm) at 298°K. Upfield shifts are given as positive numbers.

The double-helical state is much more limited as to the backbone flexibility and generally the phosphate diester conformation is limited to g,g (9,10). The P-31 spectra of double helical nucleic acids will hopefully therefore provide better information on the P-31 chemical shift of a purely g,g phosphate diester. The results in Figures 4 and 6 suggest that a difference in the averaged P-O conformation states exists even for the self-complementary dimers (ApU, GpC, d-ApT, and dpApT) which have P-31 signals 0.1-.2 ppm upfield from their respective homodimers (UpU, CpC, ApA, dApA and TpT). The self-complementary dimers are apparently more constrained to the g,g conformation since they tend to form miniature double helices (11,12).

Additionally, the P-31 signal of a 1:1 mixture of ApA + UpU shifts slightly further upfield from the signals of the homodimers alone.* (See Figure 8)

---

*Note in previous spectra (Figures 4-7) we report P-31 chemical shifts using a F-19 lock standard. Conversion between the D_2O lock system used in the remaining spectra and previous data is conveniently made by use of eqn. 1 in reference 8.

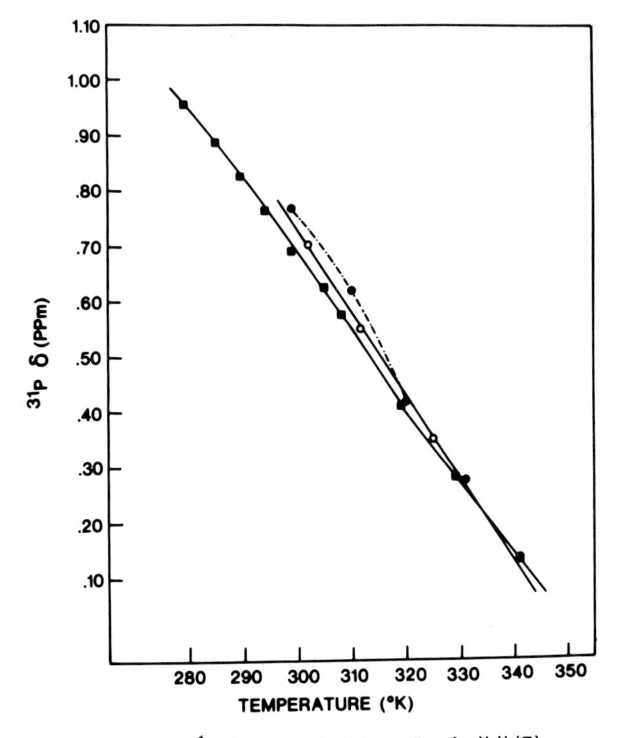

Figure 8. 3^1p chemical shift vs. temperature for UpU (□), ApA (o), ApA + UpU (1:1) (●).

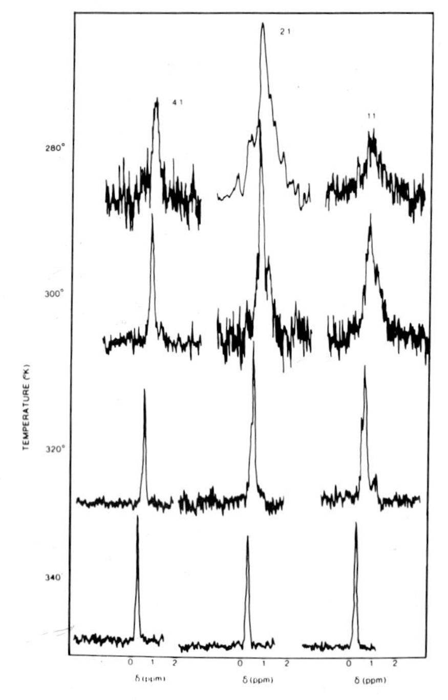

Figure 9 ^{31}P NMR spectra for poly A + oligo U mixtures, 4:1, 2:1, and 1:1 at various temperaturers.

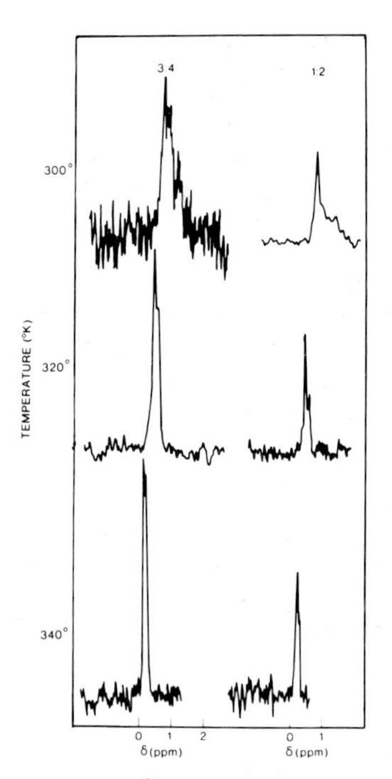

Figure 10. ^{31}P NMR spectra for poly A + oligo U mixtures, 3:4 and 1:2 at various temperatures.

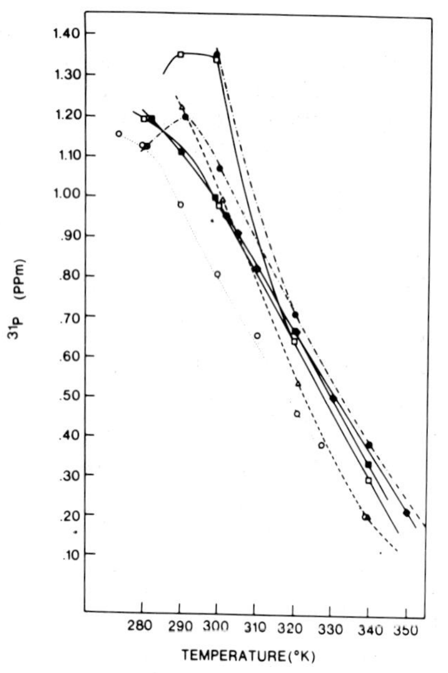

Figure 11. ^{31}P chemical shift vs. temperature for oligo U (○), poly A (◆), poly A + oligo U (1:1) (●), poly A + oligo U (2:1) (□), poly A + oligo U (4:1) (■). Both signals at T = 320° are plotted for (□) and (●).

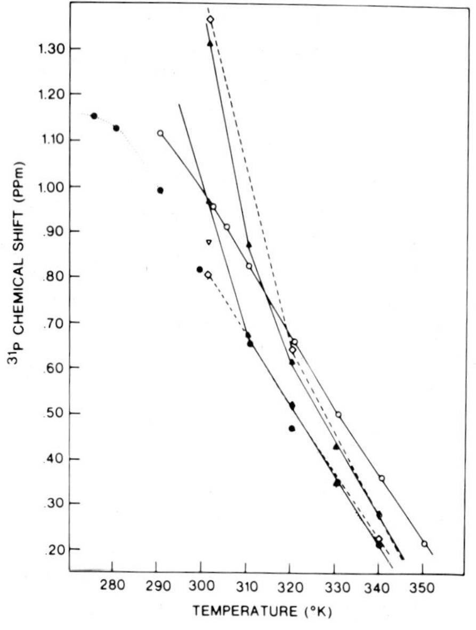

Figure 12 ^{31}P chemical shift vs. temperatures for oligo U (O), poly A + oligo U (3:4) (▲), poly A + oligo U (1:2)(◇). Both signals are plotted for (▲) and (◇).

Only a single signal was observed for the mixture even though at lower temperature, ApA and UpU P-31 shifts are noncoincident. These results indicate a slightly greater degree of helical character in the mixture presumably again due to a small fraction of miniature double helices. Unfortunately, even at low temperatures, the proportion of double helix will be small in these duplex dimers (probably <10% at 0° under our conditions; see 1).

Self-complementary polymers or mixtures of complementary polymers would eliminate this problem since melting temperatures of double-stranded polymers are 50° (13) and at room temperature exist almost entirely in the double-helix state (g,g phosphate diester conformation). Unfortunately the P-31 NMR signal of such a double-helical polymer at lower temperatures is too broad to be useful or even observable (12,14).

We have therefore chosen to study a mixture of complementary oligo U and poly A to avoid this problem since the double-helix state in this mixture will have single-strand breaks every 20 uridine bases corresponding to the length of the oligo U chain (Cal Biochem., based upon MW_{avg} 7500). This imparts considerable flexibility to the polymer-oligomer complex although not as much as present in the single-strand poly A. In addition, the oligo U/poly A mixture forms only the double helix (1 U/1 A) even at temperatures below the melting temperature and none of the triple helix (2 U/1 A) that is favored in the poly A/poly U mixture (13,15,16).

The temperature dependence to the P-31 chemical shifts of poly A and oligo U, separately and in various mixtures, is shown in Figures 9-12. At temperatures of 320°K or higher P-31 spectra correspond largely to the superposition of the component poly A and oligo U spectra. As shown in Figure 10 for the 3:4 and 1:2 (poly A:oligo U) spectra at 320°K, two separate signals are observed in the approximate ratio of the respective nucleic acids. In these spectra the downfield peak corresponds to the oligo U species although the signal occurs *ca* 0.06 ppm upfield from the oligo U species in the absence of added poly A. Similarly the poly A signal in the 3:4 and 1:2 mixtures occurs .02-.06 ppm downfield from the poly A signal in the absence of added oligo U. Although not as apparent, separate signals are likely present in the 4:1, 2:1, and 1:1 spectra at 320° K (Figure 9). The smaller proportion of oligo U in these mixtures is seen as a small downfield shoulder on the main (poly A) signal. In the 2:1 mixture, the U and A signals are shifted in the same direction as observed in the 3:4 and 1:2 solutions. The 4:1 mixture main signal corresponds exactly to the poly A signal in the absence of oligo U. In contrast to all the other spectra at 320°, the main signal in the 1:1 mixture comes .04 ppm *upfield* from the poly A signal in the absence of poly U. These effects on the P-31 nmr spectrum of one nucleic acid species upon the other indicate that some interaction between the complementary base pairs is occurring even at elevated temperatures. The separation of the signals for the U and A nucleic acids suggests that chemical exchange is rapid (on the NMR time scale) between the free, uncomplexed nucleic acids and the complexed, largely double-helical species which must be present in minor (unobservable) quantities.

The most interesting aspect to these spectra occurs when the temperature is lowered to 300°K. At this temperature in all but the 4:1 mixture a new P-31 signal is observed, ca. .3-.55 ppm upfield from the P-31 signal of the poly A or oligo U species free in solution (Figures 9 and 10). This signal always comes at 1.34 ± .03 ppm in these spectra and likely corresponds to the double helix, poly A-(oligo U)$_n$. Identification of the downfield peak as the single-strand form of the nucleic acids is based upon chemical shift comparisons. Thus, at 300° in the 1:2 spectra, the downfield peak resonates at .804 ppm, nearly identical to the P-31 chemical shift of free oligo U at the same temperature. Assuming a 1:1 double helix, poly A is limiting in this 1:2 mixture and the remaining uncomplexed nucleic acid is oligo U. (The double helix signal is seen as the small broad signal .55 ppm upfield from the sharper oligo U signal.) Similarly in the 2:1 spectra at 300°, the sharper downfield signal at .979 ppm corresponds to the chemical shift of free poly A. In this solution oligo U will be limiting in forming the double helix, leaving poly A as the uncomplexed single-strand nucleic acid. In the 4:1 spectra only one signal is observable, with the chemical shift corresponding to free poly A. In the 1:1 spectra the downfield peak is .1 ppm upfield from free poly A. In the 3:4 spectra, it appears possible to see all three species: "free" single-helix oligo U (.875 ppm), "free" single-helix poly A (1.095 ppm), and double helix (1.35 ppm). Unfortunately, poor signal-to-noise in these spectra makes assignment difficult. In these 1:1 and 3:4 spectra the downfield peaks are all .1 ppm upfield from the expected completely free, uncomplexed nucleic acid. Some intermediate chemical exchange between the uncomplexed nucleic acid and the double helix is probably partially averaging the complexed and uncomplexed signals.

At still lower temperatures, the signals continue to broaden due to the slower tumbling of these large molecules, and fine structure is lost. Total signal intensity, however, stays about the same at these lower temperatures and thus any really rigid species which would have a P-31 signal so broad as to be unobservable, is likely not present to any appreciable extent. In the 2:1 spectra at 290° a single signal at 1.35 ppm suggests largely double helix is present although because of the broadness of the signal (~1.0 ppm) a separate single-stranded poly A signal would probably not be observed. At still lower temperature (280°) only a broad single peak is observed with a chemical shift (measured at highest peak height) corresponding to free poly A. Apparently the double helix signal is so broad ($\Delta\nu_{1/2} > 35$ Hz) that only the relatively sharp "free" poly A signal ($\Delta\nu_{1/2} < 12$ Hz) stands out. Superimposed underneath the single-helix poly A signal is the broad poly A. oligo U double-helix signal. The slight upfield asymmetry to the peak probably reflects the imperfect superposition of the double-helix signal.

Similarly the 1:1 spectra at 290° shows a single, very broad, low intensity signal with a chemical shift (1.20 ppm) in between the expected single-helix and double-helix regions. At 280° the sharpest P-31 signal at 1.13 ppm corresponds to free oligo U (apparently, in preparing the 1:1 solution by weight a slight excess of oligo U was added). Again, superimposed under these sharper, single-stranded signals is a broad, featureless signal corresponding to the double helix.

The assignment of the new upfield (1.35 ppm) signal to the double-helical state is suggested by

1) the signal broadness
2) the sharp transition for its appearance and
3) its upfield position as predicted by theory.

The polymeric double helix will be much more rigid than the polymeric single helix and internal tumbling of the backbone phosphates will be greatly reduced. This shortens the P-31 transverse relaxation time and broadens the signal. Thus Patel and Canuel (12) observed only a very broad P-31 signal for the self-complementary poly(dAT) below 60°C. The oligo U/poly A double helix retains more flexibility than the polymer/polymer double helix and the P-31 signal for the former is not as broad.

The P-31 melting profile for this upfield peak at T's $< 320^{\circ}$ (Figures 11,12) suggests a sharp melting transition for this species, characteristic of a cooperatively melting double-stranded species. As shown earlier and seen for poly A or oligo U signals, the single-helix P-31 melting profiles are spread over a wider temperature interval, characteristic of a non-cooperative melting phenomenon.

5. CONCLUSIONS

The .2-.5 upfield P-31 shift for the double-helical state from the stacked single-helical state supports our previous assignments (8) and our theoretical calculations (2) that a phosphate ester in a gauche, gauche conformation should be shielded relative to other non-g, g conformations. These results, combined with our earlier demonstration that the P-31 chemical shifts of the single-helix state is >1.0 ppm upfield from the random coil state, suggests that P-31 chemical shifts may serve as a sensitive probe to the conformation of nucleic acids. Drug-nucleic acid (17,18) and Protein-nucleic acid interactions may be usefully probed by P-31 NMR. Thus as shown in Figure 11, the binding of the intercalating drug ethidium bromide to the poly A/oligo U (1:1) mixture produces perturbations in the P-31 chemical shift. Patel (17) has observed even larger downfield shifts of the internucleotide phosphates in both the dideoxyribonucleotide duplex,

2(d-pGpC), and in the hexadeoxyribonucleotide duplex, 2(d-ApTpGpCpApT), upon binding the intercalating drug, actinomycin D. Jain and Sobell (19) have proposed models for these intercalated complexes that involve partial unwinding of a specific section of the double helix. This requires significant disruption of the normal, g,g phosphate diester conformation, which must be responsible for these large downfield shifts.

The 2 ppm downfield shift observed by Griffin et al. (20) in the binding of 3'-methyleneuridylyl-3',5'-adenosine phosphonate to ribonuclease A likely reflects the binding of this dinucleotide analogue in a more open, g,t conformation as found in the x-ray crystal structure of the complex. The binding of uridylyl-2',5'-adenosine to the same enzyme also produces a downfield shift at low temperature (D. Gorenstein, unpublished observation).

P-31 studies in nucleic acids may prove to be especially significant since other spectroscopic probes largely fail to provide detailed conformational information on the phosphate ester bonds in the nucleic acids. Since it is now believed that of the six torsional angles that largely define the conformational structure of nucleic acids, the two P-O ester torsional angles provide the main conformational flexibility to the nucleic acid backbone. A probe of the solution conformation for phosphate ester-enzyme complexes becomes especially useful, since Gorenstein, et al. (23-25) and Lehn and Wipfe (26) have demonstrated that the reactivity of phosphate esters is highly dependent upon the conformation. The reactivity of phosphates is also highly dependent upon bond angle (3,24,27) and the recent observation (28-30) that the phosphate covalently bound to alkaline phosphatase is shifted 6-8 ppm downfield from the inorganic phosphate in solution has been suggested to originate from bond angle strain in the enzyme complex. With the realization that separate, modestly sharp P-31 signals may be observed for even very large phosphate complexes (such as the >200,000 MW ν body in chromatin (31) and the ATP complex of the 237,000 MW enzyme muscle pyruvate kinase;32)P-31 NMR therefore holds great promise in providing detailed solution structures of complex biological phosphates.

Acknowledgements. Support of this research by NSF, NIH, and the Alfred P. Sloan Foundation for a fellowship is gratefully acknowledged. The hospitality of the Oxford Enzyme group and the Dyson Perrins Laboratory, Oxford University and the support provided by a Fulbright Fellowship during the writing of this paper is most appreciated. The collaboration with Bruce Luxon, John Findlay, David Kar, Robert Momii and Alice Wyrwicz has been, of course, crucial to these studies.

REFERENCES

1. Gorenstein, D. G.: J. Amer. Chem. Soc. 97, 898 (1975).

2. Gorenstein, D. G. and Kar, D.: Biochem. Biophys. Res. Commun. 65, 1073 (1975).
3. Gorenstein, D. G., Kar, D., Luxon, B. A., Momii, R. K.: J. Amer. Chem. Soc. 98, 1668 (1976).
4. Gorenstein, D. G., and Kar, D.: ibid. 99, 672 (1977).
5. Gorenstein, D. G., Luxon, B. A., and Findlay, J. B.: Biochim. Biophys. Acta 475, 184 (1977).
6. Perahia, D. and Pullman, B.: ibid. 436, 282 (1976).
7. Gorenstein, D. G., Wyrwicz, A. M., and Bode, J.: J. Amer. Chem. Soc. 98, 2308 (1976).
8. Gorenstein, D. G., Findlay, J. B., Momii, R. K., Luxon, B. A., and Kar, D.: Biochemistry 14, 3796 (1976).
9. Ts'o, P.O.P.: "Basic Principles in Nucleic Acid Chemistry", Vol. I and II, New York and London, Academic Press (1975).
10. Day, R. O., Seeman, N. C., Rosenberg, J. M., and Rich, A.: Proc. Natl. Acad. Sci. U.S.A. 70, 849 (1973), and references therein.
11. Krugh, T.R., Laing, J.W., and Young, M.A.: Biochemistry 15, 1224 (1976).
12. Patel, D. J., and Canuel, L.: Proc. Natl. Acad. Sci., U. S. 73, 674 (1976).
13. Bloomfield, V. A., Crothers, D. M., and Tinoco, I.: "Physical Chemistry of Nucleic Acids," Chap. 6, Harper & Row, New York, N. Y. (1974).
14. Akasaka, K., Yamada, A., and Hatano, H.: FEBS Lett. 53, 339 (1975).
15. Michelson, A. M. and Monny, C.: Biochem. Biophys. Alta, 149, 107 (1967).
16. Porschke, D.: Biopolymers 10, 1989 (1971).
17. Patel, D. J.: Biochemistry 13, 2396, 2388 (1974).
18. Reinhardt, C. G. and Krugh, T. R.: ibid. 16, 1900 (1977).
19. Jain, S. C., and Sobell, H. M.: J. Mol. Biol. 68, 1 (1972).
20. Griffin, J. H., Schechter, A. N., and Cohen, J. S.: Ann. N. Y. Acad. Sci. 222, 693 (1973).
21. Sundaralingam, M.: Biopolymers 7, 821 (1969).
22. Ts'o, P.O.P., Kondo, N.S., Schweizer, M.P.O. and Hollis, D.C.: Biochemistry 8, 997 (1969).
23. Gorenstein, D. G., Findlay, J. B., Luxon, B. A., and Kar, D.: J. Amer. Chem. Soc. 99, 3473 (1977).
24. Gorenstein, D. G., Luxon, B. A., Findlay, J. B., and Momii, R.: ibid. 99, 4170 (1977).
25. Gorenstein, D. G., Luxon, B. A., and Findlay, J. B.: ibid. 99, 8048 (1977).
26. Lehn, J. M. and Wipfe, G.: J. Chem. Soc., Chem. Comm. 800 (1975).
27. Westheimer, F. H.: Accounts Chem. Res. 1, 70 (1968).
28. Chlebowski, J. F. and Coleman, J. E.: J. Biol. Chem., 249, 7192 (1974)
29. Hull, W. E., Halford, S. E., Gutfreund, H., and Sykes, B. O.: Biochemistry, 14, 1547 (1976).
30. Bock, J. and Sheard, B.: Biochem. Biophys. Res. Commun. 61, 24 (1975).

31. Cotter, R. I. and Lilley, D.M.J.: Febs Lett. 82, 63 (1977).
32. Gupta, R. K. and Milovan, A. S.: J. Biol. Chem. 252, 5967 (1977).

DISCUSSION

Navon: In P^{31} NMR studies of living cells it was often observed that the linewidths of the peaks of the β phosphorus of ATP are appreciably greater than those of the α and the γ phosphorus. Particulary in spectra taken with high field NMR spectrometers. It was suggested (G. Navon, R. Navon, R.G. Shulman and T. Yamane, Proc. Natl. Acad. Sci. USA, in press) that this phenomena is due to an exchange of ATP with bound state where its linewidth is determined by chemical shift anisotropy. Do you have any experimental or theoretical data on the chemical shift anisotropy of the phosphorus atoms in ATP ?

Gorenstein: Unfortunately I do not know of any chemical shift anisotropy theoretical studies on ATP (or any phosphates in fact). Some recent advances in solid state, oriented crystal, ^{31}P NMR of various nucleic acids have appeared in the literature. Large C.S.A.'s are found for mononucleoside monophosphates. Your suggestion for the line broadening in ATP is dertainly reasonable but possibly could also be explained by chemical exchange broadening (?).

Rüterjans: It seems that the influences of charges on the P31 chemical shifts of phosphate groups are only small. May be the arrangement of the H-bonds formed between water molecules or side chains of enzymes has a larger influence.

Gorenstein: Since we don't have enough experimental information on H-bonding interactions, it is certainly possible that the arrangement of H-bonds can have a large effect on P-31 shifts. However, so far all evidence suggests the effect is not very large. Thus RNase A does not significantly perturb the P-31 shifts of the fully dianionic or mono-anionic states of either 3'-, 5'-, or 2'-CMP upon binding, yet the arrangement of H-bonds is quite different.

Feeney: Do any of your calculations of ^{31}P chemical shifts explain the biphasic nature of your correlation of ^{31}P shifts with O-P-O bond angles ?

Gorenstein: Unfortunately not and the one or two data points which provide the biphasic behavior should possibly be ignored, leaving only a simple near-linear correlation. Phosphate trianion may be a very unusual species by symmetry compared to all other phosphates.

Abraham: In many n.m.c. studies of conformational equilibrium by variable temperature observations, the use of a standard which removes the intrinsic temperature dependence of the nuclear chemical shifts, thus allowing the conformational dependence to be isolated, has proved very useful. Is it possible to use the phosphate anion in this way in your analysis of the ^{31}P chemical shifts of phosphates in aqueous solution ?

Gorenstein: A priori it is not possible to say that a trianionic phosphate is a much better reference standard than a simple monoanionic diester since we know so little about intrinsic temperature dependencies to chemical shifts. Hydrogen-bonding interactions to solvent and electrostatic effects to counter-ions may be quite different between the two types of phosphates and thus I am not sure that one "standard" is necessarily better than another. What is important is the relative difference in temperature dependency to P-31 shifts between any reasonable simple model and the nucleic acids. We will though investigate your suggestion. Note that our choice of lock standards (P_2O or C_5F_6) also exerts an influence on these measurements (see discussion in ref. 8).

NMR STUDIES ON DYNAMIC PROPERTIES OF NUCLEIC ACID BASE PAIRS IN NONAQUEOUS SOLVENTS

Hideo Iwahashi and Yoshimasa Kyogoku
Institute for Protein Research, Osaka University
Suita, Osaka 565 JAPAN

In the double-stranded structures of deoxyribonucleic acid and ribonucleic acid, adenine forms specific hydrogen bonds with thymine (or uracil) and guanine with cytosine. The specific bonds are believed to be the molecular basis of information transfer in nucleic acids. Much work has been done to find the basis of this specificity by the use of synthetic poly- and oligonucleotides and single base derivatives. Some infrared (1∿8) and proton magnetic resonance studies (9∿12) clearly showed that even single base residues interact by means of specific hydrogen bonds in solution. The data obtained in these experiments revealed the strength of interaction on the amino- and imino-protons, but little is known about the acceptor sites of the hydrogen bonds. In the base pair model proposed by Watson and Crick, thymine (or uracil) pairs with adenine by using the C-4 carbonyl group. However, X-ray analyses showed that 5-bromouracil (13) and 4-thiouracil (14) derivatives through the C-2 carbonyl group. In the present experiment we observed ^{13}C magnetic resonances of single base derivatives in solution and tried to find interaction sites from the concentration dependency of chemical shifts.

In spite of much accumulation of the knowledge about the static structure of the interacted bases, little is investigated about the dynamic properties of the base pairs. We tried to measure the mobility of free bases in non-aqueous solvents and the effect of the base pairing on the mobility by measuring rotational correlation times of ^{13}C nuclei. On the other hand the spin lattice relaxation times of ^{1}H reflect the effect of the intermolecular interaction more directly. The experiment of saturation transfer by labile protons shows us the sites of exchangeable proton in the interacted system. Combining both data, here we will report the evidence of the direct proton transfer between the complementary base pairs, and quantitative treatment of the transfer. A study on the dynamic process will present us new information about the transfer mechanism of genetic information.

1. MATERIALS AND PROCEDURES

B. Pullman (ed.), Nuclear Magnetic Resonance Spectroscopy in Molecular Biology, 17-30.

1.1 Materials and instruments.

9-Ethyladenine(A), 1-cyclohexyluracil(U), 1-cyclohexylthymine(TH), 1-cyclohexyl-5-bromouracil(BU), 1-cyclohexyl-5,6-dihydrouracil(HU) and 1-cyclohexyl-4-thiouracil(TU), 9-ethylguanine(G) and 1-methylcytosine(C) were purchased from Cyclo Chemical Co., Los Angeles. Nuclear magnetic resonance spectra were measured for their chloroform-d_1 or dimethylsulfoxide-d_6 solutions.

^{1}H and ^{13}C resonance spectra were obtained at 100 and 25 MHz, respectively, with JEOL PFT-100 and FX-100 pulse Fourier transform NMR system locked on deuterium. The temperature of the sample tubes was kept at 27°C throughout the experiments. Spin lattice relaxation times were measured with the saturation recovery method by applying 180°-t-90° pulse sequences. Saturation transfer was detected in the proton spectra by irradiating at the frequency of the amino or imino proton resonances.

1.2 Procedures for the calculation of association constants and population differences

The equilibrium equation for the formation of a 1:1 complex of adenine and uracil derivatives is expressed as

$$A + U = AU, \qquad \ldots\ldots(1)$$

but in the present case, various conformations of dimers as shown in Figure 1 should be taken into account, because A and U have two proton

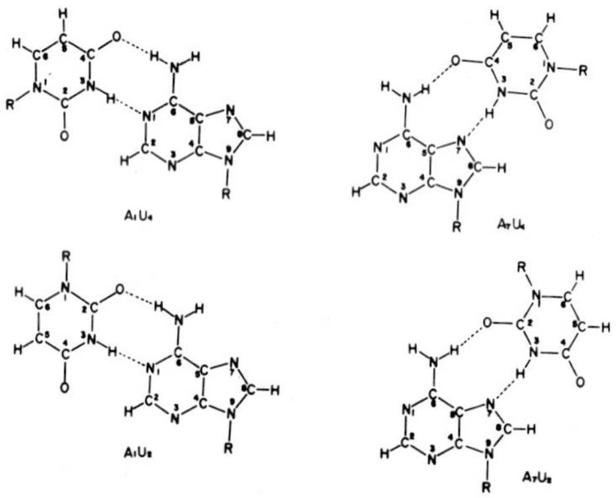

Figure 1. Structures of AU dimers

acceptor sites each. Taking all the possibilities into account, the above relation can be rewritten as follows:

$$A + U = kA_1U_2 + lA_7U_2 + mA_1U_4 + nA_7U_4. \qquad \ldots\ldots(2)$$

Here the relation of k + l + m + n = 1 holds for k, l, m and n. Then the association constant K can be defined as follows:

$$K = C^{k}_{A1U2} \cdot C^{l}_{A7U2} \cdot C^{m}_{A1U4} \cdot C^{n}_{A7U4} \,/\, C_A \cdot C_U \,, \quad \text{.....(3)}$$

where C_A and C_U are the concentrations of monomers and C_{AiUj} is the concentration of each A_iU_j complex. If δ_U, δ_{A1U2}, δ_{A7U2}, δ_{A1U4} and δ_{A7U4} are the limiting chemical shifts of U, A_1U_2, A_7U_2, A_1U_4, and A_7U_4 respectively, the observed chemical shift δ_m can be expressed by the next equation, assuming rapid exchange among complexes,

$$\delta_m = (\delta_U C_U + \delta_{A1U2} C_{A1U2} + \delta_{A7U2} C_{A7U2} + \delta_{A1U4} C_{A1U4} + \delta_{A7U4} C_{A7U4}) / (C_U + C_{A1U2} + C_{A7U2} + C_{A1U4} + C_{A7U4}) . \quad \text{.....(4)}$$

The limiting chemical shift δ_k is the shift which would be realized if all molecules in the equilibrium system were in the state k. Then we can obtain an equation which relates the binding constant K with the observed chemical shift when the initial concentration of C°_U is equal to that of A:

$$K = \frac{k^k l^l m^m n^n (\delta_m - \delta_n)(\delta_{AU} - \delta_U)}{C^{\circ}_U (\delta_{AU} - \delta_m)^2} \quad \text{.....(5)}$$

where δ_{AU} is a newly defined variable which equals to

$$k\delta_{A1U2} + l\delta_{A7U2} + m\delta_{A1U4} + n\delta_{A7U4} . \quad \text{.....(6)}$$

Uracil derivatives associate with each other, though the interactions are weak compared with that of the A-U association. Self-association is here treated in the same way as hetero-association. The eqilibrium relation for self-association is expressed as follows

$$U + U \rightleftharpoons kU_2U_2 + lU_2U_4 + mU_4U_4 \quad \text{.....(7)}$$

Each complex has the structure shown in ref. 21.

A computer program for a non-liner least square regression analysis[15] was written to calculate the limiting shifts (δ_U, δ_{UU} and δ_{AU}) and apparent binding constants ($K/k^k l^l m^m$ or $K/k^k l^l m^m n^n$). This was done by feeding in the initial concentration (C°_U) and the observed chemical shifts (δ_m), and assuming that only 1:1 dimers are present and other polymers are negligible.

Now we can define new values for Δ^2_{UU}, $\Delta^2_{U_2U_2}$, $\Delta^2_{U_2U_4}$, $\Delta^2_{U_4U_4}$, Δ^4_{UU}, $\Delta^4_{U_2U_2}$, $\Delta^4_{U_2U_4}$, $\Delta^4_{U_4U_4}$. Δ^2_{UU} means the difference between the limiting shift of the C-2 carbonyl carbon of free monomer δ_U and that of dimer δ_{UU}. $\Delta^2_{U_2U_2}$ is the contribution from the U_2U_2 dimer to the difference. $\Delta^2_{U_2U_4}$, $\Delta^2_{U_4U_4}$, $\Delta^4_{U_2U_2}$, $\Delta^4_{U_2U_4}$ and $\Delta^4_{U_4U_4}$ are also defined in a similar manner to $\Delta^2_{U_2U_2}$. Thus we can obtain the following relation:

$$\Delta^2_{UU} + k\Delta^2_{U_2U_2} + l\Delta^2_{U_2U_4} + m\Delta^2_{U_4U_4} . \quad \text{.....(8)}$$

For the C-4 signal a similar relation can be written;

$$\Delta^4_{UU} = k\Delta^4_{U_2U_2} + l\Delta^4_{U_2U_4} + m\Delta^4_{U_4U_4} \quad .....(9)$$

Subtracting eq.9 from eq.8 gives:

$$\Delta^4_{UU} - \Delta^2_{UU} = -k(\Delta^2_{U_2U_2} - \Delta^4_{U_2U_2}) + l(\Delta^4_{U_2U_4} - \Delta^2_{U_2U_4}) + m(\Delta^4_{U_4U_4} - \Delta^2_{U_4U_4}). \quad .....(10)$$

It is safe to assume that the magnitude of perturbation at the C-4 carbon on the formation of a hydrogen bond using the C-2 carbonyl group is nearly identical to that at the C-2 carbon on the formation of a hydrogen bond using the C-4 carbonyl group. It is also reasonable to assume that the C-2 carbon participating in a hydrogen bond suffers the same perturbation as the C-4 carbon in a corresponding case. The above assumptions enable us to rewrite eq.10 as:

$$\Delta^4_{UU} - \Delta^2_{UU} + (m-k)\Delta' \quad .....(11)$$

where $\Delta' = (\Delta^4_{U_4U_4} - \Delta^2_{U_4U_4}) = (\Delta^2_{U_2U_2} - \Delta^4_{U_2U_2})$.

If Δ' is the same for all the U derivatives, the value of $(\Delta^4_{UU} - \Delta^2_{UU})$ is determined by $(m-k)$. Therefore the larger value of $(\Delta^4_{UU} - \Delta^2_{UU})$ means that the C-4 carbonyl group is more often used as a proton acceptor in hydrogen bonding. We can also treat the case of association with adenine in a similar manner, and can say that the value of $(\Delta^4_{AU} - \Delta^2_{AU})$ represents semi-quantitatively the population difference between the complexes which use the C-2 and C-4 carbonyl groups as proton acceptor sites.

1.3 MO Calculation

The electron charge densities of free uracil were calculated by the CNDO/2 method using the parameters employed in reference (16). The same procedure was applied to the system where a uracil molecule forms a cyclic hydrogen bonded dimer with amino aldehyde, a model of the *cis*-amide part of uracil. The N-H bond length of free uracil was taken as 0.948 A and the bonded one as 1.043 A. The H...O distance in the dimer was 1.657 A. The other bond lengths and bond angles were taken from the crystal structure of 1-cyclohexyluracil (17).

1.4 Calculation of rotational correlation time

From the obtained spin lattice relaxation time T_1, the rotational correlation time τ_r can be calculated by the following equation, if the relaxation is dominantly contributed by dipole-dipole interaction,

$$\frac{1}{T_1} = \frac{h^2\gamma_i^2\gamma_j^2}{4\pi^2 r^6}\tau_r \quad .....(12)$$

where γ_i and γ_j are the gyromagnetic ratio of the i th and j th nucleus, respectively, and r is the distance between i and j atoms (18). This equation is only valid, when the product of resonance frequency ω and τ_r is quite small than 1. This assumption is reasonable for the bases in chloroform, since rotational correlation time is defined as (18).

$$\tau_r = 4\pi\eta a^3/3kT \qquad .....(13)$$

and the viscosity of the solvent η and the radius of the molecule a are not so big at the present time. We put the following values for each physical parameters.

γ_H : 2.68 rad.sec^{-1}.G^{-1}, γ_C : 0.673 rad.sec^{-1}.G^{-1},
r(CH) : 1.08 x10^{-8} cm, $h/2\pi$: 1.05 x 10^{-27} erg.sec

1.5 Determination of the rate of proton exchange

Let us consider the system where the nuclei X exchange between the sites A and B, and the X nuclei in the B site are saturated by irradiation. The change of magnetization of the nuclei at A in the exchanging system irradiated at the B site is expressed as follows (19,20).

$$\frac{dM_A}{dt} = -\frac{(M_A - M_A^\circ)}{T_{1A}} - \frac{M_A}{\tau_A} \qquad .....(14)$$

where τ_{1A} is given by $1/\tau_{1A} = 1/\tau_A + 1/T_{1A}$. M_A° is the equilibrium value of z magnetization at the non-irradiated condition, τ_A is the life time of X at the site A and T_{1A} is the spin-lattice relaxation time of X at A. The decay of M_A is obtained by solving eq.14,

$$M_A = (M_A^\circ/\tau)\,[1 - 2\mathrm{Exp}(-\tau t/T_{1A})] \qquad .....(15)$$

where τ is defined as

$$\tau = 1 + T_{1A}/\tau_A \qquad .....(16)$$

M_o^A/τ is proportional to the intensity of the resonance at A site during the irradiation at B site. Thus the apparent T_1 ($T_{app} = T_{1A}/\tau$) is obtained from eq.15 by applying 180°-t-90° pulse sequence.

2. RESULTS AND DISCUSSION

2.1 Detection of Proton Acceptor Sites of Hydrogen Bonding Between Bases (21).

Concentration dependences of ^{13}C chemical shifts of the 1:1 mixtures of 8-bromouracil derivative with A are shown in Figure 2. The C-2 and C-4 signals of the bromouracil derivative shifts downfield by a much larger amount than other signals. The downfield shihts are attributable to an increase in hydrogen bond formation with increase in concentration. The larger shifts of the C-2 and C-4 carbons of uracil derivatives indicate that the C-2 and C-4 carbonyl groups directly participate in hydrogen bond formation and suffer much more perturbation when bonding with hydrogen than do other carbons in the molecule. The result is consistent with the infrared absorption spectra in the carbonyl stretching region (22).

The total electron density was calculated for the free uracil derivative and the hydrogen bonded uracil derivative with a *cis*-amide

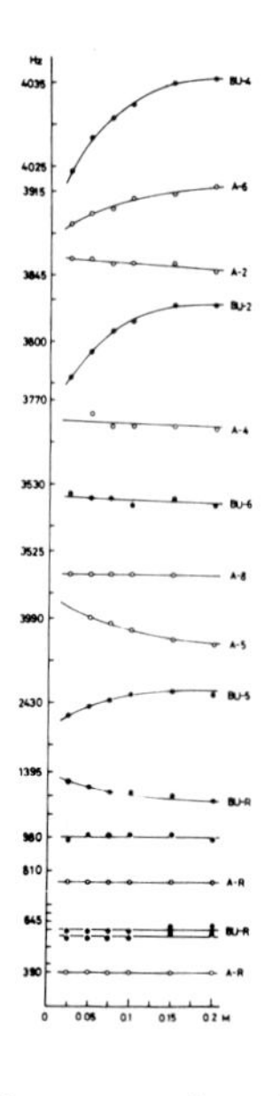

Figure 2. Concentration dependence of ^{13}C chemical shifts of the 1:1 mixture of 9-ethyladenine and 1-cyclohexyl-5-bromouracil.

Table 1. Difference in electron charge density between free uracil and hydrogen bonded uracil (calculated by CNDO/2)

Atom No.	Free Form.	3,4-complex (Δ)	3,2-complex (Δ)
N-1	5.192	0.000	-0.001
C-2	3.552	0.000	-0.010
N-3	5.247	0.028	0.028
C-4	3.628	-0.014	-0.001
C-5	4.169	0.006	-0.003
C-6	3.823	-0.006	0.004
O-2	6.373	-0.001	0.026
O-4	6.351	0.029	-0.001
H-1	0.868	0.002	0.004
H-3	-0.048	-0.048	-0.047
H-5	0.946	0.002	0.002
H-6	0.996	0.001	0.001

group using the C_2 or C_4 carbonyl and the N_3 imino group. The differences in electron charge densities are given in Table 1. The result indicates that the formation of cyclic hydrogen bonds at C_2 and N_3 withdraws electrons from the C_2 carbon and results in the downfield shift of the ^{13}C resonance of C_2. Our calculations also indicated that the formation of hydrogen bonds at the C_2 carbonyl group do not affect the electron density at the C_4 carbonyl group and vice versa. According to the empirical relationship (160 ppm/electron) proposed by Spiesecke and Schneider (23) the present calculation predicts that $\Delta^4_{U_4U_4} = 56$ Hz,

$\Delta^2_{U_2U_2} = 40$, $\Delta^4_{U_2U_2} = \Delta^2_{U_4U_4} = 0$, $\Delta^4_{U_2U_4} = 28$ and $\Delta^2_{U_2U_4} = 20$ Hz. The calculated shifts only explain half or one third of the observed chemical shift differences when we assume k=l=m=1/3. It may be due to the fact that the present calculation was not done for the complete dimer system or the shift is caused in part by other factors.

Computer analysis enables us to calculate the binding constants and limiting shifts for free monomers and associated complexes by using the plots of the relationship between concentration and chemical shift. The calculated binding constants well agree with those obtained by infrared measurements (5,6). The values obtained for the limiting shift differences $\Delta = \delta_{UU} - \delta_U$ or $\delta_{AU} - \delta_U$ are given in Table 2. The table

Table 2. Difference (Δ) between the limiting shifts of ^{13}C resonances of carbonyl carbons in free and bound states

Compounds	Carbonyl Position	$\Delta_{UU} = \delta_{UU} - \delta_U$ (Hz) (self association)	$\Delta_{AU} = \delta_{AU} - \delta_U$ (Hz) (association with adenine)
5,6-Dihydro-uracil	C-2	45.7 ± 0.5	
	C-4	45.6 ± 0.7	51.2 ± 0.7
Uracil	C-2	30.3 ± 1.2	46.1 ± 0.4
	C-4	55.2 ± 0.7	65.5 ± 0.5
Thymine	C-2	34.5 ± 0.6	45.9 ± 0.6
	C-4	41.7 ± 0.6	74.9 ± 0.5
5-Bromo-uracil	C-2	34.8 ± 0.4	50.6 ± 0.3
	C-4	32.4 ± 0.7	65.2 ± 0.2
4-Thio-uracil	C-2	66.3 ± 1.0	51.1 ± 0.8
	C-4	53.4 ± 1.2	34.0 ± 0.7

indicates that the self-association using the 4-carbonyl group is more common in uracil and the ratio of the 4-carbonyl dimer to the 2-carbonyl gradually falls in the order, thmine, 5,6-dihydrouracil, 5-bromouracil, 4-thiouracil while that of the 2-carbonyl dimer rises in the reverse order. However, with adenine association the ratio is highest in thymine and this is followed by uracil, 5-bromouracil and 4-thiouracil in descending order of magnitude. In the association of U derivatives with A, the difference in availability of the C-4 carbonyl group in hydrogen bonds can be explained by the difference in electronegativity of the substituent at the 5 position. Since the methyl group is considered an electron releasing group, the methyl group of thymine pushes electrons into the pyrimidine ring to increase the electron density at the C-4 carbonyl group. On the other hand, electron density of the pyrimidine ring of 5-bromouracil decreases, particularly at the C-4 site which is close to the strong electron attractive group, bromine. Thus electron density of the C-4 carbonyl group increases in ascending order from thymine> uracil >5-bromouracil; this is identical to the order of the relative induced shift of the C-4 carbon. Therefore we can say that the electron migration induced by substitution on the pyrimidine ring is an important factor in determing the population of the hydrogen bonded complex. The 2-carbonyl group of 4-thiouracil is more ofen hydrogen bonded than the C-4 thio-carbonyl group because S is inferior to O as a proton acceptor (24). The above discussion is valid only when we can assume that $\Delta^2_{U2U4} = \Delta^4_{U2U4}$ and $\Delta^4_{U4U4} - \Delta^2_{U4U4} = \Delta^2_{U2U2} - \Delta^4_{U2U2}$. There is no way to test it experimentally, but tentative calculation of electron charge density shows that the assumption is not far from the mark. Even when $\Delta^2_{U2U4} = \Delta^4_{U2U4}$ and $\Delta^4_{U4U4} = \Delta^2_{U2U2}$, the order of population differences is unaffected.

2.2 Change in the Spin-lattice Relaxation Times of Nucleic Acid Base by the Formation of Base Pairs.

^{13}C Spin-lattice relaxation time and rotational correlation times calculated by eq.12 are given in Table 3. The ^{13}C nucleus at the

Table 3. T_1 of ^{13}C nuclei in 1-cyclohexyl-5-Br-uracil

Position	Free		Complex with A	
	NT_1 (sec)	τ_r (10^{-11}sec)	NT_1	τ_r
C-2	16.4		9.7	
C-4	18.4		16.0	
C-5	9.5		7.9	
C-6	1.1	4.1	0.51	8.4
C-1'	1.5	3.0	0.71	6.0
C-2'	1.6	2.8	0.78	5.4
C-3'	1.6	2.8	0.72	6.0
C-4'	1.3	3.5	0.58	7.4

6-position of 5-bromouracil is dominantly relaxed by the attached proton and the obtained rotational correlation time of the nucleus should directly reflect the molecular motion of the 5-bromouracil base ring. This value changed from 4.1×10^{-11} sec to 8.4×10^{-11} sec by mixing with equimolar 9-ethyladenine. We can ascribe the increase of the rotational correlation time to the apparent increase of molecular radius, since the bromouracil molecules exchange rapidly between the associated and monomer states, but the life time of the associated dimer is larger than the rotational correlation time. We know the association constant of the system and can calculate apparent increase of the molecular radius in weight average. If η of the solvent does not affected on the mixing with adenine, the increase of the rotational correlation time of C-6 corresponds to the apparent increase of the molecular size by factor two.

The rotational correlation times of ^{13}C in the cyclohexyl group also becomes almost twice on the comlpexation with adenine. But the absolute value of the times is shorter than that of C-6. It must be due to the contribution from the internal motion of cyclohexyl group. The C-4 and C-5 carbons do not show remarkable change in the spin-lattice relaxation times on the mixing with adenine. The relaxation of these carbon nuclei is governed by the long distance protons including the intermolecular interaction. Thus the change in the relaxation times can not be simply explained by the change in the molecular radius.

Among the proton relaxation times obtained by the saturation recovery method, the decrease of the H-6 proton relaxation time from

Table 4. ^{1}H Spin-lattice relaxation times of base pairs

Compounds	Concentration(M)	A-C2H	A-C8H	A-NH_2	U-C6H	U-NH
9-ethyladenine	0.15	2.42	3.87	0.72		
1-cyclohexyluracil	0.15				2.27	2.69
9-ethyladenine + 1-cyclohexyluracil	0.15	2.60	3.42	0.52	1.14	0.59

the free to associated state can be explained in terms of the increase of the rotational correlation time of the dimer. On the other hand these of H-2 and H-8 of 9-ethyladenine do not change. It may be due to the fact that the relaxation of these protons are governed by the dipole-dipole interaction by the protons in the ethyl group which rotates freely both in the free and associated state. There is no difference in the relaxation rate in the two states. The change in the relaxation times of the imino proton of BU is remarkable. It is attributable not only to the change in rotational correlation time, but also to the exchange of the protons with the amino protons of A. The evidence will be shown later in a different way. If so, the observed relaxation times of the imino and amino protons in the mixed solution should take an average value T_1,

$$1/T_1 = p_{amino}/T_{1,amino} + p_{imino}/T_{1,imino} \quad .....(14)$$

where p_{amino} and p_{imino} are population of the protons at the amino and imino groups, respectively.

2.3 Direct Proton Exchange Between Complementary Nucleic Acid Bases(25)

The proton magnetic resonance spectra of 9-ethyladenine and 1-cyclohexyl-5-bromouracil (1:1) mixture are shown in Figure 3. When the amino proton signal of A is irradiated, the imino proton signal of BU disappears. In a similar manner, the irradiation at the imino proton of BU reduces the intensity of the amino proton of A. The result can be explained as follows. The protons are reversibly exchanging between the amino group of A and the imino group of BU, and thus the irradiated proton moves to the non-irradiated site through the formation of hydrogen bonded dimers. The exchange of the saturated protons produces

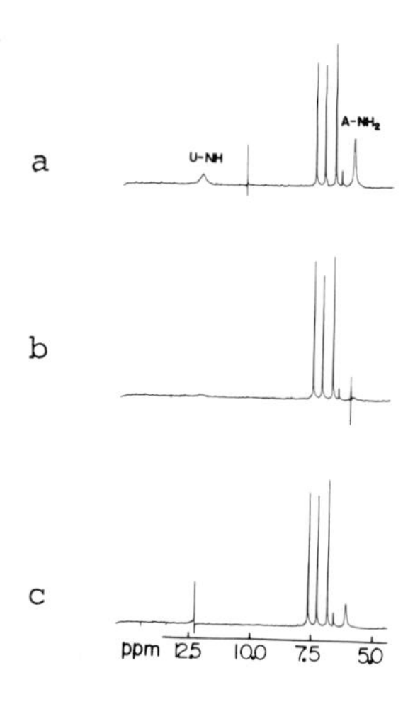

Figure 3. Proton nmr spectra of a mixture of 9-ethyladenine and 1-cyclohexyl-5-bromouracil (both at 0.2M) in chloroform. a, non-perturbed, b, irradiated at the amino proton of A, c, irradiated at the imino proton of V.

partial saturation of the non-irradiated proton site to weaken the intensity of the non-irradiated proton resonance, if the exchange rate is faster than the relaxation time (20). The irradiation at the signal of water proton contaminated in solution gives little change in the intensity of other proton resonances. This fact allows us to exclude the possibility that most of protons exchange via water contained in the solvents.

The same experiment has been done on the mixture of 9-ethylguanine (G) and 1-methylcytosine(C) in DMSO. In this system, it is interesting to see that proton exchange occurs between the imino proton of G and the amino proton of C, but the amino proton of G does not exchange with both the amino proton of C and the imino proton of G (Table 5). If the

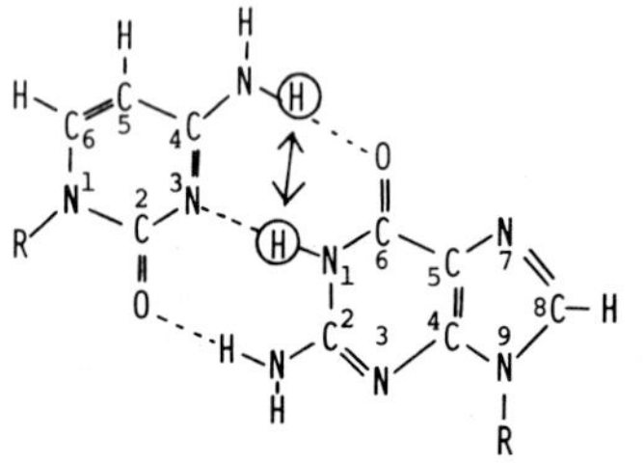

Figure 4. Exchangeable protons in the G-C pair

formation of cyclic hydrogen bonds of the G-C pair (26,27) is enough condition for the exchange of protons, the amino proton of G should exchange with the other protons participating in hydrogen bonds, judging from the distance between the protons. These results make us to consider other factors than distance for the mechanism of the intermolecular proton exchange.

The life time of the proton exchange was quantitatively estimated following the procedures given in eqs.14∿16. We obtained 0.1 sec as the life time of the exchange,i.e., the proton exchanges ten times in a

Table 5. Partial saturation induced by intermolecular exchange of saturated protons

Base pairs	Solvent	Irradiated nuclei	Decrease in the peak intensity(%)				
			G-amino	C-amino	G-imino	A-amino	U-imino
9-ethyladenine[a)] + 1-cyclohexyl-5-bromouracil	$CDCl_3$	A-amino					100
		U-imino				36	
		H_2O				0	0
9-ethylguanine[b)] + 1-methylcytosine	DMSO	G-amino		12	20		
		C-amino	12		86		
		G-imino	0	71			
		H_2O	3	5	20		

a) 0.2M, b) 0.1M

Table 6. Proton exchange rates between 9-ethyladenine and 1-cyclohexyl-uracil

Temperature(°c)	$M_A(\infty)/M_A^\circ=1/\tau$	T_{app}(sec)	T_{1A}(sec)	k(sec -1)
-5	0.855	0.12	0.14	2.41
4	0.786	0.12	0.15	3.64
15	0.611	0.12	0.20	6.45
26	0.437	0.10	0.23	11.1
44	0.257	0.06	0.23	25.0

second in A-BU system at room temperature.

Rate constants were measured at different temperatures and from the Eyring plots activation energies of proton exchanges were obtained as follows: 7.9 kcal/mole for A-BU, 12.7 for A-U, 13.6 for A-TH and 15.3 for A-DH. The order is parallel to that of pk valves of the imino protons of uracil derivatives. The more acidic protons exchange more easily.

For the exchange process formation of hydrogen bonded pair seems necessary. In the dimethylsulfoxide solution of the A-BU mixture, the extent of decrease in peak intensity diminishes compared with that of chloroform solution. Dimethylsulfoxide disturbs more effectively the formation of solute-solute hydrogen bonds. Another evidence is more effective intensity decrease of the resonance due to the bonded hydrogen of the amino group of C. The amino proton resonances appear as a singlet at room temperature, but becomes doublet at lower temperatures. The lower field resonance corresponds to the signal of the amino proton which is participated into the G-C pair and the upper field one is due to the non-bonded one. When the imino proton of G was irradiated, the lower amino proton signal of C more effectively decreased in intensity than the upper one.

The concentration dependency of the rate constants is given in Figure 5. If the proton exchange occurs only through the base pairs, the exchange rate depends on the total concentration of bases, but corrected rate ($k=k_{obs}[A°]/[AU]$) should not depend on the concentration. However the corrected values are still far from constant. It may be partly due to the exchange via water molecules contaminated in solution. Such effect is expected to be predominant in lower solute concentration region. The deviation in higher concentration region means that the formation of dimers is not enough for proton transfer. There are four kinds of A-U dimers as shown in Figure 1. It is probable that only one of the structures can contribute to the proton exchange.

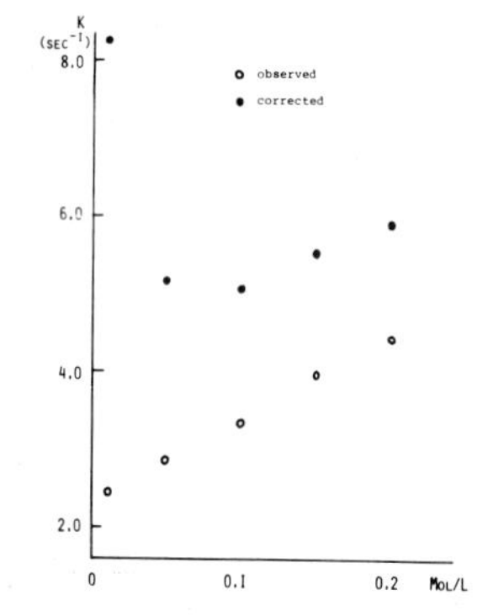

Figure 5. Concentration dependence of the rate constants of the proton exchange between A and U.

Combining these evidences one of the probable mechanisms seems to be keto-enol equilibrium in the hydrogen-bonded dimers. If so, only the imino protons and the conjugated amino proton in the cyclic hydrogen bond can exchange. Watson-Crick type base pairs contribute to it, but Hoogsteen type does not, which would result in the lower rate of exchange than that expected from total dimer concentration.

3. CONCLUSION

Although there is ambiguity over the quantitative population difference, it is clear that the C-2 carbonyl group is used with fairly high probability as a proton acceptor in dimer formation. Among the uracil derivatives uracil and thymine more often use the C-4 carbonyl group than the C-2 carbonyl group. It is reasonable to assume that the two most common bases are able to form hydrogen bonded complexes by using the C-4 carbonyl group, just like the base pair models proposed by Watson and Crick (28), and by Hoogsteen (29). However, even of the base pairs in nucleic acids are able to exist for most of the time using the C-4 carbonyl group, the possibility still exists of complexes being formed which use the C-2 carbonyl group, i.e., the reversed Watson-

Crick and Hoogsteen type base pairs. It is quite surprising that mis-reading of the base sequences during the processes of transcription and translation is quite rare in spite of the apparent arbitrariness of the hydrogen bonded site. It is known that in place of thymine, 5-bromo-uracil incorporates into DNA at the duplication process and induces mutation (30). The present experiments show that in base pairing 5-bromouracil uses the C-2 carbonyl group as a proton acceptor much more frequently than thymine. This presumably means that 5-bromouracil is more likely to give rise to mis-pairs like G-BrU, where the C-2 carbonyl group is used.

It also has become clear that the protons of the amino and imino groups of the base derivatives can exchange directly in non-aqueous atomosphere, if appropriate conditions are satisfied. Although the experiment were carried out on the monomer systems, there is no reason to deny the proton exchange in interior of nucleic acids. The exchange is possibly due to the presence of the keto-enol tautomerism, and we may estimate the frequency of tautomerism from the present date. Besides it the structure of nucleic acid bases which are realized at the time of the proton exchange are different from the normal structure. It might be possible that such altered structures would induce mis-pairing in the process of replication and translation of nucleic acids (31).

REFERENCES

1) R.M. Hamlin Jr., R.C. Lord and A. Rich, Science, 148, 1734 (1965).
2) E. Kuchler and J. Derkosch, Z. Naturf., 21B, 209 (1966).
3) Y. Kyogoku, R.C. Lord and A. Rich, Science, 154, 518 (1966).
4) J. Pitha, R.N. Jones and P. Pithova, Can. J. Chem., 44, 1045 (1966).
5) Y. Kyogoku, R.C. Lord and A. Rich, J. Amer. Chem. Soc., 89, 496 (1967).
6) Y. Kyogoku, R.C. Lord and A. Rich, Proc. Nat. Acad. Sci. U.S., 57, 250 (1967).
7) J.H. Miller and H.M. Sobell, J. Mol. Biol., 24, 345 (1967).
8) Y. Kyogoku, R.C. Lord and A. Rich, Biochim. Biophys. Acta., 179, 10 (1969).
9) R.R. Shoup, H.T. Miles and E.D. Becker, Biochem. Biophys. Res. Commun., 23, 194 (1966).
10) L. Katz and S. Penman, J. Mol. Biol., 15, 220 (1966).
11) L. Katz, J. Mol. Biol., 44, 279 (1969).
12) T. Morishima, T. Inubushi, T. Yonezawa and Y. Kyogoku, J. Amer. Chem. Soc., 99, 4299 (1977).
13) L. Katz, K. Tomita and A. Rich, J. Mol. Biol., 13, 340 (1965).
14) W. Saeuger and D. Suck, Nature, 227, 1046 (1970).
15) R.M. Thomas, "Introduction to Numerical Methods and FORTRAN Programing", John Wiley & Sons, Inc., New York, 1967.
16) O. Kikuchi, "Molecular Orbital Calculation", Kodansha, Tokyo, 1971.
17) K. Morikawa, T. Katsura and Y. Iitaka, unpublished.
18) A. Abragam, The Principles of Nuclear Magnetism, Oxford, Univ.

Press., London, Chapter 8 (1961).
19) B.E. Mann, J.Mag. Resonance, 25, 91 (1977).
20) S. Forsen and R.A. Hoffman, J. Chem. Phys., 39, 2892 (1963).
21) H. Iwahashi and Y. Kyogoku, J. Amer. Chem. Soc., 99, 7761 (1977).
22) A. D'Albis, M.P. Wickens and W.B. Gratzer, Biopolymers, 14, 1423 (1975).
23) H. Spiesecke and W.A. Schneider, Tetrahedron Lett., 14, 468 (1961).
24) G.C. Pimental and A.L. McClellan, "The Hydrogen Bond", Freeman, San Francisco, 1960.
25) H. Iwahashi and Y. Kyogoku, Nature, 271, 277 (1978).
26) E.J. O'Brien, J. Mol. Biol., 7, 107 (1963).
27) H.M. Sobell, K. Tomita and A. Rich, Proc. Nat. Acad. Sci. U.S., 49, 885 (1963).
28) J.D. Watson and F.H.C. Crick, Nature, 171, 737 (1953).
29) K. Hoogsteen, Acta Cryst., 12, 822 (1959); ibid., 16, 907 (1963).
30) W. Hayes, "The Genetics of Bacteria and Their Viruses", Blackwell, Oxford, 1964, p. 278.
31) M.D. Topal and J.R. Fresco, Nature, 263, 285 (1976).

DISCUSSION

B. McConnell: Could you comment on evidence against the possibility that saturation transfer might occur via a free intermediate or catalyst at small concentration ?

Kyogoku: We measured concentration dependency of the transfer rate constants. They gradually increased as the concentration of the base increased. Thus it can be said that the formation of dimers is necessary for the transfer and contaminated species at small concentration in the solvent like water and alcoohol are not responsible for the transfer. But I cannot deny the possibility that the transfer occurs via the intermediate species of the base and solvent molecules themselves like protonated forms.

NMR STUDIES AND MINIMUM ENERGY CONFORMATION CALCULATIONS ON THE 5'-TERMINUS OF MAMMALIAN mRNA

M. M. Dhingra* and Ramaswamy H. Sarma
Institute of Biomolecular Stereodynamics
State University of New York at Albany
Albany, New York 12222 USA.

INTRODUCTION

The 5'-terminal region of a wide variety of mRNAs of eukaryotes contain an unusual sequence of nucleosides in which the terminal is 7-methylguanosine(Rottman et al. 1974). In this structure the 5' hydroxyl of 7-methylguanosine is connected through a triphosphate bridge to the 5'hydroxyl of 2'-O-methylated nucleosides. It has been shown that this unusual sequence is necessary for mRNA to carry out in vitro translation with fidelity (Adams and Cory, 1975; Furuichi et al, 1977). To understand the structure-function relationships we have explored whether the presence of 7-methylated guanosine at the terminus endows any unique stereochemical features to mRNAs. The conformational properties of 7-methylguanosine-5'monophosphate (7-Me-5'GMP) and 5'GMP havebeen studied using NMR spectroscopy and theoretical calculations using classical potential functions.

NUCLEAR MAGNETIC RESONANCE STUDIES.

NMR spectra were recorded 270 MHz in the Fourier transform mode using systems described elsewhere (Lee and Sarma, 1976). The spectra were analysed by LACON III. The coupling constant data were translated into conformational parameters using equations developed by Lee and Sarma (1976) and the data are summarised in Table 1.

Inspection of the data in Table I shows that 7-methylation has an important influence on ribose ring conformational distribution. The ribose conformation which prefers ^{2}E sugar pucker in 5'GMP registers a decrease of 23% ^{2}E on 7-methylation, so much so, the ribose in 7-Me-5'GMP prefers ^{3}E pucker. This is the first example of a purine-5'-mononucleotide which shows preference for ^{3}E pucker in aqueous solution. The data further reveal that 7-methylation has very little influence on the population of conformers about C5'-O5' bond, but has a significant effect on the conformational preference about C4'-C5' bond. There is an increase of 20% in the gg (ψ=60°) population in

B. Pullman (ed.), Nuclear Magnetic Resonance Spectroscopy in Molecular Biology, 31-39.

going from 5'GMP to 7-Me-5'GMP. This increase in gg population is possibly the result of electrostatic interaction between the enhanced positive charges on various atoms of the base and the negative charges on phosphate oxygens. NMR studies on mono- and dinucleoside phosphates have shown that the preferred glycosidic torsion in these systems lies in the anti domain (Lee et al 1976; Ezra et al 1977; Cheng and Sarma, 1977). NMR data do not allow a quantitative evaluation of the magnitude of χ or that of the population distribution of the syn and anti conformers. However, the relative changes in χ can be monitored from the shift trends of H1' and H2' (Giessner-Prettre and Pullman, 1977ab). A comparison of the shift data for 5'GMP and 7-Me-5'GMP (Table I) indicates that 7-methylation causes downfield shift of H1' by 36 Hz and upfield shift of H2' by 24 Hz. These observed shift differences are large in magnitude and are largely due to changes in χ upon 7-methylation with small contribution from electronegativity changes. This is entirely expected because such a substitution can lead to electrostatic interactions between the positive charges on the base atoms and phosphate negative charges which in turn will bring about a concomitant change in χ and ψ. A detailed discussion of the present NMR findings has been presented elsewhere (Kim and Sarma, 1978).

Table I Conformational Properties and Chemical Shift Data on 5'GMP and 7-Me-5'GMP in D_2O, pH=5.0° and Temp 20°C*.

Nucleotide	Conformational Preferences about various bonds			Chemical Shift in Hz	
	C5'-05' g'g'$\rightleftarrows$g'/t'	C4'-C5' gg $\rightleftarrows$ g/t	Ribose ^3E	H1'	H2'
5'GMP	71 $\rightleftarrows$ 29	71 $\rightleftarrows$ 29	37	741.9	420.6
7-Me-5'GMP	74 $\rightleftarrows$ 26	91 $\rightleftarrows$ 9	60	778.8	395.8

*Data from Kim and Sarma (1978)

THEORETICAL CALCULATIONS.

Several theoretical calculations dealing with the conformational preferences about the exocyclic C4'-C5' bond have been reported (Saran et al. 1972, 1973; Yathindra and Sundaralingam, 1973ab). The primary aim has been to establish the correlation between the exocyclic C4'-C5' bond and the various other torsion angles say C5'-05', χ and sugar puckers. These calculations predict that gg conformation about C4'-C5' should be the most stable one in purine and pyrimidine systems irrespective of the nature of the sugar pucker. The conformational preference about the C5'-05' bond is predicted to be g'g' ($\phi \simeq 180°$). Investigation of the interrelation between χ and ψ predicts that the anti-gg conformation to be energetically favored for adenine mononucleotides (Yathindra and Sundaralingam, 1973a). On the other hand, for guanine mononucleotides, the most favored conformation is syn-gg ie χ=190°, ψ=80° for ^{2}E and χ=180°, ψ=50° for ^{3}E (Yathindra and Sundaralingam, 1973b).

In order to explore the effect of positive charge in7-methylgnanosine base on the conformational interrelationships between χ and ψ and to supplement the NMR data, we have undertaken a classical potential energy calculations on 7-Me-5'GMP and 5'GMP. The calculations have been done on the nucleotides in which the phosphate group is in the monoionic state. Conformational energies have been evaluated by simultaneous variation of the rotation angles about the glycosyl and C4'-C5' bonds at intervals of 10°. The conformational angle about C5'-O5' bond was kept at the most favored value of 180°. Contributions from Vander Waals, electrostatic and torsional interactions have been considered for the total potential energy and were evaluated as reported by earlier workers (Yathindra and Sundaralingam, 1973a). The electrostatic interactions have been estimated using the electronic charge distributions calculated by CNDO method. The electronic charge distribution in 7-methylguanosine and guanosine is given in Table II. The CNDO electronic charge distribution for the phosphate group were approximated to those for dimethylphosphate mono anion (Groenstein, 1978, personal communication). A dielectric constant of 4 was used in these calculations. The bond lengths and bond angles for guanosine moiety were taken from the available x-ray data on guanosine dihydrate (Thewalt et al. 1970). Calculations have been performed for C(2') endo and C(3') endo sugar geometries. The conformational energy maps have been constructed as a function of torsions about the C4'-C5' and glycosyl bonds for both type of sugar puckers. The presentation of the results on the conformational energy maps has been limited to 5 kcal/mole above the global minimum. The isoenergy curves for 7-Me-5'GMP and 5'GMP are shown in Figs. 1 and 2 respectively.

The results of the calculation carried out for ^{3}E 7-Me-5'GMP are presented in Fig. 1a. The map shows a global minimum at ψ=50° and χ=310° and this indicates the preponderance of anti-gg conformation. In addition there exists a local minimum (0.5 kcal/mole above the global one) corresponding to ψ=180° (gt) and χ in the range of 260° to 50°. The tg conformation corresponding to ψ=300° is 3 kcal/mole higher in energy compared to the global minimum. This conformation also spans a large region of χ ie from 250° to 50°. The prediction of three familiar conformations gg, gt and tg about the exocyclic C4'-C5' bond is in good agreement with those predicted by other theoretical and experimental methods. Of the three conformers, the gg is the most favored and is in excellent agreement with our experimental data discussed in the NMR section. However, the large areas of 1-3 kcal/mole (enveloping χ in the anti and syn domains) associated with the global and local minima indicate rather greater degree of flexibility about the glycosidic torsion. Such larger domains for χ have been predicted for guanine derivatives by PCILO method (Berthod and Pullman, 1971ab).

The isoenergy curves for 7-Me-5'GMP with ^{2}E sugar geometry is shown

TABLE II Electronic Charge Distribution Guanine and 7-Methylguanine

Atom	CNDO Guanine	CNDO 7-Methylguanine	RL Method* Guanine
N1	-0.2268	-0.2163	-0.184
C2	+0.4334	+0.4738	+0.556
N3	-0.3726	-0.3629	-0.636
C4	+0.2349	+0.2706	+0.311
C5	-0.1222	-0.1410	-0.302
C6	+0.4167	+0.4403	+0.370
N7	-0.2285	+0.0940	-0.540
C8	+0.1590	+0.1989	+0.224
N9	-0.0717	-0.0425	+0.057
06	-0.4336	-0.3847	-0.505
H1	+0.1137	+0.1511	+0.192
H8	-0.0098	+0.0516	+0.068
N2	-0.2440	-0.2244	-0.384
H2	+0.1207	+0.1497	+0.225
H2	+0.1409	+0.1666	+0.225
C1'	+0.3651	+0.3510	-0.140
01'	-0.3273	-0.3167	+0.105
C7		+0.1424	
H-		+0.0073	
H- CH_3		+0.0246	
H-		+0.0073	

*From Renugopalakrishnan and A. V. Lakshminarayanan (1971).

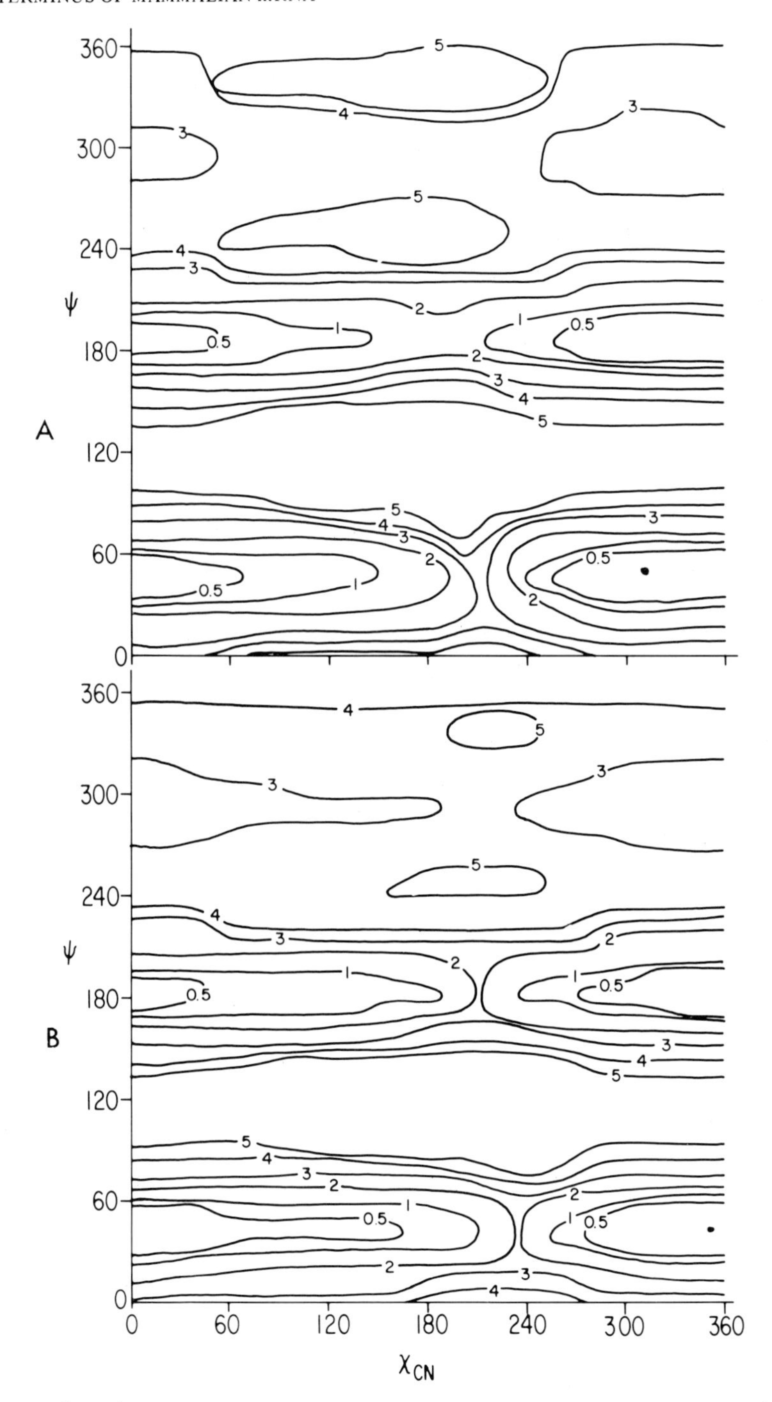

Fig. 1. (ψ-χ) conformational energy map, Isoenergy curves in kcal/mole with respect to global minimum. (a) ^{3}E (b) ^{2}E 7-Me-5'GMP

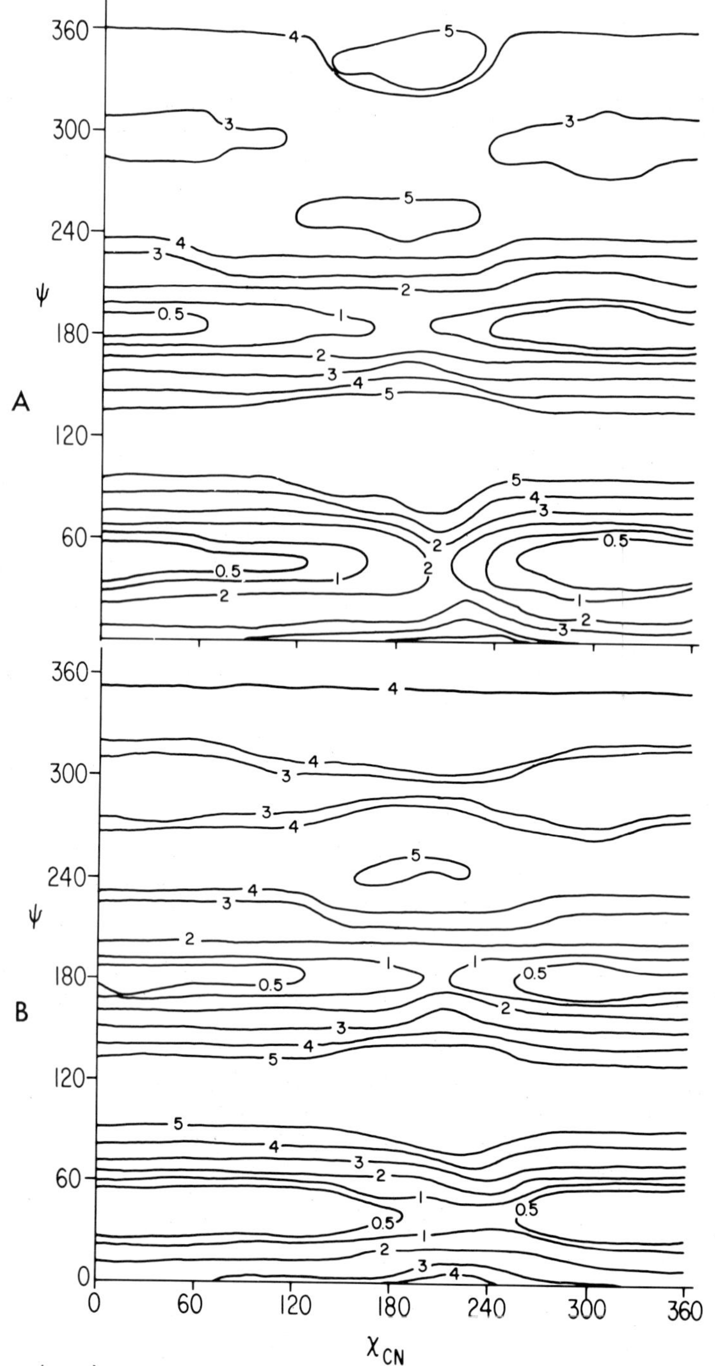

Fig 2. (ψ-χ) conformational energy map. Isoenergy curves in kcal/mole with respect global minimum. (a) ^3E (b) ^2E. 5'GMP

in Fig. 1b. This map is very similar to the one presented for ^3E sugar geometry. The global minimum is however shifted to $\psi=40^\circ$ and $\chi=350^\circ$. The glycosidic torsion for the most favored conformation is once again in the anti domain. A local minimum, 0.5 kcal above the global one, is observed for $\psi=180^\circ$ and χ almost spaning the entire range of 0°-360°. However, the areas engulfing the 1-2 kcal/mole isoenergy curves are still larger compared to corresponding isoenergy curves for the ^3E geometry. The 3 kcal/mole energy curves practically spans the whole range of χ indicating greater flexibility about the glycosidic linkage under ^2E sugar geometry conditions. This greater flexibility for ^2E sugar geometry is generally true, not only for guanine derivatives, but for other purine nucleotides.

The isoenergy curves for 5'GMP (Figs. 2ab) are very similar to those for 7-Me-5'GMP. For ^3E sugar pucker, the global minimums is at $\psi=50^\circ$ and $\chi=300^\circ$ while for ^2E sugar, it is at $\psi=40^\circ$ and $\chi=30^\circ$. In 5'GMP as well, the ^2E geometry imparts greater flexibility about χ.

A comparison of the theoretical deductions on 7-Me-5'GMP and 5'GMP suggests that the most favored conformation about C4'-C5' bond is gg in both. However, there are certain differences in the flexibility about glycosidic torsion. In 5'GMP, the glycosidic torsion has a greater flexibility and the existence of anti and syn conformers in equilibrium is probable, while in 7-Me-5'GMP, the flexibility about the glycosidic torsion is relatively smaller and hence relatively smaller populations of the syn conformer in the equilibrium blend is expected. These differences in the glycosidic torsion are corroborated by the NMR data discussed earlier.

It is pertinent to point out that the results of the potential energy calculations for 5'GMP presented here are at variance with those reported earlier by Yathindra and Sundaralingam (1973b). The results of their calculations predict syn-gg conformation to be the most stable and the potential barrier for the interconversion of syn-anti equilibrium is rather high. Secondly the observation of a syn-gg conformation is unexpected because the syn conformation in purine mononucleotides is normally associated with a tg conformer about C4'-C5' (Sarma et al. 1974). These differences in the CPF calculation results may be due to the differences in the electrostatic interactions. The electrostatic interactions which depend on the electronic charge distributions calculated by different methods. A comparison of the electronic charge distribution used by Yathindra and Sundaralingam (1973b) and the one used in the present calculations (Table II) indicate large differences in the magnitude of charges on the same atom (for example N3) and in some cases, even reversal in sign (for example O1'). We believe that the conformational energies calculated in this paper present a realistic picture of the conformational properties of guanosine derivatives because they agree with solution studies and PCILO calculations (Berthod and Pullman, 1973) and the more accurate CNDO method is used to calculate the electronic charge distribution. Further, recently Imoto et al. (1977) have shown from spin lattice relaxation data, that even 5'-adenine nucleotides have greater flexibility and that 5'AMP exists as an equilibrium blend of conformers having χ in the anti, intermediate and syn ranges.

ACKNOWLEDGMENTS

We thank the organizers of the 11th Jerusalem Symposia for inviting us to present these results. This research was supported by Grant No. CA12462 from NCI and PCM 75-16406 from NSF. This research was also supported in part by NIH Grant No. 1-P07-PR00798 from the Division of Research Resources. We thank G. Govil for several of the computer programs which were of use in this study. The authors thank the University Computer Center, particularly Mr. J. Quinn, for assistance in the course of this work.

REFERENCES

Adams, J., and Cory, S. (1975), Nature. 255, 28

Berthod, H. and Pullman, B. (1971a), Biochim. Biophys. Acta 232, 595.

Berthod, H., and Pullman, B. (1971b), Biochim. Biophys. Acta 246, 359.

Berthod, H., and Pullman, B. (1973), FEBS Lett. 30, 231.

Cheng, D. M., and Sarma, R. H. (1977), J. Amer. Chem. Soc. 99, 7333.

Ezra, F. S., Lee, C. H., Kondo, N. S., Danyluk, S. S., and Sarma, R. H. (1977) Biochemistry 16, 1977.

Furuichi, Y., LaFiandra, A., and Shatkin, A.J., (1977), Nature 266, 235.

Giessner-Prettre, C., and Pullman, B. (1977a), J. Theor. Biol. 65, 189.

Giessner-Prettre, C., and Pullman, B. (1977b), J. Theor. Biol. 65, 171.

Imoto, T., Shibata, S., Akasaka, K. and Hatano, H, (1977) Biopolymers 16, 2705.

Kim, C. H., and Sarma, R. H., (1978) J. Am. Chem. Soc. 100, 1571.

Lee, C. H., and Sarma, R. H., (1976) J. Am. Chem. Soc., 98, 3541.

Lee, C. H., Ezra, F. S., Kondo, N. S., Sarma, R. H. and Danyluk, S.S. Biochemistry 15, 3627.

Renugopalakrishnan, V. and Lakshminaraynan (1971), Biopolymers 10, 1159.

Rottman, F., Shatkin, A. J., and Perry, R. P. (1974), Cell 3, 197

Sarma, R. H., Lee, C. H., Evans, F. E., Yathindra, N., and Sundaralingam, M. (1974), J. Am. Chem. Soc. 96, 7337.

Saran, A., Pullman, B., and Perahia, D. (1973), Biochem. Biophys. Acta 299, 497.

Saran, A., Pullman, B., and Perahia, D. (1972), Biochem. Biophys. Acta 287, 211.

Thewalt, U., Bugg, C. E., and Marsh, R. E., (1970), Acta. Cryst. B26, 1089.

Yathindra, N., and Sundaralingam, M., (1973a) Biopolymers 12, 297.

Yathindra, N., and Sundaralingam, M., (1973b) Biopolymers 12, 2075.

* Permanent address: Tata Institute of Fundamental Research
Homi Bhabha Road, Bombay, India

SOLUTION CONFORMATIONS OF PURINE(β)NUCLEOSIDES AND ANALOGS

H.-D.Lüdemann and E.Westhof*
Institut für Biophysik und Physikalische Biochemie
Universität Regensburg - Postfach 397
D-8400 Regensburg, Germany

* present address: Department of Biochemistry
University of Wisconsin, Madison 53706, U.S.A.

INTRODUCTION

The purinenucleosides and their natural or synthetic analogs are a class of compounds with a broad spectrum of physiological effects. Modification of a single side group at the base or ribose moiety can produce substances with antibiotic, antiviral, or antitumor activity. The biological activity of these analogs can only in part be explained through their different behaviour in the complex purinenucleoside metabolism (1,2). The study of the solution conformations of these derivatives could reveal purely sterical arguments for various modes of biological action of these substances, and might in addition help to understand the complex behaviour of the purinenucleosides in preparative chemistry (3). By application of the rich choice of CW and FTNMR-techniques to this class of compounds the preferred solution conformations can be derived.

It is obvious that the conformations, which these compounds can acquire in aqueous solution are of prime importance to their biological activity, and thus water should be the solvent of choice. However, the investigation of aqueous solutions of the purine(β)nucleosides by NMR techniques has two very severe experimental limitations: First, many of the nucleosides are only very sparingly soluble in neutral water at room temperature. Second, the aqueous solutions of these compounds do show pronounced self association at the concentrations necessary for NMR-experiments (4) and this association renders the application of proton-T_1 and NOE studies to the quantitative determination of conformations nearly impossible, since it is very difficult to separate intramolecular effects from the contributions of neighbouring molecules.

In the studies presented here, liquid deuteroammonia was used as a solvent. In this liquid base-stacking is not observed (5) and the ability of the ammonia molecule, to

B. Pullman (ed.), Nuclear Magnetic Resonance Spectroscopy in Molecular Biology, 41-51.

participate like water as a donor and as an acceptor in hydrogen bonds to the nucleoside, holds the promise that similar conformations will be preferred in these two solvents. Whereever the solubility of the nucleosides in water permitted a comparison between the two liquids, no significant differences were found (6,7). These results are in marked contrast to the behaviour of the purine(β) ribosides in dimethylsulfoxide (8). In this solvent pronounced differences in the conformational preferences of adenosine and inosine are observed, which are neither found in the aqueous solutions (9) nor in liquid deuteroammonia (5). Furthermore, deuteroammonia is an excellent solvent for purinenucleosides and their analogs. All compounds tried were at least to a concentration of 0.5 molal soluble in this solvent, and carefully dried and oxygen free ammonia is a remarkably inert solvent. Some of the nucleosides studied below have been kept in ammonia solutions for several years without any observable change in the ^{1}H-NMR spectra.

CONFORMATIONS OF THE PURINE(β)RIBOSIDES

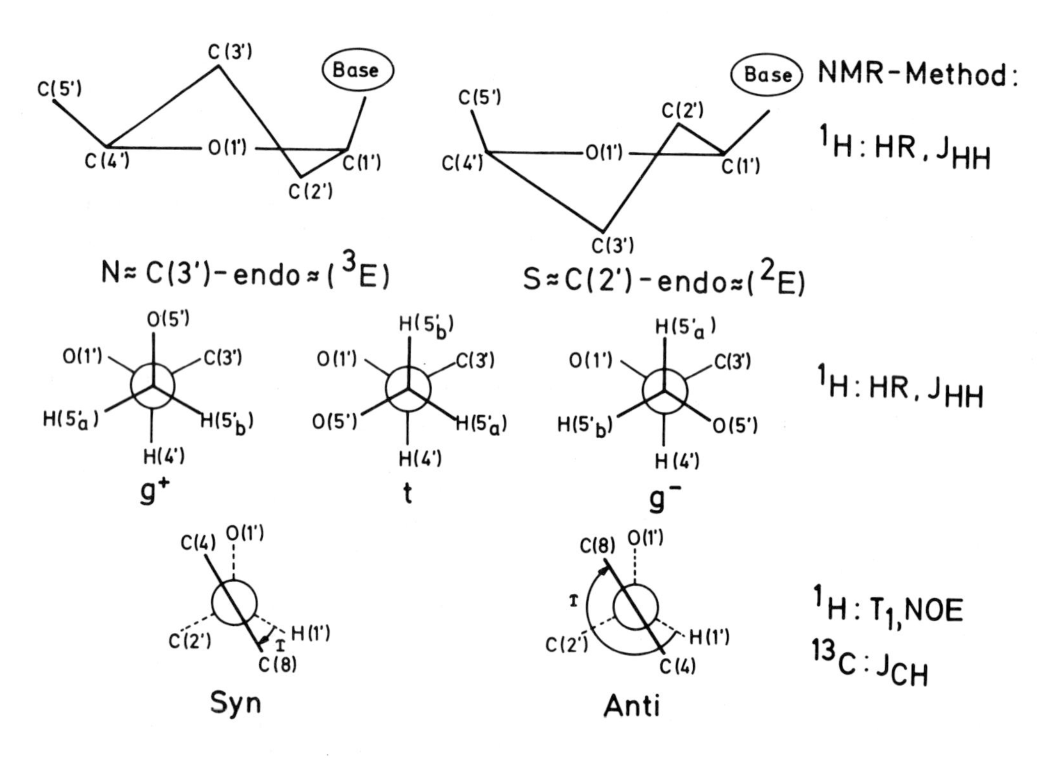

Fig. 1: Schematic representation of the major modes of internal motion in purine(β)ribosides

Fig. 1 gives the description of the conformational equilibria of the purine(β)ribosides. The representation of the furanoside ring by the two state N ⇄ S model and the interconversion of these two forms by pseudorotation was introduced by Altona and Sundaralingam (10,11). Details of the conformational equilibria can be deduced from the vicinal proton-proton coupling constants of the ribose protons H(1') to H(4'). Following a proposal by Hruska (12) the vicinal couplings between H(4'), H(5'A) and H(5'B) can be used to determine the equilibrium between the three rotamers of the 5'hydroxymethylgroup. Finally the position of the base towards the ribose ring can be determined by proton relaxation studies (8,13,14) and by proton-carbon coupling constants (15). However, the description of a nucleoside as a blend of well defined and energetically separated conformers requires that the energy barriers between the conformers are significantly higher than the thermal energy of the molecules. The activation energy for the syn ⇄ anti equilibrium of the purine(β)ribosides was determined by Rhodes and Schimmel to 26 kJ·mole^{-1} (16). At least a lower limit of ~ 15 kJ·mole^{-1} can be given for the rotation of the 5'hydroxymethylgroup from comparison with results obtained for stereochemically similar ethane fragments. For the pseudorotation of the furanose ring of the common purine(β)ribosides our group (17) derived from the analysis of the temperature dependence of the ^{13}C longitudinal relaxation rates activation energies of 20 ± 2 kJ·mole^{-1}. Taken together one can deduce from these data that all three energy barriers are large compared to the thermal energy and the nucleosides can be described as a mixture of the different conformers.

DETERMINATION OF THE ACTIVATION ENERGY FOR THE N ⇄ S TRANSITION FROM ^{13}C-RELAXATION RATE MEASUREMENTS.

In Fig. 2 the longitudinal relaxation rates R_1 of the single, proton-noise-decoupled tertiary carbon-13-atoms in adenosine (A), isopropylideneadenosine (iA), and the 5'methylester of 5'adenosinemonophosphate (5'CH_3AMP) are given. The results for A and iA have been published previously (17) and are included for comparison. In iA the longitudinal relaxation rates of C-2, C-8 and C-1' to C-4' are identical in the whole temperature range covered. These results allow two important conclusions: First, all C-H bond length within the base and the sugar are identical. Second, the diffusion of the molecule can be described by a single correlation time. The isotropic rotating sphere appears a valid description for the nucleosides in solution. The introduction of the isopropylidenegroup into the ribose moiety gives additional rigidity to the furanose ring.

In the unblocked adenosine the relaxation rates of the ri-

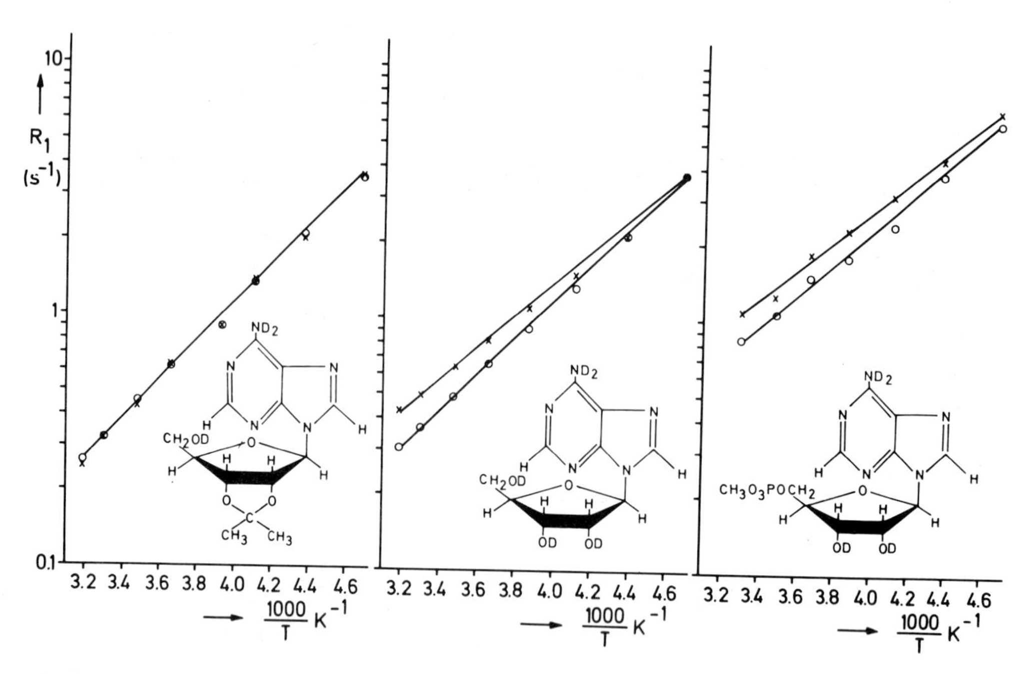

Fig. 2: Longitudinal C-13-relaxation rates vs reciprocal temperature for isopropylideneadenosine, adenosine and 5'adenosinemonophosphatemonomethylester. (x)R1 [C2,C8] (o)R1 [C1' to C4']

bose carbons C-1' to C-4' are at high temperatures smaller than the rates found for C-2 and C-8. This proves that the pseudorotation occurs at a rate comparable to the rotatoric diffusion. From the differences in the relaxation rates as function of temperature one can derive the activation energy for this process. Two models for the calculation of the correlation time of a small rotating side group fixed to an isotropic tumbling molecule are given in the literature: One describing jumps of 120° around a fixed axis (18) and the second for the case of small step diffusion (19). Both models are certainly a simplification for the motion of the single carbon-proton vectors in the process of pseudorotation and the correlation times derived from the relaxation rates by the two models differ by a factor of three. The temperature dependence of τ_c is however the same in the two sets of results and one arrives by both models at identical activation energies. The details of the calculation have been published (17). Table I compares the activation energies for the unmodified purine(β) ribosides with some recently obtained results for nucleoside analogs (20). The data found are remarkably independent on modifications at the base and the ribose moiety. In formycin B and 8-bromoadenosine the S-conformation dominates with [S] ≈ 0.8 (compare table II) and one would expect an increased rigidity of the furanose ring, which is not found in the experiments. The removal of the

Table I: Activation Energies for the N ⇄ S Pseudorotation of some Purine(β)ribosides and Analogs in Liquid Deuteroammonia

Compound	ref.	E_G(kJ·mol^{-1})
2'3'Isopropylideneadenosine	17	> 30(not measurable)
Adenosine	17	21
Inosine	17	19
Guanosine	17	20
Xanthosine	17	19
Formycin B	20	23
8 Bromoadenosine	20	21
5'Deoxyadenosine	20	22
5'Adenosinemonophosphatemono-methylester	20	18

5'hydroxylgroup in 5'deoxyadenosine eliminates the possibility of hydrogen bond formation between N(3) and the 5' hydroxymethylgroup and reduces [S] to ~ 0.5, but this modification also does not influence the flexibility of the ring. With the introduction of the monomethylphosphategroup at C(5') a neutral nucleotide analog is formed, and the furanose ring is bound between two bulky substituents. Even this drastic alteration does not lead to a restriction of the mobility. All these results are in marked contrast to the data found for uridine and cytidine (17). In the latter compounds the carbons of the ribose moiety and the base yield identical correlation times, and the activation energies for the pseudorotation in the pyrimidine (β)nucleosides must be raised by at least 6 kJ·mole^{-1} compared to the purinederivatives.

DETERMINATION OF THE RIBOSE CONFORMER EQUILIBRIA FROM THE VICINAL PROTON-PROTON COUPLING CONSTANTS.

The results derived for this equilibria are contained in table II. The position of the N ⇄ S equilibrium and structural details of the N and S conformer can be deduced from the vicinal coupling constants J(1'2'), J(2'3') and J(3'4') following the statistical approach and the concept of pseudorotation proposed by Altona and Sundaralingam (10,11). Comparison of the coupling constants and the crystal conformations of a large number of β-nucleosides permitted the determination of a Karplus equation for the vicinal coupling constants of the protons H(1') to H(4'). Following this procedure we derived for our data in liquid ammonia the Karplus equation given below (21):

(1) $J(ij) = 10.0 \cos^2\varphi(ij) - 0.95 \cos\varphi(ij)$

(2) $J(obs.,i,j) = [N]J(N,i,j) + [S]J(S,i,j)$

Table II Solution Conformation of the Ribose Moiety in Purine(β)ribosides and Analogs.

II.I Influence of Base Modifications

Solvent:	ND_3(-60°C)				D_2O(+40°C)			
Compound	[N]	[g⁺]	[t]	ref.	[N]	[g⁺]	[t]	ref.
A	.43	.70	.20	21	.37	.71	.18	22
G	.37	.75	.14	21	-	-	-	-
I	.37	.77	.14	21	.41	.64	.23	22
X	.41	.77	.12	21	-	-	-	-
Pr	.44	.71	.20	21	.42	.65	.20	23
T	.38	.68	.20	21	-	-	-	-
2zA	.47	.65	.23	23	-	-	-	-
3cA	.32	.76	.14	24	.40	.58	.40	24
Bza	.31	.77	.06	24	.40	.55	.42	24
3c8zA	.31	.50	.21	24	.54	.47	.42	24
8zA	.38	.19	.41	22	.52	.56	.31	22
8zI	.37	.27	.40	22	.51	.54	.32	22
8zG	.37	.32	.40	22	.48	.52	.34	22
Fo	.28	.75	.11	23	.26	.65	.22	23
FoB	.16	.80	.10	21	.26	.65	.23	23
8BrA	.17	.22	.47	21	.25	.83	.12	23
8BrI	.12	.64	.21	23	.28	.63	.27	23
8BrG	.12	.72	.14	21	-	-	-	-
8BrX	.12	.74	.12	23	-	-	-	-
II.II Influence of Ribose Modifications								
2'dA	.36	.56	.21	25	.30	.54	.30	26
3'dA	.95	.86	.06	25	.75	.59	.30	25
5'dA	.51	-	-	25	.55	-	-	25
8Br5'dA	.45	-	-	25	.56	-	-	25
2'amA	.13	.52	.25	7	-	-	-	-
3'amA	.94	.93	.07	7	.77	.74	.12	7
3'ClA	.61	.76	.15	27	-	-	-	-
3'thioA	1.0	1.0	0	27	-	-	-	-

II.III 5'Adenosinemonophosphatemonomethylester (5'CH_3AMP)

ND_3	[N]	[g⁺]	[t]	[g']	ref.
-60°C	.43	.86	.11	.93	20

Abbreviations: A:adenosine; G:guanosine; I:inosine; X:xanthosine; Pr:purineriboside; T:tubercidin; 2zA:2-azaadenosine; 3cA:3-deazaadenosine; Bza:benzimidazoleriboside; 3c8zA:3-deaza-8-azaadenosine; 8zA:8-azaadenosine; 8zI:8--azainosine; 8zG:8-azaguanosine; Fo:formycin; FoB:formycin B; 8BrA:8-bromoadenosine; 8BrI:8-bromoinosine; 8BrG:8-bromoguanosine; 8BrX:8-bromoxanthosine; 2'dA:2'deoxyadenosine; 3'dA:3'deoxyadenosine; 5'dA:5'deoxyadenosine; 8Br5'dA:8-bromo-5'-deoxyadenosine; 2'amA:2'aminoadenosine; 3'amA: 3'aminoadenosine; 3'ClA:3'chloroadenosine; 3'thioA: 3'thioadenosine.

The constants of equation (1) differ slightly from the one in the original paper and are derived from compounds with unmodified ribose moieties only. Replacement of any of the hydroxylgroups at C-2', C-3' or C5' by a substituent with a different electronegativity will change the constants of eq. (1). The exact value for this electronegativity correction for the different substituents introduced is unknown. It could be obtained experimentally by the same statistical approach as used for the ribosides. Neither the solid state data nor sufficient solution conformation data for this approach are available at present. Consequently the Karplus equation with the constants given above was used for the modified ribose moieties also. The additional uncertainty introduced by this procedure into the determination of the N and S mole fractions should not exceed ± 5 %.
Hruska (12) developed a procedure to determine the rotamer populations of the 5'CH_2OH-group from the coupling constants between H(5'A), H(5'B) and H(4'). With the Karplus parameters given in eq. (1) the equations used are:

$$P(g^+) = 1.46 - \frac{J_A(4'5') + J_B(4'5')}{8.9}$$

$$P(t)[P(g^-)] = \frac{J_A(4'5')}{8.9} - .23 \qquad (3)$$

The determination of the two conformers t and g^- rests on an unequivocal assignment of the two protons H(5'A) and H(5'B) in the spectra. This has recently been achieved by stereospecific partial deuteration of C(5') in adenosine (28). The spectrum of the aqueous solution of A has been assigned with the result, that the signal of H(5'B) is found at higher field. With this assignment the spectra of all aqueous nucleoside solutions yield the result that the g^+ rotamer is the most populated with a mole fraction of ~ 0.7, while the t-rotamer has a population of 0.2 to 0.3 and g^- being the rotamer with the lowest probability. This distribution of populations is also found in the solid state (29). In order to retain this sequence of populations also for the deuteroammonia solutions the assignment of the H(5') has to be reversed for the latter solvent. Alternativly one could argue, that specific interactions between the ammonia molecules and the 5'hydroxylgroup destabilize t relative to g^-, since regardless which assignment is chosen, the distribution between t and g^- rotamer is more balanced in ammonia than in water.
In table II,I the influence of base modifications in the purine(β)ribosides upon the conformations preferred by

the ribose are collected. In the common nucleosides the S-state dominates with a mole fraction ~ 0.6. Substitution of H(8) by a bulky substituent restricts the base to the syn range (29,30) and increases the population of S to ~ 0.85. A similar preference is observed for the C-nucleosides formycin and formycin B, where the mole fraction of the ribose in the S-state is ~ 0.80. The summary of these data is that all significant changes observed resulting from a base modification, lead to a stabilisation of the S-state.
Table II,II collects the data for the purine(β)ribosides with chemical modifications at the 2', 3' or 5'hydroxylgroup. The removal of the 5'hydroxylgroup in 5'dA leads to an increase of the N-conformer. The increase is even more pronounced, when 8 BrA and 8 Br5'dA are compared. This result is indirect evidence for the stabilisation of the S-syn-g^+ conformation by a hydrogen bond between N(3) and O(5'). Substitution of the 2'hydroxylgroup by a hydrogen in 2'dA or an aminogroup in 2'amA stabilizes, compared to the unsubstituted A, the S-conformer. This effect is most pronounced for the amino group. In all 3'-substituted adenosine(β)ribosides the N-g^+ conformer is strongly favoured. 3'dA, 3'amA and 3'thioA occur at -60°C almost exclusively in this conformation. The substituents replacing the 3'-hydroxylgroup in the ribose moiety do have no common feature, to distinguish them from the hydroxylgroup. They do represent a wide variation in van der Waals radii as well as in polarities, and with our present state of knowledge it appears impossible to give a convincing explanation for the experimental results.
The description of the ribose by a small number of possible conformers separated by sufficiently high energy barriers permits the calculation of the dipole-dipole relaxation rates ρ(d,s) between the individual protons from:

$$\rho(obs,d,s) = [S]\rho(S,d,s)+[N]\rho(N,d,s)$$

Only the average relaxation rates between H(8) and the ribose protons do have to be extracted from relaxation studies (5,6).
Among the analogs presented in table II 2'amA and 3'amA are a pair of compounds with a high conformational purity and their spectra are sufficiently resolved to allow the determination of nuclear Overhauser enhancements and longitudinal relaxation rates for the single ribose protons. These two compounds were therefore chosen for a critical check of a previously proposed correlation between the conformational states of the ribose and the syn ⇄ anti equilibrium (5,6,7). The proton relaxation data of the common purine(β)ribosides A,G,I and X can be quantitatively reproduced by a correlation between N and anti and S and syn (5,6). Since the four nucleosides represented mixtures of **comparable** concentrations of the N and S state, it ap-

peared desirable, to analyse compounds consisting practically of one conformer, in order to reduce the number of adjustable parameters. The results obtained from 3'amA confirmed the strict correlation between anti and N, also observed in the solid state (29). The data for 2'amA can be explained by an exclusive preference of the S ribose for the syn conformation of the base. However, the scatter of the results is significantly reduced, if one calculates the individual relaxation rates under the assumption, that approximately one third of the molecules with a S-ribose have the adenine in the anti-position (7).
In table II,III the results from the conformational analysis for the 5'adenosinemonophosphate-monomethylester are compiled. The close correspondence between these results and the data obtained for 5'AMP in aqueous solution (14, 31,32) give additional support for the similarity of the conformations of the nucleotides in water and ammonia. A preliminary analysis of the proton relaxation data (20) leads to the conclusion that this nucleotide does have the same conformational preferences and correlations as the purine(β)ribosides. The relaxation data can only be explained by a distribution of the base between the syn and anti position.

SUMMARY

The purine(β)ribosides represent a class of compounds with a high conformational variability. Replacement of the 2'hydroxylgroup by a hydrogen or an aminogroup increases the population of the S-state of the ribose moiety. Exchanging the 3'hydroxylgroup by a hydrogen, a chlorine, an amino- or a thiogroup leads to strong stabilisation of the N-conformer. Proton relaxation studies on 2'amA ([S] = 0.87) and 3'amA ([N] = 0.94) corroborate the correlation between the N-state of the ribose and the anti position of the base and between S and syn. Preliminary studies on the monomethylester of 5'AMP can be explained by the same correlation.
The activation energies for the N ⇄ S equilibrium have been determined for some nucleoside analogs and 5'CH_3AMP. Neither the removal of the 5'hydroxylgroup in 5'dA nor the introduction of the 5'methylphosphate group in 5'CH_3AMP has a significant influence upon the mobility of the furanose ring.

REFERENCES

1. A.Bloch
   Ann.N.Y.Acad.Sci. Vol. 255, 1975
2. R.J.Suhadolnik
   Nukleoside Antibiotics, Wiley Intersience N.Y., 1970
3. A.Williamson ed.
   Chemistry of Nucleic Acids Components Nucleic Acid Research, Spec.Publ.Nr. 1, 1975
4. P.O.P.Ts'o
   Basic Principles in Nucleic Acid Chemistry, Vol. 1, 526, 1974
5. H.-D.Lüdemann, O.Röder, E.Westhof, E.v.Goldammer, A.Müller
   Biophys.Struct.Mechanism 1, 121, 1975
6. H.-D.Lüdemann, E.Westhof, O.Röder
   Europ.J.Biochem. 49, 143, 1974
7. H.Plach, E.Westhof, H.-D.Lüdemann, R.Mengel
   Europ.J.Biochem. 80, 295, 1977
8. J.P.Davis, P.A.Hart
   Tetrahedron, Vol. 28, 2883, 1972
9. H.-D.Lüdemann, E.v.Goldammer
   Zeitschrift für Naturforschung 28c, 361, 1973
10. C.Altona, M.Sundaralingam
    J.Amer.Chem.Soc. 94, 8205, 1972
11. C.Altona, M.Sundaralingam
    J.Amer.Chem.Soc. 95, 2333, 1973
12. F.E.Hruska
    The Jerusalem Symposia on Quantum and Biochemistry, Vol. V, 345 (E.D.Bergmann, B.Pullman eds.) 1973
13. C.Chachaty, G.Langlet
    FEBS Letters, 68, 181, 1976
14. K.Akasaka, T.Imoto, S.Shibata, H.Hatano
    J.Magn.Res. 18, 328, 1975
15. D.B.Davis
    studia biophysica, 55, 29, 1976
16. L.M.Rhodes, P.R.Schimmel
    Biochem. 10, 4426, 1971
17. O.Röder, H.-D.Lüdemann, E.v.Goldammer
    Europ.J.Biochem. 53, 517, 1975
18. D.E.Woessner
    J.Chem.Phys. 36, 1, 1962
19. A.Allerhand, R.Komorski
    J.Amer.Chem.Soc. 95, 8228, 1973
20. H.Plach
    Dr.-Thesis, Universität Regensburg, 1978
21. E.Westhof, O.Röder, I.Croneiss, H.-D.Lüdemann
    Zeitschrift für Naturforschung 30c, 131, 1974
22. H.-D.Lüdemann, E.Westhof, I.Cuno
    Zeitschrift für Naturforschung 31c, 135, 1976
23. H.-D.Lüdemann, E.Westhof
    Zeitschrift für Naturforschung 32c, 528, 1977

24. H.-D.Lüdemann, H.Plach, E.Westhof, L.B.Townsend
to be published
25. E.Westhof, H.Plach, I.Cuno, H.-D.Lüdemann
Nucleic Acid Research 4, 939, 1977
26. K.N.Slessor, A.S.Tracey
Carbohydrate Res. 27, 407, 1973
27. H.-D.Lüdemann, R.Mengel
to be published
28. R.G.S.Ritchie, A.S.Perlin
Carbohydrate Res. 55, 121, 1977
29. M.Sundaralingam
The Jerusalem Symposia on Quantum Chemistry and Biochemistry, Vol. V, 417 (E.D.Bergmann and B.Pullman eds.), 1973
30. S.S.Tavale, H.M.Sobell
J.Mol.Biol. 48, 109, 1970
31. F.E.Evans, R.H.Sarma
J.Biol.Chem. 249, 4754, 1974
32. C.Chachaty, T.Zemb, G.Langlet, Tran-Dinh Son, H.Buc, M.Morange
Europ.J.Biochem. 62, 45, 1976

NITROGEN-15 NUCLEAR MAGNETIC RESONANCE SPECTROSCOPY OF ^{15}N-LABELED NUCLEOTIDES
Investigations of Ribonuclease A - Nucleotide Interactions

P.Büchner, F.Blomberg and H.Rüterjans
Institut für Physikalische Chemie der Universität
D-44oo Münster, Germany

We here report on ^{15}N NMR investigations of ribonucleoside-3'-phosphates, biologically enriched with the ^{15}N isotope at all nitrogen positions. The studies are carried out in water over a large pH range up to alkaline pH values resulting in a complete pH dependence of the ^{15}N chemical shifts. With the ^{15}N-enriched compound we are able to study the pH dependence of ^{15}N-coupling constants. Together with the various ^{13}C-^{15}N couplings which are currently being investigated in our laboratory, these couplings should allow certain insights into the electronic structure of the nucleobases at different pH values. We also demonstrate that ^{15}N NMR spectroscopy of the ^{15}N amino groups is a good technique for the study of the exchange of amino protons with the protons of the solvent water. The pH dependence of the structure of the ^{15}N amino resonance signal reflects the change of the exchange rate of the amino protons in the course of titration. In addition the interactions between pancreatic ribonuclease A and ^{15}N-labeled pyrimidine-3'-phosphates are investigated using ^{15}N NMR spectroscopy.

EXPERIMENTAL

The ribonucleotides cytidine-3'-phosphate, uridine-3'-phosphate, adenosine-3'-phosphate and guanosine-3'-phosphate were labeled by growing Escherichia coli MRE 6oo in a minimal medium containing $(^{15}NH_4)_2SO_4$ as the only nitrogen source. The isolation of ribosomes and the preparation of ribosomal RNA was carried out according to the method of Boulton (1). After incubation of the RNA with a mixture of bovine pancreatic ribonuclease A (Worthington) and ribonuclease T_2 (isolated from Aspergillus oryzae) the 3'-nucleotides were separated chromatographically.
The H_2O solutions of the nucleotides were treated with Chelex 1oo (Biorad) in order to remove paramagnetic impurities.

B. Pullman (ed.), Nuclear Magnetic Resonance Spectroscopy in Molecular Biology, 53-70.

The final nucleotide concentrations of the samples were o.o8 M as determined by spectrophotometric methods. (The pH was adjusted with a Radiometer PH meter (PHM 26) in conjunction with an Ingold type 4o5-M 3/2 combined electrode.) Bovine pancreatic ribonuclease A (phosphate free) was obtained from Worthington Biochemical Corporation (Freehold, N.J., USA) as a lyophylized powder and used without further purification. ^{15}N NMR spectra were recorded on a Bruker HFX-9o high-resolution spectrometer (^{15}N resonance frequency: 9.12 MHz). (The instrument was equipped with a Bruker Fast Fourier transform unit, a Nicolet 1o8o data handling system, a Bruker B-SV 2 proton decoupler, and a Bruker deuterium lock system.) The probe was kept at 43 $\pm$ 2°C with a flow of temperature controlled air (Bruker temperature control unit B-ST 1oo/7oo). Field frequency lock was obtained from D_2O in a 2 mm capillary inserted into the 1o- or 15-mm sample tube. 27.36 MHz ^{15}N NMR spectra of the ribonuclease-nucleotide complexes were recorded on a Bruker WH-27o spectrometer in the Fourier transform mode. The ^{15}N chemical shift data are given in ppm relative to the external standard (4 M solution of 95%-enriched $^{15}NH_4NO_3$ in 2 M HNO_3).

RESULTS AND DISCUSSION

In Fig.1 the pH dependence of the ^{15}N chemical shifts of uridine-3'-phosphate and cytidine-3'-phosphate are shown. At pH=6 the chemical shift values of the two ^{15}N resonances of uridine-3'-phosphate are -137.5 and -124.8 ppm, respectively. These values are similar to those found for the ^{14}N resonance of pyrrole (2).
The assignment of the observed signals of uridine-3'-phosphate to the two nitrogen atoms is based on a pH dependence of chemical shifts (Fig.1). One of the two nitrogen resonances (-137.5 ppm) at pH=6 is shifted by more than 4o ppm to lower field on deprotonation of the uracil ring. The observed change in the chemical shift is supposed to be caused by the deprotonation of the N3 nitrogen atom and may be explained by a change of the paramagnetic term of the shielding constant, which should be increased in going from neutral to alkaline pH values. The paramagnetic term of the shielding constant is assumed to be a predominant one for nitrogen in heterocyclic compounds (3). According to our assignment of the signal at -137.5 ppm, pH=6.o to the N3 atom, the remaining resonance at -124.8 ppm has to be assigned to the N1 nitrogen. The signal also exhibits a low-field shift on deprotonation of the N3 nitrogen, although to a much smaller extent (Fig.1). The change at N3 is supposed to affect the whole heteroaromatic ring, implying the observed change of the chemical shift of the glycosidic nitrogen. In addition, a small change of the ^{15}N1 resonance is observed around

pH=6, reflecting the ionization of the phosphate group.

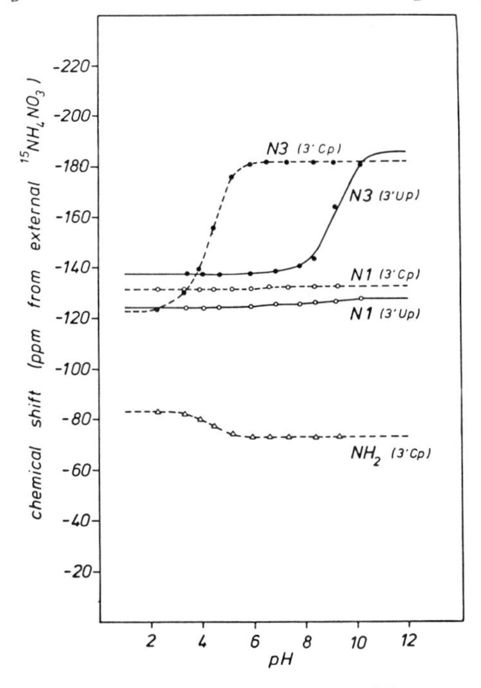

Fig.1 pH dependence of the ^{15}N chemical shifts of uridine-3'-phosphate (solid line) and cytidine-3'-phosphate (dashed line). For the extremes of pH the chemical shifts have been determined by calculating the titration curves using the Henderson-Hasselbalch equation.

The pH dependence of the ^{15}N chemical shift of cytidine-3'-phosphate is also illustrated in Fig.1. The resonance which is shifted furthest downfield at neutral pH values (-18o.4 ppm) reveals a chemical shift value comparable to the resonance signal of the N3 nucleus in uridine-3'-phosphate at high pH. Hence, we assign this signal to the N3 atom, which is not protonated at neutral pH values. At pH=7.o the resonance of the glycosidic nitrogen (N1) is shifted by -132.o ppm relative to the reference. The diamagnetic contribution to the screening constant of the amino nitrogen is dominating. Consequently this resonance signal should be shifted less to lower field, compared with the other nitrogen signals, although the value of the chemical shift (-73.2 ppm) does not fully correspond to the chemical shift of an sp^3-hybridized amino nitrogen atom (4).
The observed deviation of the chemical shift of about 3o ppm further downfield from the observed region of an sp^3-hybridized amino nitrogen is presumably due to the more sp^2-hybridized state (4). Such a partial double-bond character

diminishes the shielding by the nitrogen lone pair electron and should result in a shift to lower field. Protonation of the pyrimidine ring causes a drastic high-field shift (more than 6o ppm) of the N3 resonance (Fig.1). The proton is attached to the lone pair electron of N3, which results in a smaller paramagnetic contribution to the chemical shift (3,5).
The ^{15}N resonance of the amino nitrogen is shifted to lower field upon protonation of the cytosine ring (Fig.1). We assume that protonation at N3 nucleus induces the almost sp^2-hybridized state of the amino nitrogen, present already at neutral pH, to be more extensive in acid solution. This may lead to a decrease in the shielding of the amino nitrogen. The more sp^2-hybridized state of the amino nitrogen is consistent with the results of Becker et al. (6,7), who observed a hindered rotation of the amino group in ^{15}N-labeled 1-methyl-cytosine due to a partial double-bond character of the bond between the C4 and the amino nitrogen. The ^{15}N resonance signal of N1 is shifted by o.3 ppm to lower field during the ring protonation; protonation of the phosphate group results in a small upfield shift (o.6 ppm).

Table 1 : ^{15}N-^{15}N and ^{15}N-^{1}H Coupling Constants (in Hz) of Nucleoside-3'-phospahtes at different pH-values.

Compound	Number of bonds	Nuclei	pH=3	pH=7	pH=1o
3'Gp	2	$^{15}N1$-$^{15}NH_2$	2.2	2.2	6.o
	2	$^{15}N3$-$^{15}NH_2$	6.o	6.o	6.o
	2	$^{15}N9$-$^{15}N3$	3.7	3.7	3.7
	2	$^{15}N7$-$^{1}H8$		1o.o	1o.o
	1	$^{15}NH_2$-^{1}H		9o.7	
3'Ap	2	$^{15}N1$-$^{15}NH_2$	1.o	5.2	
	2	$^{15}N3$-$^{15}N9$	1.5	2.2	
	2	$^{15}N1$-$^{1}H_2$	12.5	14.o	
	2	$^{15}N3$-$^{1}H2$	14.o	15.5	
	2	$^{15}N7$-$^{1}H8$	1o.5	1o.5	
	2	$^{15}N9$-$^{1}H8$	9.5		
	1	$^{15}NH_2$-^{1}H		88.2	
3'Cp	2	$^{15}N3$-$^{15}NH_2$	1.5	5.8	
	1	$^{15}NH_2$-^{1}H		86.o	
3'Up	2	$^{15}N1$-$^{15}N3$	2.2	2.2	

From the proton broad-band decoupled ^{15}N spectra of 3'CMP the two bond ^{15}N3-^{15}N7 coupling is obtained (Table 1). During the protonation of the pyrimidine ring the absolute value of this coupling constant decreases (Table 1). Values of ^{15}N-^{1}H, ^{13}C-^{15}N or ^{15}N-^{15}N two-bond couplings are supposed to be dependent on the orientation of the lone electron pair at the nitrogen (8-12). In the case of the cisoid configuration large coupling constants are usually observed whereas in the transoid configuration small values are found. Goto et al. (14) and Tori et al. (13) demonstrated that the protonation of the nitrogen lone pair results in a decrease of these coupling constants similar to the observed dependence of the ^{15}N3-^{15}NH$_2$ two-bond coupling on protonation of N3. The decrease in the ^{15}N3-^{15}NH$_2$ two-bond coupling constant is a further indication that protonation of the pyrimidine ring of cytidine-3'-phosphate occurs at the N3.
In Fig.2 the ^{15}N NMR spectrum 3'AMP is shown at pH=5.o with and without proton decoupling.

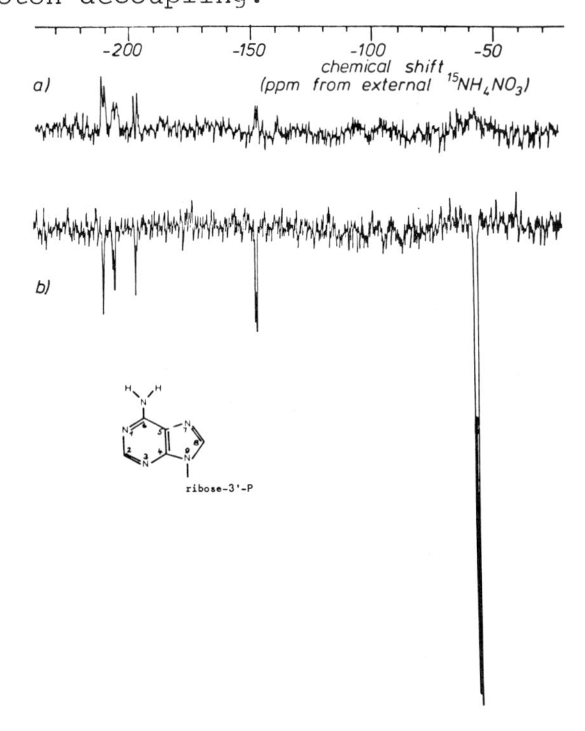

Fig. 2 Proton-coupled (a) and proton-decoupled (b) ^{15}N NMR spectra of fully ^{15}N-labeled adenosine-3'-phosphate at pH=5.o.

The assignment of the five resonances was possible either by the coupling of the ^{15}N nuclei with the neighbouring protons or by comparing of a partially labeled adenosine derivative (15). The amino nitrogen resonance can be readily assigned according to the one-bond coupling $^{15}N-^{1}H$ (Tab.1). As for the amino nitrogen signal of 3'CMP, this resonance is not found in the normal absorption region of sp^3-hybridized amino nitrogen resonances. The signal (-58.o ppm) is shifted further downfield relative to the amino nitrogen signal of amino pyridine (8). The extent of this shift is not as large as that observed for the corresponding resonances of 3'CMP, but we can assume a partial double-bond character for the C6-NH_2 bond of 3'AMP, although less pronounced. The resonance of the glycosidic nitrogen N9 is supposed to resemble that of a pyrrole nitrogen (2), and hence the resonance at -147.5 ppm (Fig.2) can be assigned to the glycosidic nitrogen. The pyridine type nitrogen resonances of N1, N3, and N7 are shifted downfield considerably. The N1 resonance (-2o3.3 ppm) can be readily assigned by comparison to the ^{15}N NMR spectrum of adenosine-labeled with ^{15}N only at N1 (11). The N7 resonance can be recognized from the coupling of $^{15}N7$ to $^{1}H8$. The corresponding coupling constant is found to be 1o.5 Hz at pH=7 (Table 1). A similar value of two-bond $^{15}N-^{1}H$ coupling for a two-bond nitrogen in the imidazole part of a ^{15}N-labeled adenine derivative was found by Leonard and Henderson (16), whereas the corresponding two-bond $^{15}N-^{1}H$ coupling for a two-bond nitrogen in the pyrimidine part of the ^{15}N-labeled adenine derivative was found to be about 16 Hz. In addition, guanosine, ^{15}N-labeled only in the imidazole part, revealed a $^{15}N7$ resonance at -214 ppm which is very similar to the chemical shift value of -21o.4 ppm found for the corresponding nitrogen resonance of 3'AMP (17) (compare also Fig.3). Hence, the assignment of the resonance at -21o.4 ppm to N7 seems to be justified, and the remaining resonance at -196.7 ppm should belong to the N3 nitrogen.

A different assignment of the three low-field ^{15}N resonances of the adenine ring is proposed by Happe and Morales (18). These authors investigated ^{15}N-labeled ATP and assigned the N1 and N3 nitrogen resonances in a reverse manner. In view of our comparison of the ^{15}N NMR spectra of the totally ^{15}N-labeled 3'AMP with a ^{15}N NMR spectra of adenosine, labeled at the N1 position, our assignment seems to be conclusive. Furthermore, it is generally assumed that the N1 nitrogen of the adenine ring is protonated at pH=3.8. The ^{15}N resonance at -2o3.3 ppm shifts upfield by about 7o ppm to higher field. This change in signal position is certainly a further proof for a correct assignment.
The drastic upfield shift of the N1 resonance upon protonation can be interpreted as a conversion of a pyridine-type nitrogen to a pyrrole-type nitrogen in a way similar to that

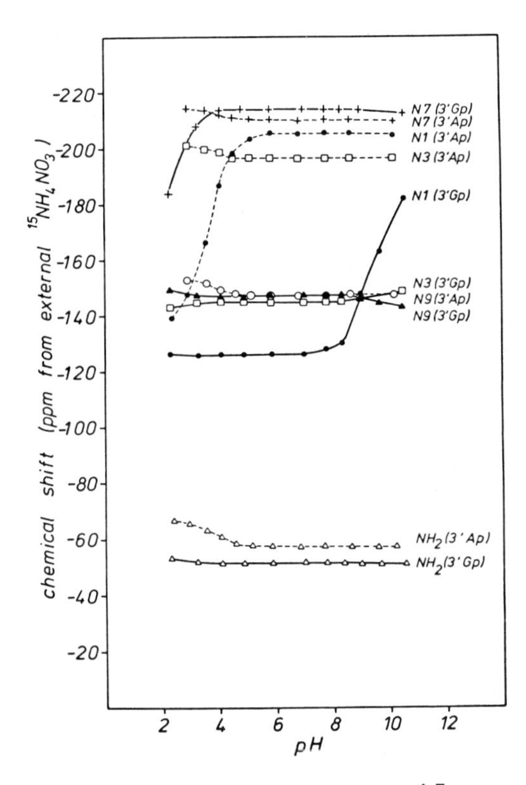

Fig. 3 pH dependence of the ^{15}N chemical shifts of adenosine-3'-phosphate (dashed line) and guanosine-3'-phosphate (solid line).

discussed for N3 of 3'CMP. The paramagnetic part of the chemical shift decreases again.
All other ^{15}N resonances of 3'AMP are shifted downfield during protonation of N1 (Fig.3). It seems that the aromatic character of the adenine ring increases, which induces a deshielding of all nitrogen atoms except N1. In addition, the rather strong downfield shift of the amino nitrogen resonance indicates that a partial double-bond character of the C6-NH_2 bond also increases when a proton is bound to N1.

The two-bond ^{15}N-^{15}N coupling constants of 3'AMP are in most cases small and hardly detectable (Table 1). The absolute value of the two-bond ^{15}N1-$^{15}NH_2$ coupling decreases from about 5.2 Hz to about zero upon protonation of N1. We were not able to determine the sign of this J value with our experimental device. However, from similar two-bond ^{15}N-^{1}H coupling constants it can be concluded that at neutral pH values the two-bond ^{15}N-^{15}N coupling constant is positive owing to the orientation of the lone electron pairs at either nitrogen. With protonation of N1 one of the lone

pairs disappears, which is reflected in the change of the two-bond $^{15}N1-^{15}NH_2$ coupling value. It may be that this coupling reverses its sign below pH=3.8; the absolute value could not be detected within the resolution of our instrumentation. The pH dependences of the ^{15}N resonances of guanosine-3'-monophosphate are also shown in Fig.3.
are shown in Fig.3.
The assignment of the amino nitrogen resonance (51.6 ppm) is possible in the same way as that described for 3'CMP or 3'AMP, i.e. from the direct coupling of a nitrogen nucleus with the amino protons at neutral pH values. The remaining four resonances can be assigned by comparison of the ^{15}N spectra with the corresponding spectra of guanine and guanosine, both ^{15}N-labeled either in the imidazole part (N7 and N9) or the pyrimidine part (N1, N3 and NH_2) (17). Since N1 is known to be deprotonated at high pH values, the ^{15}N resonances of the pyrimidine part can be distinguished. The $^{15}N1$ resonance (-126.o ppm, pH=6.o) is shifted downfield by about 6o ppm upon deprotonation (Fig.3). This drastic change in the paramagnetic term might be explained in the same way as that discussed above for the deprotonation of the N3 of uridine-3'-phosphate. The direct coupling of the proton attached to N1 at neutral pH values is not observed because the exchange of this proton with the solvent protons is too rapid on the NMR time scale. The chemical shift values indicate that the electronic structure of the N1 nitrogen changes from a pyrrole-type nitrogen below pH=8 to a pyridine-type nitrogen at higher pH values. The N3 resonance reveals a chemical shift value which resembles that of a pyrrole-type nitrogen (-145.1 ppm, pH=6.o, Fig.3)(2). Relative to the N3 signal of 3'AMP, the N3 resonance of 3'GMP is shifted by about 51 ppm to higher field (Figs.2 and 3). The aromatic character of the guanine ring, especially at the N3 position, seems to be much less pronounced than that of the adenine ring. Apparently the electron-donating substituents in ortho- and para-position to N3 induce the upfield shift of the N3 nitrogen of the pyrimidine part.
The two-bond $^{15}N1-^{15}NH_2$ coupling constant value changes from 6.o Hz at pH=1o to about 2.2 Hz at pH=7, which is very similar to the change observed for the two-bond $^{15}N3-^{15}NH_2$ value of 3'CMP at pH=7 and pH=3, respectively (Table 1). Again, because of the steric arrangement of the lone-pair electron at N1 of 3'GMP at high pH values, the absolute value of the two-bond $^{15}N1-^{15}NH_2$ coupling is relatively large, whereas upon protonation at N1 this value decreases. The two-bond $^{15}N3-^{15}NH_2$ coupling value is found to be about 6.o Hz, and, indeed, this large value indicates that there is a lone pair electron at N3 instead of a proton attached to N3. This lone pair electron is supposed to be cisoid to the amino nitrogen following the interpretation of large absolute values of two-bond couplings between $^{15}N-^{1}H$ or $^{15}N-^{13}C$ (8-12). In

addition, the two-bond $^{15}N3-^{15}NH_2$ coupling does not change in the pH range of 2.2 to 1o, thus indicating that no protonation process occurs at N3 in this pH range. Also, the two-bond $^{15}N9-^{15}N3$ coupling is constant in that pH range. The chemical shift of the N9 resonance of 3'GMP (-147.1 ppm, pH=6.o) corresponds to that of a pyrrole-type nitrogen, whereas the chemical shift of the N7 resonance (-213.8 ppm, pH=6) is that of a more sp^2-hybridized character (Fig.3). The N7 resonance shifts upfield below pH=4 because of the protonation of the N7 nitrogen. Since there has been some uncertainty in the literature whether the N7 or the N3 nitrogen is protonated at low pH values, this shift of the N7 resonance seems to be a clear evidence for a protonation at N7. However, a very small upfield shift of the N3 resonance below pH=4 indicates, that, to a much lesser extent, the N3 nitrogen is also protonated (Fig.3). Apparently at pH values below pH=3 there exists an equilibrium between a large number of 3'GMP molecules protonated at N7 and a small number protonated at N3. Protonation at N7 converts the more sp^2-hybridized state of that nitrogen to a more sp^3-hybridized state.

It should be mentioned that the assignment of the ^{15}N resonances of all nucleotide-3'-monophosphates described above are essentially in agreement with those of Markowski et al. (19) reported for some nucleosides and nucleoside-5'-monophosphates.

EXCHANGE OF AMINO PROTONS

Chemical exchange of the protons attached to the nitrogen of amino groups with water protons occurs at just the proper rate to affect the $^{15}N-^{1}H$ scalar coupling (2o).

As is demonstrated for 3'CMP and 3'GMP in Fig.4 broadening of the triplet resonances of the amino nitrogens is observed when the pH of the solution is raised or lowered.

At high pH values the one-bond $^{15}N-^{1}H$ coupling collapses, resulting in a singlet structure of the ^{15}N resonances. For some limiting conditions, line widths of the observed resonances can be used to obtain values of the exchange rate $1/\tau_e$ (21,22). These limiting conditions are fulfilled at all pH values for the amino group resonances.

At some pH values the ^{15}N resonances have completely disappeared. In this case the simple relation

$$\tau_e = 1/2\pi J$$

according to the line shape calculation procedure of Gutowsky et al. (23) and Kaplan (24) may apply. The exchange rates determined following this procedure agree with the exchange rates at nearby pH values evaluated from the line widths. It should be mentioned that the multiplet structure due to the

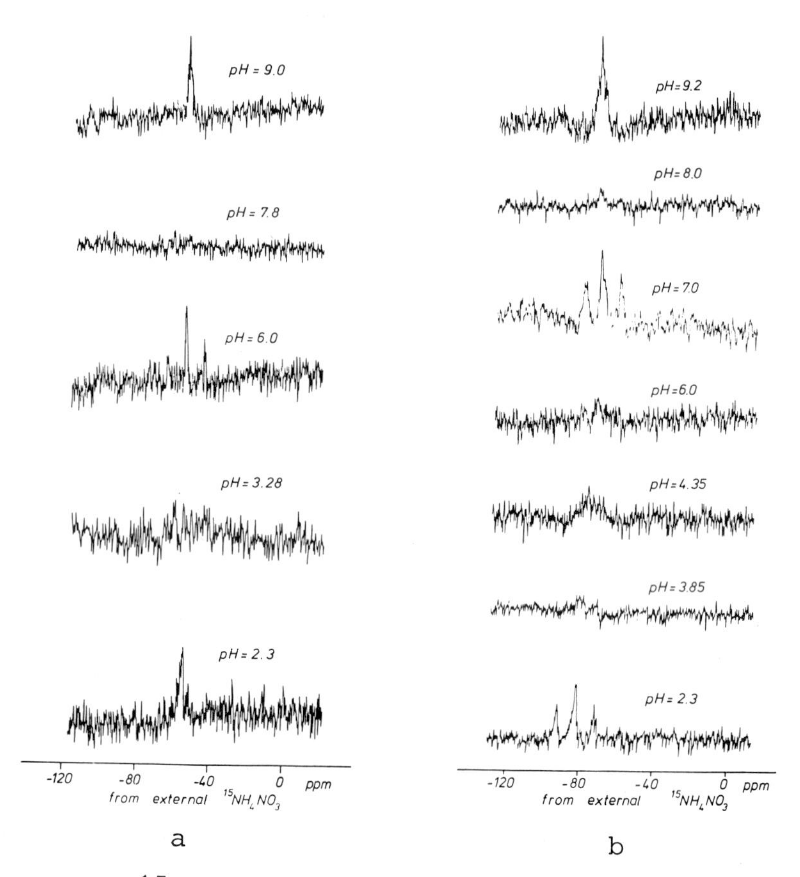

Fig.4 ^{15}N amino resonances of the proton-coupled ^{15}N NMR spectra of guanosine-3'phosphate (a) and cytidine-3'-phosphate (b) at different pH values.

coupling of the amino nitrogen with neighbouring ^{15}N nuclei or ^{1}H nuclei is considered in the line shape.
Because of the relatively slow exchange the direct coupling of the amino protons with the ^{15}N nuclei is observed at neutral pH values for 3'CMP, 3'AMP and 3'GMP. The direct ^{15}N-^{1}H coupling constants are found to be 86.o Hz (3'CMP), 88.2 Hz (3'AMP) and 9o.7 Hz (3'GMP). The exchange rates of the amino group protons of the three nucleotides investigated increase at high pH values in a similar way (Fig.5). At pH=8 the triplet structure collapses, so that the ^{15}N resonance is no longer observed. Above pH=8 a singlet structure appears, reflecting a further increase of the exchange rate. This behaviour of the exchange rate indicates that the exchange of the amino protons is a base catalysed reaction dependent on the concentration of OH^- ions (19,25-27). One possible

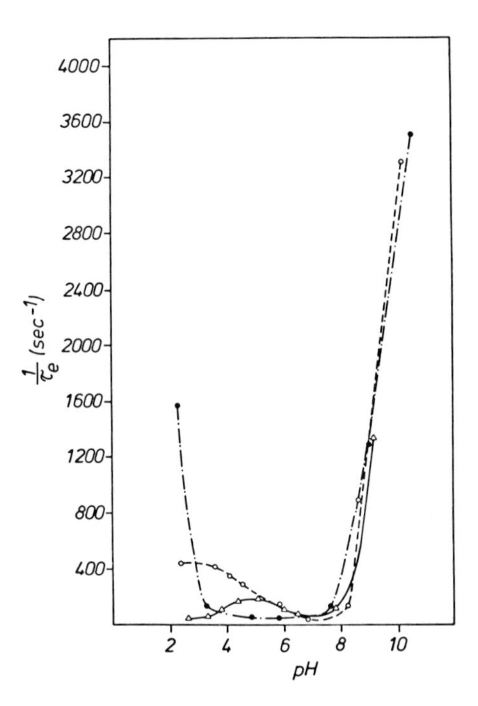

Fig. 5 pH dependence of the exchange rates $1/\tau_e$ for protons attached to the ^{15}N amino nitrogen with water protons. △——△ Cyd-3'-P; ○- - -○ Ado-3'-P; —●—·—·●— Guo-3'-P.

mechanism of the base catalysed proton exchange is depicted in Scheme 1.

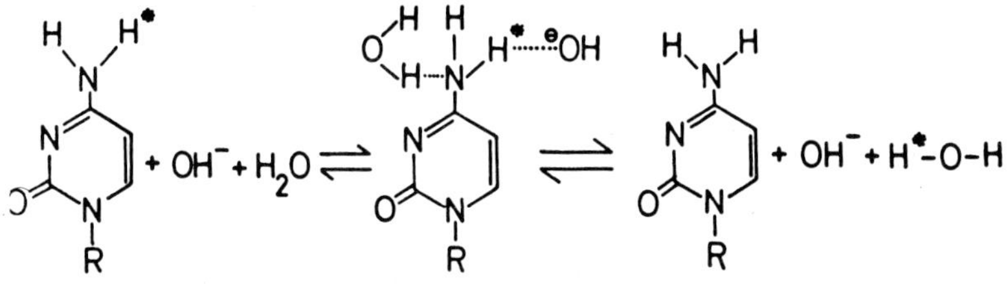

Base catalyzed enhancement of the rate of the exchange of the amino protons with the protons of the solvent water

Scheme 1

Marked differences between the nucleotides are found for the exchange behaviour of the amino group protons at acid pH values. The exchange rate is enhanced with decreasing pH in the case of 3'GMP (Fig.5). The triplet structure of the ^{15}N amino resonance is lost, and at pH=2 a singlet is observed. Line width broadening is found for the ^{15}N resonance of 3'CMP when the pH is lowered (Fig.4). Around pH=6 the triplet structure is lost. A broad singlet appears with a minimal line width at pH=5. Correspondingly, the exchange rate reveals a maximum at this pH value. Below pH=4 the broad lines of the triplet reappear, revealing a decrease in the exchange rate. The exchange rate of the amino protons of 3'AMP increases in going from pH=6.5 to pH=4 (Fig.5). Below pH=4 it levels off to a constant or slightly increasing value.
Various mechanisms for the acid-catalysed exchange of the amino protons are possible. The most commonly discussed mechanism is the direct protonation-deprotonation mechanism (26,28,29). Direct protonation of the amino group should generally be enhanced at pH values not too far from the acid pK value. Although it seems unlikely that the amino group of 3'GMP is protonated at very acid pH values - the guanine ring being protonated at N7 - it is certainly possible that the enhancement of the exchange rate at low pH is induced by protonation or deprotonation processes. However, additional effects must be considered in order to explain the peculiar pH dependence of the exchange rate of the amino protons of 3'CMP or 3'AMP at low pH values.
In addition to the mechanism of direct protonation and abstraction of a proton the exchange of the amino group protons with water protons may be mediated through the surrounding nitrogens of the heteroaromatic ring systems (2o). If one considers that a complex may favourably be formed between a hydronium ion and the guanine ring of 3'GMP, as depicted in Scheme 2, the N3 nitrogen may act as an auxiliary base.

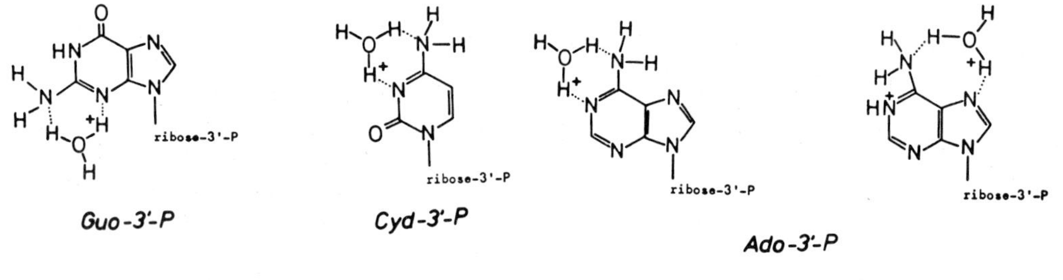

ribose-3'-P

Guo-3'-P

ribose-3'-P

Cyd-3'-P

ribose-3'-P

ribose-3'-P

Ado-3'-P

Acid catalyzed enhancement of the rate of the exchange of the amino protons of various nucleotides with the protons of the solvent water assuming neighbouring nitrogens to be an auxiliary base

Scheme 2

Abstraction of a proton from the amino group of the complex of Scheme 2 should be facilitated and hence, the exchange rate should be dependent on the concentration of the complex or on the concentration of hydronium ions. The sudden increase of the exchange rate of the amino protons of 3'GMP below pH=3 and even the relatively fast exchange of all nucleotide amino protons at neutral pH values may be explained in this way. It was pointed out previously that it is difficult to explain the exchange rate at neutral pH values with the assumption of a direct protonation mechanism or a base catalysis mechanism (25,29). If some of the ring nitrogens act as an auxiliary base the relatively high exchange rate at pH=7 would be plausible.
The N3 nitrogen of the pyrimidine ring is supposed to take this role in the case of 3'CMP (Scheme 2). Consequently the exchange rate significantly decreases when the N3 nitrogen is protonated.
The N1 nitrogen of 3'AMP seems to be the auxiliary base at neutral pH values (Scheme 2). When the N1 position is protonated the exchange rate should decrease in the same way as that observed for 3'CMP. Since the exchange rate does not decrease below pH=3.8 the N7 position may take over the predominant role of the auxiliary base (Scheme 2); or the relatively high exchange rate may be due to a direct protonation mechanism.
In general, it should be said, that it is difficult to give a definite answer with regard to a mechanism of exchange. Further studies with other amino-group-bearing compounds are necessary. Since the dynamic range for these exchange processes is usually such that the ^{15}N-^{1}H scalar coupling is affected , ^{15}N NMR seems to be very useful for this type of investigation.

RIBONUCLEASE A - NUCLEOTIDE INTERACTIONS

In a further investigation the interaction of pyrimidine-3'-phosphates with pancreatic ribonuclease A has been studied using ^{15}N NMR spectroscopy. It has been concluded from an X-ray analysis of various complexes that specific hydrogen bonds are formed between some amino acid side chains of the active site of RNase A and the uracil or cytosine ring of corresponding nucleotides (3o-34). It should be possible to detect the proposed H-bonds between the N3 of 3'-UMP or 3'CMP and the OH group of Thr 45 or the suggested hydrogen bond between the amino group of 3'CMP and the OH of Ser 123 either from shifts of the N3 and amino nitrogen resonances of ^{15}N labeled nucleotides or from the direct coupling of the hydrogen bond proton with the ^{15}N nuclei. In Fig.6 the chemical shifts of the ^{15}N resonances of 3'UMP and 3'CMP are plotted in dependence of the ratio of enzyme concentration to nucleotide concentration at pH=5.6. Nearly all ^{15}N

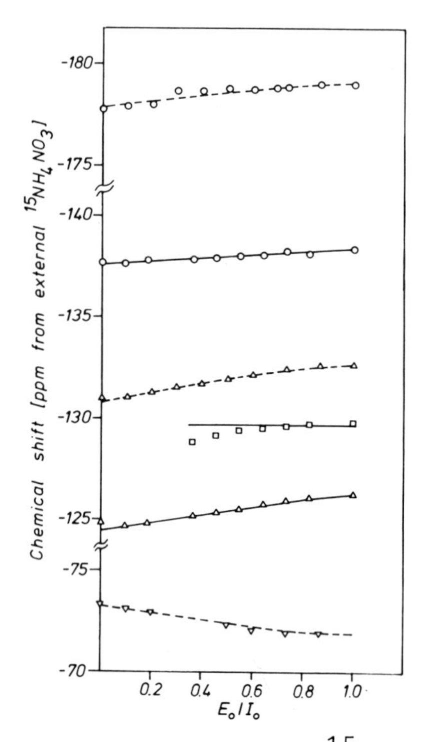

Fig. 6 Chemical shifts of ^{15}N resonances of 3'CMP (dashed lines) and 3'UMP (solid lines) in dependence of the ratio of RNase A concentration to nucleotide concentration at pH= 5.6. $^{15}N3$ resonances: ○ ; $^{15}N1$ resonances: △ ; amino nitrogen: ▽ $^{15}N3$ resonance (complex): □

resonances are changing with the RNase A concentration. The N3 resonance of 3'CMP is shifting to lower field. However, no characteristic line width broadening is observed due to a slow exchange or due to the appearance of a typical doublet structure attributable to a direct $^{15}N-^{1}H$ coupling. The shift to lower field may be produced by the ring current of the phenyl ring of Phe 12o which is known to be adjacent to the base in the complex. The amino nitrogen resonance of 3'CMP is slightly shifting to higher field. The direct coupling of the amino protons with the ^{15}N nucleus can be detected at all investigated concentration ratios. In addition the line widths of the triplet signals do not change. Apparently even the exchange rates of the amino protons are unchanged. From this result a hydrogen bond between the amino group of 3'CMP and the OH of Ser 123 seems to be unprobable. Also the H-bond between N3 of 3'CMP and the OH group of Thr 45 is unlikely, since the shift of the N3 resonance on complex formation is to lower field instead of higher field which should be expected for a protonation process.
Similar shifts of the ^{15}N resonances of 3'UMP are observed

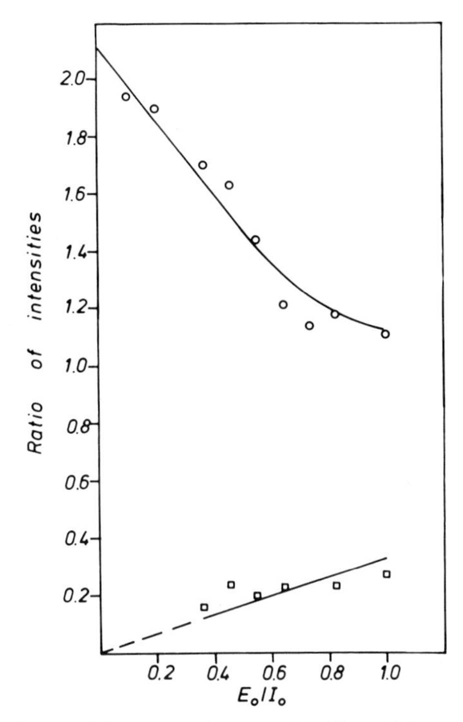

Fig. 7 Intensity ratios of the N3 resonances (N3 at 129 ppm to N3 at 138 ppm) □—□ and of N3 to N1 (N3 at 138 ppm, N1 at 125 ppm) ○—○ of 3'UMP in dependence of the ratio of RNase A concentration to nucleotide concentration.

on complex formation with RNase A(Fig.6). However, the intensity of the N3 resonance is decreasing with increasing concentration of RNase A (Fig.7), an additional resonance appears at about 129 ppm (Fig.6). Apparently this resonance has to be assigned to a second complex different from the first one and different from the RNase A-3'CMP complex. The average life time of this complex is long with respect to the NMR time scale, since we observe one separate N3 resonance of 3'UMP in this complex. The ratio of the two different RNase A-3'UMP complexes is not dependent on the RNase A concentration as can be concluded from the plot of the intensity ratios of the observed two N3 resonances versus E_o/I_o (Fig. 7). If we follow the generally accepted assumption that there is only one binding site for pyrimidine bases(B_1 according to ref. 35) we would like to suggest that 3'UMP is bound to RNase A in the anti conformation and in the syn conformation thus forming two different complexes. In both complexes the exchange of the proton at N3 of 3'UMP is fast. The direct coupling of this proton with ^{15}N3 is not observed. Hence, a hydrogen bond between the N3-H of 3'UMP and the OH

group of Thr 45 seems to be unlikely.
Further studies of other complexes of nucleotides with RNase A are currently investigated in order to get a better understanding of the specificity of this type of protein nucleic acid interaction.

Acknowledgement: We would like to thank the Deutsche Forschungsgemeinschaft for a grant.

REFERENCES

1. E.T.Bolton, in "Procedures in Nucleic Acid Research" (S.L.Cantoni and D.R.Davies,Eds.),p.437, Harper&Row, New York,1966.
2. D.Herbison-Evans and R.E.Richards, Mol.Phys.7,9(1964).
3. G.A.Webb and M.Witanowski, in "Nitrogen NMR" (M.Witanowski and G.A.Webb,Eds.),p.1, Plenum,London/New York, 1973.
4. W.M.Litchman, J.Magn.Resonance 12, 182 (1973).
5. E.W.Randall and D.G.Gillies, in "Progress in Nuclear Magnetic Resonance Spectroscopy" (J.W.Emsley, J.Feeney, and L.H.Sutcliffe,Eds.), Vol.6,p.119,Pergamon, Oxford, 1971.
6. H.T.Miles, R.B.Bradley, and E.D.Becker, Science 142, 1569 (1963).
7. E.D.Becker, H.T.Miles, and R.D.Bradley, J.Am.Chem.Soc. 87, 5575 (1965).
8. R.L.Lichter, D.E.Dorman, and R.Wasylishen, J.Am.Chem. Soc. 96, 93o (1974).
9. D.Crépaux, J.M.Lehn, and R.R.Dean, Mol.Phys. 16, 225 (1969).
1o. D.Crépaux and J.M.Lehn, Org.Magn.Reson. 7, 524 (1975).
11. J.P.Kintzinger and J.M.Lehn, J.Che.Soc.Chem.Commun. 66o (1967).
12. F.Blomberg, W.Maurer, and H.Rüterjans, J.Am.Chem.Soc. 9o, 8149 (1977).
13. K.Tori, M.Ohtsuru, K.Aono, Y.Kawazoe, and M.Ohnishi, J.Am.Chem.Soc. 89, 2765 (1967).
14. T.Goto, M.Isobe, M.Ohtsuru, and K.Tori, Tetrahedron Lett. 1511 (1968).
15. G.Grenner and H.-L.Schmidt, Chem.Ber. 11o, 373 (1977).
16. N.J.Leonard and T.R.Henderson, J.Am.Chem.Soc. 97,499o (1975).
17. F.Blomberg, H.Rüterjans, and W.Guschlbauer, unpublished results.
18. J.A.Happe and M.Morales, J.Am.Chem.Soc. 88, 2o77 (1966).
19. V.Markowski, S.R.Sullivan, and J.D.Roberts, J.Am.Chem. Soc. 99, 714 (1977).
2o. F.Blomberg, W.Maurer, and H.Rüterjans, Proc.Natl.Acad. Sci. USA 73, 14o9 (1976).

21. E.Grunwald, A.Loewenstein, and S.Meiboom, J.Chem.Phys. 27, 6.. (1957).
22. C.S.Johnson,Jr., in: "Advances in Magnetic Resonance" (J.S.Waugh,Ed.), p.33, Academic Press, New York, 1965.
23. H.S.Gutowsky, D.W.McCall, and C.P.Slichter, J.Chem. Phys. 21, 279 (1953).
24. J.Kaplan, J.Chem.Phys. 28, 278 (1958).
25. B.McConnell and P.C.Seawell, Biochemistry 11, 4382(1976).
26. B.McConnell and P.C.Seawell, Biochemistry 12, 4426(1973).
27. M.Raszka, Biochemistry 13, 4616 (1974).
28. D.S.Cross, A.Brown, and H.F.Fisher, Biochemistry 14, 2745 (1975).
29. B.McConnell, Biochemistry 13, 4515 (1974).
3o. H.W.Wyckoff, K.D.Hardman, N.M.Allewell, T.Inagami, L.N. Johnson, and F.M.Richards, J.Biol.Chem. 242, 3984 (1967).
31. H.W.Wyckoff, D.Tsernoglov, A.W.Hanson, J.R.Knox, B.Lee, and F.M.Richards, J.Biol.Chem. 245, 3o5 (197o).
32. F.M.Richards, H.W.Wyckoff, and N.M.Allewell, in: "The Neurosciences: Second Study Program", ed.F.O.Schmitt, Rockefeller Press, New York 197o.
33. F.M.Richards, H.W.Wyckoff, W.D.Carlson, N.Allewell, B.Lee, and Y.Mitsui: Cold Spring Harbor Symp. Quant. Biol. 36, 35 (1971).
34. S.Y.Wodak, M.Y.Liu, and H.W.Wyckoff, J.Mol.Biol. 116, 855 (1977).
35. F.M.Richards and H.W.Wyckoff, in: "The Enzymes" (ed. P.D.Boyer), 3rd. ed.,4, Academic Press, New York and London.

DISCUSSION

Gorenstein: We have also found evidence in the CMP complexes with RNase A that the first complex undergoes an isomerization to a second distinct complex with a rate ≈ 200 sec^{-1}, and that this isomerization is pH dependent. Do you find any pH effects to your N-15 spectra of the UMP-RNaseA complex ? Do you have any evidence that these two isomers represent syn/anti conformational isomers ?

Rüterjans: The answer is no. The N-15 resonances of 3'CMP in the complex with RNase A do not change in the pH range of pH = 3 to pH = 6. A conformational change of the enzyme in this pH range is not reflected in the behaviour of the N-15 resonances. In addition we can make sure from the pH dependance of the ^{15}N chemical shift of the 3'CMP-RNase A complex that the N3 position is not protonated around pH = φ in the complex. This result is in agreement with calorimetric studies on the binding of 3'CMP to RNase A we performed recently.

McConnell: The pH maximum for cytosine at pH4 can be accounted for by cytosine N-3 acting as an acceptor of the amino of the protonated form of cytosine. The product of protonated and neutral cytosine goes through a maximum at the pK of C(N-3). For adenine, the acidic plateau is due to H_2O (pK = -2) as an acceptor of the amino of protonated adenine (at N 1). Therefore, normal diffusional processes account for this exchange of A and C and the assertion of a concerted hydrogen bonded hydronium complex is not necessary in either case. It would not account for the pH4 maximum in cytosine. Your N^{15} data and my proton data are in excellent agreement. Because of this I would be most interested in your opinion about the dramatic low pH exchange increase for the guanine amino, since in this case your method is superior to the measurement of proton resonances in the estimate of lifetimes and your concerted mechanism could be valid.

Rüterjans: The dramatic increase of the exchange rate of the amino group protons of 3'GMP at low pH values may be due to protonation at N7 resulting in a change of pK of the amino group itself or by assuming a stable complex between an hydronium ion and the amino nitrogen and N3 position of the guanine , as we have proposed in the paper. I agree that the direct protonation mechanism may explain the exchange rates of all nucleotides at all pH values. However, I think that other mechanisms cannot be excluded.

CO-OPERATIVE CONFORMATIONAL PROPERTIES OF NUCLEOSIDES, NUCLEOTIDES AND NUCLEOTIDYL UNITS IN SOLUTION

DAVID B. DAVIES
Department of Chemistry, Birkbeck College,
Malet Street, London WC1E 7HX, UK.

Nuclear magnetic resonance spectroscopy has made an important contribution to an understanding of the conformations and interactions of nucleic acids derivatives in solution (Davies 1978a). Methods are available for determining detailed descriptions of sugar ring conformations (Altona and Sundaralingam 1972, 1973) and the conformational properties of four out of the six bonds of the sugar-phosphate backbone utilising $^{1}H-^{1}H$, $^{1}H-^{31}P$ or $^{13}C-^{31}P$ vicinal coupling constants (Davies, 1978a). Many nmr methods have been used to determine glycosidic bond conformations in a qualitative manner. No method provides an unequivocal description of the *syn* and *anti* conformational ranges or the equilibrium between them but methods based on relaxation measurements {n.O.e (Guéron *et al*, 1973) and T_1 measurements (Lüdemann *et al*, 1975; Chachaty *et al*, 1976)} and $^{3}J(^{13}C,H1')$ observations (Lemieux, 1973; Davies, 1976) offer the possibility of quantitative determinations of glycosidic bond conformations. Each single bond is taken to be flexible and appropriate nmr parameters are analysed in terms of the time-averaged equilibrium of contributions from a number of conformers for each bond. The available conformational regions are defined, as far as possible, by those observed in the solid state by X-ray crystallography (Sundaralingam, 1973; Jack *et al*, 1976). It is pertinent to investigate the principles that govern the fundamental conformational properties of nucleosides and nucleotides in solution by comparison of results of purine with pyrimidine derivatives, ribose with deoxyribose molecular fragments and momomers with oligomers.

A demonstration of the need for further understanding of the conformational state of nucleotides is the paradox between the concept of a "rigid" nucleotidyl unit postulated from X-ray crystallographic studies (Sundaralingam, 1973) and semi-empirical potential energy calculations (Yathindra and Sundaralingam, 1974) and the apparent existence in solution of a similar highly stable conformational state at the monomer (Davies and Danyluk, 1974) and dimer levels (Altona, 1975; Lee *et al*, 1976; Ezra *et al*, 1977) even though nmr parameters are analysed in terms of "flexible" conformations. This paradox can be resolved by considering the limits that can be placed on the term "flexible" as

B. Pullman (ed.), Nuclear Magnetic Resonance Spectroscopy in Molecular Biology, 71-85.

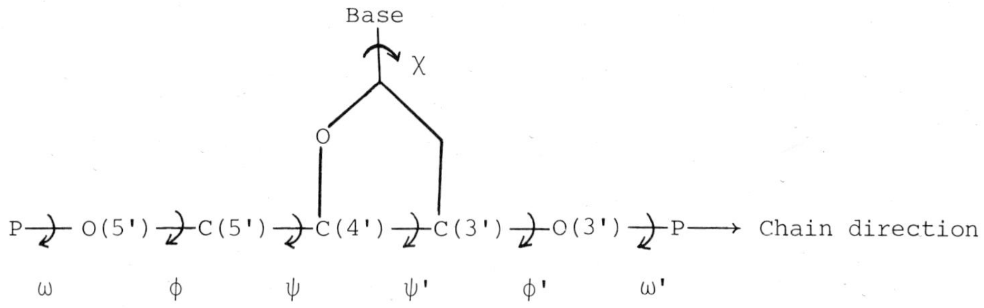

Base

χ

O

P — O(5') — C(5') — C(4') — C(3') — O(3') — P ⟶ Chain direction

ω φ ψ ψ' φ' ω'

FIG. 1. Conformational Nomenclature for Nucleotidyl Units (Sundaralingam *et al*, 1973)

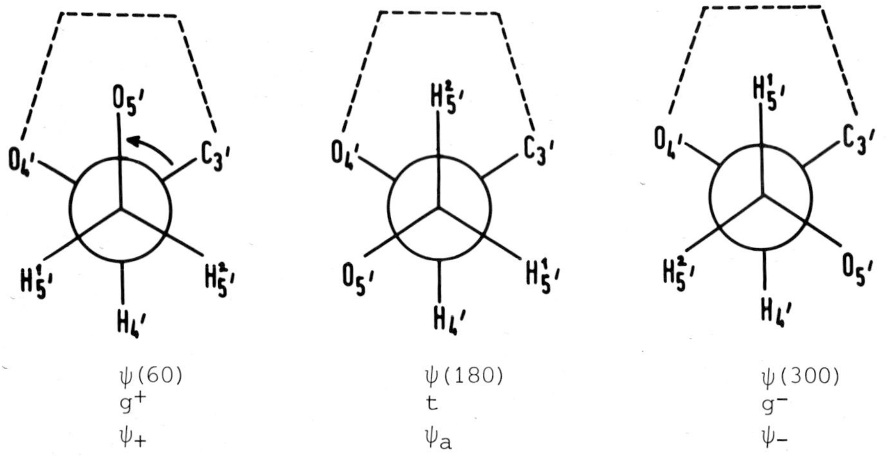

$O_{5'}$ $O_{4'}$ $C_{3'}$ $H^1_{5'}$ $H^2_{5'}$ $H_{4'}$

$H^2_{5'}$ $O_{4'}$ $C_{3'}$ $O_{5'}$ $H^1_{5'}$ $H_{4'}$

$H^1_{5'}$ $O_{4'}$ $C_{3'}$ $H^2_{5'}$ $O_{5'}$ $H_{4'}$

ψ(60)	ψ(180)	ψ(300)
g^+	t	g^-
ψ_+	ψ_a	ψ_-

FIG. 2. Notation for classical staggered rotamers for C(5')-C(4') bonds, ψ.

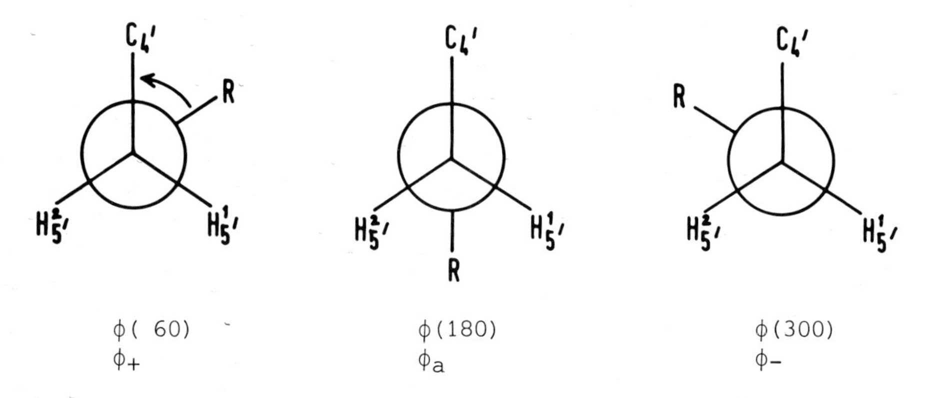

FIG. 3. Notation for classical staggered rotamers for O(5')-C(5') bonds, φ.

applied to nucleotides in solution. A nucleotidyl unit consists of eleven bonds about which rotation can occur. Five of these bonds comprise the sugar ring which, for convenience, is represented by the conformation of one bond, C(4')-C(3'), as this bond is also part of the sugar-phosphate chain. The problem reduces to considering the seven bonds which are shown in Figure 1. If the conformational properties about each bond are independent and if each bond has an overwhelming preference for one conformation, say 80%, then the overall probability of observing a particular conformational state is only 0.21 ($.8^7$) and the nucleotide can be considered "flexible", *ie*, the conformational state has a "persistence" of about 20%. If the conformational properties of each bond are interdependent, the probability of observing one state could increase to a maximum of 0.8, in this example; although the molecule could not be considered "rigid", it would have a "persistence" of 80%. Hence the idea of a nucleotidyl unit with a stable conformation that is maintained in a polynucleotide makes the tacit assumption that the conformational properties of each bond are interrelated and that there is an approximate 1:1 correspondence between the relative proportions of the most stable conformers for adjacent bonds or for the conformational properties that are interrelated.

The interrelation between conformational properties of nucleosides and nucleotides in solution is investigated, as far as possible, by plots of appropriate nmr parameters from results that are available in the literature. Only schematic representations of the plots are presented in this work so that the salient features can be discussed. The investigation provides methods by which results for nucleosides, mononucleotides and oligonucleotides can be rationalised. It is also shown that the conformational properties of adjacent bonds of nucleotidyl units are interdependent and analysis of the interrelationships provides insight into the cooperative nature of the conformational properties of nucleosides and nucleotides in solution.

1. NOMENCLATURE AND NOTATION

The system of nomenclature for describing nucleotide conformations recommended at a previous Jerusalem Symposium (Sundaralingam *et al*, 1973) is used in this work and shown in Figure 1. A number of notations have also been used to describe particular conformations and nmr descriptions based on the terms *gauche* and *trans* (*viz.* ψ bond: *gg*, *gt*, *tg* and ϕ bond: *g'g'*, *g't'* and *t'g'*) differ from those used in X-ray crystallographic studies (ie, g^+, t, g^- for each bond). An alternative description of these conformers which is based on the nomenclature of Klyne and Prelog (1960) and which can be used to specify conformational properties about any bond in an unequivocal and convenient way, has been suggested (Davies, 1978a). The notation is described in terms of the nomenclature suggested by Sundaralingam *et al* (1973) but it is consistent with any conformational nomenclature :

The symbols ϕ, ψ etc. represent the O(5')-C(5') and C(5')-C(4') bonds etc. and the subscripts +, a, - represent the 60° (+sc), 180° (ap) and 300° (-sc) conformations, respectively.

The descriptions of staggered conformations for ψ and ϕ bonds are shown in Figures 2 and 3, respectively. The terms *gauche* and *trans* may still be used to denote spin-coupling relationships.

2. ψ, ψ' RELATIONSHIP (Σ *v.* $J_{3'4'}$ PLOT)

The interrelation between sugar ring (represented by ψ') and backbone conformations of both nucleosides and nucleotides has been investigated by plots of $\Sigma(=J_{4'5'} + J_{4'5''})$Hz. *v.* $J_{3'4'}$Hz as shown schematically for pyrimidine and purine molecular fragments in Figures 4 and 5, respectively. Linear correlations (slope ~ -1) were observed both for nucleosides (including 3'-nucleotides and Lp-fragments of dimers and trimers) and 5'-nucleotides (including -pM fragments of dimers and trimers). It was found (Davies 1978b) that such correlations correspond to a proportional dependence of the fractional population of the most stable C(5')-C(4') bond conformer, $p(\psi_+)$, and the relative population of the sugar ring *N* conformer, ${}^{N}X$, according to equation (1) where *m* and *c* are constants. Making appropriate substitutions

$$p(\psi_+) = m.{}^{N}X + c \qquad (1)$$

for $p(\psi_+)$ and ${}^{N}X$, it was shown (Davies 1978b) that the linear relationships between Σ and $J_{3'4'}$ could be represented by equation (2)

$$\Sigma = (13-10c) - m.J_{3'4'} \qquad (2)$$

The observed slopes (~ -1) indicate that there is a 1:1 correspondence between the sugar ring and backbone conformational properties of nucleosides and nucleotides, ie, $m \sim 1$ in equation (1). This observation is a prerequisite of the existence in solution of a highly stable nucleotidyl conformational state.

A detailed analysis of the Σ *v.* $J_{3'4'}$ behaviour in Figures 4 and 5 leads to greater understanding of the conformational properties of nucleosides and nucleotides in aqueous solution. The behaviour of pyrimidine nucleosides (Figure 4, unbroken line, $c \sim 0.05$) is within "experimental" error of that predicted by the direct correlation between $p(\psi_+)$ and ${}^{N}X$ (dotted line, $m = 1$, $c = 0$) but quite different from that of -pPyr molecular fragments (dashed line, $c \sim 0.25$) which have stabilised backbone ψ_+ and sugar ring *S* conformations compared to the corresponding nucleosides. The correlation observed for *anti*-type pyrimidine nucleosides quantifies the behaviour previously found by Hruska (1973) and the correlation observed for pyrimidine nucleotides is in line with the prediction that 5'-phosphorylation causes stabilisation of S,ψ_+ conformations (Sundaralingam, 1973; Yathindra and Sundaralingam, 1973). Such stabilisation by 5'-phosphorylation has already been noted from

FIGURE 4.

FIGURE 5.

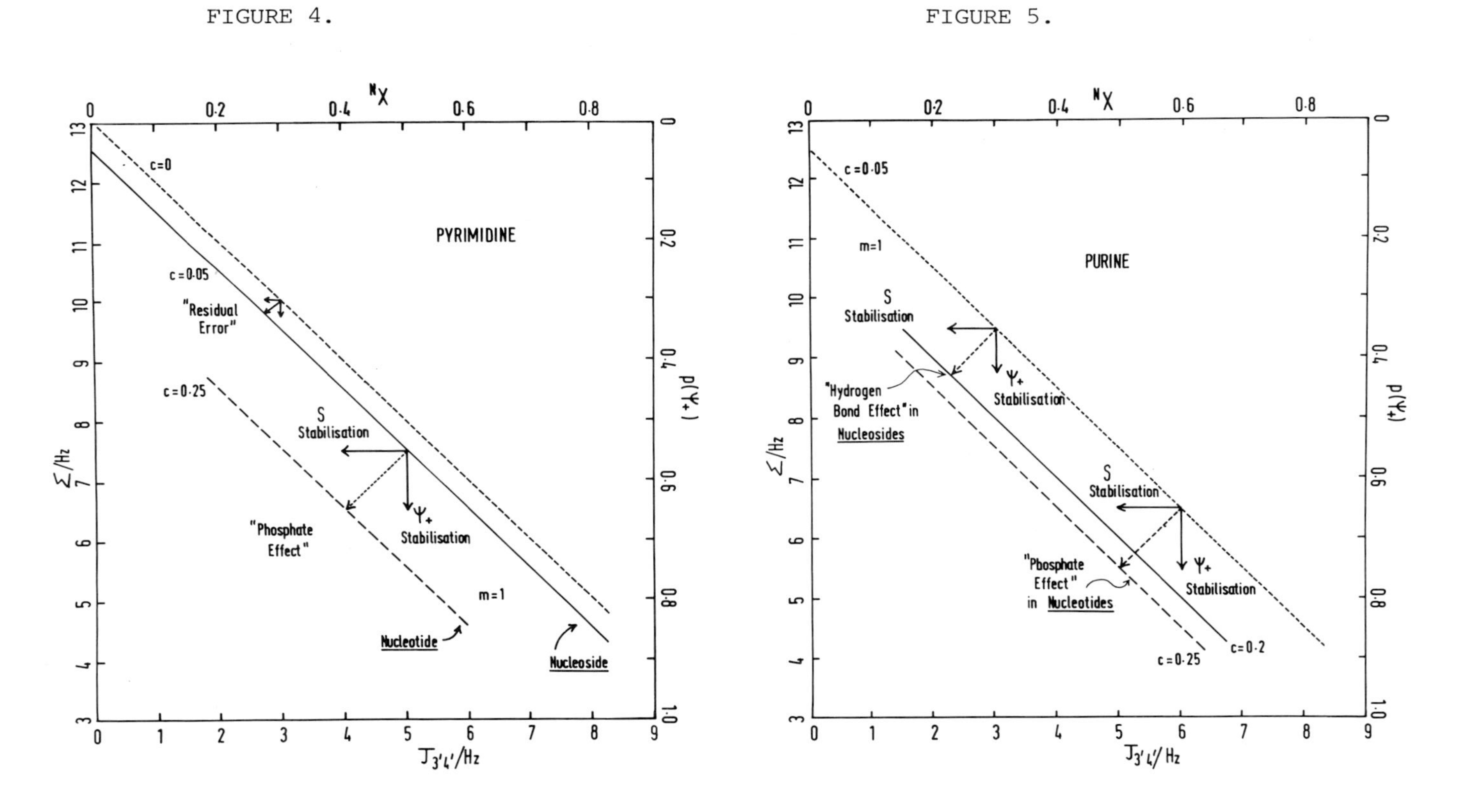

Relationship between backbone C(5')-C(4') bond (ψ_+) and sugar ring *N* ($^N\chi$) conformational properties of pyrimidine (Figure 4) and purine (Figure 5) nucleosides and 5'-nucleotides. The correlations between Σ and $J_{3'4'}$ (slope ~ -1) observed for nucleosides (which includes results for 3'-nucleotides and Lp-molecular fragments of oligomers) and nucleotides (which includes results for -pM molecular fragments of oligomers) are described by equation (2), with $m = 1$ and different magnitudes of c.

nmr studies of pyrimidine nucleosides and nucleotides in solution by Evans and Sarma (1976) and the displacement of the two curves in Figure 4 shows that the magnitude of the "phosphate effect" is equivalent to 2-3kJ mole^{-1} stabilisation of the ψ_+ conformer of nucleotide compared to nucleoside at the same sugar ring conformation (Davies, 1978b).

The Σ *v.*$J_{3'4'}$ behaviour of purine nucleosides (Figure 5, unbroken line, $c \sim 0.2$) is similar to that observed for purine nucleotides (dashed line, $c \sim 0.25$) but the slope (~ -1) of the former is opposite to that suggested by Hruska *et al* (1976) using a limited set of results (slope ~ 1). The discrepancy between these two conclusions may be explained by taking into account possible effects of CH_2OH....N(3) intramolecular hydrogen bonding on sugar ring and backbone conformations of purine derivatives in solution (Davies, 1978b). Indeed, the difference in Σ *v.*$J_{3'4'}$ behaviour of purine and pyrimidine nucleosides is equivalent to $\sim$ 2kJ mole^{-1} stabilisation of backbone ψ_+ conformers of purine compared to pyrimidine nucleosides at the same sugar ring conformational equilibrium. An alternative explanation of this difference in behaviour in terms of destabilisation of ψ_+ conformers by the base ring of pyrimidine compared to purine derivatives is unlikely because purine and pyrimidine-5'-nucleotides exhibit the same Σ *v.*$J_{3'4'}$ behaviour and intramolecular hydrogen bonding is not expected to occur in these molecules. The present analysis also provides an explanation of the apparent difference in behaviour of purine and pyrimidine nucleotidyl units; it was concluded by Evans and Sarma (1976) that 5'-phosphorylation does not increase the "rigidity" of the nucleotide backbone conformation of purine derivatives as it does for pyrimidine derivatives. An alternative explanation is that 5'-phosphorylation does stabilise the *S*, ψ_+ conformational state of nucleotides compared to nucleosides in line with the predictions of the "rigid" nucleotidyl unit (Sundaralingam, 1973) *but* the effect in purine derivatives is masked by the stabilising effect, probably by intramolecular hydrogen bonding, operating for purine nucleosides in solution (Davies, 1978b).

3. ϕ, ψ RELATIONSHIP (Σ' *v.* Σ PLOT)

The interrelation between O(5')-C(5'), ϕ, and C(5')-C(4'), ψ, bond conformations of nucleotidyl units has been investigated by plots of $\Sigma'(= J_{P5'} + J_{P5''})$Hz *versus* Σ Hz which monitor the populations of the ϕ_a and ψ_+ conformers respectively. A schematic representation of the results shown in Figure 6 indicates that the behaviour of 5'-monoribonucleotides (*anti*-type, pyrimidine O, purine ●; *syn*-type, pyrimidine Δ purine ▲) is markedly different from that of homo(py-py □ , pu-pu ■) and heterodimers (py-pu ◩ , pu-py ◪). Faint lines connect results for molecules observed at different temperatures. The dashed line is similar to that observed by Wood *et al* (1973) for *syn*- and *anti*-type 5'-mononucleotides and the slope ($\Delta\Sigma'/\Delta\Sigma \sim 0.54$) shows that the effect of changing the glycosidic bond from predominantly *anti* to predominantly *syn*-type conformations is to destabilise the ψ_+ conformer about four times as readily as the ϕ_a conformer, ie, $\Delta p(\phi_a)/\Delta p(\psi_+) \sim 0.25$. This

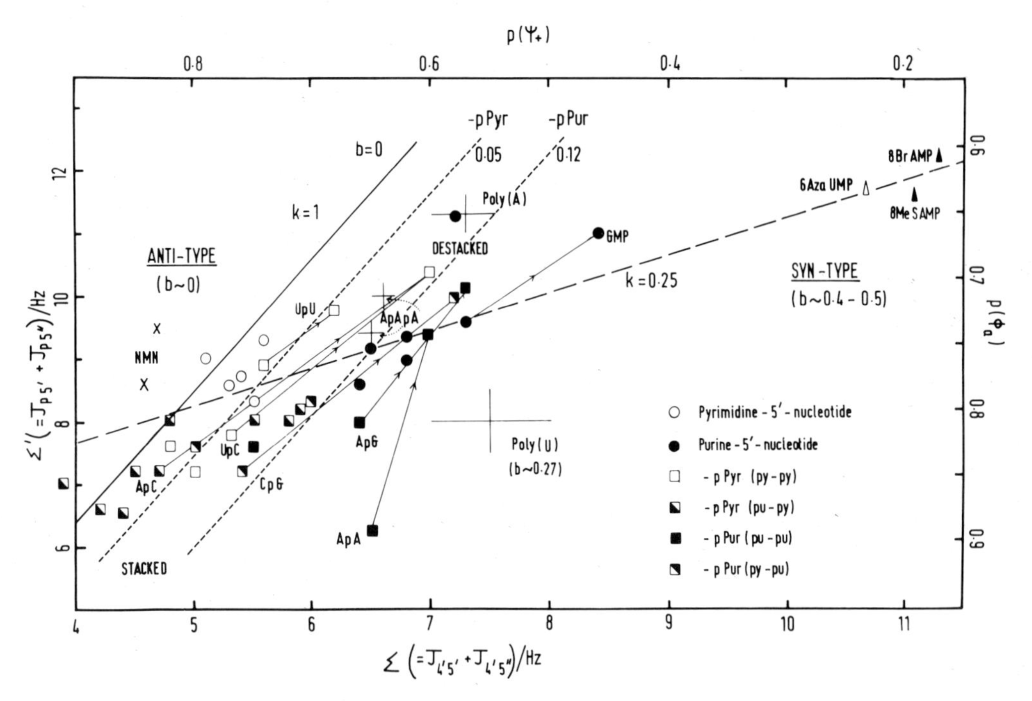

FIGURE 6. Nucleotide Backbone Conformational Map; ϕ, ψ. Plot of Σ' *v.* Σ for 5'-mononucleotides, dinucleoside phosphates, oligo- and polynucleotides.

effect is expected in view of the proximity of the base ring to O(5') for the C(5')-C(4') bond in the ψ_+ conformation. Hence, it was proposed that the population of the stable ψ_+ conformer of mononucleotides reflects, primarily, the base ring *syn* $\rightleftharpoons$ *anti* equilibrium (Davies, 1978c). The results for dinucleoside phosphates, a trinucleoside diphosphate and two polynucleotides do not conform to the monomer behaviour; in particular, the effect of increasing temperature of dinucleoside phosphates (ApA, ApG, UpU, UpC, ApC, CpG) corresponds to about equal destabilisation of both ϕ_a and ψ_+ conformers, ie, slope $\sim$ 2-3. It was suggested (Davies, 1978c) that the backbone conformations of normal oligonucleotides depend, primarly, on the sugar ring conformation, ie, base-stacking on the Altona model (1975).

Assuming a proportional dependence of the relative populations of ϕ_a and ψ_+ conformers, it was shown (Davies, 1978c) that the $\Sigma' v. \Sigma$ behaviour could be represented by equation (3) where k and b are constants defined by equation (4).

$$\Sigma' = 2.1\, k.\Sigma + (25-27k-21b) \quad (3)$$

$$p(\phi_a) = k.p(\psi_+) + b. \quad (4)$$

A number of solutions of equation (3) are compatible with the observed band of results but a reasonable interpretation can be made assuming a 1:1 correlation between the ψ_+ and ϕ_a populations (ie, $k = 1$) and values of b represent, primarily, the effect of the glycosidic bond *syn* $\rightleftharpoons$ *anti* equilibrium on nucleotide backbone conformations, ie, $b = 0$ (unbroken line) represents the behaviour of molecules with predominantly *anti*-type conformations. It is found for mononucleotides that $b \sim 0.02$(*anti*-type pPyr) $< b \sim 0.16$(*anti*-type pPur) and both magnitudes are substantially less than b(*syn*-type) $\sim$ 0.4-0.5. The behaviour at the monomer level is reflected at the oligomer level, ie, $b \sim 0.05$(-pPyr) $< b \sim 0.12$(-pPur). For monomers and most dimers {except Apa and ApG, (Lee *et al*, 1976)} the magnitude of b increases with temperature in agreement with the expected increase in population of glycosidic bond *syn*-type conformers at elevated temperatures. The important result for the present discussion of nucleotide flexibility is that there is an approximate 1:1 correspondence between the fractional populations of the backbone ψ_+ and ϕ_a conformers of oligonucleotides and both conformer populations decrease at higher temperatures.

4. ψ', ϕ' RELATIONSHIP (ϕ' BOND CONFORMATIONAL MODEL)

The conformational properties of nucleotide C(3')-O(3') bonds, ϕ', may be determined from observations of $^3J(POCH_{3'})$ magnitudes and from ^3J(CCOP) magnitudes for both C(2') and C(4') nuclei. Magnitudes of ^3J were initially analysed by Smith and co-workers (1973) in terms of three classical staggered rotamers but little correspondence was found between ^{1}H and ^{13}C nmr results (Davies and Danyluk, 1975) and the conformational analysis has been revised (Lee *et al*, 1976; Ezra *et al*, 1977) in the light of evidence from X-ray crystallographic studies and potential energy calculations (Tewari *et al*, 1974). X-ray evidence (Sundaralingam 1973; Jack *et al*, 1976) indicates ϕ' conformations in the range 160-280 with two narrow ranges of conformations symmetrically placed with respect to ϕ'_{240} ie, the P-O(3') bond coplanar with C(3')-H(3'). It was also found for tRNA that the sugar ring usually adopts the *N* conformation for lower magnitudes of ϕ' (200-220) and the *S* conformer for $\phi' > 240$ (Jack *et al*, 1976). Using such results it was suggested that the P-O(3') bond is restricted to librational motion between two domains in which the phosphate group is at an angle $\pm\theta$ to H(3'), the angle θ being determined from the approximately constant ^3J(HCOP) magnitudes of 3'-nucleotides and Lp- molecular fragments rather than assume classical *gauche* ($\theta = 60^o$) domains (Davies, 1978a). For example, magnitudes of ^3J(HCOP) of ribo- and deoxyribonucleotide derivatives vary at the oligomer level but not at the monomer level (*viz. ribo*; monomer 7.6 ($\pm$ 0.2), oligomer 8.3 ($\pm$ 0.3)Hz and *deoxyribo*; monomer 7.7 ($\pm$ 0.1) and oligomer 6.5 ($\pm$ 0.3)Hz) and correspond to conformational domains that are similar for monomers (ribo- and deoxyribo-, $\theta \sim 37^o$) and that differ slightly for dimers (ribo- $\theta \sim 35^o$, deoxyribo- $\theta \sim 42^o$).

A ϕ' bond conformational model that is related to the sugar ring equilibrium (ie, ψ') was suggested by Ezra *et al* (1977) in the form of equation (5) where N and S represent the classical $_2E$, ^3_2T, ^3E and ^2E, ^2_3T, $_3E$ conformational regions, respectively, and ϕ'_a and ϕ'_- represent the appropriate *gauche* domains.

$$N\phi'_- \rightleftharpoons N\phi'_a \rightleftharpoons S\phi'_- \rightleftharpoons S\phi'_a \qquad (5)$$

Using the ^1H nmr data for ribo-dimers a simplified ϕ' bond conformational model shown in equation (6) can reconcile some of the ^{13}C results for monomers and dimers (Davies, 1978a).

$$N\phi'_{205} \rightleftharpoons S\phi'_{275} \qquad (6)$$

This model also shows that the conformational properties of the sugar ring (summarised by ψ') and the exocyclic C(3')-O(3') bond (ϕ') are directly related which is a necessary precondition for observation of a persistent conformation for nucleotides in solution.

5. χ, ψ RELATIONSHIP (ΔJ *v.* $J_{3'4'}$ PLOT)

A number of X-ray crystallographic studies have shown that both purine and pyrimidine base rings exist in the *anti* or high *anti* conformational region (χ in range 180-300) for 5'-nucleotides (Sundaralingam, 1973) and all 76 bases in yeast $tRNA^{Phe}$ (Jack *et al*, 1976). These studies established a correlation between sugar pucker and χ, with χ in the range 180-210 for the sugar ring in the C(3')-*endo*(N) conformation and χ some 50° higher for the C(2')-*endo*(S) conformation in order to maintain the weak interaction between the base {C(8) in purines and C(6) in pyrimidines} and the exocyclic group O(5').

Glycosidic bond conformations of nucleosides and nucleotides have been investigated by $^3J(^{13}C,H1')$ measurements (Lemieux, 1973; Schweizer and Kreishman, 1973; Davies, 1976). It was shown that observation of vicinal coupling constants between H(1') and both C(2) and C(6) of pyrimidine rings is necessary to determine either *syn*- or *anti*-type conformations unless $^3J(^{13}C, H1') > \sim 5Hz$ (Davies, 1976). It is expected that similar considerations apply to purine derivatives. The glycosidic bond conformational equilibrium has been determined by comparison of ΔJ magnitudes with those observed for free rotation where ΔJ is defined by equations (7) and (8) for purine and pyrimidine derivatives.

$$\Delta J(pu) = {}^3J(C8,H1') - {}^3J(C4,H1') \qquad (7)$$

$$\Delta J(py) = {}^3J(C6,H1') - {}^3J(C2,H1') \qquad (8)$$

Magnitudes of ΔJ(*free*) determined from N-substituted derivaties vary for purine and pyrimidine rings, ie, U $\sim$ 0.6Hz, C $\sim$ 0.9Hz, A, G $\sim$ 1.2Hz (Davies and Sadikot, unpublished results).

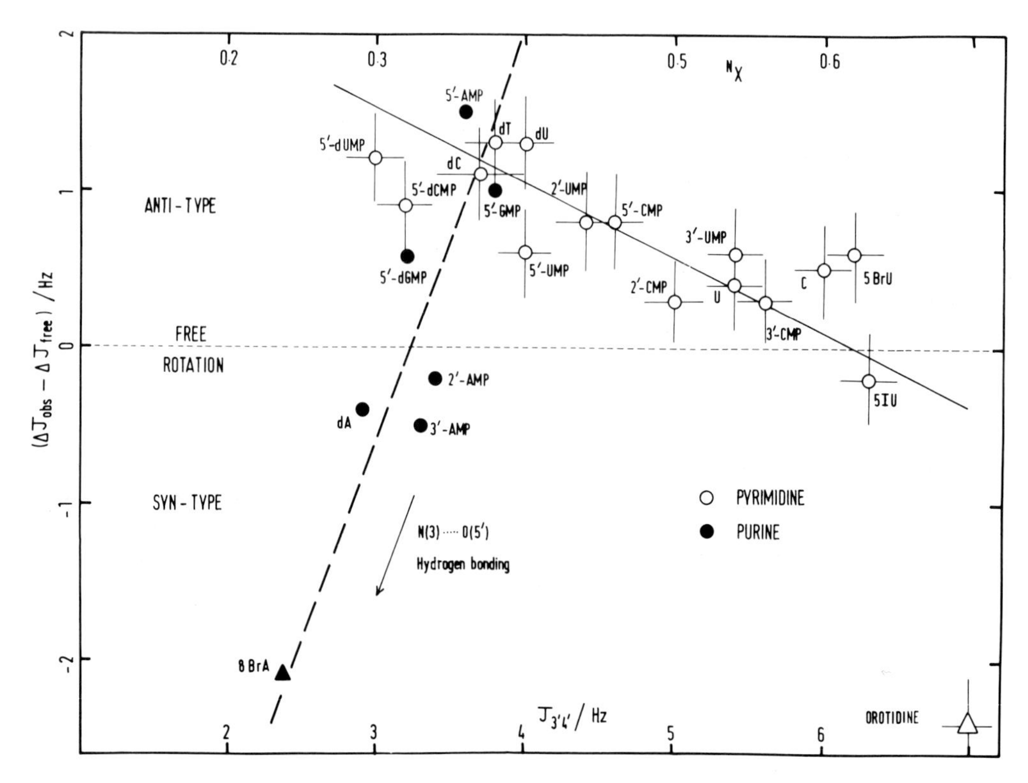

FIGURE 7. Relation between glycosidic bond *syn* and *anti* conformation ($\Delta J_{obs.}-\Delta J_{free}$)Hz and sugar ring conformation ($J_{3'4'}$Hz) for purine and pyrimidine nucleosides and nucleotides.

A plot of ($\Delta J_{obs}-\Delta J_{free}$) *v.* $J_{3'4'}$Hz monitors the relationship between glycosidic bond and sugar ring conformation. Preliminary results for a number of nucleosides and nucleotides plotted in Figure 7 demonstrate not only a similarity in behaviour for purine derivatives (A and G, so far) and for pyrimidine derivatives (U,C,dT) within experimental error but also a marked difference in behaviour between them. The results for pyrimidine derivatives indicate an increase in the population of the *anti* conformer with increasing sugar ring *S* conformation and the linear correlation (slope ∿ - 0.48) corresponds to an approximate 1:1 dependence between them. The linear correlation (slope ∿ 2.5) observed for purine derivaties is consistent with the demands of N(3)...HO(5') intramolecular hydrogen bonding in purine nucleosides in which there is a concomitant stabilisation of the glycosidic bond *syn* conformation and the sugar ring *S* conformation. A relationship between χ, ψ' bond conformers is further evidence of the cooperative nature of the conformational properties of nucleosides and nucleotides.

6. RELATION BETWEEN BASE-STACKING AND BACKBONE CONFORMATIONS.

The cooperative nature of nucleotidyl conformational properties is demonstrated vividly by the relation between the fraction of base-stacked conformers, $p(S)$, and the ψ, ϕ backbone conformations of -pPyr (Figure 8) and -pPur (Figure 9) molecular fragments of dimers and trimers. The proportion of base-stacked conformers was calculated from sugar ring conformations according to equation (9) by the method of Altona (1975) who assumed that the unstacked state is represented by the corresponding monomer conformation and that stacked conformations have exclusively *N*-type ribose rings.

$$p(S) = (J_{unstack} - J_{obs.})/(J_{unstack} - J_{stack}) \qquad (9)$$

Magnitudes of $p(S)$, the fraction of the stacked form, have been calculated from sugar ring $J_{1'2'}$ (Altona, 1975) and $J_{3'4'}$ magnitudes (Lee *et al*, 1976; Ezra *et al*, 1977). Using average values of $p(S)$ determined from $J_{1'2'}$ and $J_{3'4'}$ magnitudes, a sequence of stacking proclivities of bases in dinucleoside monophosphates at ambient temperatures has been determined (Ezra *et al*, 1977; Davies, 1978a)

$$\text{pu-py} \sim \text{pu-pu} > \text{py-py} \sim \text{py-pu} \qquad (10)$$

$$34(\pm 8) \quad 31(\pm 5) \quad 22(\pm 8) \quad 18(\pm 6)$$

6.1 Pyrimidine: py-py (□) v pu-py (◩)

Linear correlations are observed between $p(S)$ and both Σ (dashed line) and Σ' (unbroken line) for -pPyr molecular fragments as shown in Figure 8. Assuming that $p(S)$ is directly proportional to the preferred conformations for C(5')-C(4'), ψ_+, and O(5')-C(5'), ϕ_a, bonds, it was shown (Davies, 1978d) that these linear correlations were predicted by equations (11) and (12) where D,D' and F,F' are constants.

$$\Sigma = (13 - 10D) - 10F.p(S) \qquad (11)$$

$$\Sigma' = (25 - 21D') - 21F'.p(S) \qquad (12)$$

D and D' represent the relative populations of ψ_+ and ϕ_a conformers at zero-base stacked conformations (ie, similar to monomers) whereas F and F' reflect the effect of base-stacking on stabilising the backbone conformation.

The effect of base-stacking on -py backbone conformational properties of homo- (py-py) and hetero-dimers (pu-py) is the same within experimental error as shown in Figure 8. This suggests that the -pPyr molecular fragment controls not only the adjacent backbone conformational properties but also the base-stacking tendency of the molecules; these properties may, in turn, be controlled by the overwhelming preference for *anti*-type conformations of pyrimidine base-rings in 5'-nucleotides and, hence, -py fragments in oligomers (Davies, 1978d).

FIGURE 9

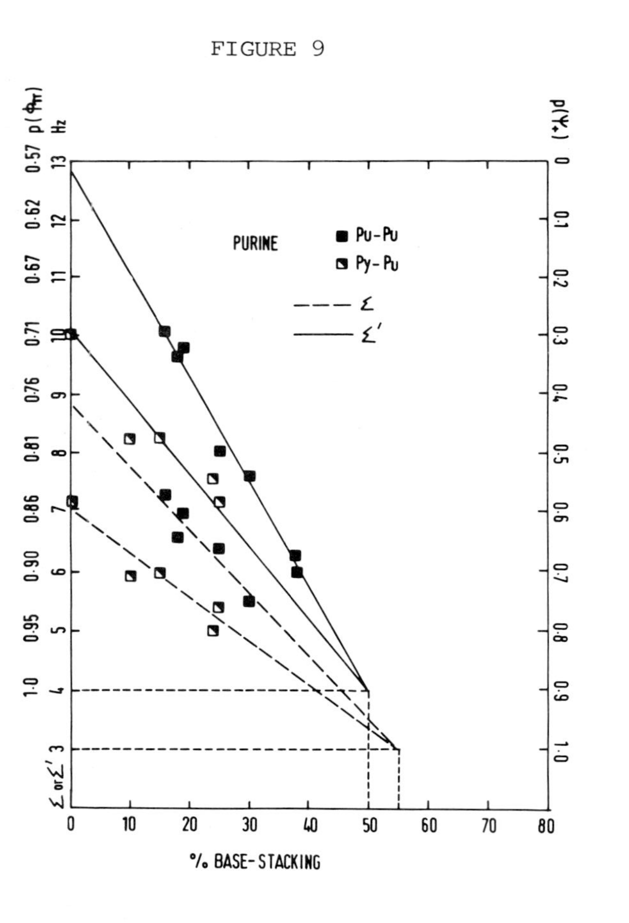

FIGURE 8

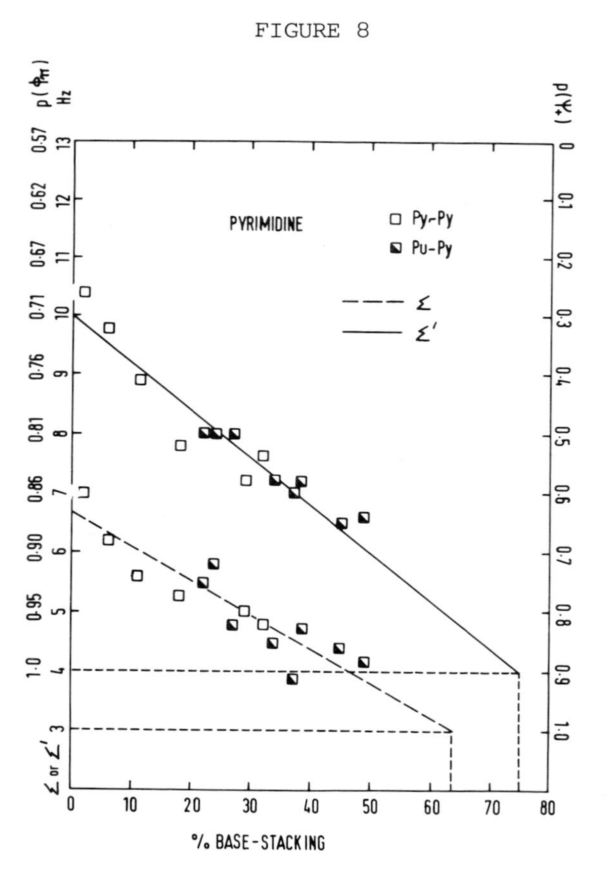

Correlation between % base-stacking and proportions of backbone O(5')-C(5') bond, ϕ_a, (plotted as Σ') and C(5')-C(4') bond, ψ_+, (plotted as Σ) conformers for pyrimidine (Figure 8) and purine (Figure 9) 5'-nucleotidyl units in homo- and hetero-dimers.

6.2 Purine : pu-pu (■) *v.* py-pu (◪)

Linear correlations are observed between $p(S)$ and both Σ and Σ' for -pPur molecular fragments as shown in Figure 9 and they can be described by equations (11) and (12) respectively. The effect of base-stacking on backbone conformations of purine derivatives *appears* to be greater for homo- (pu-pu) than for hetero- (py-pu) dimers. Extrapolation to conditions for observation of the "rigid" nucleotidyl unit (ie, $p(\psi_+) = p(\phi_a) = 1$; Σ *ca* 3Hz and Σ' *ca* 4Hz) indicates that this conformational state is given by 50-55% base-stacked conformations for -pu derivatives compared to 65-70% base-stacked conformers for -py dimers in line with the expected greater base-stacking tendencies of purine compared to pyrimidine rings.

The effect of base-stacking on backbone conformation depends, to some extent, on the amount of overlap of bases involved in the stacking process and the angle between two successive base planes. Although such detailed information on base-stacking is not yet available for these molecules in solution, the effect of base-stacking on nucleotide backbone conformations can be compared by the ratio of the slopes (F and F') for each molecular fragment with those for -py fragments which generates the sequence of behaviour summarised in equation (13) together with the averaged values of the ratio (F and F') (Davies, 1978a,d). These results suggest that sequence may be important for determining local conformations in polyribonucleotides.

$$\begin{array}{ccccccc} \text{pu-pu} & > & \text{py-pu} & > & \text{pu-py} & \sim & \text{py-py} \\ 2.0 & & 1.4 & & \sim 1.0 & & 1.0 \end{array} \qquad (13)$$

There is an approximate 1:1 correspondence between base-stacking and backbone ψ_+ and ϕ_a conformations for pu-pu dimers (F 1.05, F' 0.84) which decreases in step for py-pu (0.73, 0.58), pu-py and py-py (0.57, 0.38) dimers; in each case the effect on ψ_+ is greater than on ϕ_a. These results show that base-stacking is a cooperative process and that the equilibrium between stacked and unstacked forms is accompanied by corresponding changes in the populations of backbone ψ_+, ϕ_a conformations, sugar ring N conformations (ψ'_N) and, presumably, the glycosidic bond *anti* conformation (χ_a) and exocyclic bond ϕ'_{205} conformation. It is expected that the fundamental nucleotidyl conformation in the base-stacked helical form is consistent with that of the "rigid" unit observed by X-ray studies (Sundaralingam, 1973). This fundamental unit can also give rise to bends, loops and extended conformations depending on the ω', ω conformations. However, it is not expected that ω' and ω bonds are appreciably more flexible than other nucleotide bonds as the proportion of base-stacked conformers (on the Altona model, 1975) is roughly in line with the "persistence" of the stable conformational unit eg, GpC, 293K, Ezra *et al*, 1977; 45% base-stacking; ψ_+ 0.9, ϕ_a 0.9, ψ'_N 0.75 (hence ϕ'_{205} 0.75) give a "persistence" of 0.46 assuming χ_a 1.0, ω' 1.0 and ω 1.0 .

7. CONCLUSIONS

(i) Methods have been developed for rationalising the conformational properties of nucleosides and nucleotides in solution and for comparing results of ribose with deoxyribose molecular fragments, purine with pyrimidine derivatives and monomers with oligomers.

(ii) Conformational properties of adjacent bonds of nucleotides (excluding ω' and ω) have been shown to be cooperative in nature with approximate 1:1 relationships between the populations of the most stable conformer for each bond. These results resolve the paradox between the concept of a "rigid" nucleotidyl unit and the existence of a "flexible" nucleotide in solution by showing that a particular conformational state (χ_a, ϕ_a, ψ_+, ψ'_N, ϕ'_{205}) has a significant "persistence" in solution, especially at the oligomer level.

(iii) Base-stacking of oligonucleotides in solution has been demonstrated to be a cooperative process with populations of backbone ψ_+ and ϕ_a conformations increasing in line with proportions of base-stacked conformers. It was shown that the base-sequence of oligoribonucleotides alters the equilibrium between base-stacking and backbone conformational properties at the dimer level which suggests that base-sequence might affect the local backbone conformation at the polymer level.

ACKNOWLEDGEMENTS

I thank Professor B. Pullman for the invitation, the organisers for their hospitality and the Royal Society for a travel grant to attend the Symposium.

REFERENCES

Altona, C.: 1975, *Structure and Conformations of Nucleic Acids and Protein-Nucleic Acid Interations* (Sundaralingam, M. and Rao, S. T. eds.), Univ. Park Press, Baltimore, pp. 613-629

Altona, C. and Sundaralingam, M.: 1972, *J. Amer. Chem. Soc.*, 94, 8205

Altona, C. and Sundaralingam, M.: 1973, *J. Amer. Chem. Soc.*, 95, 2333

Chachaty, C., Zemb, T., Langlet, G. and Tran-Dinh, S.: 1976, *Eur. J. Biochem.*, 62, 45

Davies, D. B. and Danyluk, S. S.: 1974, *Biochemistry*, 13, 4417

Davies, D. B. and Danyluk, S. S.: 1975, *Biochemistry*, 14, 543

Davies, D. B.: 1976, *Studia Biophysica*, 55, 29

Davies, D. B.: 1978a, *Progress in NMR Spectroscopy* (Emsley, J. W., Feeney, J. and Sutcliffe, L. H.) Pergamon Press, Oxford, in Press

Davies, D. B.: 1978b, *Eur. J. Biochem.*, submitted to Press

Davies, D. B.: 1978c, *J. Chem. Soc.*, submitted to Press

Davies, D. B.: 1978d, *Biochem. Biophys. Acta*, submitted to Press

Evans, F. E. and Sarma, R. H.: 1976, *Nature,* 263, 567
Ezra, F. S., Lee, C-H., Kondo, N. S., Danyluk, S. S. and Sarma, R. H.: 1977, *Biochemistry,* 16, 1977
Guéron, M., Chachaty, C. and Tran-Dinh, S.: 1973, *Annals. NY Acad. Sci. (USA)* 222, 307, and references therein.
Hruska, F. E.: 1973, *Conformations of Biological Molecules and Polymers* (Bergmann, E. D. and Pullman, B., eds.) Academic Press, New York, p.345
Hruska, F. E., Wood, D. J. and Singh, H.: 1976, *Biochim. Biophys. Acta,* 474, 129
Jack, A., Ladner, J. E. and Klug, A.: 1976, *J. Mol. Biol.,* 108, 619
Klyne, W. and Prelog, V.: 1960, *Experentia,* 16, 521
Lee, C-H., Ezra, F. S., Kondo, N. S., Sarma, R. H. and Danyluk, S. S.: 1976, *Biochemistry,* 15, 3627
Lemieux, R. U.: 1973, *Annals. NY Acad. Sci.,* 222, 915
Lüdemann, H-D., Röder, O., Westhof, E., v Goldammer, E., and Müller, A.: 1975, *Biophys. Struct. Mechanism,* 1, 121
Schweizer, M. P. and Kreishman, G. P.: 1973, *J. Mag. Res.,* 9, 334
Smith, I. C. P., Mantsch, H. H., Lapper, R. D., Deslauriers, R. and Schleich, T.: 1973, *Conformations of Biological Molecules and Polymers* (Bergmann, E. D. and Pullman, B., eds.) pp. 381-402, Academic Press, New York
Sundaralingam, M.: 1973, *Conformations of Biological Molecules and Polymers* (Bergmann, E. D. and Pullman, B., eds.) pp. 417-455, Academic Press, New York, and references therein
Sundaralingam, M., Pullman, B., Saenger, W., Sasisekharan, V. and Wilson, H. R.: 1973, *Conformations of Biological Molecules and Polymers* (Bergmann, E. D. and Pullman, B., eds.) pp. 815-820 Academic Press, New York
Tewari, R., Nanda, R. K. and Govil, G.: 1974, *Biopolymers,* 13, 2015
Wood, D. J., Mynott, R. J., Hruska, F. E. and Sarma, R. H.: 1973, *Fed. Eur. Biochem. Soc. Letters,* 34, 323
Yathindra, N. and Sundaralingam, M.: 1973, *Biopolymers,* 12, 297, 2075, 2261

THE QUANTITATIVE SEPARATION OF STACKING AND SELF-ASSOCIATION PHENOMENA IN A DINUCLEOSIDE MONOPHOSPHATE BY MEANS OF NMR CONCENTRATION-TEMPERATURE PROFILES: 6-N-(DIMETHYL)ADENYLYL-(3',5')-URIDINE.

C. Altona, A.J. Hartel, C.S.M. Olsthoorn, H.P.M. de Leeuw and C.A.G. Haasnoot
Department of Organic Chemistry, Gorlaeus Laboratories of the University, Wassenaarseweg 76, Leiden, Netherlands.

INTRODUCTION

The question whether short oligonucleotides are really suitable as models for longer regions in nucleic acids has been raised (1), but will not be answered until many careful systematic studies are carried out with experimental probes at atomic resolution using nuclear magnetic resonance (NMR), supplemented with thermodynamic information from independent techniques such as UV-hypochromism and circular dichroism (CD), and guided by information on geometrical details provided by X-ray methods.

It should be realized that the shortest nucleic acid fragments available for studies of these properties, the dinucleoside monophosphates or dimers (XpY) represent at the same time 3'- and 5'- terminal fragments of RNA (DNA); thus constitute a limiting case. Therefore attempts at straightforward transfer of any conformational property of these dimers to longer fragments must be viewed with suspicion. Indeed, optical studies on trimers (XpYpZ) (2) purport to show that at least in some particular sequences of purines and pyrimidines the sum of the properties of the constituent dimers differs dramatically from the experimental property of the trimer, measured at the same temperature. This finding has now been substantiated by an independent technique. Recent NMR-analyses of trimers (all proton signals assigned) at a series of temperatures, recently carried out in our laboratory, have shown that in three cases studied quite dramatic selective conformational transmission effects occur (3,4).

In view of the need to understand these new effects and their implications for helix and loop formations in RNA and DNA the conformational ramifications of the constituent dimers themselves should be sorted out as thoroughly as possible. Notwithstanding the large number of studies that have appeared recently (vide infra), general agreement on the thermodynamics and geometry of helix formation has not been reached.

Early NMR studies on dinucleoside monophosphates and conformational conclusions were reported in the years 1968-1970 (5-9). The only data that could be extracted from the spectra at that time were the chemical shifts of the base and H1' protons (as well as the coupling constant J1'2'

B. Pullman (ed.), Nuclear Magnetic Resonance Spectroscopy in Molecular Biology, 87-101.

in first-order approximation) at various temperatures. It was deduced (5-7) that the ribose ring pucker in stacked dimers is "more" 3'-endo and that this "changes toward" 2'-endo on unstacking. A righthanded helical model for vertical base-base stacking was proposed (6-9) but more quantitative conclusions (e.g. pertaining to ΔH^{o} and ΔS^{o} for the stack-destack equilibrium as measured from NMR) as well as insights into the conformational properties of the sugar-phosphate backbone remained lacking. On the basis of the Altona-Sundaralingam (AS) pseudorotation model for the five-membered sugar ring (10) the interrelationships between the proton couplings J1'2', J2'3' and J3'4' were worked out by the same authors (11) and for the first time couplings were used to calculate quantitatively the amount of 3'-endo(N) and 2'-endo(S) conformers present in several dinucleotides in solution. Interestingly, the simple essence of this method, the evaluation of percentage S of N forms from either J1'2' or J3'4', taking the sum J1'2' + J3'4' as reference point, was soon adopted even by its severest critics (12) and is now commonly applied. (20-24)

Selective deuteration allowed Kondo and Danyluk (13) to report assignments of chemical shifts of ApA as early as 1972, but the quality of the spectra obtained allowed no firm conformational analysis. The advent of improved NMR instrumentation allowed Altona et al to publish the first complete NMR data sets for two ribodinucleotides in 1974*; m^6ApU (14) and ApU (15). The coupling constants of these compounds and of i^6ApU and m_2^6ApU (as well as the assignment and line-shape simulation of the ribose spectrum of the latter dinucleotide) were discussed at the 4th Steenbock Symposium held in June 1974 at the University of Madison Wisconsin, U.S.A. (16). A year later, preliminary papers on the NMR assignments of ApA and dApdA appeared (17,18), followed by an extensive discussion on the conformations of ApA by Kondo and Danyluk in 1976 (19). By that time HF-proton spectroscopy of dimers in general has become practically routine, witness the publication of spectral data of fifteen dinucleoside monophosphates (10,21), of two 2'-O-methyl substituted dimers (22,23) and of seven ethenoadenosine containing dinucleotides (24). However, the various rationalizations given (17-24) of the observed chemical shift effects appear rather confusing.

Our original simple model (11,14,16) of the conformational behavior of dimers in aqueous solution was based on evidence from X-ray analyses, CD-spectra, NMR coupling constants and shift evidence from various literature sources (5-11, 25-26 and references therein) and from our own observations (14-16) and consisted of: (a) the proposed existence of a single well-defined righthanded basestacked conformer (from 3'-unit to 5'-unit in recent backbone nomenclature**: anti, N, $\phi' = 215^{o}$, ω'-, ω-, ϕa, ψ+, N, anti) in conformational equilibrium with (b) a relati-

* Actually, the first public announcement of the complete assignment of m^6ApU at 270 MHz was made at the Stockholm Symposium on the Structure of Biological Molecules, July 11, 1973.

** D.B. Davies, private communication

vely large number of unstacked states characterized by main degrees of conformational freedom about the P-O ester bonds ω' and/or ω as well as by freedom of the sugar rings approximately equal to that enjoyed by the constituent monomeric phosphates. No outspoken preference for one or the other unstacked state was envisaged. Of course, it should be realized that the conformational preferences about the remaining degrees of freedom in the unstacked molecules (C3'-O3', Φ'; C5'-O5', Φ; C4'-C5', Ψ) are, however outspoken, certainly less than 100%. Therefore, rotation about these bonds constitute minor degrees of freedom of the unstacked states. It is for this reason that we stated: "the unstacked molecules exist in a relatively large number of states, the analysis of which poses challenging and formidable problems" (16). Recent theoretical work (27) seems to indicate that some combinations of ω',ω and 3' - 5' combinations of ribose pucker (N-N, N-S, S-N and S-S) are more favored than others.

Later, Kondo and Danyluk (19) postulated the existence of two stacked conformations, I and II, I depicted as the regular right-handed helix (ω',ω/330°,320°), II as a "more loosely" base-stacked left-handed loop structure with ω',ω/50°,80°. Shortly thereafter (20), II was described as ω',ω/80°,80°, a linear form having ω',ω/270°,90° was also postulated. Subsequently, the same authors (21) reversed the usual assignment of the 5' and 5" signals of the Xp- moiety. This reversal has now been recalled (24), concomitant with the introduction of a third stacked conformation; stacked forms I, II and III supposedly in equilibrium with a single (!) unstacked or extended form in solution. These three stacked states are supposed to be characterized by:
I A righthanded helix, ω',ω/300°, 290°; II a lefthanded helix, ω',ω/30°, 100°; III An unorthodox stacked form having the following unusual features: (1) the Xp-ribose ring in S-conformation (2) the Φ' angle in the g^+ region (260°), (3) ω',ω/50°, 220°, and (4) a high χ-angle (100°) for the base in the -pY moiety.

In our opinion, too little is known at present about the many factors which influence the composite shielding or deshielding effects on the various protons upon helix formation to warrant far-reaching generalizations of the kind described above. An interesting discussion about the sources of uncertainty was recently given by Borer et al (1) who pointed out that it is very important that a more exhaustive theoretical and empirical assessment of second-order effects should be made. Only then can small chemical shift effects be properly associated with geometrical perturbations (1). Explicit specification of angles that cannot be monitored directly by coupling constants should be regarded with special caution. Even the better understood ring current shifts exerted by the purine bases (28) do not constitute infallible guides. Even within the variation of angles allowed by X-ray crystallography the effect of an adenine base on the protons of the neighboring residue is extremely sensitive to the exact magnitude of the backbone angles adopted and no exact solution of the dinucleotide problem along these lines has as yet been offered. A complicating factor in the interpretation of observed chemical shifts is to be found in the well known, but too little heeded, tendency of the dimers to associate intermolecularly, due to vertical stacking of the bases (29). Observations seem to indicate that those bases that favor intramolecular stacking also appear to favor intermolecular association.

A side-step on nomenclature is now in order: The term "dimerization shifts" has generally been taken to denote differences between observed chemical shifts of dinucleotides and observed shifts of the constituent nucleoside monophosphates (the latter corrected for the phosphate electronic charge by measurements at the appropriate pH). Therefore, "dimerization shifts" reflect the effect of an a priori unknown amount of stacking on chemical shifts and then only if the properties of the unstacked states are correctly represented by the respective monomer models; a demand that needs strict proof for every dimer examined, especially those that are known to be strong "stackers". Surprisingly, no such proof has been forthcoming. Therefore, in this paper we propose to introduce the term "stacking shift" for the 100% net effect of intramolecular (conformational) change or changes in dinucleotides induced by complete stacking; and "association shift" for the effects due to the 100% formation of an intermolecular complex. It is a popular misconception that association phenomena can only be studied by measuring shifts over a large range of concentrations. We hereby propose to utilize a new method based on differential concentration-temperature profiles.

We start off from the following well-known experimental facts: (a) Most dimers display significant upfield (and sometimes downfield) chemical shift effects in aqueous solution when the temperature is decreased; the magnitude of these effects varies from proton to proton monitored and also depends on the individual bases present and on their sequence. These shifts are generally ascribed to "stacking".
(b) Significant (always upfield) shift effects are seen when, at a given temperature, the concentration of the dimer is increased. Here also, the magnitude varies from proton to proton and is especially noticeable at the lower end of the temperature range. These shift effects are generally ascribed to intermolecular association and are considered troublesome. Implications have appeared to the effect that association shifts can be considered minimal at a concentration of 38 mM (19) or can be neglected for 10-30 mM solutions (20) or are negligible at 5-10 mM (24). Sometimes elimination of association shifts is attempted by simple extrapolation of the measured shifts to infinite dilution (19). The latter method may give rise to appreciable errors (vide infra) whereas complete neglect needs thorough experimental checking, especially for good stackers.

The question is: can we describe stacking and association effects on a quantitative basis by adopting a simple model with a few adjustable parameters? If so, can the result be checked by means of an independent method? Clearly, one has no recourse but to thermodynamic methods first, instead of starting off from a specific molecular model with predicted properties (NMR or otherwise) at the outset. Simplifying assumptions made should, if false, show up automatically in the course of the analysis. We propose to show for the first time that, at least in the case of

m_2^6ApU, the question can be positively answered and that stacking and association shifts can be separated quantitatively by measuring chemical shifts at a few well-chosen concentrations over the full range of accessible temperatures (-4° to 98°C), without taking *a priori* recourse to information gleaned from spectra of constituent monomers. Indeed, one could put to test the usual assumption that the monomeric phosphates are the perfect models for the unstacked states (with respect to either chemical shifts or coupling constants).

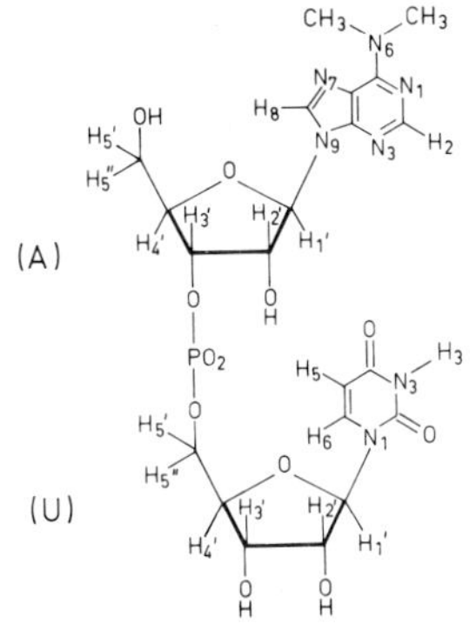

Fig. 1. Structure of 6-N-(dimethyl)adenylyl-(3',5')-uridine, m_2^6 ApU.

Our choice of model compound for this research (6-N-(dimethyl)adenylyl-uridine, m_2^6ApU) was dictated by a number of considerations: (a) the compound shows quite strong stacking characteristics, rather stronger than shown by its analogs ApU and m^6ApU (16), (b) Perusal of the relevant literature (29) showed that self-association of N^6-dimethyladenosine at 25°C is described by a relatively large K-value (22 M^{-1}), follows a single step equilibrium, and displays a large association shift for H8, H2 and H1'. The latter was ascribed (30) to an overall time-average phenomenon, associated and free species rapidly forming and breaking. (c) Self-association of the m_2^6Ap-moiety with the -pU part of another molecule, was judged to be smaller by at least one order of magnitude. This obviates the complications of triplex or higher aggregates and is well borne out by our final results.

METHOD

Fourier-transform (FT) NMR spectra were recorded on a Bruker HDX-360 (16K data points) and on a Jeol PFT-100 system (16K data points). The experimental material used consists of: 23 360 MHz spectra of 63 mM solutions and 9 of 10.3 mM solutions at temperatures spaced from -4° to 95°C; 15 100 MHz spectra of 5.7 mM and 24 of 34.3 mM solutions, temperatures spaced from -4° to 98°C (a surprising amount of supercooling was easily achieved). In a number of cases water elimination (WEFT) techniques were employed in order to detect signals otherwise hidden underneath the residual HDO peak. Small coupling constants were detected by applying a sinusoidal window (30). All 360 MHz spectra were computer simulated (program LAME) to extract accurate shifts and coupling constants, see e.g. Figure 2. In all cases the pD was adjusted to 7.4-7.8. Shift reference was 0.1-5 mM internal tetramethylammonium chloride (TMA). Special attention was

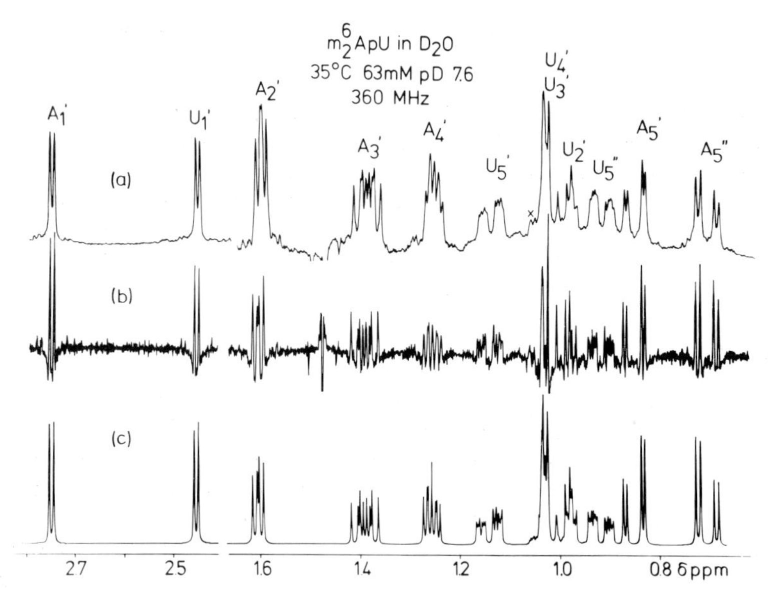

Fig. 2. NMR spectra and assignment (a) Experimental, (b) After digital filtering (sine bell), (c) simulated.

paid to the calibration of the temperature scale; by means of standard methanol and ethylene glycol samples a TMA in HDO sample was measured over the full temperature range. It was found that a reliable temperature scale could be based on δ HDO–δ TMA and the spectra have, in fact, a built-in thermometer (31).

Bimolecular association and intramolecular stacking are given by:

$$2\ \mathrm{XpY} \overset{K_A}{\rightleftarrows} (\mathrm{XpY})_2 \qquad (1)$$

$$(\mathrm{XpY})_u \overset{K_x}{\rightleftarrows} (\mathrm{XpY})_x \qquad (2)$$

(the subscript x for helix was chosen to avoid confusion with S of entropy or H of enthalpy. It also signifies that no information about this state is presupposed). The following simplifying assumptions are introduced: (a) Stacking and association in first approximation occur independently of each other (i.e. the associated molecules do not "know" whether they are both unstacked, or only one of them, or neither of them) and the chemical shift effects of stacking and association are taken to be additive.

(b) Both equilibria may be described within the boundaries of a two-state model. At the (theoretical) high-temperature limit ($T\infty$) the solution contains exclusively non-associated and unstacked species. The unstacked molecules exist in a large number of conformations in rapid equilibrium (16), but because the enthalpy differences between these species are relatively small the total blend may be regarded operationally as a single state.

(c) The usual Van 't Hoff conditions apply in the 269 - 371 K temperature range studied (ΔH^o, ΔS^o and the properties of the species present are independent of temperature).

In the following the concentrations C, the mole fractions P and the physical property monitored α of the non-associated species carry the subscript F, those of the associated species the subscript A. Similarly, subscripts u and x denote unstacked and stacked states respectively. For the high-temperature limit $P_F = P_u = 1$ and always $\alpha_u = \alpha_F$ (assumption b). Then, at any chosen temperature, the property monitored is related to the molar fractions P_A and P_x as follows:

$$\alpha_{obs} = \alpha_F + P_A\Delta_A + P_x\Delta_x \tag{3}$$

where $\Delta_A = \alpha_A - \alpha_F$ and $\Delta_x = \alpha_x - \alpha_u$. The problem boils down to the determination of seven independent parameters: ΔH^o_A, ΔS^o_A, ΔH^o_x, ΔS^o_x, α_A, α_x, α_F ($= \alpha_u$). For computational convenience our least-squares procedures iterate on Tm ($= \Delta H^o/\Delta S^o$) and ΔS^o. These seven parameters (or a group of four) can be extracted independently for each proton shift for which Δ_A and/or Δ_x is sufficiently large and the final thermodynamic parameters can be checked afterwards for overall consistency. For the association equilibrium we have:

$$K_A = C_A/(C_F)^2 = \exp|\Delta S^o/R(1-Tm/T)| \tag{4}$$

The stochiometric concentration of dinucleotide $C_o = 2C_A + C_F$; the associated proton fraction (32, 33) is given by:

$$P_A = 2C_A/C_o = [1 + 4K_AC_o - (1 + 8K_AC_o)^{\frac{1}{2}}] / 4K_AC_o \tag{5}$$

Because an automatic least-squares (LS) iteration on all seven parameters simultaneously presented difficulties, we preferred to split the problem into three parts for the time being:
I. First, our program DIMASS (31) is used to obtain (by means of standard Newton-Raphson procedures) LS estimators of the three parameters pertaining to the association equilibrium. Temperature profiles (9 - 24 data points each) of a given chemical shift δ_n of an association-sensitive proton n are measured at two (or more) different concentrations, denoted (1) and (2). The difference between the two shift vs temperature curves mirrors the temperature dependence of the association equilibrium.
At a given T:

$$\delta_n(1) - \delta_n(2) = (P_A(1) - P_A(2))\ (\delta_A - \delta_F) \tag{6}$$

By means of equations 4-6, estimators for the parameters Tm(A), ΔS^o_A, $\delta_A - \delta_F$ were extracted from the H2, H8 and A1' proton signals. Fig. 3 was prepared from the A1' curves of m^6_2ApU with $C_o(1) = 5.7$ mM and $C_o(2) = 63$ mM, with the aid of the parameters extracted from the shifts of this proton. The graphs illustrate clearly that a maximum (or minimum) should be expected on the difference curve (though not necessarily within the accessible temperature range). This occurs because at the limiting high

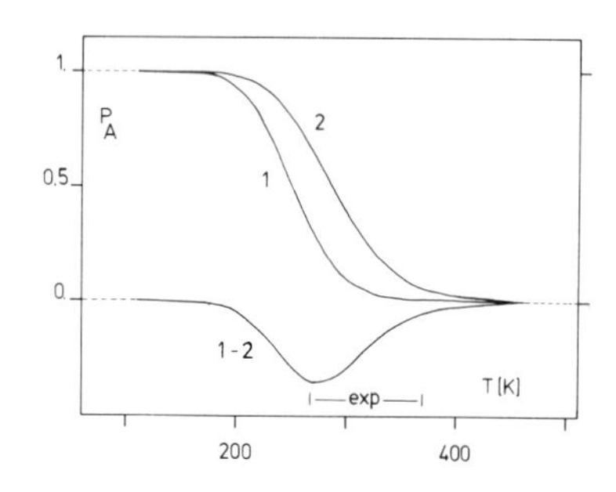

Fig. 3. Plots of the molal fraction of associated species P_A in m_2^6 ApU vs temperature (1) 63 mM solution, (2) 5.7 mM solution.

temperature both P_A and P_x approximate to zero, whereas at the low-temperature limit both approximate to 1; at intermediate temperatures the balance between the two phenomena in general will be unequal and gives rise to an extremum (+ or -) on the difference curve.

Of some practical interest is the finding (Fig. 4) that in cases where K_A is large (m_2^6ApU below room temperature) a reliable extrapolation of association-sensitive proton shifts to infinite dilution from shift vs concentration curves requires the careful recording of NMR spectra at extreme dilutions (1 mM or less), which is not usually done. At concentrations of 5-10 mM the association effects, if neglected, may give rise to errors up to 0.15 ppm or more, depending on K_A and Δ_A.

II. For the second step of the analysis proton signals were selected that showed least association shifts on the one hand, and displayed appreciable stacking shifts on the other: U5, U1', U2', U3', U5', U5", A3', A5'. By means of standard thermodynamic procedures, incorporated into our program STAKKER (34) an iterative LS analysis of the temperature profiles of this set of protons measured at 10.3 mM was carried out. The results showed gratifying internal consistency, the spread in calculated Tm and ΔS values for the individual proton shifts being within the estimated 90% confidence limits for each determination. Convergence was rapid in all cases.

III. In the third step all the available data points (covering a temperature range of -4° to 98^oC) were simultaneously used as input: a total of 70 observables for each of the low-field protons (A2, A8, U5, U6, A1', U1') and 30 observables for each of the remaining ones, except A4' and U4' (these were not considered, because of their general insensitivity toward stacking and association). Starting with the averaged thermodynamic parameters and with the Δ-values obtained from steps I and II the LS program ASSTAK (31) refined in turn three association and three stacking parameters, in the final stages δF was included as an adjustable for each proton. The final results of the procedure outlined above are shown in Tables 1 to 4.

The thermodynamic parameters ΔS_x^o and ΔH_x^o can also be obtained from an analysis of changes in CD spectra as function of temperature (25) at concentrations where association effects are indeed negligible. For this

Table 1. Thermodynamic parameters for self-association of m_2^6ApU.

Proton monitored	$-\Delta S^o$ (e.U.)	$-\Delta H^o$ (Kcal/mole)
A8	20	7.3
A2	15	5.9
A1'	25	8.7
A2'	25	8.6
A3'	26	7.6
NMR best value	24.8	8.6[7]

Table 2. Thermodynamic parameters for intramolecular stacking in m_2^6ApU.

Proton monitored	Tm (K)	$-\Delta S^o$ (e.u.)	$-\Delta H^o$ (kcal/mole)
A3'	317	23.1	7.3
A5'	314	22	6.8
A5"	322	25	8.2
U6	307	20.0	6.1
U5	311	20.9	6.5
U1'	305	19.8	6.1
U2'	314	21.8	6.8
U3'	299	18.3	5.5
U5'	308	20	6.3
U5"	286	17	4.8
NMR best value	309.3	20.7	6.4
CD value	307	20.1	6.2

Table 3. Association and stacking shifts of m_2^6ApU.

Proton	$-\Delta_A$	$-\Delta_x$	$\delta_F = \delta_u$ *
A8	0.27	-0.34	5.03
A2	0.38	0.14	5.08
A1'	0.35	0.06	2.90
A2'	0.28	-0.07	1.66
A3'	0.09	0.35	1.62
A5'	0.02	-0.23	0.73
A5"	0.06	-0.07	0.68
U6	0.11	0.26	4.61
U5	0.08	0.71	2.71
U1'	0.09	0.56	2.75
U2'	0.05	0.32	1.17
U3'	0.05	0.24	1.14
U5'	0.05	-0.22	1.05
U5"	0.04	0.18	0.99

* Reference: tetramethylammonium chloride.

purpose the CD spectra of m_2^6ApU were measured at 20 temperatures in the range 0 - 69 ^oC, Fig. 5. The spectra show four isosbestic points; this fact by itself is an excellent indication for the validity of our contention that only one intramolecularly base-stacked species exists in solution. The spectra, converted into digital form, were further processed by means of a computer. The thermodynamic analysis by means of program STAKKER (vide supra) was carried out, using integrated band intensities, for each of the peak-trough regions as well as for the complete spectrum from 210-320 nm. Within narrow error limits each analysis yielded the same parameters. The mean CD-values (Tm, ΔH_x^o, ΔS_x^o) are shown in Table 2.

DISCUSSION

Owing to space limitations, this discussion necessarily covers only the barest essentials. The results presented bring home the following major points:
(a) The calculated association parameters (Table 1) show satisfactory agreement for the 5 protons* monitored. Those obtained from A1' and A2' are least affected by possible residual errors from stacking shifts and are considered most accurate. Those from A8 and A2 may reflect a tendency for the bases to associate slightly better in the open states than in the stacked one, but further work is necessary to clear up this point. The association shifts $\delta_A - \delta_F$ for the m_2^6Ap - residue roughly parallel shift differences found (35) for two concentrations of Ado-3'-P: A2>A8>A1'>A2'>A3'. Of interest is the finding that association induces a small, but fairly constant, long range shielding of the protons of the -pU residue. This is interpreted to mean that the m_2^6A- bases stack with the two -pU units pointing in opposite directions. A more detailed model for the associated species will be discussed elsewhere. A note of warning should be issued here. If, as seems likely, the intermolecular association tendency of a given base is positively correlated to its capacity to stabilize intramolecular stacking, then published interpretations of shift data of many dimers should be eyed with suspicion (especially of adenine or guanosine containing dimers). Even at concentrations as low as 5-10 mM ring current shift effects due to association may dominate over stacking shifts in the lower temperature range. Tables 3 and 4 combined serve to illustrate this point.

(b) Let us now inspect Table 2. Only four out of fourteen protons of m_2^6ApU display stacking shifts ($\delta_x - \delta_u$) that are too small for a meaningful thermodynamic analysis. Keeping in mind the magnitude of the stacking shift and the number of available data points (U5 and U1' yielding maximum accuracy) we draw the important conclusion that the results obtained independently from ten different proton signals are mutually consistent. Moreover, the "best values" from the NMR-analysis (from A3', A5', U5,6, 1',2', 3', 5') show excellent agreement with those obtained from the CD-spectra.

* In the following the protons of the m_2^6Ap- and -pU nucleotidyl units will be designated by A8, A2, A1', U6; the coupling constants by A1'2', A2'3' and so forth.

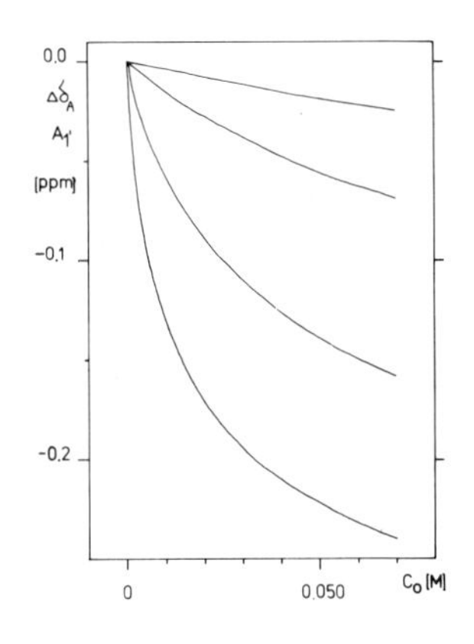

Fig. 4. Net effect of association on the chemical shift of A1' at four temperatures (278, 298, 333, 368 K), calculated from data in Tables 1 and 3.

We conclude that CD and NMR methods, if applied with care to ensure maximum accuracy, do indeed measure the same equilibrium, most likely that between a right-handed single-helical stacked conformation and an assembly of unstacked microstates. Our combined NMR and CD data definitely rule out the presence of lefthanded (base-base overlapping) stacks in pu-py dimers in relative amounts large enough to cause the observed upfield shifts of U2' and A3' (24), or the downfield shifts of A5' and A5" (21). If a left-handed ($\omega^{\underline{t}},\omega+$) stacked species occurs in pu-py dimers, it must exist as partner in the blend of microstates.

This conclusion is at variance with current interpretations of observed dimerization shifts in pu-py dimers (19-24) and prompts us to propose new (but tentative) interpretations for some of the stacking shifts $\delta_x - \delta_u$, remembering that δ_u represents a time-average shift of microstates. The surprisingly large deshielding of A8 is ascribed to its close proximity to O5' and O1' in the pure ribose-N-base-anti geometry in the stacked conformation. This effect is analogous to that experienced by U5' from O2' of the neighboring residue (19-21). The stacking shifts of A2, A1', A2' and A5" are small and presumably reflect the interplay of several opposing factors. The large downfield shift of A5' that has puzzled many workers is presumably due to several factors, one of which is the close proximity of the A-base at a low χ-angle, which puts A5' (in Ψ_+) nearly in the deshielding plane of the base (37). From the fact that the relevant coupling constants A4'5' and A4'5" behave in a perfectly normal fashion (A4'5' < A4'5") over a large stack-unstack range (31) we conclude that the original Remin and Shugar (36) assignment remains correct for the Ap-residue. The upfield stacking shift of A3', a quite general phenomenon in dimers which has received relatively little attention thus far, has been attributed to a left-handed stacked form (24), or to a decrease in χ-angle of the Ap-base in ApA (19). However, stacking entails a shift of the torsional equilibrium about the C3'-O3' bond (38): ϕ'_a ($\sim 215^o$) $\rightleftarrows$ $\phi'^{\underline{}}$ ($\sim 260^o$) to yield 100% ϕ'_a which radically changes the relative orientations of the pro S and pro R oxygens of the phosphate group (as well as that of the P-O5' ester bond) with respect to A3' and this change may well induce a large upfield shift. Perhaps a locking in of rotational freedom about ω' on stacking also plays a similar role (23). The upfield shifts experienced by U6, U5, U1', U2' and U3' are mainly ascribed to the shielding effect exerted by the overlying purine base. The downfield shift of U5' has already been commented upon. The upfield shift of U5" may reflect the locking in of the backbone (ω',ω) torsional freedom.

Table 4. Populations of self-associated molecules (100 P_A) and of intramolecularly stacked species (100 P_x) calculated from best NMR thermodynamical parameters of m_2^6ApU, Comparison with CD.

t(°C)	100 P_A		100 P_x	
	5.7 mM	63 mM	NMR	CD
0	22.2	61.4	79.7	77.8
10	15.2	52.5	72.2	70.2
20	10.1	43.7	63.8	61.8
30	6.7	35.3	55.0	53.3
50	3.0	21.4	38.8	37.8
90	0.7	6.8	17.4	17.4

For the first time we have an independent check on our original conclusion (11, 14-16) that the coupling constants J1'2' and J3'4' provide a rough but valid estimate of the stack-unstack equilibrium. From the best set of thermodynamic parameters the population of stacked species (P_x) pertaining to each spectrum was calculated. Plots of the ribose proton couplings vs P_x are shown in Fig. 6. As expected, a near-perfect linear correlation is found for both A1'2' and A3'4'. A reliable extrapolation of the observed couplings to values pertinent of the fully stacked state is now possible for the Ap-residue. We find: A1'2'= 0.8 ± 0.1 Hz, A2'3' = 4.4 ± 0.1 Hz, A3'4' = 9.0 ± 0.1 Hz, and for the sum A1'2' + A3'4' 9.8 ± 0.15 Hz. Noteworthy is the linear decrease of A2'3' (from about 5.3 Hz to 4.4 Hz), accompanied by a small increase in the sum A1'2' + A3'4', as the equilibrium shifts toward the pure stacked form. The possibility of this trend was already predicted in the original AS treatment for several combinations of the pseudorotation parameters (P_N, τ_N, P_S, τ_S). Of interest is also the finding that A1'2' for the stacked state (100% 3'N-ribose) differs from zero (footnote 4, ref. 16).

The plot of U1'2' vs P_x shows a small, but perhaps significant, curvature. If real, the deviation from linearity might indicate the presence of a varying, but minor amount (<6%) of 3'N-5'S stacked state (16). If this were the case the corresponding U3'4' plot should show an opposite curvature. Unfortunately, U3'4' can be measured only over part of the temperature range and this point remains unsettled. For the time being, coupling constants of 5'-residues should not be used to calculate P_x.

ACKNOWLEDGEMENT

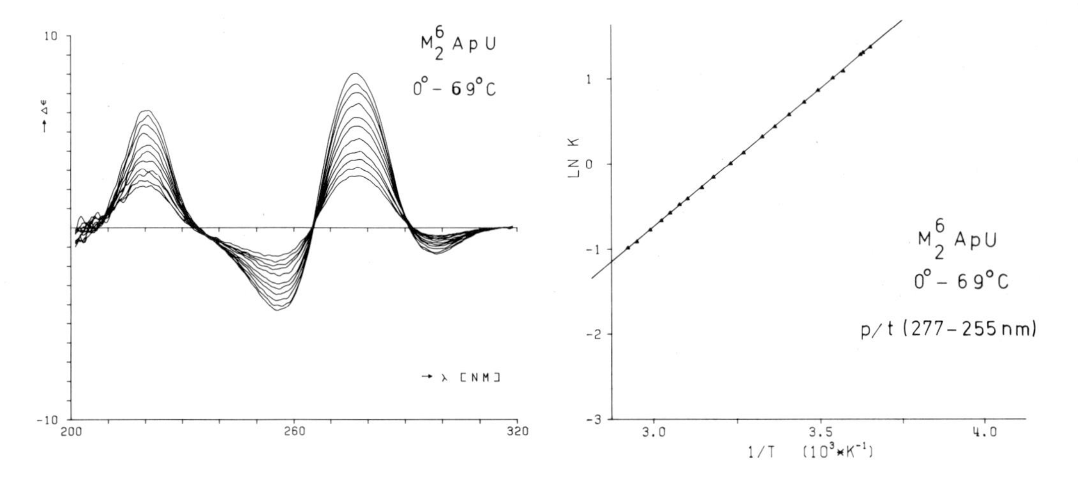

Fig. 5. Left: the CD spectrum of m_2^6 ApU in H_2O, pH 7.3, 3.8 x 10^{-5} M dinucleotide at varying temperatures.
Right: Van 't Hoff plot of the integrated intensity of the CD bands from 277 to 255 nm.

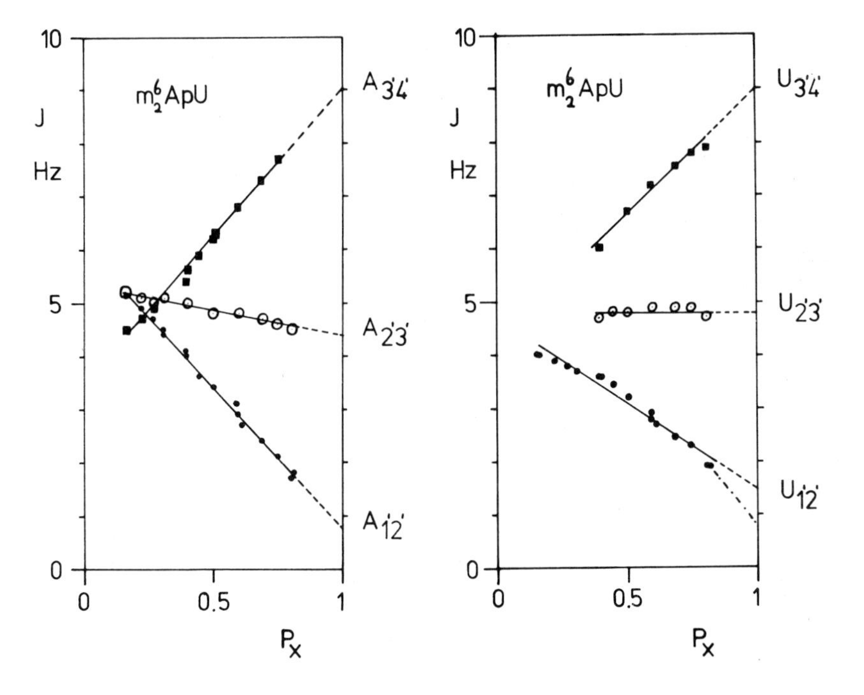

Fig. 6. Ribose coupling constants J1'2', J2'3' and J3'4' vs P_x, see text.
Left: m_2^6 Ap- residue, right: -pU residue.

This research was supported by the Netherlands Foundation for Chemical Research (S.O.N.) with financial aid from the Netherlands Organization for the Advancement of Pure Research (Z.W.O.). We wish to thank Dr. R. Kaptein, Mr. K. Dijkstra and Mr. C. Erkelens for technical aid and Dr. J.H. van Boom and his group for generous advice in the field of synthesis and purification. We are indebted to Mr. P.P. Lankhorst and to the other students of our group as well for their participation in this work.

REFERENCES

1. P.N. Borer, L.S. Kan, and P.O.P. Tso, Biochemistry, 14 4847 (1975).
2. D.M. Gray, I. Tinoco and M.J. Chamberlin, Biopolymers, 11 1235 (1972).
3. C. Altona, J.H. van Boom and C.A.G. Haasnoot, Eur. J. Biochemistry, 71, 557 (1976), and unpublished.
4. H.P.M. de Leeuw and C. Altona, unpublished.
5. F.E. Hruska and S.S. Danyluk, J. Amer. Chem. Soc 90 3266 (1968).
6. S.I. Chan and J.H. Nelson, J. Amer. Chem. Soc. 91 168 (1969).
7. B.W. Bangerter and S.I. Chan, J. Amer. Chem. Soc. 91 3910 (1969).
8. P.O.P. Ts'o, N.S. Kondo, M.P. Schweizer, and D.P. Hollis, Biochemistry, 8 997 (1969).
9. N.S. Kondo, H.M. Holmes, L.M. Stempel, and P.O.P. Ts'o, Biochemistry, 9 3749 (1970).
10. C. Altona and M. Sundaralingam, J. Amer. Chem. Soc., 94 8205 (1972).
11. C. Altona and M. Sundaralingam, J. Amer. Chem. Soc., 95 2333 (1973).
12. F.E. Evans and R.H. Sarma, J. Biol. Chem., 249 4754 (1974).
13. N.S. Kondo and S.S. Danyluk, J. Amer. Chem. Soc., 94 5121 (1972).
14. C. Altona, J.H. van Boom, J.R. de Jager, H.J. Koeners, and G. van Binst, Nature, 247 558 (1974).
15. C. Altona, H.J. Koeners, J.R. de Jager, J.H. van Boom, and G. van Binst, Rec. Trav. Chim., 93 169 (1974).
16. C. Altona in "Structure and Conformation of Nucleic Acids and Protein-Nucleic Acid Interactions", M. Sundaralingam and S.T. Rao, Univ. Park Press, Baltimore, p. 613 (1975).
17. F.E. Evans, C.H. Lee, and R.H. Sarma, Biochem. Biophys. Res. Commun., 63, 106 (1975).
18. C.H. Lee, F.E. Evans, and R.H. Sarma, FEBS Lett., 51 73 (1975).
19. N.S. Kondo and S.S. Danyluk, Biochemistry, 15 756 (1976).
20. C.H. Lee, F.S. Ezra, N.S. Kondo, R.H. Sarma, and S.S. Danyluk, Biochemistry, 15 3627 (1976).
21. F.S. Ezra, C.H. Lee, N.S. Kondo, S.S. Danyluk, and R.H. Sarma, biochemistry, 16 1977 (1977).
22. H. Singh, M.H. Herbut, C.H. Lee, and R.H. Sarma, Biopolymers, 15 2167 (1976).
23. D.M. Cheng and R.H. Sarma, Biopolymers, 16 1687 (1977).
24. C.H. Lee and I. Tinoco, Biochemistry, 16 5403 (1977).
25. J.T. Powell, E.G. Richards, and W.B. Gratzer, Biopolymers, 11 235 (1972).
26. J. Brahms, J.C. Maurizot, and A.M. Michelson, J. Mol. Biol. 25 481 (1967).

27. D. Perahia, B. Pullman, D. Vasilesca, R. Cornillon, and H. Broch, Biochim. Biophys. Acta, 478 244 (1977).
28. C. Giessner-Prettre, B. Pullman, P.N. Borer, L.S. Kan and P.O.P. Ts'o, Biopolymers, 15 2277 (1976).
29. P.O.P. Ts'o in "Basic Principles in Nucleic Acid Chemistry", vol. 1 chap. 6, vol. 2 chap. 5, Acad. Press, New York (1974).
30. A. de Marco and K. Wühthrich, J. Magn. Res. 24 201 (1976).
31. A.J. Hartel and C. Altona, unpublished.
32. K.G. Wagner and R. Lawaczeck, J. Magn. Res., 8 164 (1972).
33. Y-F Lam and G. Kotowycz, Canad. J. Chem. 55 3620 (1977).
34. C.A.G. Haasnoot and C. Altona, unpublished.
35. T.D. Son and C. Chachaty, Biochim. Biophys. Acta 335 1 (1973).
36. M. Remin and D. Shugar, Biochem. Biophys. Commun., 48, 636 (1972).
37. C. Giessner-Prettre and B. Pullman, J. Theor. Biol. 65 189 (1977).
38. J.L. Alderfer and P.O.P. Ts'o, Biochemistry, 16 2410 (1977).

THE CONTRIBUTIONS OF THE PURINE NITROGENS TO STACKING ASSOCIATION

Karl G. Wagner, Hans-Adolf Arfmann, Rüdiger Lawaczeck,
Karin Opatz, Ida Schomburg and Victor Wray
Gesellschaft für Biotechnologische Forschung
D-3300 Braunschweig-Stöckheim, West Germany

SUMMARY. The supposition that stacking affinity is primarily governed by the polarization energy, arising from the planar overlap of two polar aromatic molecules, was proven with a series of compounds that have the same atomic skeleton as purine but up to three of its nitrogens replaced by CH. By proton NMR measurements association constants have been determined in D_2O. The bond moments, generated by the presence of nitrogens in the compounds mentioned, were evaluated, with respect to their polarizing power, by complex formation with tryptamine (indole derivative) which has a weak polar ring system but a large polarizability. On the other hand the polarizability of these compounds was evaluated by complex formation with caffein, which is a strong polar compound. The results obtained clearly show that stacking affinity is strictly dependent upon the magnitude of the polarization energy estimated from the available polarizing bond moments and the polarizability. Data on the effect of exocyclic bond moments and on the temperature dependence of stacking are in accord with this idea.

1. INTRODUCTION

The phenomenon of stacking, i. e. the close and planar association of polar ring molecules possessing aromatic properties, has been well established as an important affinity contribution in the interaction within or between various biomolecules. The discussion of the origin of this interaction, however, is not completed and there is no general accord which of the known basic physical contributions is of most importance. Some people emphasize the hydrophobic origin while others emphasize London dispersion forces or charge transfer interactions.

The importance of the polarization energy gained by planar overlapping of polar aromatic rings as the main contribution to stacking has been deduced from different experimental evidence. In a comparative synopsis on the mode of coordination of bases in single crystals of nucleic constituents Bugg et al. (1971) stressed the dipole-induced dipole interaction as the driving force for the generation of the peculiar pattern of nucleobase overlap. This supposition is supported by the fact that nonpolar aromatic hydrocarbons

B. Pullman (ed.), Nuclear Magnetic Resonance Spectroscopy in Molecular Biology, 103-110.

crystallize with no overlap between parallel molecules.

In a further synopsis on affinity data from the association in aqueous solution Lawaczeck and Wagner (1974) explained stacking specificity through the magnitude of polarization energy. This energy is thought to be generated by polarizing endo- and exocyclic bond moments and the polarizability of the Π-electron ring system involved. The failure in the detection of stacking in the selfassociation of tryptamine (indole ring), described in that paper, and the well-known stacking behavior of purine prompted an investigation of the contribution of the nitrogens within the 5- and 6-membered rings of purine to the stacking association. This investigation was possible, as purine analogues with less nitrogens (benzimidazole, 7-azaindole and indole) are available, and as an analogue with three nitrogens, 1-deazapurine, was synthesized.

2. MATERIALS AND METHODS

7-Azaindole, benzimidazole and indole were purchased from Merck AG, Darmstadt; purine was obtained from Merck AG and Sigma Chemical Co., St. Louis. Caffeine was purchased from Sigma Chemical Co. and tryptamine HCl from Merck AG and Serva Feinbiochemica, Heidelberg.

To prepare 1-deazapurine a method described for the synthesis of substituted compounds (Temple et al. 1973) was modified. 2.3-Diamino-pyridine (3.4 g) was suspended in 420 ml ethyl orthoformate. After addition of 2.0 ml concentrated HCl the reaction mixture was stirred at room temperature for 72 h. The product was collected by filtration, washed with diethyl ether and dried over phosphorous pentoxide. Chromatography on a CM-Sephadex column and elution with 0.01 M HCl yielded pure 1-deazapurine. The structure was confirmed by TLC, MS, and NMR.

All ^{1}H spectra were recorded in 5 mm-tubes on a Varian XL-100-12 spectrometer operating in the Fourier transform mode at 100.06 MHz and locked to the deuterium resonance of the solvent, deuterium oxide. Pulse conditions were chosen such as to give an 8 K transform for sweep widths of 1 000 Hz which resulted in a resolution of better than 0.25 Hz with the use of a centroid interpolation routine (Ernst and Lincoln, 1974). Temperature control was performed with a Varian variable temperature accessory and was better than $\pm 1^{o}$. Chemical shifts were determined relative to internal sodium 3-(trimethylsilyl) propionate-d_4.

Experiments were performed by holding the concentration of one component constant, whose chemical shift was measured, and varying the concentration of the second component. Apparent association constants were evaluated from the shift data, on the assumption of 1:1 complex formation, by the direct fit of the mass equation to the experimental curve of the shift against the variable component concentration. A computer program, similar to the one described previously (Wagner and Lawaczeck, 1972; Lawaczeck, 1972), evaluated the association constant and shift of the complex by iteratively minimising the root-mean-square

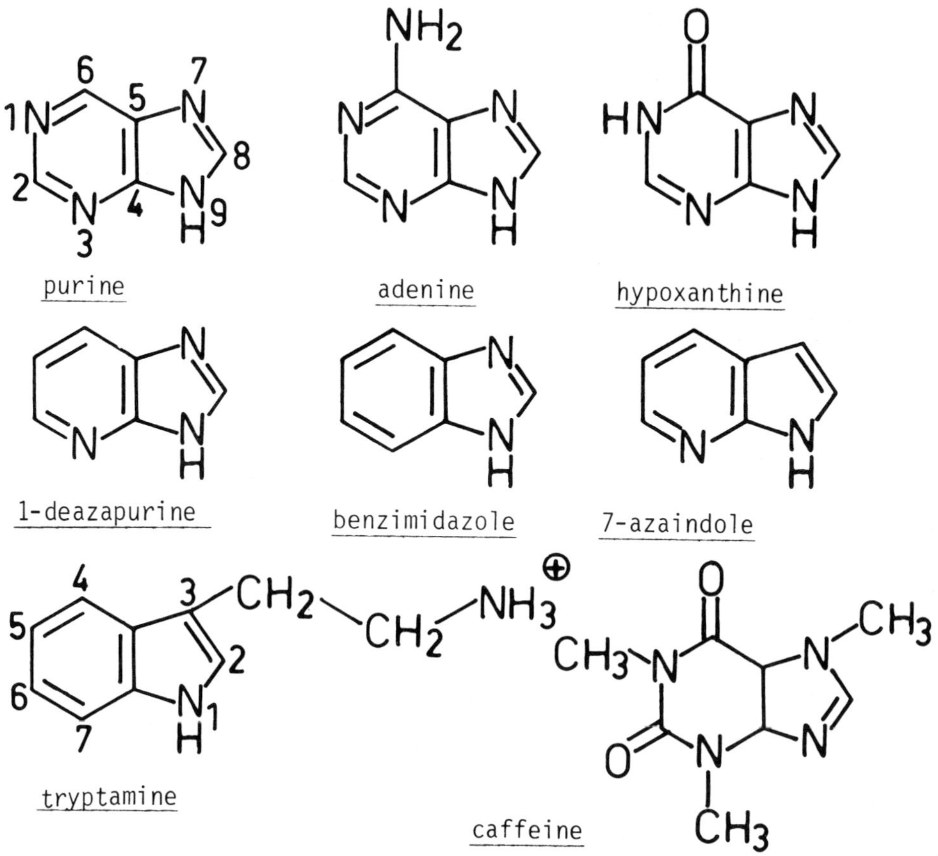

Fig. 1. Formulae

deviation from the observed curve. This procedure was repeated for each proton or group of protons in the molecule and the reported values constitute an average of these.

3. RESULTS AND DISCUSSION

According to our concept polarization energy generated by a planar overlap of two nucleobases depends both upon the polarizability and the polarizing power of the two molecules. Polarizability is assumed to have its origin in the density of mobile Π-electrons which is related to aromaticity and is also evident in the ring-current effect. Polarizing power is thought to originate in the bond moments; in nucleobases these

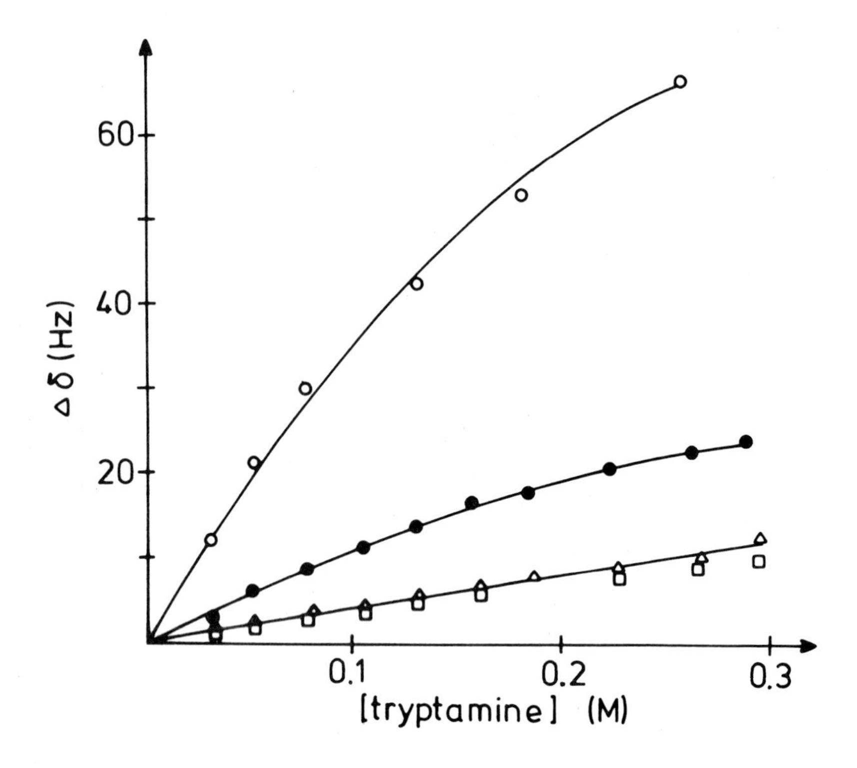

Fig. 2. Influence of purine nitrogens upon the association with tryptamine. Chemical shifts (upfield) are plotted versus tryptamine concentration.
(o) purine; (●) 1-deazapurine; (Δ) benzimidazole; (□) 7-azaindole. Experiments were performed at 7°; the concentrations of purine and its analogues was 0.01 M.

exist in the exocyclic keto and amino substituents, and they are also caused by N atoms of the base ring system.

In order to evaluate the influence of the endocyclic bond moments, caused by the four N atoms of purine, upon stacking affinity, molecules with a reduced number of nitrogens and with the same ring system as purine were chosen: 1-deazapurine, benzimidazole and 7-azaindole (Fig. 1). The contribution of these bond moments to stacking affinity is indicated best when the association is determined in complexes where other bond moments are not available. This could be done by determination of the selfassociation of these compounds or by choosing a partner, which has no bond moments but possesses a large polarizability. For this reason, and also because of its rather good solubility, tryptamine was chosen. Its indole moiety possesses a large polarizability but only a rather weak polarizing power as it has only one nitrogen.

TABLE I

Dependence of stacking affinity upon the replacement of purine nitrogens by CH groups. Complexes with caffeine determined at 7°.

Base	purine	1-deazapurine	benzimidazole	7-azaindole	indole
$K(M^{-1})$	11.7	20.6	24.9	26.7	27.3

Fig. 2 shows the upfield shifts of selected protons of the four compounds mentioned with increasing tryptamine concentration. The compounds with four (purine) and three (1-deazapurine) nitrogens reveal a behavior which is consistent with stacking, as it is observed for a series of investigations with nucleic acid constituents and also with tryptamine and different nucleotides (Wagner and Lawaczeck, 1972; Wray and Wagner, 1977). However, compounds with only two nitrogens i. e. benzimidazole and 7-azaindole are obviously incapable of forming stacking associates with tryptamine; the steady increase in the chemical shifts rather resembles those of a solvent effect.

These results clearly indicate the importance of polarizing bond moments for the formation of stacking complexes. The affinity constants derived for purine (4.3 M^{-1}) and for 1-deazapurine (1.8 M^{-1}) reveal a clear dependence upon the number of the nitrogens.

From previous work it was shown that the indole moiety supports stacking when the partner molecule provides polarizing bond moments. Thus stacking of tryptamine with nucleotides and tryptophane with nucleosides and nucleotides has been reported (cf. Lawaczeck and Wagner, 1974 and citations herein). In a second investigation the compounds of the present work were evaluated with respect to their capacity to form stacking complexes with a strong polarizing partner molecule. We chose caffeine because of its good solubility and to avoid the contribution of the ribose or ribose phosphate moiety of nucleosides or nucleotides. The purine derivative caffeine is a strong polarizing molecule because of its two exocyclic keto groups.

Stacking affinity of the different compounds with caffeine is significantly enhanced relative to tryptamine as is indicated in Table 1. Obviously with purine-tryptamine the available endocyclic bond moments are limited in their polarizing effect. It will be shown below that exocyclic keto substituents are more effective. The stacking affinity with caffeine of the purine analogues, possessing one to four endocyclic nitrogens, now inversely parallels the number of the nitrogens; purine has the lowest and indole the highest affinity. This reflects the supposition that with a strongly polarizing partner stacking affinity of a compound increases with the magnitude of its polarizability.

TABLE II
Influence of exocyclic bond moments upon stacking affinity. The K values were determined with tryptamine at 7°.

Base	purine	adenine	hypoxanthine	caffeine
$K(M^{-1})$	4.3	3.9	6.7	20.6

Polarizability can be estimated from calculated ring current effects. Giessner-Prettre and Pullman (1970) reported the greatest shielding value in a plane 3.4 Å distant from the molecular surface of 1.28 ppm for adenine, which should be about the same for purine, and a value of 1.45 ppm for indole (Giessner-Prettre and Pullman, 1971). Preliminary data on the shielding of caffeine protons in complexes with either tryptamine or purine indicate a significant larger ring current for the indole moiety of tryptamine than for purine. Obviously the replacement of nitrogens in purine by CH enhances the polarizability of its π-electron system.

In a further investigation we estimated the influence of exocyclic bond moments on purine (Fig. 1). Table II shows affinity constants of adenine, hypoxanthine and caffeine with tryptamine. Whereas an amino substitution on C-6 (adenine) does not increase the affinity, keto substitutions have a strong effect. The large affinity observed with caffeine may in addition be caused by hydrophobic contributions of its three methyl substitutents. Recent investigations with pyrimidine derivatives also elucidated the contribution of methyl groups (Plesiewicz et al., 1976 a and b) to stacking affinity. The results with adenine and hypoxanthine in Table II are consistent with calculated electron charge densities (Giessner-Prettre and Pullman, 1968; Renugopalakrishnan et al., 1971), which show a significant larger bond moment for the keto substitution. The amino substitution, however, may increase the density of the ring π-electrons by a mesomeric effect.

TABLE III
Influence of the ribose moiety upon the scattering of association constants (M^{-1}) determined for different protons. Complexes with tryptamine at 7°.

	H-2	H-8	H-1'	Average
adenine	3.9	3.9	-	3.9
adenosine	5.6	3.7	5.0	4.8
hypoxanthine	6.7	6.7	-	6.7
inosine	8.6	7.1	9.4	8.4

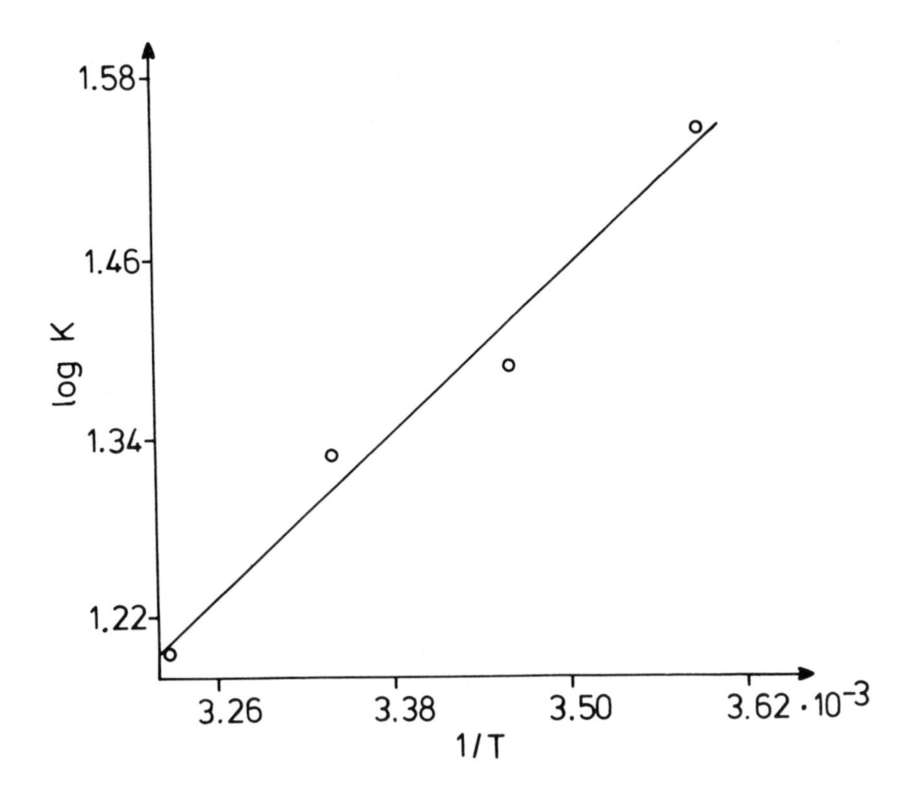

Fig. 3. Van't Hoff plot of the association of caffeine and 7-azaindole.

Table III contains data on the influence of the ribose moiety upon the stacking association with tryptamine, by comparing bases and the respective nucleosides. In both cases, i. e. with adenosine and inosine, there is an increase in the affinity which may be explained by an additional bond moment generated through the attachment of the ribose. However, there may be other contributions of the ribose, related to the interaction with the solvent.

Table III further illustrates an observation which was made in previous studies with nucleotides and tryptamine or serotonine (Wagner and Lawaczeck, 1972; Lawaczeck and Wagner, 1974). Apparent affinity constants obtained from the chemical shift of non-equivalent protons revealed considerable variations which were beyond the experimental error. We proposed an explanation for this (Lawaczeck and Wagner, 1974) by assuming complexes of minor probability with different stereochemistry and/or stoichiometry. The results in Table III indicate variation in the apparent affinity constants only with the nucleosides. Apparently complex formation of tryptamine with the bases is more homogeneous than with the nucleosides.

With suitable components, i. e. with caffeine and 7-azaindole, possessing a large affinity constant and also a high enough solubility, the temperature dependence of the affinity was elaborated. Fig. 3 shows a van't Hoff plot with an enthalpy of interaction of -4.1 kcal/mol or -17.2 kJ/mol. With a Δ G value of -1.8 kcal/mol (-7.5 kJ/mol) at 26.5^{o} a Δ S value of -9.6 cal/^{o}mol (40 J/^{o}mol) is extracted. These data are in accord with other thermodynamic parameters reported from nucleobases, e. g. for the selfassociation of N-6, N-9 dimethyladenine (Pörschke and Eggers, 1972). However, they are in contrast to the concept of hydrophobic interaction developed for unpolar compounds. Further they support the idea that hydrophobic affinity is not primarily involved (cf. also Scruggs et al., 1972; Marenchic and Sturtevant, 1973; Alvarez and Biltonen, 1973), and favour the concept of polarization energy as the primary driving force in this kind of interaction.

ACKNOWLEDGEMENTS

This work was supported by the Deutsche Forschungsgemeinschaft (Wa 91) and the Fonds der Chemischen Industrie, Frankfurt. We are very grateful to Mrs R. Jähne for typing the manuscript.

REFERENCES

Alvarez, J. and Biltonen, R.: 1973, Biopolymers 12, 1815 - 1828.
Bugg, C.E., Thomas, J.M., Sundaralingam, M. and Rao, S.T.: 1971, Biopolymers 10, 175 - 219.
Ernst, L. and Lincoln, D.N.: 1974, J. Magn. Res. 16, 190 - 191; (cf. ibid 17, 272)
Giessner-Prettre, C. and Pullman, A.: 1968, Theor. Chim. Acta 9, 279 - 287
Giessner-Prettre, C. and Pullman, B.: 1970, J. Theor. Biol. 27, 87 - 95.
Giessner-Prettre, C. and Pullman, B.: 1971, J. Theor. Biol. 31, 287 - 294.
Lawaczeck, R.: 1972, Thesis, Technische Universität, Braunschweig.
Lawaczeck, R. and Wagner, K.G.: 1974, Biopolymers 13, 2003 - 2014.
Lawaczeck, R. and Wagner, K.G.: 1974, Z. Naturforsch. 29c, 488 - 492.
Marenchic, M.G. and Sturtevant, J.M.: 1973, J. Phys. Chem. 77, 544 - 548.
Plesiewicz, E., Stepien, E., Bolewska, K. and Wierzchowski, K.L.: 1976, Biophys. Chem. 4, 131 - 141.
Plesiewicz, E., Stepien, E., Bolewska, K. and Wierzchowski, K.L.: 1976, Nucleic Acids Res. 3, 1295 - 1306.
Pörschke, D. and Eggers, F.: 1972, Eur. J. Biochem. 26, 490 - 498.
Renugopalakrishnan, V., Lakshminarayanan, A.V. and Sasisekharan, V.: 1971, Biopolymers 10, 1159 - 1167.
Scruggs, R.L., Achter, E.K. and Ross, P.D.: 1972, Biopolymers 11, 1961 - 1972.
Temple Jr., C., Smith Jr., B.H. and Montgomery, J.A.: 1973, J. Org. Chem. 38, 613 - 615.
Wagner, K.G. and Lawaczeck, R.: 1972, J. Magn. Res. 8, 164 - 174.
Wray, V. and Wagner, K.G.: 1977, Z. Naturforsch. 32c, 315 - 320.

METHYLATION EFFECTS ON NUCLEIC ACID CONFORMATIONS*

Steven S. Danyluk, Clinton F. Ainsworth, and Malcolm MacCoss,
Division of Biological and Medical Research, Argonne National
Laboratory, Argonne, Illinois 60439

INTRODUCTION

Chemical modification of nucleic acids by alkylating agents continues to be extensively explored from the perspective of chemical mutagenesis and carcinogenesis. Largely through the pioneering *in vitro* and *in vivo* studies of Lawley (1961), Lawley and Brookes (1963), Singer (1972), and Reese and coworkers (1965), coupled with theoretical predictions of Pullman (1974), a detailed profile has emerged of the principal alkylation sites for purine and pyrimidine nucleotides. Generally, alkylation of nucleic acids produces a multiplicity of reaction products dependent upon factors such as nature of base ring, electrophilicity of the alkylating agent (Lawley, 1976), and presence of activating enzymes (Magee and Barnes, 1967). Thus, for example, alklation of adenine nucleotides can occur at six sites, Figure 1. Strong electrophiles such as N-methyl-N-nitrosourea (MNUA) and N-methyl-N^1-nitro-N-nitrosoguanidine (MNNG) couple preferentially with available oxygen donors, e.g., phosphate group, 2' oxygen in ribonucleotides, while weakly electrophilic agents dimethyl sulfate and methyl methane sulfonate tend to show a higher proportion of ring nitrogen methylation at N1, N3, and N7 of purines.

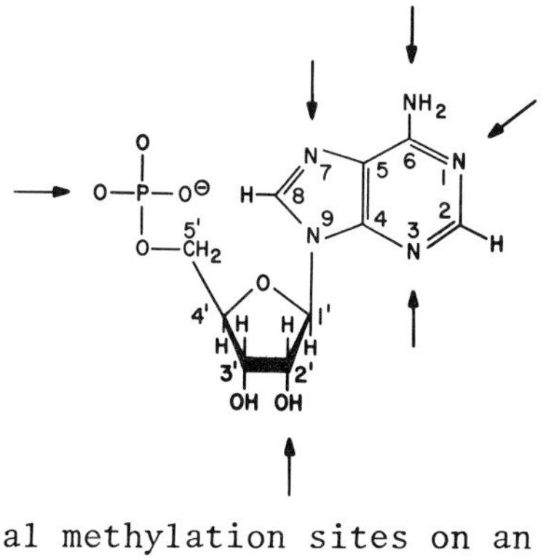

Figure 1. Potential methylation sites on an adenine nucleotide.

*Work supported by the U.S. Department of Energy.

B. Pullman (ed.), Nuclear Magnetic Resonance Spectroscopy in Molecular Biology, 111-124.

A facet of this problem only marginally investigated to the present is concerned with the impact of carcinogen modification upon structural/conformational features of nucleic acids. Yet, methylation effects on structure may be quite substantial, notably in instances where charge redistribution and/or electrostatic charges are altered. In the present communication, we explore this aspect in some detail using dinucleoside monophosphates as simple models. Of specific interest was the effect of N1 methylation in purines on base stacking and overall conformation. Since purine-rich regions of RNA and DNA are presumptive nucleating sites for ordered stacked structures, introduction of positive charges into the base ring by methylating agents can have a profound influence on total biopolymer structure.

MATERIALS AND METHODS

Methylation at the adenine N1 position of a series of adenylate dinucleoside monophosphates, ApA, ApU, and UpA, was carried out following procedures described by Reese et al. (1965). Figure 2 shows a schematic illustration of the overall process. In a typical reaction, dimethyl sulfate was added slowly to an aqueous solution of the dimer under conditions of strict pH control (pH = 7.0). Following ether extraction of the mixture, the products were separated by column chromatography and characterized further by electrophoresis, UV, and NMR spectra. Yields of the N1 methylated dimers ranged from 20-35%.

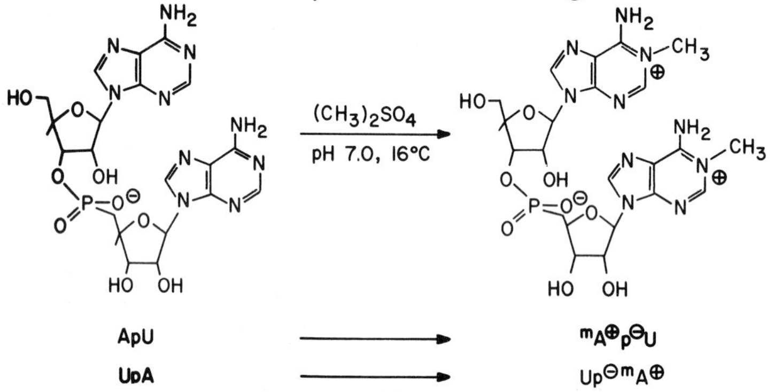

$(CH_3)_2SO_4$

pH 7.0, 16°C

ApU ⟶ $^{m}A^{\oplus}p^{\ominus}U$

UpA ⟶ $Up^{\ominus}{}^{m}A^{\oplus}$

Figure 2. Schematic illustration of adenylyl-3'-5'-adenosine, ApA, methylation reaction. Similar reactions can be written for methylation of adenylate fragments in ApU and UpA.

Samples for NMR measurements were lyophilized from 99.8% D_2O, then dissolved in 100% D_2O and the solutions adjusted to the desired concentration, 0.004-0.007 M and pD. A trace amount of TSP added to the sample served as an internal reference. Proton spectra were recorded in the FT mode on a Varian 220 MHz spectrometer equipped with a Nicolet FT accessory and 20 K data system. All of the spectra were measured at 20 ± 2°C and no sample degradation occurred in the time span of the measurements. Spectral simulations were made on a Nicolet 1080 computer using appropriate simulation programs.

RESULTS AND DISCUSSION

Spectral Assignments, Analysis, and Comparisons

Figure 3A illustrates a typical spectrum for a methylated dimer, ^{m}ApU,† recorded at pD = 7.0 and 20°C. Except for differences in chemical shifts, the overall pattern of signal multiplets conforms to that observed for the corresponding parent dimer ApU, (Ezra et al., 1977) and related N6 methylated ApU derivatives (Altona, 1975). Similar correlations between parent and methylated derivatives are seen for the other compounds. Assignment of signals to specific protons followed procedures developed in earlier dimer studies (Kondo and Danyluk, 1976; Lee et al., 1976; Ezra et al., 1977). Because of the near equivalence in methyl shifts, no assignment is possible for these signals in ^{m}Ap^{m}A. Final best-fit values of chemical shifts and coupling constants are summarized in Tables I and II, respectively.

Table I

NMR Parameters for Methylated Dinucleoside Monophosphates*

Chemical Shifts† ppm	^{m}Ap^{m}A		^{m}ApU		Up^{m}A	
	^{m}Ap-	-p^{m}A	^{m}Ap-	-pU	Up-	-^{m}pA
2	8.540	8.540	8.543	-	-	8.551
8	8.540	8.609	8.561	-	-	8.623
^{m}Ap	3.903	-	3.966	-	-	-
p^{m}A	-	3.936	-	-	-	3.927
5	-	-	-	5.772	5.818	-
6	-	-	-	7.861	7.859	-
1'	6.141	6.177	6.168	5.868	5.795	6.200
2'	4.864	4.818	4.882	4.309	4.386	4.745
3'	4.773	4.573	4.759	4.327	4.514	4.577
4'	4.391	4.391	4.468	4.277	4.227	4.395
5'	3.836	4.282	3.895	4.291	3.773	4.191
5''	3.814	4.220	3.968	4.150	3.818	4.123

*pD = 7.0; T = 20°C; Concentration = 0.005 M.

†Shifts are accurate to ± 0.005 ppm.

---

†Abbreviations used: ApA, adenylyl-3'-5'-adenosine; ApU, adenylyl-3'-5' uridine; UpA, uridylyl-3'-5'-adenosine; ^{m}Ap^{m}A, N-1 methyl adenylyl-3'-5'-N-1 methyl adenosine, etc.; m = CH_3 group.

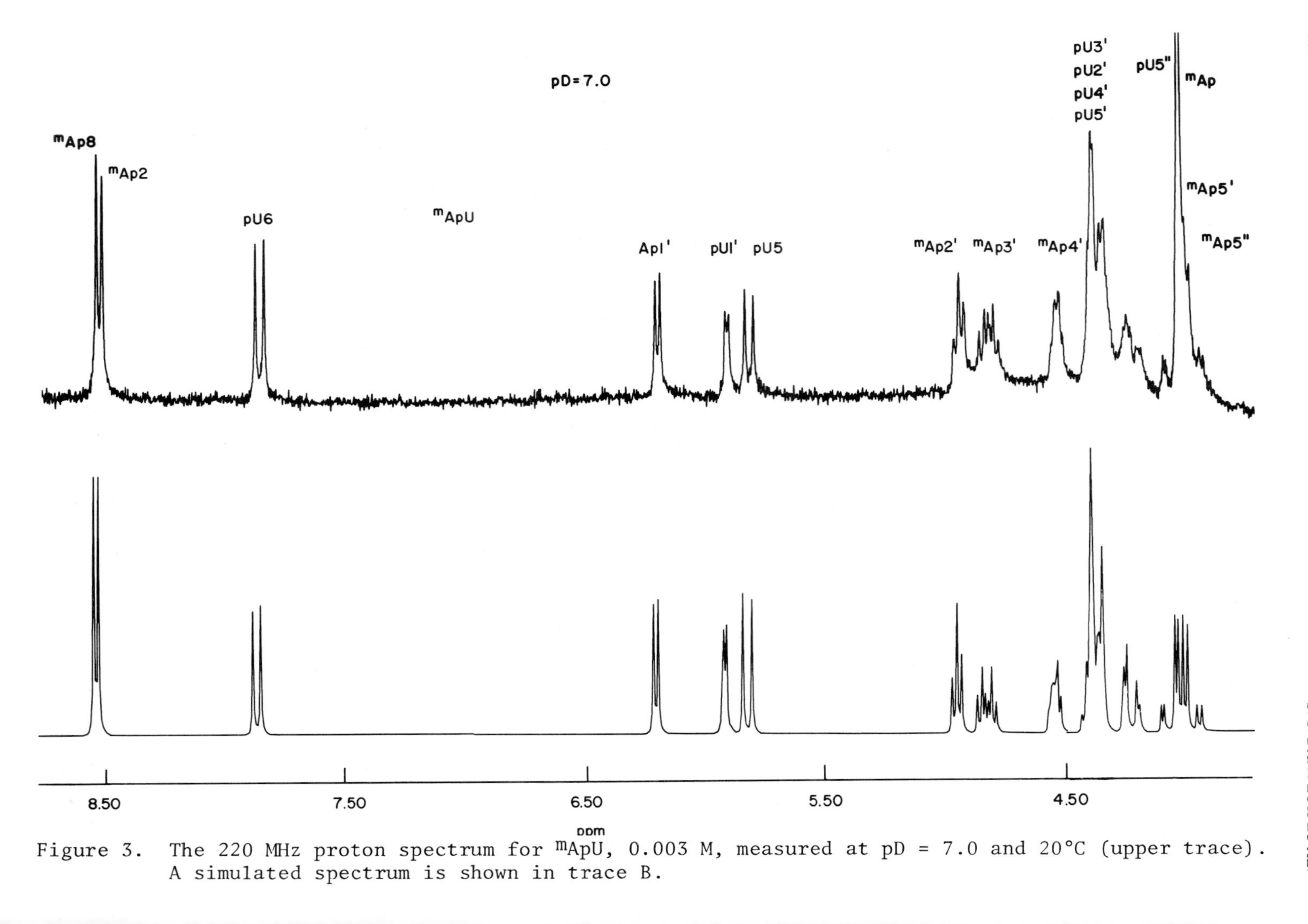

Figure 3. The 220 MHz proton spectrum for ^{m}ApU, 0.003 M, measured at pD = 7.0 and 20°C (upper trace). A simulated spectrum is shown in trace B.

Table II

NMR Parameters for Methylated Dinucleoside Monophosphates*

Coupling Constants Hz†	$^{m}Ap^{m}A$		^{m}ApU		$Up^{m}A$	
	^{m}Ap-	-$p^{m}A$	^{m}Ap-	-pU	Up-	-^{m}pA
1'2'	5.7	5.2	4.4	3.6	4.5	4.5
2'3'	5.2	5.1	5.2	5.1	5.2	5.1
3'4'	3.8	4.0	4.8	5.6	5.1	4.9
4'5'	2.5	2.8	3.8	2.2	3.5	(3.5)
4'5"	2.5	3.3	2.6	2.8	2.7	3.5
5'5"	-13.1	-12.1	-13.1	-11.4	-13.0	-11.0
5'P	-	3.7	-	3.5	-	(4.5)
5"P	-	3.7	-	4.6	-	4.5
3'P	8.2	-	8.2	-	8.0	-
4'P	-	(2.0)	-	(2.0)	-	(2.0)
56	-	-	-	8.0	8.1	-

*pD = 7.0; T = 20°C; Concentration = 0.005 M

†Couplings are accurate to ± 0.2 Hz except for values in parentheses.

Comparison of the shift data with values for parent dimers reveals extensive deshielding of base and ribose protons in all three methylated compounds, Table III, the changes being most pronounced for H2 and H8 protons of ^{m}Ap A, Figure 4. This behavior is not surprising since methylation at N1 introduces a net positive charge into the base ring. Qualitatively the deshielding arises from a combination of base-ring charge deficiency, and base destacking due to repulsive interactions between adjacent positively charged adenines in ^{m}Ap^{m}A. Both methylation and destacking affect H2 to a greater extent than H8;† also, the effects are, as expected, less marked in hetero-dimers.

As with the shifts, methylation causes alterations in magnitudes of coupling constants; however, the effects are more selective with $J_{1'2'}$,$J_{3'4'}$ of the ^{m}Ap^{m}A ribose rings showing the largest changes. The increase in $J_{1'2'}$ and decrease of $J_{3'4'}$ closely parallel the behavior seen on base protonation of corresponding adenylate mononucleotides (Danyluk, Nelson, and Ezra, unpublished results).

Conformational Properties

The utilization of chemical shifts and coupling constants for conformational analyses of short-chain nucleic acid segments has been extensively detailed previously (Altona and Sundaralingam, 1973; Kondo

†Protonation of adenine is formally analogous to methylation. The former occurs preferentially at N1 and produces deshieldings in order $\delta 2 > \delta 8$ (Danyluk and Hruska, 1968).

Table III

Methylation Effects on Chemical Shifts

$\Delta\delta^{\dagger}$ (ppm)

	^{m}ApU		^{m}Ap^{m}A		Up^{m}A	
Proton	^{m}Ap-	-pU	^{m}Ap-	-p^{m}A	Up-	-^{m}pA
2	0.362	-	0.659	0.487	-	0.326
8	0.250	-	0.364	0.364	-	0.212
5	-	0.186	-	-	0.056	-
6	-	0.107	-	-	0.103	-
1'	0.114	0.123	0.306	0.220	0.070	0.100
2'	0.044	0.070	0.190	0.233	0.131	0.035
3'	0.111	0.035	0.105	0.061	0.042	0.037
4'	0	0.028	0.032	0.013	0.038	0.020
5'	0	-0.048	-0.064	-0.098	0.025	-0.073
5''	-0.046	0.014	-0.027	0.035	0.020	-0.032

*Data at 20°C and pD = 7.0.

†$\Delta\delta = \delta_{CH_3} - \delta_0$; (values for δ_0 obtained from Ezra et al., 1976 and Lee et al., 1977).

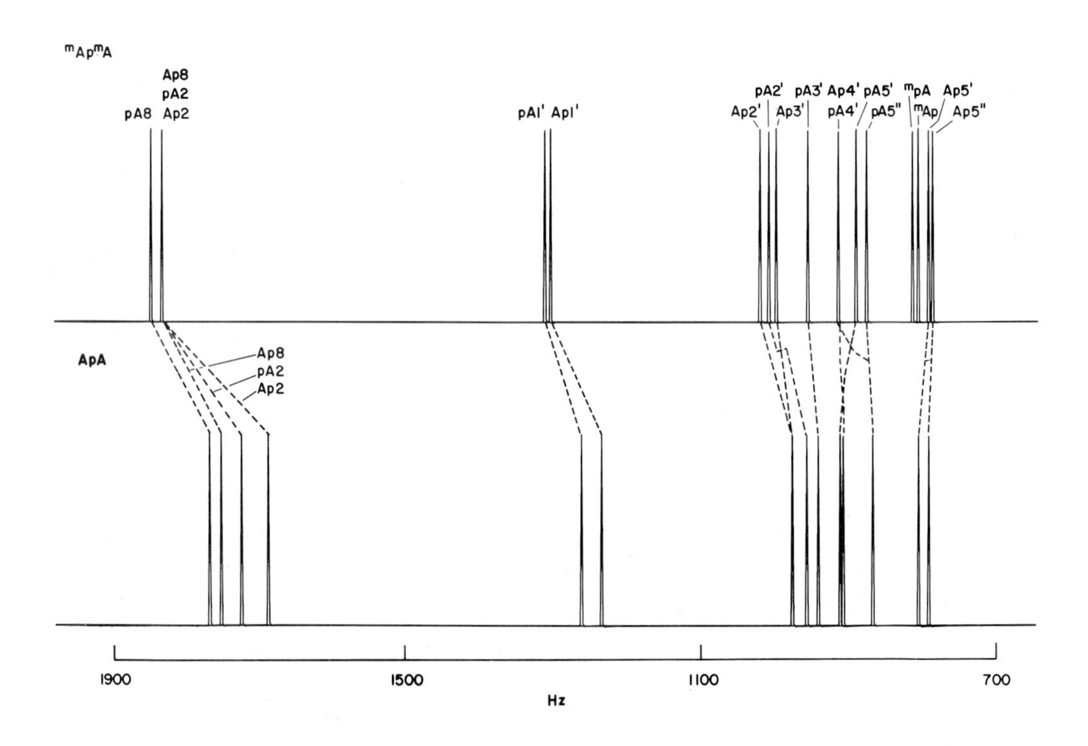

Figure 4. Comparison of methylation effects upon the proton spectrum of ApA at pD = 7.0 and 20°C.

and Danyluk, 1976; Altona, 1975; Ts'o and coworkers, 1969; Lee et al., 1976; Ezra et al., 1977). Briefly the ribofuranose ring can be represented by two puckered modes ^2E and ^3E undergoing rapid interconversion via pseudorotation, i.e., $^2E \rightleftharpoons {}^3E$. Since $J_{2'3'}$ (5.2 Hz) and $J_{1'2'}$ + $J_{3'4'}$ (9.4 ± .2 Hz) are identical to values for dinucleoside monophosphates, no significant change in pseudorotational angle or degree of pucker takes place on methylation. Equilibrium populations were accordingly evaluated as for regular dimers and yielded ^3E populations listed in Table IV. A marked shift is noted from a favored ^3E form in both nucleotidyl residues of ApA (% E values are given in parentheses,(Table IV) to an ^2E conformation in $^mAp^mA$. Values in the latter are in fact close to those for corresponding monomers (33 and 38% ^3E for 3'-AMP and 5'-AMP, respectively (Davies and Danyluk, 1974,1975), and ApA at elevated temperatures (Lee et al., 1976). This trend is qualitatively indicative of decreased base stacking in $^mAp^mA$ versus ApA. In contrast, introduction of a positive charge into the adenine ring of ^mApU and Up^mA has minimal direct or indirect effect on ^3E populations in either dimer.

Table IV

Conformational Properties for Methylated Dinucleoside Monophosphates

	$^mAp^mA$		^mApU		Up^mA	
Property	^mAp-	^{m}-pA	^mAp-	-pU	Up-	$-p^mA$
Ribose Ring						
Stacked[a]	8(38)		27(37)		10(15)	
^3E[b]	40(58)	42(61)	50(57)	59(59)	54(53)	51(51)
Backbone						
$gg(\psi_+)$[c]	85(79)	78(74)	75(89)	87(95)	77(74)	70(79)
$g'g'(\phi_+)$[c]	-	85(90)	-	81(85)	-	77(80)
θ PH	± 35°		± 35°		± 33°	

[a]Computed using $J_{1'2'} + J_{3'4'}$ = 9.5 Hz (dimers) and 9.3 Hz (monomers); % stacked = $J_{3'4'}$(dimer) - $J_{3'4'}$(monomer)/9.5 - $J_{3'4'}$(monomer).
[b]% $^3E = J_{3'4'}/J_{1'2'} + J_{3'4'}$
[c]Rotomer equations: $\psi_+ = (13.7 - \Sigma)/9.7$; $\phi_+ = (25 - \Sigma^1)/20.8$.

Rotamer populations along the phosphodiester backbone, C4'-C5', ψ_+; C5'-O5'; ϕ_+; and C3'-O3', can be evaluated using expressions, Table IV, footnote, derived by the usual staggered rotamer formalism (Wood et al., 1973; Sarma et al., 1974; Evans and Sarma, 1974; Lee and Sarma, 1976). The values summarized in Table V are not significantly different from those for unmodified dimers (parentheses) further exemplifying the extraordinary conformational integrity of these three conformational bonds across a wide spectrum of charged and uncharged dimers.

As with all short chain nucleic acid fragments, the overall three-dimensional solution structure is predominantly determined by two factors, the extent of base-base interactions and the swivel nature of

P-O3' and P-O5' bonds. Although there is no single method for precise evaluation of base stacking, a reasonable semi-quantitative estimate can be made from magnitudes of $J_{3'4'}$ (or $J_{1'2'}$). The underlying rationale assumes that ribofuranose rings adopt an ^{3}E form (100%) in fully-stacked dimers and a monomer $^{3}E/^{2}E$ distribution in unstacked states (Hruska and Danyluk, 1968; Altona, 1975; Kondo and Danyluk, 1976). With this approach, stacked populations are estimated to range from 8% in $^{m}Ap^{m}A$ to 27% in ^{m}ApU, Table V. It is apparent that methylation produces striking changes in the homo-adenylate and somewhat smaller but not insignificant alterations in hetero-dimers. Whereas stack populations increase in order, UpA < ApU < ApA, methylation shifts the trend to $^{m}Ap^{m}A$ < $Up^{m}A$ < ^{m}ApU. This behavior is clearly linked to more pronounced repulsive forces between two adjacent positively charged adenine rings in $^{m}Ap^{m}A$. Charge neutralization at high pH removes this interaction and shifts the equilibrium towards base stacking.

Two additional stacking indices of a somewhat more qualitative nature are dimerization-induced shift changes for pyrimidine H5 signals and relative differences in shift non-equivalence for 5',5" protons of 5'-nucleotidyl fragments (-pX). In each instance, increased stacking correlates with increased dimerization shifts and 5'5" nonequivalence (Ezra et al., 1977). Table V summarizes relevant data for a series of hetero-dimers including ^{m}ApU and $Up^{m}A$. Both methylated hetero-dimers follow the general patterns observed for normal dimers and are consistent with a diminished stacking in methylated derivatives.

Table V
Variation in Stacking Parameters of Dimers

Dinucleotide	$\Delta\delta H5^{a}$	$\Delta J_{1'2'}$ [b]		% Stack	Δ^{c}
		Pyrimidine	Purine		
GpC	0.476	1.9	3.2	45	0.190
ApC	0.449	1.6	2.6	38	0.143
ApU	0.350	1.4	2.4	34	0.133
GpU	0.234	1.4	1.4	27	0.111
CpA	0.280	1.1	1.5	24	0.152
CpG	0.215	1.0	1.8	25	0.129
UpA	0.153	0.6	1.1	15	0.109
UpG	0.122	0.1	1.1	10	0.068
^{m}ApU	0.200	1.3	1.7	27	0.127
$Up^{m}A$	0.097	0.5	1.1	10	0.035

[a] $\delta_{H(5)}$(monomer) - $\delta_{H(5)}$(dimer), ppm.
[b] $J_{1'2'}$(monomer) - $J_{1'2'}$(dimer), Hz.
[c] $[(\delta_{5'} - \delta_{5''})_{dimer} - (\delta_{5'} - \delta_{5''})_{monomer}]$ for the 5'-nucleotidyl unit.

Base proton and H1' dimerization shifts have been used in the past to confirm existence of base-stacking (Ts'o et al., 1969; Chan and

Nelson, 1969; Danyluk and Hruska, 1968) and, in combination with theoretical isoshielding profiles (Giessner-Prettre and Pullman, 1970; Giessner-Prettre and Pullman, 1977), to fix relative orientations of bases in stacks. As noted earlier, a more complex situation prevails in methylated dimers where base protons are influenced by at least two major factors, the presence of a delocalized positive charge (deshielding), and base stacking (shielding increase). Nevertheless, a qualitative measure of base overlap is possible from a consideration of dimerization shifts† $^{m}\delta$(dimer) - $^{m}\delta$(monomer). Table VI summarizes $\Delta\delta_d$ values for protons of all three methylated dimers. Several features are apparent when one compares $\Delta\delta_d^m$ values in Table VI with corresponding $\Delta\delta_d$ for parent dimers (Ezra et al., 1977; Lee et al., 1976). In each instance, the absolute magnitudes of dimerization shifts are substantially less for methylated compounds, particularly for base and H1' protons. However, the general trend of $\Delta\delta^m_{H2(5)} > \Delta\delta^m_{H8(6)}$ follows the previous pattern. From this it can reasonably be concluded that base-base orientations are not radically different in normal and methylated dimers, even though stacked populations are reduced sharply, as in $^{m}Ap^{m}A$. A further quantitation of base orientations does not seem warranted given the relatively small magnitudes of $\Delta\delta_d^m$. Similarly, no assessment can be made of glycosyl torsion angles, χ_{CN}, in methylated derivatives, though any major change, viz. anti → syn domains is quite unlikely on methylation or dimerization.

Table VI

Dimerization Shifts* for Methylated Dimers

	$^{m}Ap^{m}A$		^{m}ApU		$Up^{m}A$	
Proton	^{m}Ap-	-$p^{m}A$	^{m}Ap-	-pU	Up-	-$p^{m}A$
2(5)	0.026	0.126	0.023	0.166	0.097	0.115
8(6)	- .008	.135	- .013	.130	.046	.121
1'	.059	.134	.032	.122	.165	.111
2'	.015	.024	- .003	.054	.073	.097
3'	- .061	.021	- .047	.065	.058	.017
4'	.058	.061	- .019	- .003	.073	.057
5'	.070	- .091	.011	- .131	.075	0
5''	.142	- .048	- .012	- .060	.110	.049
CH_3	0	0	0			- .009

*$\Delta\delta_d^m$ = δ(monomer) - δ(dimer) in ppm.

From the preceding discussion a generalized conformational model for methylated dimers emerges that is similar to that of normal dimers except for a higher population of extended forms. Both the shift and coupling data are consistent with a rapidly interconverting mixture of stacked right helical, loop stack, and extended forms (Kondo and

†It is assumed that ring current distributions are unaltered in positively charged monomers and dimers.

Danyluk, 1976; Lee et al., 1976; Ezra et al., 1977). Whether the blend of these forms differs from normal dimers cannot be ascertained from the present data.

Effects of pD Change

Previous optical studies by Macon and Wolfenden (1968) found an apparent pK_a value of the conjugate acid of 1-methyladenosine, $N1^mAdo^+ \rightleftharpoons N1^mAdo$, to be 8.25 at 25°C. Since the conjugate base is uncharged, it is of interest to test whether neutralization of positive charges in both adenylate rings of $^mAp^mA$ leads to enhanced base

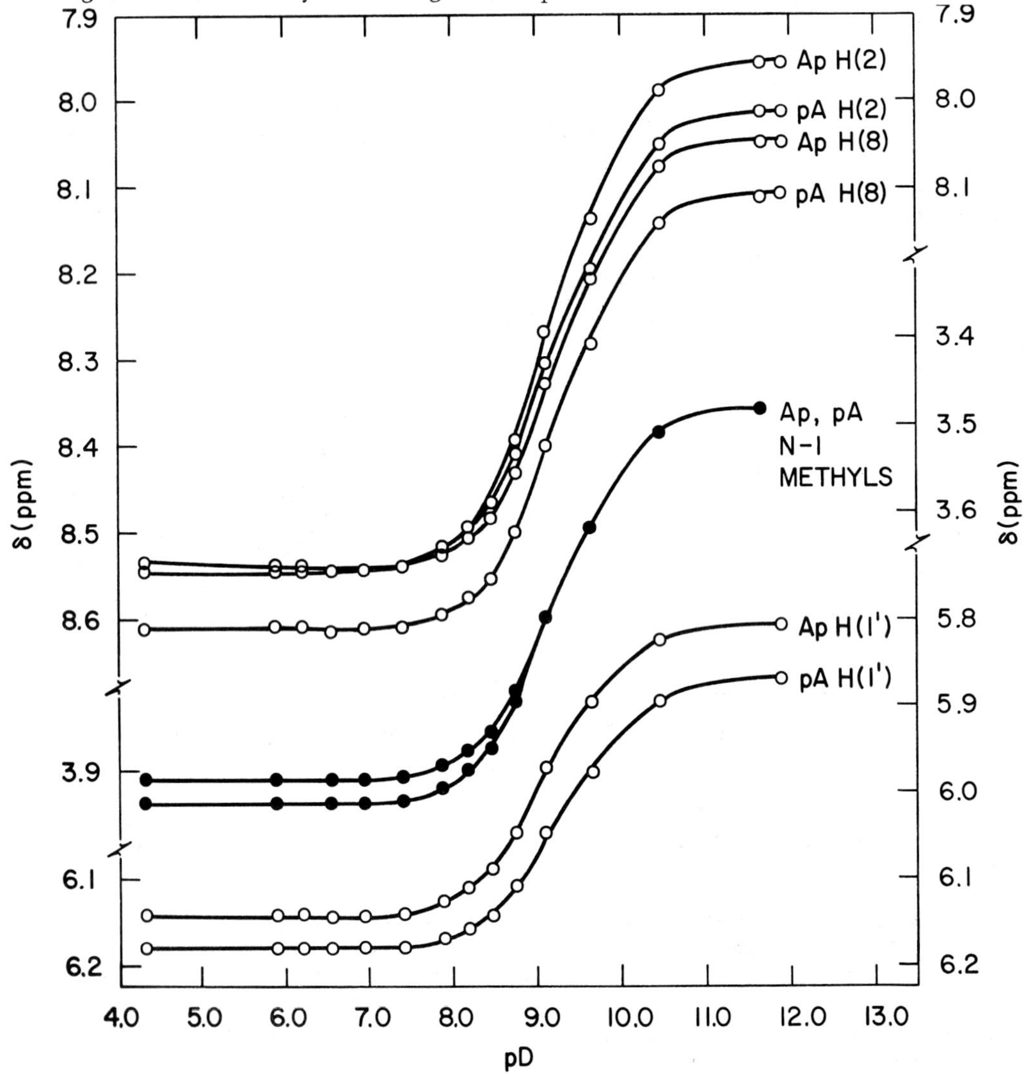

Figure 5. Variation of selected base proton shifts of $^mAp^mA$ with pD at 20°C.

stacking. Titration of $^{m}Ap^{m}A$ with base produces striking upfield shifts of base and H1' signals, Figure 5, and a decrease in $J_{1'2'}$ magnitudes, Figure 6 (the latter are accompanied by mirror-image increases of $J_{3'4'}$). Both sets of parameters plateau above pD $\sim$ 11.0 at values very close to those for ApA (Figure 4), e.g., δH8 Ap- = 7.88 ppm versus 7.90 ppm for ^{m}Ap-; $J_{1'2'}$ Ap- = $J_{1'2'}$ ^{m}Ap = 3.6 Hz. Interestingly, pK_a is slightly lower, 9.0 ± 0.05 for ^{m}Ap- than for -$p^{m}A$, 9.25 ± 0.05; both are stronger bases than corresponding mononucleotides.

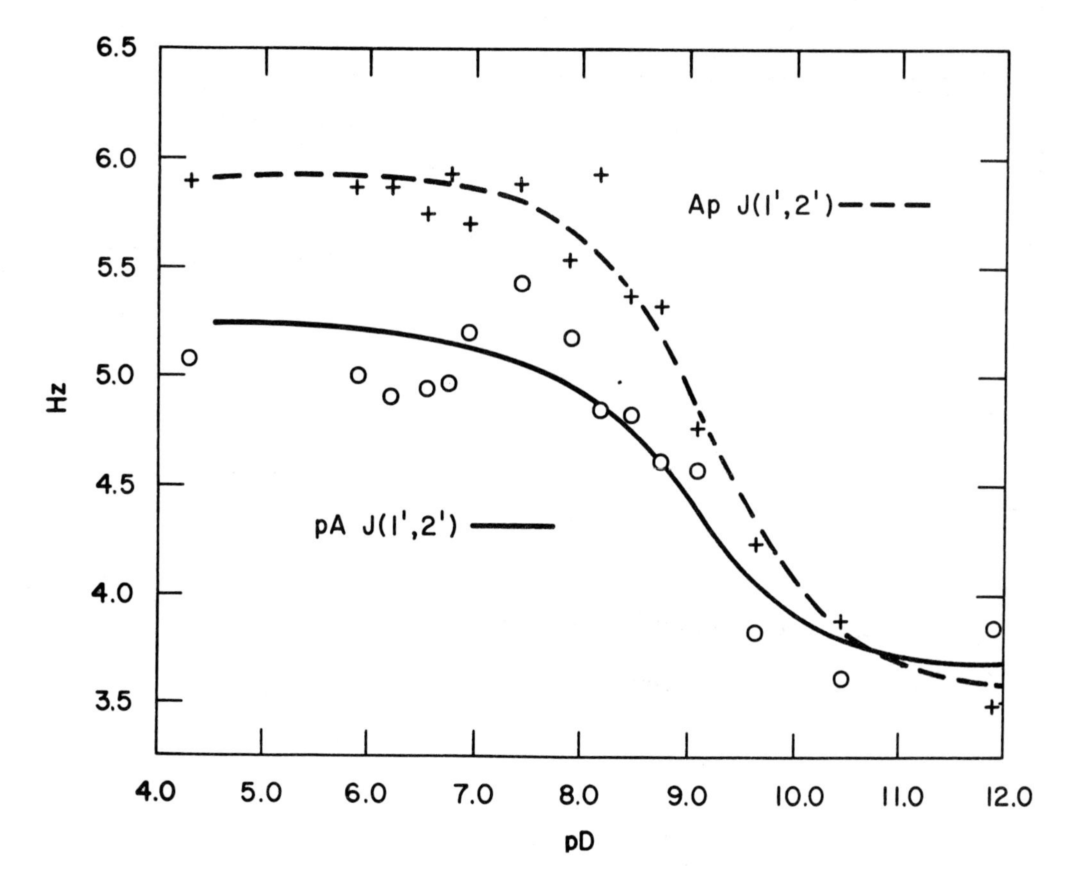

Figure 6. Variation of $J_{1'2'}$ couplings for $^{m}Ap^{m}A$ with pD at 20°C.

Clearly, the transition from a zwitterionic form for adenylate fragments to a mono-anion (bases uncharged) greatly enhances base stacking by all NMR criteria. Somewhat surprisingly, the total stacked population (right helical + loop forms) is effectively the same as in ApA, whereas purine aggregation studies suggest an increase in the former.

Relevance to Polynucleotide Structure

Base modified nucleosides are fairly common in t-RNA and m-RNA (Hall, 1967; Nishimura, 1972; Rich and RajBhandary, 1976) and are generally believed to play important and specific structural roles. Amongst the most prevalent are methylated bases such as N3 methylcytidine, N7 methylguanosine, and N1 methyladenine (Dunn, 1961; Hall, 1967; Rich and RajBhandary, 1976). N1 Methyladenylate residues are found in nearly all t-RNA molecules for which accurate sequence information is available. Significantly, the recent crystallographic structure determinations for $tRNA^{Phe}$ (Jack, Ladner, and Klug, 1976; Quigley et al., 1975; Kim et al., 1975; Brennan et al., 1975) show $N1^{m}$ Ado at position 58 as the linking nucleotidyl residue between TψC arm (stem) and TψC loop regions. At this point, the ordered stacked structure of the stem region alters into a more open loop conformation. Generalizations from solution conformational data for dimers to expectations in polynucleotide structures are risky but it is nevertheless of interest that the highly disordered nature of N1 methylated adenylate dimers is not inconsistent with the crystallographic results. It is tempting to assign a role for methylated adenylate residues (and by inference $N7^{m}$Guo) as conformational mediators between ordered and open domains of t-RNA and m-RNA molecules. From this perspective, methylation of available base sites by carcinogens produces irreversible conformational changes and concommitant alterations of biological function.

REFERENCES

Altona, C., and Sundaralingam, M., J. Amer. Chem. Soc. 95, 2333 (1973).

Altona, C., Proceedings of the Fourth Annual Harry Steenbock Symposium, Madison, WI, M. Sundaralingam and S. T. Rao, ed., University Park Press, 1975, p. 613.

Brennan, T., McMullan, R. K., Rubin, J., Ichikawa, T., Rao, S. T., and Sundaralingam, M., Fourth Annual Steenbock Symposium, Madison, Wi, M. Sundaralingam and S. T. Rao, ed., University Park Press, 1975, p. 39.

Brinacombe, R. L. C., Griffin, B. E., Haines, J. A., Haslam, W. J., and Reese, C. B., Biochemistry 11, 2452 (1965).

Chan, S. I., and Nelson, J. H., J. Amer. Chem. Soc. 91, 168 (1969).

Danyluk, S. S., and Hruska, F. E., Biochemistry 7, 1038 (1968).

Davies, D. B., and Danyluk, S. S., Biochemistry 13, 4417 (1974).

Davies, D. B., and Danyluk, S. S., Biochemistry 14, 543 (1975).

Dunn, D. B., Biochim. Biophys. Acta 46, 198 (1961).

Evans, F. E., and Sarma, R. H., FEBS Letters 41, 253 (1974).

Ezra, F. S., Lee, C-H, Kondo, N. S., Danyluk, S. S., and Sarma, R. H., Biochemistry 16, 1977 (1977).

Giessner-Prettre, C., and Pullman, B., J. Theor. Biol. 27, 87 (1970).

Giessner-Prettre, C., and Pullman, B., J. Theor. Biol. 65, 189 (1977).

Hall, R. H., Methods Enzymol. 12, 305 (1967).

Hruska, F. E., and Danyluk, S. S., J. Amer. Chem. Soc. 90, 3266 (1968).

Jack, A., Ladner, J. E., and Klug, A., J. Mol. Biol. 108, 619 (1976).

Kim, S. H., Suddath, F. L., Quigley, G. J., McPherson, A., Sussman, J. L., Wang, A. H. J., Seeman, N. C., and Rich, A. Fourth Annual Steenbock Symposium, Madison, WI, M. Sundaralingam and S. T. Rao, ed., University Park Press, 1975, p. 7.

Kondo, N. S., and Danyluk, S. S., Biochemistry 15, 757 (1976).

Lawley, P. D., J. Chem. Phys. 1011 (1961).

Lawley, P. D., and Brookes, P., Biochem. J. 89, 127 (1963).

Lawley, P. D. Screening Tests in Chemical Carcinogenesis, International Agency for Research on Cancer, 1976, p. 181.

Lee, C-H., Ezra, F. S., Kondo, N. S., Sarma, R. H., and Danyluk, S. S., Biochemistry 15, 3627 (1976).

Lee, C-H., and Sarma, R. H., Biochemistry 15, 2460 (1976).

Macon, J. B., and Wolfenden, R., Biochemistry 10, 3453 (1968).

Magee, P. N., and Barnes, J. M., Advance in Cancer Res. 10, 163 (1967).

Nishimura, S., Progr. Nucl. Acids Res. Mol. Biol. 12, 50 (1972).

Pullman, A., Chemical Carcinogenesis, P. O. P. Ts'o and J. A. DiPaolo, ed., Part A, Marcel Dekker, Inc., 1974, Chapt. 16.

Quigley, G. J., Wang, A. H. J., Seeman, N., Suddath, F. L., Rich, A., Sussman, J. L., and Kim, S. H., Proc. Nat. Acad. Sci. U.S.A. 72, 4866 (1975).

Rich, A., and RajBhandary, U. L., Ann. Rev. Biochem. 45, 805 (1976).

Sarma, R. H., Lee, C-H., Evans, F. E., Yathindra, N., and Sundaralingam, M., J. Amer. Chem. Soc. 96, 7337 (1974).

Singer, B., and Kusmierek, J. T., Biochemistry 15, 5052 (1976).

Ts'o, P. O. P., Kondo, N. S., Schweizer, M. P., and Hollis, D. P., Biochemistry 8, 997 (1969).

Wood, D. J., Hruska, F. E., Mynott, R. J., and Sarma, R. H., Can. J. Chem. 51, 2571 (1973).

STRUCTURAL STUDIES ON THE YEAST $tRNA^{Phe}$ BY THE ^{1}H NMR OF MODIFIED BASES. EFFECT OF REMOVAL OF THE Y BASE AND ANTICODON BINDING.

by P. DAVANLOO-MALHERBE, M. SPRINZL and F. CRAMER
Max-Planck-Institut für Experimentelle Medizin,
D-3400 Göttingen, West Germany

1. INTRODUCTION

Recent studies of Kan et al. [1] and Robillard et al. [2] monitoring the methyl and methylene resonances of yeast $tRNA^{Phe}$ in the high field region of the spectra have demonstrated that these resonances are very sensitive to structural changes which occur upon heating or chemical modification. Furthermore the assignment of these resonances is less ambiguous than is the assignment of H-bonded resonances. In this communication we report the assignments of the modified bases in yeast $tRNA^{Phe}$ at low temperature, and the effect of removal of the Y base from the anticodon loop of yeast $tRNA^{Phe}$ on the resonances of minor nucleosides.

The information provided by the high field resonances of minor nucleosides about $tRNA^{Phe}$ structure in solution can be further applied to approach the question concerning the conformational change in tRNA upon interaction with other macromolecules during protein biosynthesis. Schwarz et al. [3] and Schwarz & Gassen [4] suggested that tRNA may undergo a conformational change in the TΨC loop region as the result of codon-anticodon interaction. In our present studies we attempted to determine such a conformational change. We investigated the effect of the formation of the yeast $tRNA^{Phe}$ and E.coli $tRNA^{Glu}$ complex on the ^{1}H NMR signals of their respective modified bases in the high field region of the spectrum; these two tRNAs have complementary anticodons which leads to formation of strong complex via anticodon-anticodon interaction [5]. This interaction between two anticodons mimics the codon-anticodon interaction during ribosomal translation.

2. MATERIALS AND METHODS

Highly purified $tRNA^{Phe}$ was obtained from baker's yeast by a previously published procedure [6]. $tRNA^{Glu}$ from E.coli was prepared in accordance with Nishimura's procedure [7]. $tRNA^{Phe}$ lacking the Y base was prepared according to Thiebe and Zachau [8]. Yeast $tRNA^{Phe}$ (-Y) was then purified by chromatography on benzoylated DEAE-cellulose, eluting

B. Pullman (ed.), Nuclear Magnetic Resonance Spectroscopy in Molecular Biology, 125-136.

with 1M NaCl in acetate buffer, pH 6. D_2O, 99.9 % was purchased from Merck, Darmstadt, Germany. The NMR samples were prepared by dialysis several times against 10 mM $MgCl_2$, 100 mM KCl, 10 mM K-phosphate, pH 6.6. The samples were then lyophilized, dissolved in 99.9 %D_2O, again lyophilized, redissolved in 0.3 ml of 99.9 % D_2O and transferred to a Wilmed 5 mm NMR tube.

^1H NMR spectra were recorded using the Fourier transform (90° pulse, 2.7 s acquisition time,5 s pulse delay and Quadrature phase detection) procedure on a Bruker WH-270-MHz spectrometer located at Max-Planck-Institute for Biophysical Chemistry, Göttingen. Temperatures were measured before and after each run using the chemical shift difference between the methylene and hydroxyl protons of ethylene glycol [9]. The field was locked to D_2O in the solvent. Chemical shifts are reported with respect to DSS (2,2-dimethyl-2-silapentane-5-sulfonate).

3. RESULTS

3.1. Assignments of Yeast $tRNA^{Phe}$ Resonances at Low Temperature.

The assignment of the 12 methyl and methylene resonances originating from the minor nucleosides in yeast $tRNA^{Phe}$ was performed by Kan et al. [1]. These authors compared the high temperature spectra of $tRNA^{Phe}$ with corresponding mononucleotides or oligonucleotides. However, there are still some ambiguities in the assignment of the resonances at low temperature since a strong shift of the resonances takes place upon formation of the secondary and tertiary structure of tRNA. Monitoring the spectra in a wide range of temperatures does not lead to unambiguous interpretation since in some cases the signals of several methyl groups collapse at certain temperatures to a single peak. In such cases, if by further decrease of the temperature the peaks again separate, the assignment may be difficult.

We have tried to resolve these ambiguities in the low temperature assignments by using tRNAs other than yeast $tRNA^{Phe}$ having the same modified nucleoside in the same position of the primary sequence [10], but a simpler spectrum. This approach relies on the assumption that the tertiary structure of different tRNA species is similar [11]. Therefore a particular modified nucleotide located in the defined position of the sequence should have a similar chemical shift regardless of the tRNA species.

Fig. 1 shows a 270-MHz fourier transform high field NMR spectra of yeast $tRNA^{Phe}$ measured at different temperatures. The dependence of the chemical shifts on the temperature and the assignments of the resonances is given in Fig. 2. Under the applied conditions the melting of $tRNA^{Phe}$ occurs around 75°C. The assignment of the resonances above the melting temperature (95°C) is identical with that of Kan et al. [1]. Below the melting temperature, e.g. 60°C spectrum in Fig. 1, the peak at -2.77 ppm was assigned to the m^2G-10 resonance which is super-

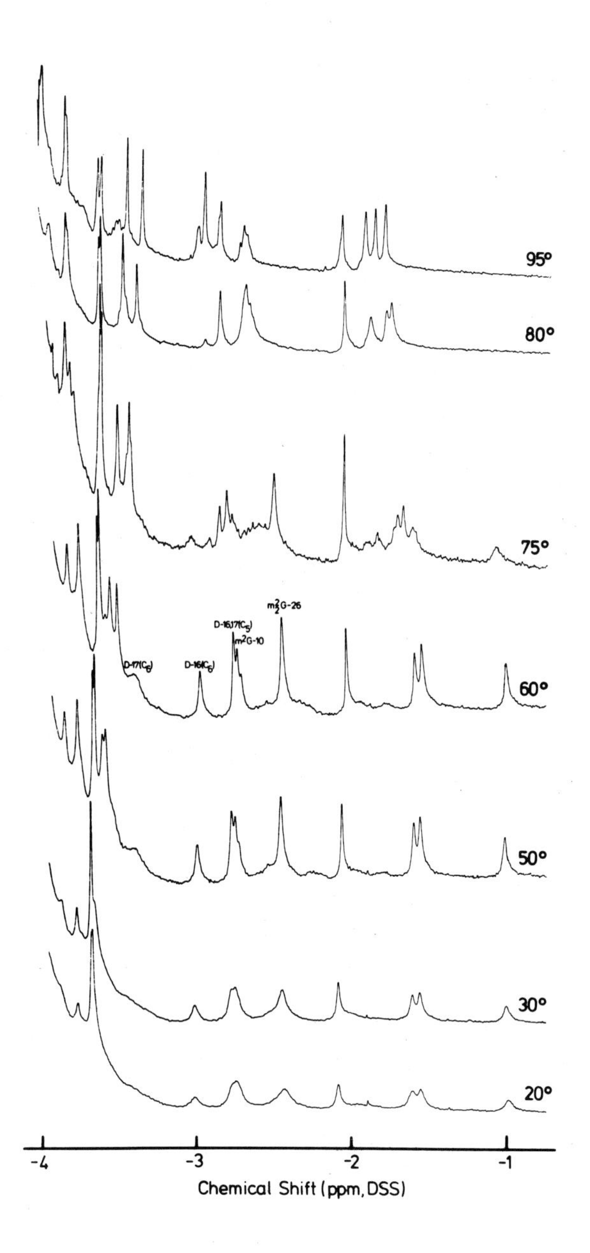

Figure 1. 270-MHz proton NMR spectra of 0.55 mM yeast tRNA^{Phe} in 10 mM $MgCl_2$, 100 mM KCl, 10 mM K-phosphate, pH 6.6 as a function of temperature.

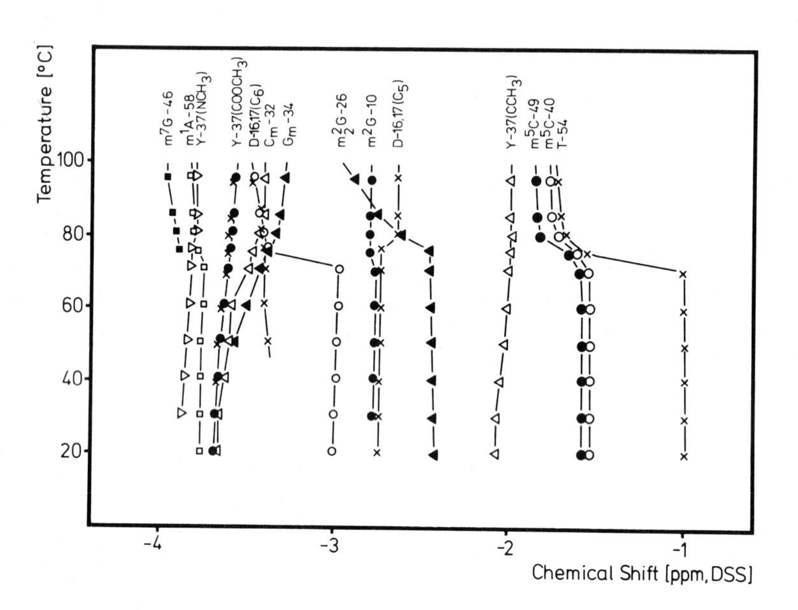

Figure 2. Plot of the chemical shift as a function of temperature for the resonances in Figure 1.

imposed by a triplet on the right edge of the peak belonging to the C_5 methylene resonances of D-16 and D-17. This assignment was made by comparison of the spectra of yeast $tRNA^{Phe}$ with that of $tRNA^{fMet}$ from Thermus thermophilus [12]. $tRNA^{fMet}$ from Thermus thermophilus has in this part of the spectrum only resonances belonging to the C_6 and C_5 methylene groups of D-21 which appear at low temperature at -3.0 ppm and -2.84 ppm, respectively (to be published). The peak at -2.75 ppm in the high field spectrum of yeast $tRNA^{Phe}$ (60°C) was recently assigned also by Robillard et al. [2] to the C_5 methylene resonances of D-16,17 which is in accordance with our finding.

Kan et al. [1] have assigned the peak at -3.0 ppm to the C_6 methylene protons of D-16,17. We have observed that at 40°C beside the peak at -3.0 ppm, a broad peak at -3.4 ppm appears and its intensity increases until 70°C. At 75°C, the peaks at -3.4 and -3.0 ppm moved downfield and superimposed under G_m-34 resonance. At 95°C, these resonances appear as a resolved triplet at -3.48 ppm. This observation leads us to speculate that in intact $tRNA^{Phe}$ possessing its native tertiary structure, the

C_6 methylene protons of D-16 and D-17 have different environments; one methylene being highly shielded ($\Delta\delta$ from 40° to $95^{\circ}C$ = 0.495 ppm) and the other being less shielded ($\Delta\delta$ from 40° to $95^{\circ}C$ = 0.089 ppm). The peak at -3.4 ppm (at $40^{\circ}C$) is tentatively assigned to the C_6 methyene of base D-17. In support of this assignment it is interesting to note that E.coli $tRNA^{Val}$, which contains only D-17, has in the NMR spectrum taken at $27^{\circ}C$ only the peaks at -3.65 and -2.75 ppm which were assigned to the C_6 and C_5 methylene resonances of D-17, respectively [13].

C_m-32 and G_m-34 resonances are at $20^{\circ}C$ superimposed under the Y-37 ($2COOCH_3$). As the temperature is increased they shift upfield and at $60^{\circ}C$ are well separated. We observed that in the presence of Mg^{2+}, through a temperature range from $20^{\circ}C$ to $95^{\circ}C$, the C_m-32 and G_m-34 resonances shifted upfield by 0.25 and 0.36 ppm, respectively. This effect was not observed in the absence of Mg^{2+} [2], indicating that in the presence of Mg^{2+} the C_m and G_m bases in the anticodon loop are strongly stacked, and that the Mg ion plays an important role in stabilization of the anticodon conformation. In the crystal structure of yeast $tRNA^{Phe}$, a strong Mg^{2+} binding site in the anticodon loop was identified [14].

In E.coli $tRNA^{fMet}$ lacking m^1A-58 we observed a peak at -3.77 ppm at low temperature which moved to -3.98 ppm at the melting temperature. This peak was assigned to the m^7G-46. It is therefore conceivable that in yeast $tRNA^{Phe}$, the m^7G-46 resonance is superimposed with the resonance of m^1A-58 (-3.78 ppm) through a temperature range of 20-$70^{\circ}C$. At the melting temperature the m^7G-46 is shifted to -3.99 ppm, which represents a downfield shift of 0.2 ppm upon unfolding. The assignment of m^5C-40 and m^5C-49 at the low temperature is still not resolved. All other assignments are in accordance with the previously reported work [1,2].

3.2. The Effect of the Removal of Y Base on ^1H NMR of the Modified Base in Yeast $tRNA^{Phe}$.

The profile of chemical shifts vs. temperature of each modified base in $tRNA^{Phe}$(-Y) is shown in Fig. 3. The major changes as compared to native $tRNA^{Phe}$ are as follows:

(a) Among the modified bases in the anticodon loop, the temperature dependence of G_m-34 peak is affected the most when the Y-base is removed. The G_m-34 resonance in the $tRNA^{Phe}$(-Y) at $50^{\circ}C$ has been shifted by 0.195 ppm to the higher field as compared to native $tRNA^{Phe}$. The excision of Y-base has, on the other hand, very little effect on the temperature dependence of the chemical shift of C_m-32 resonance. This observation may indicate that the G_m-34 in the native $tRNA^{Phe}$ participates in the stack involving the Y-37 base formed by the anticodon and its 3' neighbors. Hence as a result of the excision of the Y-base, this stacking interaction is disturbed.

(b) m^5C-49 and m^5C-40 resonances at low temperature (below $40^{\circ}C$) are superimposed and are less shielded in $tRNA^{Phe}$(-Y) than in native $tRNA^{Phe}$.

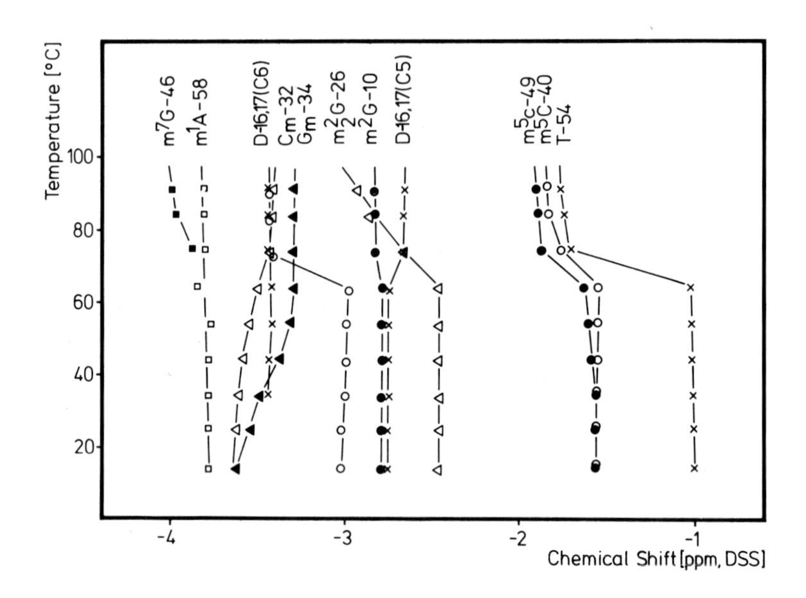

Figure 3. The chemical-shift data of the methyl proton resonances from the yeast $tRNA^{Phe}$ (-Y) (0.55 mM) vs. temperature (10 mM $MgCl_2$, 100 mM KCl, 10 mM K-phosphate, pH 6.6).

(c) The resonances of T-54, m^2_2G-26 and D-16(C_6) in $tRNA^{Phe}$ (-Y) show temperature-induced transitions of 3°C, 5°C and 3°C, respectively, lower than in the native $tRNA^{Phe}$ (Fig. 3). In general, the thermal denaturation of $tRNA^{Phe}$ (-Y) takes place at a lower temperature than that of $tRNA^{Phe}$.

3.3 Peak Assignments in E.coli $tRNA^{Glu}$.

The 50°C spectrum of $tRNA^{Glu}$ from E.coli (Fig. 4) contains three major peaks in the region between 0 and -4 ppm from DSS. The peak at -1 ppm arises from the methyl protons of 5-methyluridine (T-54). The peaks at -2.33, -4.0 and -2.68 ppm, which are not sensitive to temperature variations between 20°-95°C were assigned to 2-methyladenosine (m^2A-38) and to the CH_2 and $N-CH_3$ resonances of 2-thio-5-(N-methylaminomethyl)uridine (s^2mam^5U-35), respectively. These assignments are based on the spectra of free nucleotides in D_2O.

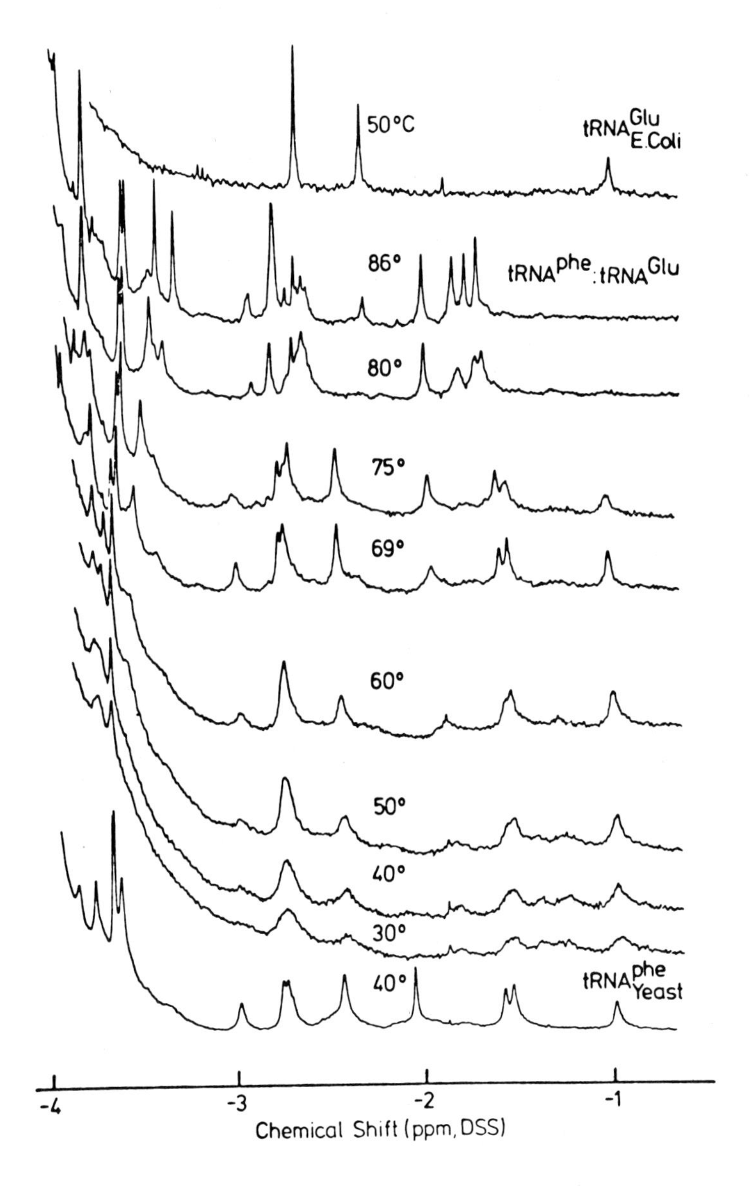

Figure 4. Temperature dependence of upfield 270-MHz NMR spectra of yeast tRNA^{Phe} · E.coli tRNA^{Glu} complex each 0.57 mM in 10 mM $MgCl_2$, 100 mM KCl, 10 mM K-phosphate, pH 6.6.

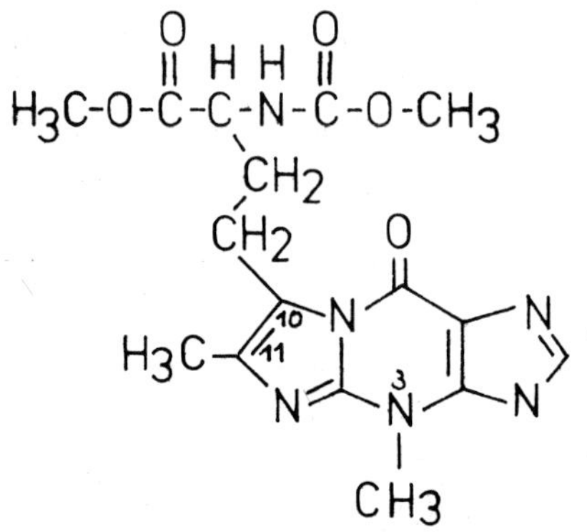

Figure 5. Molecular structure of the Y base

3.4 E.coli $tRNA^{Glu}$·Yeast $tRNA^{Phe}$ Complex.

In equimolar concentration, a dimer of these tRNA species is formed due to complementarity of their anticodons [5]. The 270-MHz spectra of the complex at various temperatures is shown in Fig. 4. As compared to the spectrum of single tRNAs the methyl and methylene resonances of all modified nucleosides show a strong broadening upon E.coli $tRNA^{Glu}$ and yeast $tRNA^{Phe}$ dimer formation. As expected, the resonances belonging to the nucleosides which are a part of the anticodons or in their vicinity are influenced most strongly.

A very broad peak at -1.83 ppm in the 30°C spectra (Fig. 4) which gradually shifts downfield with increasing temperature belongs to C_{11}-CH_3 resonance of Y-37 (Fig. 5) of $tRNA^{Phe}$. Above the melting temperature of the $tRNA^{Phe}$·$tRNA^{Glu}$ complex, which is at 85°C, this resonance appears as a sharp peak at -2.0 ppm, corresponding to the chemical shift of Y-37(C_{11}-CH_3) resonance of free $tRNA^{Phe}$. Hence, the upfield shift of about 0.25 ppm for Y-37(C_{11}-CH_3) can be used to monitor the $tRNA^{Glu}$·$tRNA^{Phe}$ complex formation.

Due to the high density of the peaks in the region between -4 to -3.5 ppm and to a strong broadening at low temperature, it is difficult to perform an unambiguous assignment of resonances originating from other modified nucleosides from the anticodon loop. However, the peak

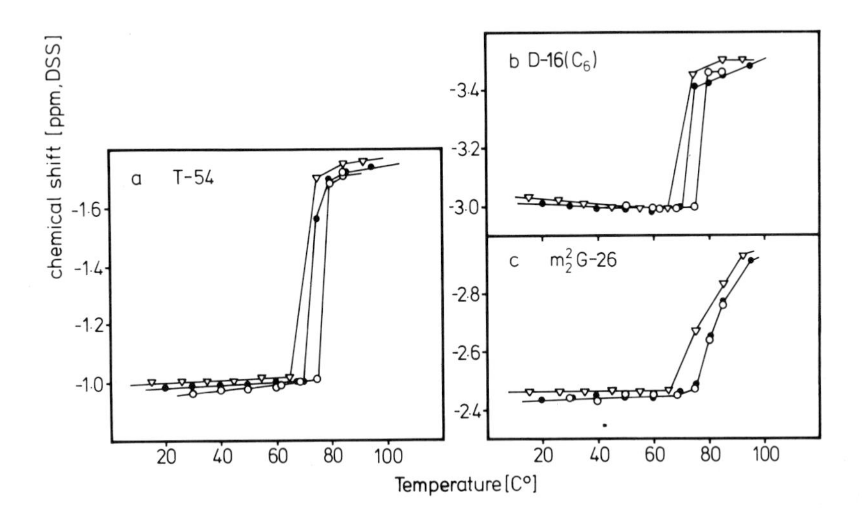

Figure 6. Effect of the removal of Y base and effect of the yeast $tRNA^{Phe}$·E.coli $tRNA^{Glu}$ dimer formation on the melting temperature of: a) T-54; b) D-16(C_6); c) m_2^2G-26. (-●-) is the yeast $tRNA^{Phe}$ (0.55 mM), (-Δ-) is the yeast $tRNA^{Phe}$ (-Y) (0.55 mM) and (-O-) is the yeast $tRNA^{Phe}$.E.coli $tRNA^{Glu}$ complex (each 0.57 mM).

at -3.74 ppm in the 50°C spectrum of the dimer (Fig. 4), which moves by 0.16 ppm upfield as the temperature increases, probably belongs to one of the ($COOCH_3$) methyl groups of Y-37. This assignment is based on the observation that in the 69°C spectrum of the dimer this peak is at -3.66 ppm, a position where the Y-37 ($COOCH_3$) resonance appears in the spectrum of $tRNA^{Phe}$ from yeast.

The N-methyl resonance of s^2mam^5U-35 of $tRNA^{Glu}$ moves by 0.08 ppm downfield upon complex formation and is superimposed over the C_5-methylene protons of D-16,17 of yeast $tRNA^{Phe}$. The resonance of the m^2A-38 (C_2-CH_3) which is visible as a sharp peak in the 50°C spectrum of free $tRNA^{Glu}$ completely disappears from the spectrum of the $tRNA^{Phe}$·$tRNA^{Glu}$ dimer. This can be due either to superimposition with some other resonance or to strong broadening which makes the identification impossible. This resonance reappears in the 86°C spectrum of the dimer at which temperature the anticodon·anticodon complex starts to dissociate.

This behaviour indicates the involvement of the m^2A residue in the interaction between the tRNAs and in addition a sharp monomer-dimer transition.

Broadening of the peaks upon dimer formation was observed also for methyl and methylene resonances of the modified bases from other parts of the tRNA than anticodon. For these modified nucleotides however no significant change in the chemical shift was detected upon dimer formation. The resonance of the T-54 methyl group is shifted by only 0.015 ppm upfield but the peak is much broader in the dimer at 50 ^{o}C as compared to the corresponding spectra of single $tRNA^{Glu}$ or $tRNA^{Phe}$, respectively (Fig. 4).

The temperature vs. chemical shift dependence of some resonances is given in Fig. 6. The melting of TΨC regions of $tRNA^{Phe} \cdot tRNA^{Glu}$ dimer is reflected in a very sharp cooperative transition at 78 ^{o}C. The melting monitored by the change of the chemical shift of T-54 of yeast $tRNA^{Phe}$ takes place in the dimer at a temperature 5 ^{o}C higher than in the case of free $tRNA^{Phe}$ (Fig. 6a). Similarly a higher thermal stability of the dimer is reflected by the D-16(C_6) chemical shifts vs. temperature dependence. Whereas in the case of the free $tRNA^{Phe}$ a sharp transition occurs at 73 ^{o}C, melting in the dimer is observed only at 78 ^{o}C (Fig. 6b). No difference in the melting of free $tRNA^{Phe}$ and $tRNA^{Phe} \cdot tRNA^{Glu}$ was observed by monitoring the temperature dependence of m_2^2G-26 resonance. However, in this case the melting occurs at a higher temperature as compared to the T-54 or D-16 resonances already in free $tRNA^{Phe}$ (Fig. 6c).

4. DISCUSSION

The main aim of this work was to study the effects on the resonances of methyl and methylene groups in the high field NMR spectra of chemical modification in the anticodon loop and of the binding of a complementary sequence to the anticodon.

The removal of the Y-37 hypermodified nucleoside from yeast $tRNA^{Phe}$ strongly influences the structure of the anticodon. The stacking of the anticodon to the 3'-part of the anticodon loop, as apparent from the X-ray structure of $tRNA^{Phe}$ [14] probably takes place also in the intact $tRNA^{Phe}$ in solution. This can be deduced from the strong changes observed in the G_m-34 resonance of $tRNA^{Phe}$(-Y) as compared to native $tRNA^{Phe}$, which can be interpreted as a disturbance of the stacking of the anticodon to its 3'-neighbours in the anticodon loop.

Dimer formation between the $tRNA^{Glu}$ from E.coli and $tRNA^{Phe}$ from yeast via their complementary anticodons was studied by Grosjean et al. [5] using the temperature-jump technique. These authors pointed out the importance of the modified bases adjacent to the 3' end of the anticodon for stabilization of the anticodon·anticodon interaction. Our observation of the 0.25 ppm upfield shift of the Y-37(CCH_3) resonance as well as the effects on the m^2A-38 resonance upon $tRNA^{Phe} \cdot tRNA^{Glu}$ dimer formation are in good agreement with these studies.

Surprisingly not only the resonances belonging to the modified bases from the anticodon are influenced by excision of the Y base, but the resonances belonging to the modified bases far distant are also affected. The temperature-induced melting of the $tRNA^{Phe}$ (-Y), as monitored by the changes of the chemical shift of D-16(C_6), T-54 and m_2^2G-26 resonances, occurs at lower temperature as compared to native $tRNA^{Phe}$. This indicates that a structural disturbance in the anticodon loop may result in long range effects in the remote parts of the molecule. Whereas by excision of the Y base the thermal stability in the TΨC, D-16 and m_2^2G regions are decreased, an opposite effect was observed in the case of the $tRNA^{Phe} \cdot tRNA^{Glu}$ dimer. The thermal unfolding, as monitored by T-54 and D-16(C_6) resonances, takes place at a higher temperature as compared to free $tRNA^{Phe}$ (Fig. 6).

The lower melting temperature for the T-54 and D-16(C_6) resonances due to the removal of Y base and the higher melting temperature for the above bases as the result of $tRNA^{Phe} \cdot tRNA^{Glu}$ complex formation illustrate the phenomenon of transmission of stability in tRNA. The concept of such telestability in DNA in which one region of a DNA helix may influence the dynamic structure of a contiguous but remote portion of the helix, was clarified by Wells and his colleagues [15,16] using a synthetic DNA polymer in their model studies. They observed a strong stabilization of an AT region of a polymer by an adjacent CG region even in the case when the AT block was separated from a CG block by 15 base pairs (about 50 Å).

In our present study we were unable to prove a hypothesis that the codon anticodon interaction may trigger an unfolding of the tertiary structure of tRNA and induce accessibility of the TΨC region [3,4]. This may however, be due to limitations in our model system using tRNA species with complementary anticodons. On the other hand, it is also possible that other components involved in the ribosomal decoding process as well as the proper codon are necessary to induce a conformational change to allow the proper binding of the aminoacyl-tRNA to the ribosomal A-site.

Very recently, a report appeared [17] on the NMR study of the interaction of the anticodon of yeast $tRNA^{Phe}$ with the oligonucleotide UpUpCpA. In good agreement with our results, the hydrogen-bonded proton spectra of the yeast $tRNA^{Phe}$ indicated no loss of hydrogen-bonded proton resonances. This implies that the hydrogen bonds between the D and TΨC-loops which are necessary for stabilization of the tertiary structure of $tRNA^{Phe}$ remain intact after the binding of this oligonucleotide to the anticodon.

Acknowledgement
This work was supported by the grant of the Deutsche Forschungsgemeinschaft Cr 8/45.

REFERENCES

1 L.S. Kan, P.O.P. Ts'O, M. Sprinzl, F. von der Haar & F. Cramer (1977) Biochemistry, 16:143.
2 G.T. Robillard, C.E. Tarr, F. Vosman & B.R. Reid (1977) Biochemistry, 16:5261.
3 U. Schwarz, H.M. Menzel & H.G. Gassen (1976) Biochemistry, 15:2484.
4 U. Schwarz & H.G. Gassen (1977) FEBS Lett., 78:267.
5 H. Grosjean, D.G. Söll & D.M. Crothers (1976) J.Mol.Biol., 103:499.
6 D. Schneider, R. Solfert & F. von der Haar (1972) Hoppe Seyler's Z. Physiol.Chem., 535:1330.
7 S. Nishimura, in: Procedures in Nucleic Acids Research (1971) eds. G.L. Cantoni & D.R. Davies, (Harper and Row, N.Y.) pp. 342-564.
8 R. Thiebe & H.G. Zachau (1968) Eur.J.Biochem., 5:546.
9 A.L. Van Geet (1968) Anal.Chem., 40:2227.
10 M. Sprinzl, F. Grüter & D.H. Gauss (1978) Nucleic Acids Research, in press.
11 A. Rich & U.L. RajBhandary (1976) Ann.Rev.Biochem., 45:805.
12 K. Watanabe, T. Oshima & S. Nishimura (1976) Nucleic Acids Research, 3:1703.
13 R.V. Kastrup & P.G. Schmidt (1978) Nucleic Acids Research, 5:257.
14 G.J. Quigley, M.M. Teeter & A. Rich (1978) Proc.Natl.Acad.Sci., U.S.A. 75:64.
15 J.F. Burd, R.M. Wartell, J.B. Dodgson & R.D. Wells (1975) J.Biol.Chem., 250:5109.
16 J.F. Burd, J.E. Larson & R.D. Wells (1975) J.Biol.Chem., 250:6007.
17 H.A.M. Geerdes, J.H. Van Boom & C.W. Hilbers (1978) FEBS Lett., 88:27.

SPECTROSCOPIC STUDIES OF DRUG-NUCLEIC ACID INTERACTIONS

THOMAS R. KRUGH
Dept. of Chemistry, University of Rochester, Rochester, New York, 14627, U.S.A.

At a preceding Jerusalem Symposium I had the pleasure of presenting the initial results of our development of the use of oligonucleotides as models to study the interaction of drug-nucleic acid complexes (Krugh, 1974). The main thrust of this early work was the study of the sequence preferences observed in the interaction of actinomycin D with deoxydinucleotides as studied by visible absorption spectroscopy (Krugh, 1972) and the determination of the geometry of the complexes formed using proton magnetic resonance spectroscopy (Krugh and Neely 1973a,b). Several interesting developments have occurred in our understanding of drug-nucleic acid complexes during the past few years: some of these will be discussed below. Figure 1 shows the chemical structures of actinomycin D (1a), ethidium bromide (1b), daunorubicin and adriamycin (1c) and actinomine (1d). Actinomycin D, daunorubicin and adriamycin are important drugs in the area of cancer chemotherapy (e.g. see Meienhofer and Atherton, 1973, and Sartorelli, 1976, ethidium bromide is a trypanocidal drug (e.g. see LePecq et. al., 1974), while actinomine is an analog of actinomycin in which the uncharged cyclic pentapeptides are replaced by the positively charged (at neutral pH) groups shown in Figure 1d. Actinomycin D, actinomine, and ethidium bromide have been shown to intercalate into double stranded DNA (e.g. see Reinhardt and Krugh, 1978; Krugh et. al., 1977, and references therein) while daunorubicin and adriamycin are also believed to intercalate (e.g., see Henry, 1976).

An example of the combined use of different spectroscopies to study the binding of drugs to oligonucleotide systems can be seen in our recent experiments with ethidium bromide and the ribodinucleoside monophosphate CpG and its deoxy analogue dCpdG, as illustrated in Figures 2 and 3 (e.g. see Krugh et. al. 1975; Krugh and Reinhardt, 1975; and Reinhardt and Krugh, 1978). Ethidium bromide is an optically inactive molecule which becomes optically active when it binds to the asymmetric double helix (Figure 2). The characteristic bands which are observed at 308, 330, and 380 nm in the ethidium-DNA complexes are also observed when ethidium bromide forms complexes with the self-complementary deoxyribodinucleosides (Figure 2A) or mixtures of complementary dinucleosides (e.g., Figure 2B). The similarity between the circular

B. Pullman (ed.), Nuclear Magnetic Resonance Spectroscopy in Molecular Biology, 137-146.

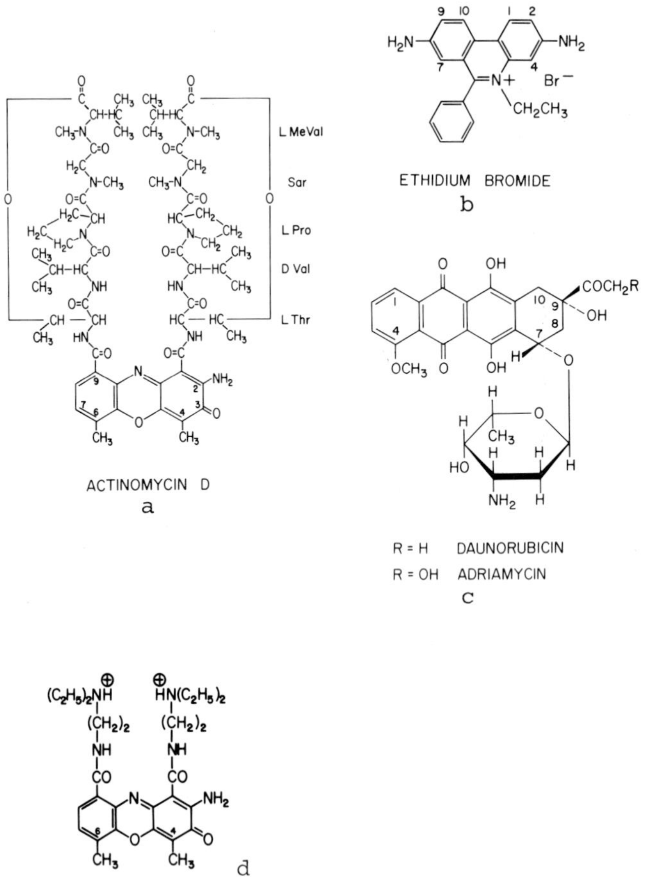

Fig. 1. The chemical structures of: (a), actinomycin D, (b), ethidium bromide, (c), daunorubicin and adriamycin, and (d), actinomine.

dichroism data on the model systems with corresponding spectra on the polymers illustrates that the oligonucleotides are good model systems for studying drug-nucleic acid complexes. The proton magnetic resonance data for ethidium bromide and the 2.0:1 (dC-dG):ethidium bromide complex shown in Figure 3 provides a means of obtaining information on the geometry of the complexes. The corresponding ethidium resonances in the two spectra are indicated by a dashed line in Figure 3. The upfield shifts of the ethidium resonances that accompany complex formation result from the ring current effects of the adjacent nucleotide bases (e.g., see Giessner-Prettre et. al., 1976; and references therein). The unusually large upfield shifts of the aromatic resonances of ethidium (~0.6 to 1.0 ppm) provides convincing evidence that ethidium bromide is forming a miniature intercalated complex with two of the

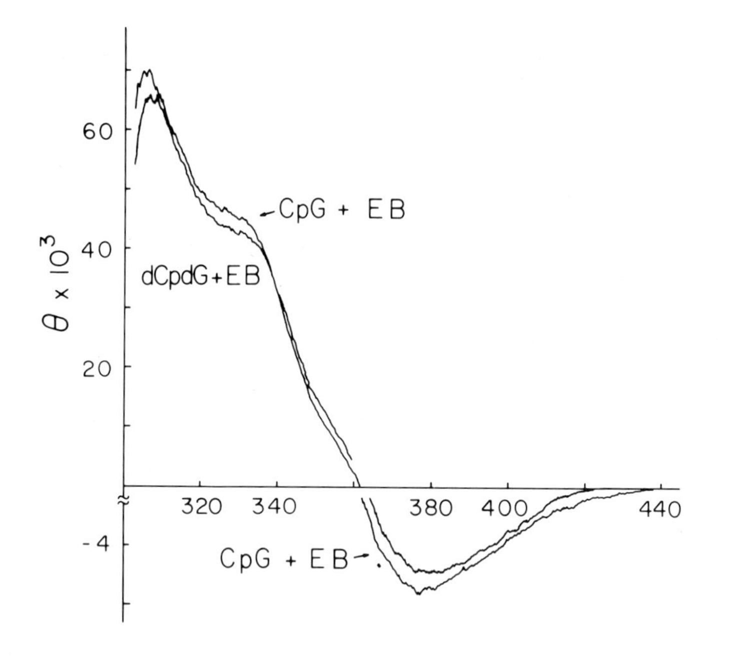
CpG + EB
dCpdG+EB
CpG + EB
θ x 10^3
60
40
20
-4
320
340
380
400
440

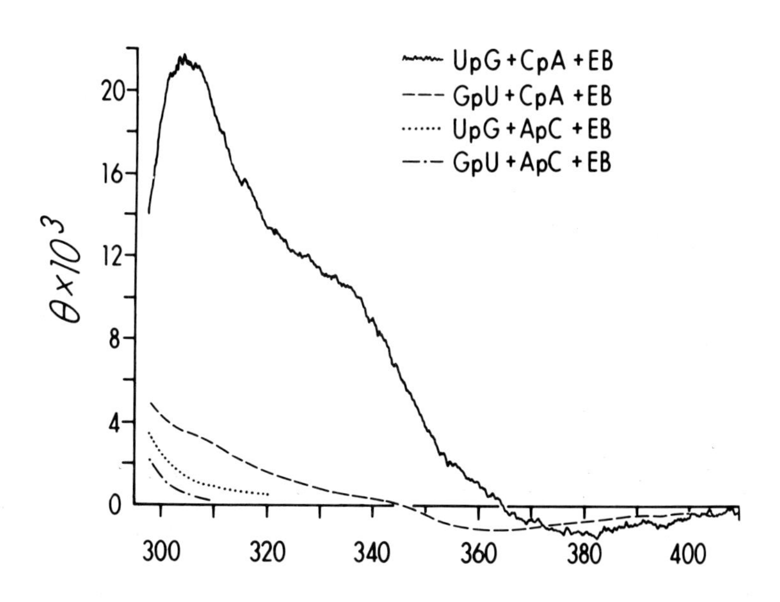
UpG+CpA+EB
GpU+CpA+EB
UpG+ApC+EB
GpU+ApC+EB
θx10^3
20
16
12
8
4
0
300
320
340
360
380
400

Fig. 2 (A, top). Induced CD spectra of ethidium bromide upon addition of CpG and dC-dG in a 1-mm thermostatted cell at 3°C. The samples were 0.75 mM EB and 2.3 mM CpG and 0.75 and EB and 3.2 mM dC-dG. Both samples were dissolved in a D_2O potassium phosphate buffer (5 mM) pD = 7.4 (pH meter + 0.4). The θ values are direct readings from the recorder. Note that the scale of the ordinate has been expanded below the zero level. (B, bottom figure). Induced CD spectra of ethidium bromide in the presence of equimolar mixtures of the complementary (UpG + CpA and GpU + ApC) and non-complementary ribodinucleoside monophosphates at ~3°C. Note that only the complementary mixture of pyrimidine-purine dinucleosides UpG + CpA gives rise to the characteristic induced CD spectrum under the present experimental conditions.

self-complementary dinucleoside monophosphates (either ribose or deoxyribose) in which the long axis of ethidium is oriented parallel to the base pairs. By using short pathlength cells it is also possible to record the circular dichroism spectrum of similar solutions which thus provides a link between the geometry of the complex deduced from proton NMR data on the oligonucleotides and the binding of the drug to polynucleotides.

Sobell and his coworkers (Tsai et. al., 1975; Jain et. al., 1977) utilized our initial experiments with ethidium bromide as guidance in selecting to cocrystallize ethidium with the ribodinucleoside monophosphates. The pyrimidine-purine sequence dinucleosides UpA and CpG were easily cocrystallized with ethidium and formed 2:2 complexes. The crystal structures provide a visualization of an intercalated complex and contain a wealth of stereochemical information. Figure 4 (top) shows the overlap of the U·A base pairs with the phenanthridinium ring of ethidium which was observed in the crystalline complex, whereas Figure 4 (bottom) shows a hypothetical overlap if the base pairs are interchanged to generate a possible complex in which ethidium is intercalated at an ApU sequence. The sequence dependent stacking is a result of the helical aspect of the complex. Of course, other factors such as hydrogen bond formation, the solvation energy of the drug, and the stacking energies of the nucleotide bases at the intercalation site will also contribute to the overall stability of the complexes.

While the determination of whether or not a drug intercalates is an interesting question in itself, the more important questions concern the preferential binding of the drugs to the various sequences available on DNA, the distortion of the double helix which accompanies intercalation, the transmission of this conformational information along the DNA molecule, and the cooperative or competitive binding of two or more drugs to DNA. A dramatic example of all of these phenomena is our recent observation that daunorubicin (or adriamycin) facilitates the binding of actinomycin D to poly(dA-dT)·poly(dA-dT) as illustrated by the circular dichroism data in Figure 5 (from Krugh and Young, 1977). A comparison of the CD data in Figure 5a for the binding of the drugs to DNA with the data in Figure 5b leaves no doubt that

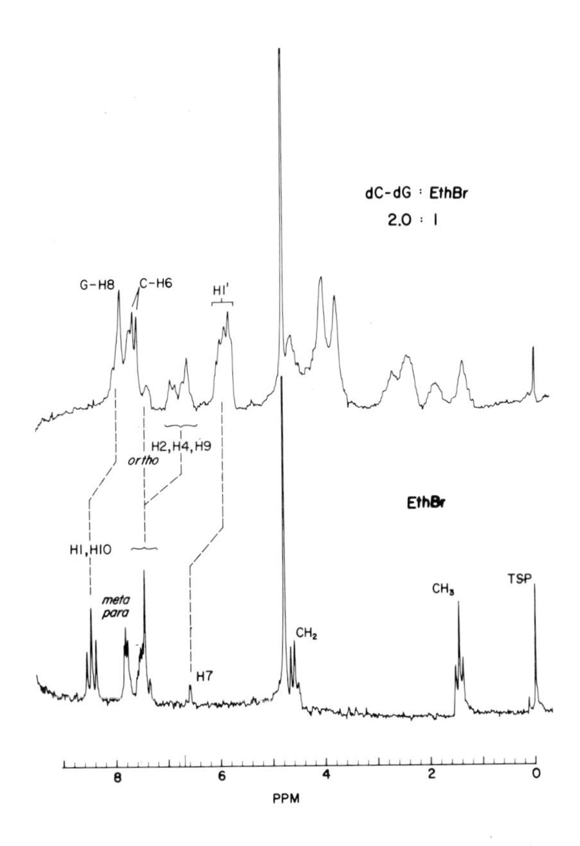

Fig. 3. 100 MHz Proton FT-NMR Spectra of a Solution of Ethidium Bromide and a 2.0:1 dC-dG:Ethidium Bromide Solution (25°C). The concentration of ethidium bromide was 1.6 mM in a 5 mM D_2O phosphate buffer (pD = 7.4) in each spectrum. EthBr, ethidium bromide, TSP, sodium 3-trimethylsilylpropionate-2,2,3,3-d_4.

the binding of daunorubicin to poly dA-dT·poly(dA-dT) facilitates the binding of actinomycin D to this synthetic polynucleotide. The shape of the approximate CD spectrum of bound actinomycin D (the ······ curve in Figure 5b) when compared to that observed in model systems (e.g., Auer et. al., 1978) is a good indication that the phenoxazone ring is intercalated between two A·T base pairs. However, nuclear magnetic resonance experiments are currently in progress on drug-oligonucleotide solutions in order to determine the structure of the complexes.

An interesting aspect of the facilitated binding of actinomycin D to poly(dA-dT)·poly(dA-dT) is the question of whether the bound actinomycin D would inhibit RNA polymerase, since it has been well documented that actinomycin D very effectively inhibits RNA polymerase

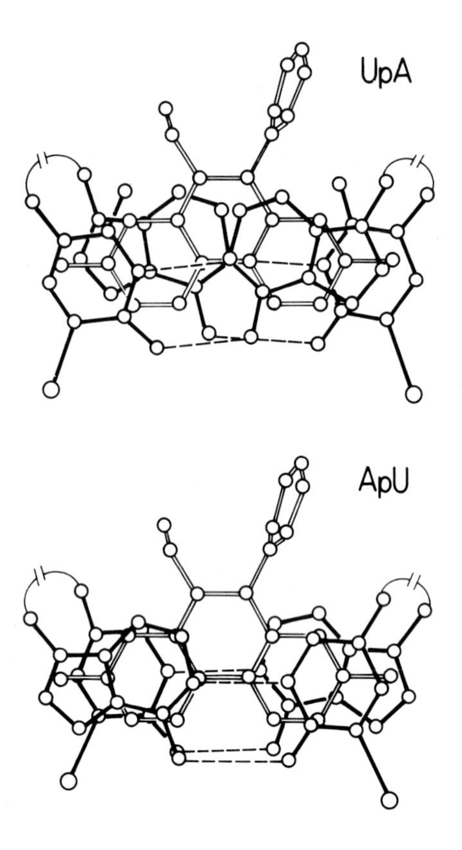

Fig. 4. Schematic illustration of the overlap of the phenanthridinium ring in the ethidium-5-iodo-UpA crystalline complex (top) and a hypothetical complex of ethidium with ApU (bottom). The ApU structure was generated by simply interchanging the two U·A base pairs in the UpA-EB complex shown as the top figure.

when DNA is used as a template. We have performed a variety of experiments to answer this question and to investigate the possible synergistic inhibition of RNA polymerase when daunorubicin and actinomycin D are used in combination. The details of this work will be published separately, but the most interesting result is that the addition of actinomycin D to a poly(dA-dT)·poly(dA-dT) + daunorubicin system did not result in any significant increase in the inhibition of RNA polymerase over that produced by daunorubicin alone. In 1968 Müller and Crothers proposed that the very slow dissociation of actinomycin D was coupled to the effective inhibition of RNA polymerase and we therefore measured the duponol induced dissociation rate constants of actinomycin D from the actinomycin D-daunorubicin-poly(dA-dT)·poly(dA-dT) complex. We observed (Krugh et. al., in preparation) that actinomycin D

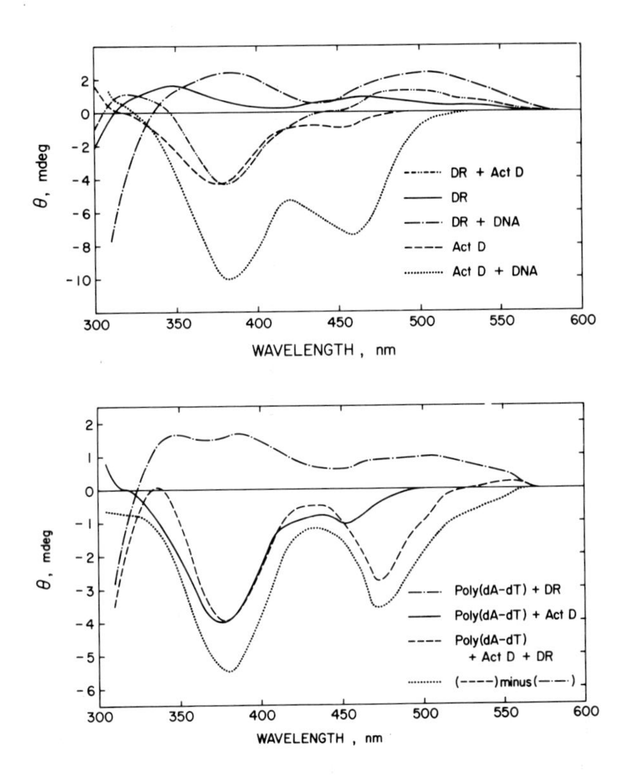

Fig. 5. Circular dichroism spectra of solutions of: a, 8.5×10^{-6} M daunorubicin (DR) and actinomycin D (Act D) alone and in the presence of 8.5×10^{-5} M DNA; b, 8.5×10^{-6} M daunorubicin plus 8.5×10^{-5} M poly(dA-dT)·poly(dA-dT); 8.5×10^{-6} M actinomycin D plus 8.5×10^{-5} M poly(dA-dT)·poly(dA-dT); 8.5×10^{-6} M actinomycin D plus 8.5×10^{-6} M daunorubicin plus 8.5×10^{-5} M poly(dA-dT)·poly(dA-dT). The curve ········ was calculated by subtracting the poly(dA-dT)·poly(dA-dT) + DR spectrum from the poly(dA-dT)·poly(dA-dT) + DR + Act D spectrum which, to a first approximation, is an estimate of the circular dichroism spectrum of actinomycin D when bound to poly(dA-dT)·poly(dA-dT). All spectra were recorded on a Jasco J-40 circular dichroism instrument in a 4-cm path length cell at 20°C. All solutions contained 10 mM potassium phosphate buffer, pH 7.0. Essentially similar results were obtained at lower drug-to-phosphate ratios, as well as in the presence of 0.1 M NaCl. The spectra in which adriamycin were used in place of daunorubicon gave qualitatively similar results.

dissociates much more rapidly and binds less tightly when poly(dA-dT)·poly(dA-dT) is used than with DNA. This result provides an independent confirmation of the relationship between the slow dissociation of actinomycin D and its pharmacological activity in the inhibition of RNA polymerase.

Müller and Crothers (1968) reported the presence of three slow dissociation constants for the dissociation of actinomycin C_3 from DNA complexes and proposed a coupled kinetic scheme to account for these observations. We have studied the dissociation (and association) kinetics of actinomycin D from poly dG· poly dC and poly(dG-dC)·poly(dG-dC) and we have observed two new phenomena: (1) the dissociation of actinomycin D from poly(dG-dC)·poly(dG-dC) is characterized by a single exponential decay and, (2) the magnitude of this dissociation constant is a function of the ratio of the nucleotides/bound drug. The latter observation is undoubtedly another manifestation of a site-site interaction which results from the transmission of conformational information along the DNA double helix. We therefore would not be surprised to observe drug-drug effects in mixed drug-DNA complexes, which may occur over rather long regions of the double helix.

The dissociation of actinomycin D from poly(dG)·poly(dC) is characterized by two slow dissociation constants, which contrasts with the single exponential decay observed for the dissociation of actinomycin D from poly(dG-dC)·poly(dG-dC). Segments of these two polynucleotides are illustrated in Figure 6 where we note that in poly(dG)·poly(dC) all

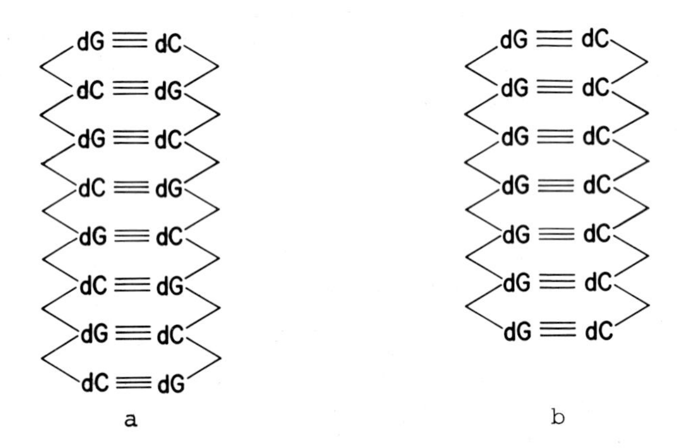

Fig. 6. Schematic illustrations of a portion of (a), poly(dG-dC)·poly(dG-dC) and, (b), poly(dG)·poly(dC).

the intercalation sites are identical and yet two slow dissociation constants are observed. In poly(dG-dC)·poly(dG-dC) there are two types of intercalation sites (i.e., (dC-dG)·(dC-dG) and (dG-dC)·(dG-dC)) but each site is characterized by a twofold symmetry. It is important to note that the phenoxazone ring of actinomycin D is asymmetric and thus there are two distinct ways of intercalating actinomycin D into the poly(dG)·poly(dC) duplex (Figure 6b) whereas the twofold symmetry of the intercalation sites in poly(dG-dC)·poly(dG-dC) (Figure 6a) results in only one type of complex. It is also interesting to note that in model studies with mixtures of the complementary deoxydinucleotides pdG-dG + pdC-dC the proton magnetic resonance data showed that a miniature intercalated complex was formed in which there was an approximate 50:50 distribution of the two predominant complexes as schematically illustrated in Figure 7. (Krugh et. al., 1977).

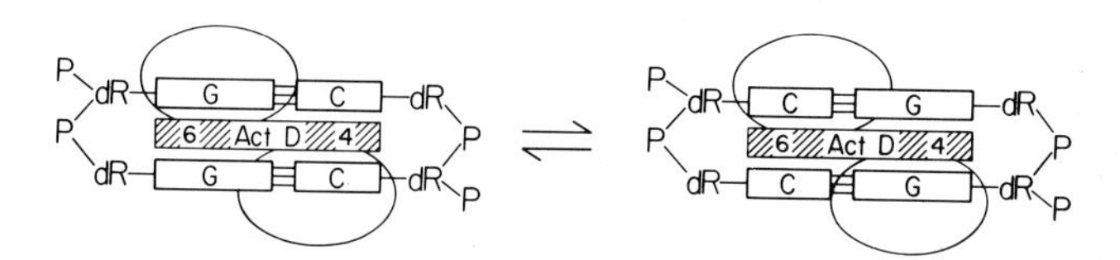

Fig. 7. Schematic illustrations of the two possible intercalated complexes which may be formed by actinomycin D in the presence of the complementary deoxydinucleotides pdG-dG and pdC-dC (from Krugh et. al., 1977).

A plausible explanation for all of this data is that actinomycin D exhibits separate dissociation constants for each type of sequence at the various intercalation sites. If we assume a preferential binding at the (dG-dC)·(dG-dC) intercalation sites in poly(dG-dC)·poly(dG-dC) then a single dissociation is not surprising. On the other hand, there are two possible ways of orienting the phenoxazone ring when actinomycin D intercalates into poly(dG)·poly(dC), which is consistent with the two slow dissociation constants observed. Extrapolating these results to DNA where there are three slow dissociation constants observed leads to the conclusion that there are three different *types* of intercalation sites, each of which exhibits a characteristic slow dissociation constant, and presumably therefore, different physiological activities (if the apparent correlation between the magnitude of the dissociation constants and the inhibition of RNA polymerase is reliable). The preliminary dissociation data at low nucleotide/drug levels with poly(dG-dC)·poly(dG-dC) suggests that this polynucleotide has a relatively long conformational memory and that it does not rapidly relax back to its original conformation as the actinomycin D is dissociating. We anticipate that all of these interesting phenomena will be observed at the oligonucleotide level where magnetic resonance techniques will provide valuable structural information. In anticipation of this, we have developed a method of assigning the ^{31}P resonances of oligonucleotides using shift reagents and paramagnetic relaxation probes. These same approaches should also prove useful in assigning the proton resonances of the oligonucleotides and in locating the site or sites of intercalation in the drug-oligonucleotide complexes.

This research was supported by research grants CA-14103 and CA-17865 and Research Career Development Award CA-00257 from the National Cancer Institute, DHEW, and an Alfred P. Sloan Fellowship.

REFERENCES

Auer, H.E., Pawlowski-Konopnicki, B.E., Chiao, Y-C.C., and Krugh, T.R. (1978), *Biopolymers*, in press.
Giessner-Prettre, C., Pullman, B., Borer, P.N., Kan, L.S., and Ts'o, P.O.P. (1976), *Biopolymers* 15, 2277-2286.
Henry, D.W. (1976), in Cancer Chemotherapy, Sartorelli, A.C., ed., Amer. *Chem. Soc. Symp.* 30, *Amer. Chem. Soc.*, Washington, D.C., pp. 15-57.
Jain, S.C., Tsai, C-C., and Sobell, H.M. (1971), *J. Mol. Biol.* 114, 317-331.
Krugh, T.R. (1972), *Proc. Natl. Acad. Sci. U.S.A.* 69, 1911-1915.
Krugh, T.R. (1974), Molecular and Quantum Pharmacology, Bergmann, E. and Pulman, B. eds. Reidel Pub. Co. Dordrecht, Holland 7, 465.
Krugh, T.R., Mooberry, E.S., and Chiao, Y-C.C. (1977), *Biochemistry* 16, 740-747.
Krugh, T.R., and Neely, J.W. (1973a), *Biochemistry* 12, 1775-1782.
Krugh, T.R., and Neely, J.W. (1973b), *Biochemistry* 12, 4418-4425.
Krugh, T.R., and Reinhardt, C.G. (1975), *J. Mol. Biol.* 97, 133-133-162.
Krugh, T.R., and Young, M.A. (1977), *Nature* 296, 627-628.
Meienhofer, J., and Atherton, E. (1973), *Adv. in Applied Microbiology* 16, 203-300.
Müller, W., and Crothers, D.M. (1968), *J. Mol. Biol.* 35, 251.
Reinhardt, C.G., and Krugh, T.R. (1978), *Biochemistry*, in press.
Sartorelli, A.C. ed. (1976), Cancer Chemotherapy, *Amer. Chem. Soc. Symp.* 30, *American Chem. Soc.*, Washington, D.C.
Tsai, C-C., Jain, S.C., and Sobell, H.M. (1975), *Proc. Natl. Acad. Sci. U.S.A.* 72, 628.

A MODEL FOR THE SPECIFIC SITE MELTING OF DNA _IN VIVO_

Bruce McConnell
University of Hawaii, Honolulu, Hawaii

While there is evidence that strand separation in DNA is an important control event for both transcription and replication (1,2) there is no specific model that would provide a satisfactory mechanistic account as to how localized helical destabilization might be initiated by a protein and be restricted to a specific DNA sequence under the highly stabilizing conditions found _in vivo_ (3). An hypothetical model for sequence-specific destabilization can be formulated quite naturally on the basis of proton exchange mechanisms in DNA to be described in this report. This model provides for the exclusive destabilization of G-C rich sequences under the influence of polycationic sequences of proteins. Superficially, these features are contradictory to the well known stability of G-C regions and the ionic stabilization of DNA by polycations. In addition, the model contains the apparently self-contradictory notion that an increase in H-bond strength will establish the condition of destabilization. Therefore, justification of such a model will be based on close examination of a rational basis for its formulation and second on a comparison of its predictive features with several general observations on DNA melting reported in the literature.

Proton Exchange and Hydrogen Bonding

The model is based on consideration of mononucleotide-proton interactions that represent natural regulatory mechanisms for controlling the lifetimes of exchangeable hydrogens in DNA. The amino protons of adenine, guanine and cytosine and the N-1 imino proton of guanine provide discrete pmr signals in aqueous solution, whose line widths are direct measurements of the proton lifetimes (4,5). Plots of line width as a function of pH, exemplified in Figure 1 (bottom) for the amino protons of cytosine, demonstrate two general exchange characteristics: First, that the ability of OH^- to act as a proton acceptor (alkaline pH broadening) is anomalously slow and second, that the low pH broadening reflects increased exchange due to protonation of the weakly basic endocyclic nitrogen of the nucleobase (5,6). The maximum at pH 4 arises from amino proton transfer from (N-3) protonated cytosine to the N-3 site of the conjugate pyrimidine base. The involvement of the

B. Pullman (ed.), Nuclear Magnetic Resonance Spectroscopy in Molecular Biology, 147-160.

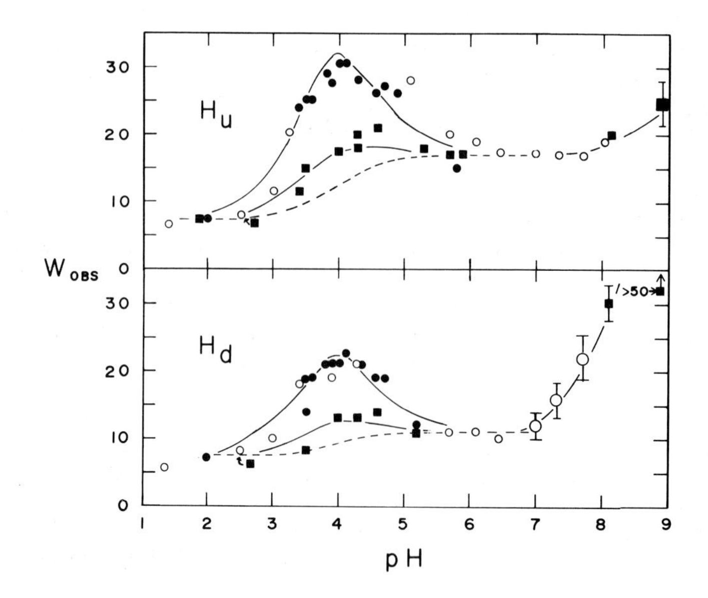

Figure 1. Pmr Line Width vs pH for the Amino Protons of 2',3'-cCMP. Observed line widths of the separate upfield (H_u)(top) and downfield (H_d)(bottom) $-NH_2$ resonances obtained at nucleotide concentrations of 0.4 M (circles) and 0.1 M (squares). Data were obtained at 360 MHz (closed) and 100 MHz (open) at a probe temperature of 3 $\pm$ 1°C. Solid curves are calculated from eq. 1 and dashed curves represent rotational broadening corrections.

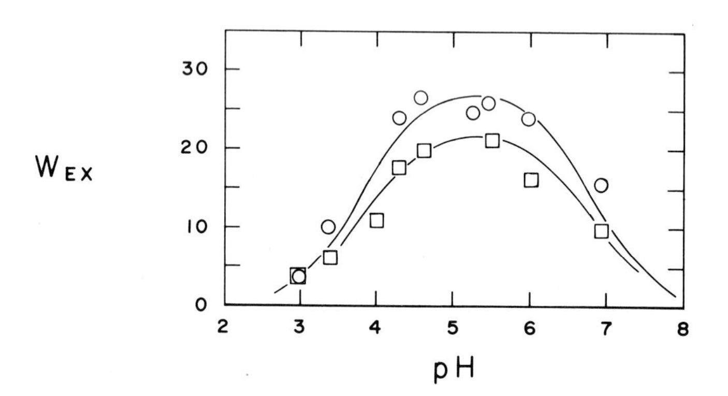

Figure 2. Phosphate Catalysis of Amino Proton Exchange vs pH for 2',3'-cCMP. Additional pmr line broadening (W_A) due to (Na) phosphate addition (0.05 M) was measured at 3 $\pm$ 1°C at 360 MHz for the separate $-NH_2$ resonances. H_u (upfield)(circles) and H_d (downfield)(squares). Solid curves were calculated from eq. 1.

protonated nucleobase as an important intermediate in exchange was confirmed by the addition of buffer exchange catalysts (5,6). As shown in Figure 2 line broadening of the amino proton resonances induced by the addition of buffer typically exhibits a maximum midway between the pK of the nucleobase protonation site and the pK of the buffer (phosphate in this case). Although low pH data for guanylic acid are not available, the gross features of the pH dependence of line width with and without catalyst establish the protonated intermediate for the purine amino protons, also (Figure 3). The mechanism that accounts for these pH broadening patterns is shown for G and C in Figure 4. Here the buffer conjugate base (A) acts as an acceptor of the exchangeable protons slowly from the neutral nucleobase (U mechanism) and much more rapidly from the protonated form (P mechanism). There is an additional intermediate possible for the guanine amino, i.e., the zwitterion protonated at G(N-7) and deprotonated at G(N-1). In any case an N-7 protonated form of the guanine base, whether the conjugate acid or the zwitterion, accounts for the pH dependence of buffer catalysis. Here we see what amounts to a natural regulatory mechanism for control of proton lifetimes by an interaction at a different site on the molecule. Protonation renders the H-bonding protons much more labile to transfer, i.e., much more acidic and thus more able to form stable hydrogen bonds (32).

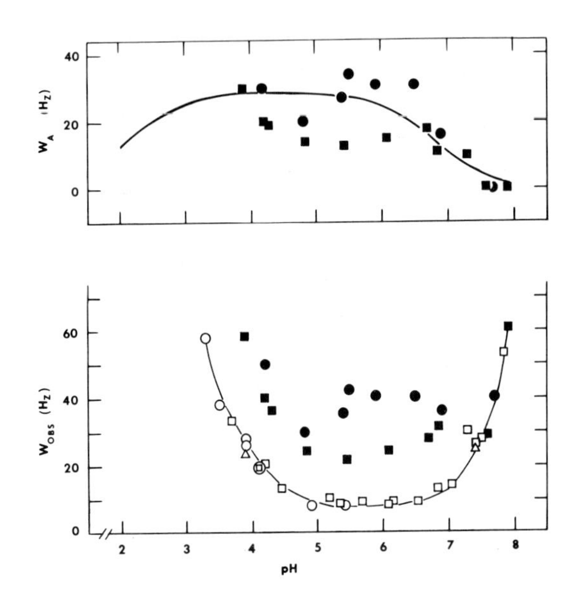

Figure 3. Pmr Line Broadening vs pH for the Amino Protons of 2',3'-cGMP. Observed line widths (W_{obs})(bottom) were measured from 360 MHz spectra obtained at 26° for nucleotide alone (open figures) and in the presence of (Na) phosphate at 0.5 M (closed squares) and 0.75 M (closed circles). Phosphate broadening (W_A)(top figure) was obtained by data subtraction with and without phosphate. Solid curves are calculated from eq. 1.

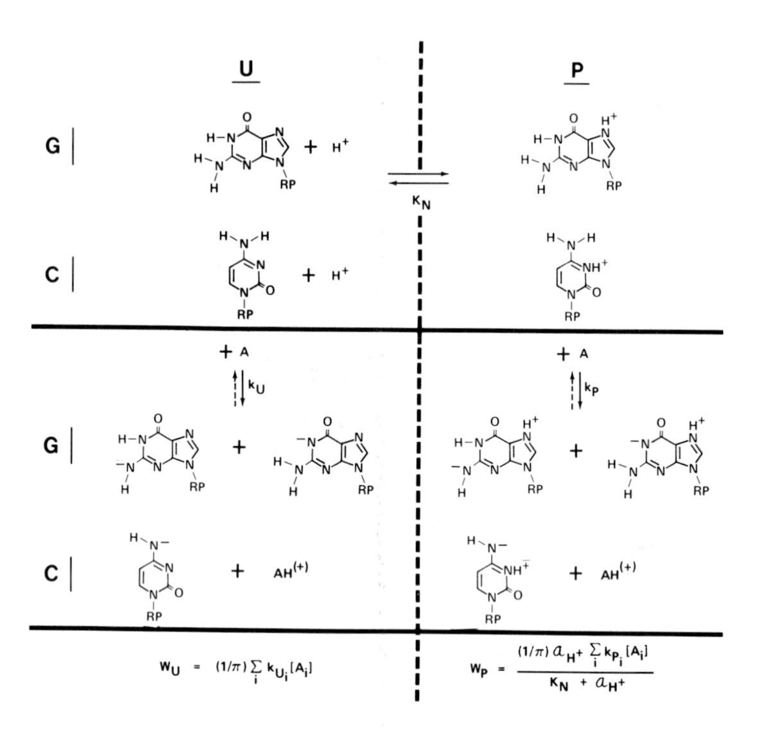

Figure 4. Mechanisms of Amino and Imino Proton Exchange in Guanine and Cytosine. The exchangeable proton is transferred to an acceptor A from the neutral base (U mechanism) or from the protonated base (P mechanism). Kinetic expressions for each are shown at the bottom.

Can we estimate the acidity increase that occurs when the nucleobase is protonated? In the case of guanine N-1 the calculation is straight-forward (5). The relation between kinetic parameters and line width is described by equation 1,

$$W_{obs} = W_u + W_p = \left(\frac{1}{\pi}\right) \sum_i k_{u_i} [A_i] + \frac{\left(\frac{1}{\pi}\right) a_{H^+}}{K_N + a_{H^+}} \sum_i k_{p_i} [A_i] \qquad (1)$$

where W_u and W_p are the respective broadening contributions of the U and P mechanisms, K_N is the acid dissociation constant of the nucleobase protonation site, a_{H^+} the proton activity defined by the pH electrode and A is the concentration of acceptor. The acceptor is the conjugate base of any acid-base pair established by acid dissociation of a buffer, the nucleotide itself or water. As shown in eq. 1 the pmr line widths provide catalytic rate constants for transfer of the G(N-1) proton from

Table I
Rate Constants[a] for Proton Transfer From Exchange Sites of 2',3'-cGMP and 2',3'-cCMP

Acceptor	pK_A	Guanine $-NH_2$ [b]		Guanine (N-1) [c]		Cytosine $-NH_2$ [d]	
		k_p	k_u	k_p	k_u	k_p	k_u
H_2O	-1.7	60[e]		100		10	
Chloracetate	3.0			$5x10^4$	50		
Acetate	4.5			$1x10^6$	$3x10^3$		
Cytosine (N-3)	4.0			$9x10^4$	$2x10^3$	$1x10^3$	
Phosphate	6.8	$6x10^6$		$2x10^8$	$5x10^5$	$1x10^6$	
Imidazole	7.1	0				$2x10^6$	
2-Methylimidazole	8.0					$1x10^7$	
OH^-	15.8		10^8		$2x10^{10}$		$3x10^7$, $3x10^8$

(a) Dimensions of k_u and k_p are M^{-1} sec^{-1}. (b) Data collected at 26°C. (c) Data collected at 3 ± 1°C. (d) Separate values for each $-NH_2$ proton resonance are shown for k_u, but are averaged for k_p, due to their similarity in the latter case. Data collected at 3 ± 1°C.
(e) This value is valid between pH 3.5 and 4.5 only.

the neutral and protonated nucleobase designated k_u and k_p, respectively. Some of these values are listed in Table I. These constants are used to obtain an exchange site pK value (pK_u or pK_p) by means of equation 2,

$$k_u \text{ (or } k_p) = \frac{k_d \times 10^{pK_A - pK_u \text{ (or } pK_p)}}{1 + 10^{pK_A - pK_u \text{ (or } pK_p)}} \tag{2}$$

This equation is a kinetic simplification for proton transfers of the "normal" type (7,31), in which rate processes defining proton transfer within the donor-acceptor transition complex are not considered (not rate-limiting). It provides a determination of the exchange site pK (pK_u or pK_p) if the diffusion rate constant for formation of the transition complex (k_d) is known in addition to the pK of the acceptor (pK_A). From a Brönsted plot (Figure 5) we can guess k_d to be approximately 10^8 M^{-1} sec^{-1}, because it is safe to assume that water, whose k_p value is 2 log units above a line of unit slope defined by the other catalysts, acts as a proton acceptor with a k_d of 10^{10} M^{-1} sec^{-1} (31). For the U mechanism k_d is the only unknown in eq. 2 and is equal to 10^8 M^{-1} sec^{-1} as well. The result is a pK calculation showing that the pK of the G(N-1) group falls 2 to 3 units from 9.4 to 6 or 7 when the G(N-7) site is protonated. At physiological pH the G(N-1) site should be ionized for the protonated purine.

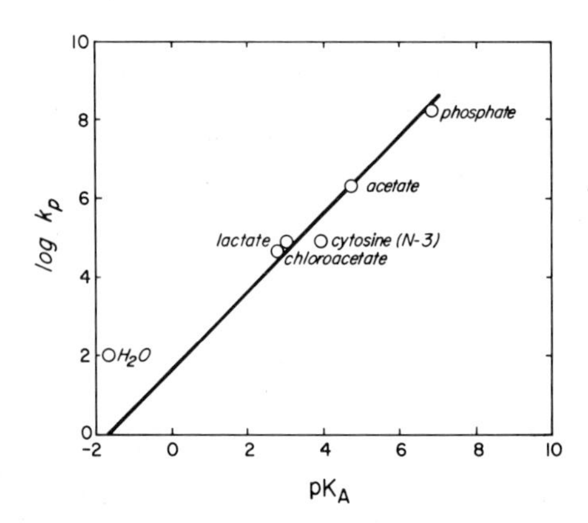

Figure 5. Brönsted Plot of Exchange Catalysis for the N-1 Proton of 2',3'-cGMP. P mechanism proton transfer rate constants were obtained from catalytic broadening data (360 MHz, 3 $\pm$ 1°C). Catalyst pK_A values were obtained by direct potentiometric titration of experimental mixtures and by extraction from the kinetic data (eq. 1). The pK_A of the $H_3O^+ \leftrightarrow H_2O + H^+$ reaction is -1.7.

A similar calculation for the amino protons gives a strikingly different result. The amino pK_p for guanine ranges between 8 and 10, while a similar calculation for the neutral base (A = OH^-) gives a pK_u value of about 18. Surely, this huge difference in pK_u and pK_p cannot be due to a change in the intrinsic acidity of the amino group in view of the 3 pK unit change in the G(N-1) imino, which reflects the inductive effect of the G(N-7) protonation. This calculation merely illustrates the fact that eq. 2 is not valid for the neutral base amino, as it does not account for rate processes that would decrease the lifetime of the transition complex below that value required for completion of proton transfer (7). The formation of a stable transition complex requires "normal" H-bond formation between the donor and acceptor. Slow exchange of the neutral nucleobase does not reflect a pK_A-pK_u term (eq. 2), is not "normal" and is a reflection of the profound reluctance of these protons to form stable H-bonds with the highly solvated hydroxyl group and other acceptors generally. Since the protonated nucleobase qualifies as a "normal" acid (buffer catalysis can be measured), we can form the interesting conclusion that there is a qualitative increase in hydrogen bonding ability of the amino proton when the nucleobase binds a proton at its ring nitrogen. There is a transformation from a poor H-bond donor to a "normal" donor when the amino group experiences an increase in acidity of a few orders of magnitude through the inductive effect of nucleobase ring protonation.

In summary, two things happen when the basic endocyclic nitrogen of guanine is protonated. First, there is anion formation at neutral pH by dissociation of the imino site. Second, the amino protons are transformed from poor H-bond donors to "normal" donors. This latter point applies to adenine and cytosine as well as guanine and leads to the suggestion of natural mechanisms involving this transformation that could affect the functional dynamics of the double helix. Since guanine is unique in having its proton binding site exposed to solvent in the large groove of the double helix, a model for the effect of G(N-7) protonation on H-bonding in the helical G-C base pair can be visualized.

The Model

In Figure 6 are shown six states of the helical G-C base pair, three in the unprotonated form and three protonated at G(N-7). For the unprotonated states 1, 2 and 3 there is an initial process of H-bond breakage, which allows access of water to the H-bonding sites. Thus, two equilibria are shown; primary breakage of H-bonds ($1 \rightleftharpoons 2$) and the competitive formation of water H-bonds with the interbase protons ($2 \rightleftharpoons 3$). From the above considerations the amino-to-water H-bonds would be weak or short-lived and the ratio of state 3 to state 1 would be small (as is known)(8,9). Indeed, it is likely that in the helix reaction $1 \rightleftharpoons 2$ must occur many times before an amino proton is transferred to water, while the rate limiting reaction for the chemically fast G(N-1) proton

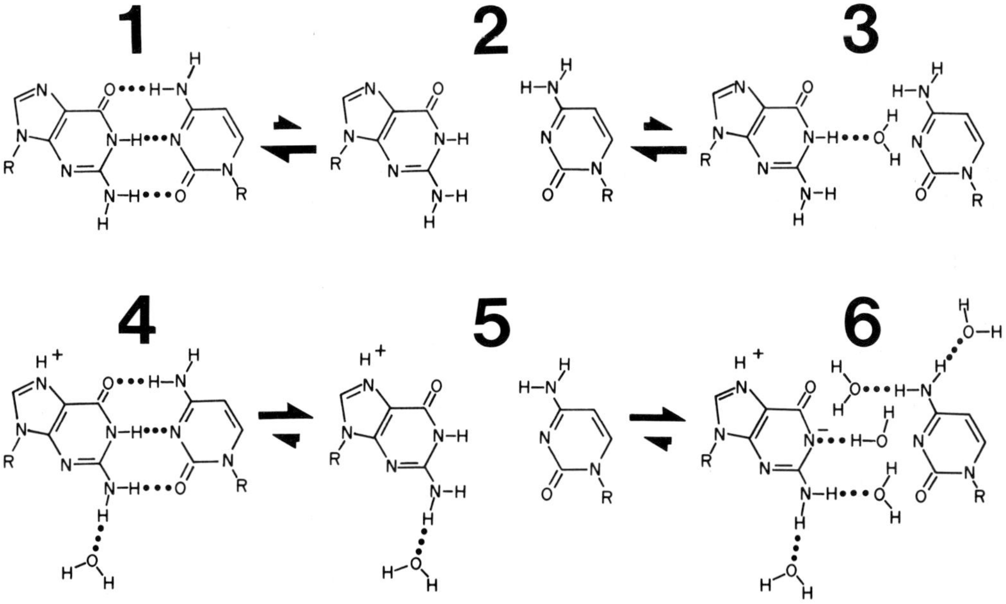

Figure 6. A Mechanism of Helical G-C Destabilization by Protonation of Guanine N-7. H-bonding water interactions of three unprotonated states of the G-C base pair (1, 2 and 3) are compared with that of three corresponding states of the G-C pair protonated at G(N-7)(4, 5 and 6).

exchange is 1 → 2 (8). Therefore, water is shown bonded only to the G(N-1) proton in state 3 and the overall equilibria are in the direction of state 1.

Corresponding reactions of the three protonated states (4, 5 and 6) would lead to a different result. The ionization of G(N-1) and the transformation of the amino protons to H-bond donors will allow water to compete successfully with the cytosine acceptor sites. These events would greatly increase the ratio of state 6 to state 4 as compared to 3/1. An obvious question is if the H-bond strength is increased so dramatically by protonation, why wouldn't this decrease the equilibrium production of state 5 from state 4? If we view this G-C base pair as part of the double helix, then we are considering a structural change that is not as drastic as melting, but which can act as a nucleus for melting (9). Both this structural transition of the helical strands and the resultant access of water to exchange sites is well documented by observations of tritium loss from H-bonding sites equilibrated with this isotope (8,9). There is an "open" state of the helical strands (2 and 3), which occupies many base pairs (8) such that the opening reaction rates (1 → 2 and 4 → 5) are determined predominantly by inertial effects in the macromolecule and negligibly by the H-bond strength. While the rates for 1 → 2 and 4 → 5 should be roughly the same the reverse rates should be quite different. It is important to note that as soon as the G(N-7) is protonated the non-H-bonding amino proton in the small groove can now form a strong water H-bond previous to the reaction 4 → 5. In effect, the solubility of the H-bonded guanine in the external solvent has been increased. This event would satisfy a well known thermodynamic condition for the denaturation of macromolecules, since the previously lipophylic residues no longer maintain conformation by excluding solvent and seeking the hydrophobic interior of the macromolecule (10). This increase in solvent interaction would represent a profound decrease in stacking energy, which is the predominant stabilizing force in DNA (11) and depends to a large extent on the solvent exclusive property of the nucleobases (10). Thus, the rate 5 → 4 would be decreased and the accumulated non-bonded state 5 would be further trapped as state 6, due to the well established access of water to the helix interior and its superior position (55 M) for H-bond competition. With the opening of the G-C pair as in state 6, the intrinsic basicity of the more exposed cytosine N-3 site (pK = 4) can now be expressed and would increase by ca. factor of ten its extent of protonation and the resulting amino-to-water H-bonding. Overall, protonation at G(N-7) would increase greatly the solvent interaction of the monomeric units and would impose a distortion in the helix to accommodate the length of the base-to-water H-bonds and the water molecule itself.

The extension of this model from destabilization to strand separation demands consideration of three requirements. First, the proton has to be held in place for a time exceeding its normally short lifetime on G(N-7). Second, this extended lifetime would occur at several adjacent G-C pairs simultaneously to satisfy the minimum condition for cooperative nucleation. Third, required G(N-7) interaction would be accomplished by proton donors other than hydronium ion and would include protonated amino groups (lysine and spermine), guanidino groups (arginine) and imidazole groups (histidine). It is apparent that all three requirements would be met by a tightly bound protein that could position several proton donor sidechains at G(N-7) sites in the large helical groove. Sterically, the pleated sheet structure is appropriate for this positioning, since the alpha carbon residue distance (3.54 Å) is suited to the vertical base stacking distance (3.4 Å) when the twist of the helical groove is considered. Less conceivable due to lower charge density is the replacement of the proton by metal cations or tetraalkylamines at the required G(N-7) interaction. Studies to demonstrate increased mononucleotide exchange by G(N-7) interaction with such cations are being undertaken (12).

By accepting the plausibility of satisfying these requirements a number of predictive features can be seen about the condition and elements of DNA destabilization *in vivo* and *in vitro*, which account for many general observations described in the literature. These are listed as "Predictions."

"Predictions"

1. G-C clusters would be necessary components of control regions for transcription. The relationship between the number of transcription sites and G-C clusters in phage and bacterial DNA is well established through the work of Szybalski and co-workers (13). In addition, sequences of promotor regions in bacterial DNA show a thermally unstable A-T-rich region flanked by two G-C-rich regions (14). Dickson and co-workers have suggested the possibility of destabilization of this region by the action of the catabolite activator protein.

2. The overall stabilizing effect of cations and polycations in melting studies should appear to favor stabilization of A-T over G-C. As the ionic stabilization approaches saturation at sufficiently high salt concentrations (∿ 1 M) the destabilizing effects exclusive to G-C through cation binding to G(N-7) will be observed as a decrease in the width of the melting transition. In the literature "preferential A-T stabilization" appears to be the rule with no exception in aqueous systems. Melting studies performed in high salt (16), with spermine (17) and with cationic polypeptides (18) all demonstrate sharper melts and preferential A-T stabilization, which has not been accounted for by preferential binding. Preferential G-C destabilization added algebraically to the overall ionic stabilization is an equivalent description of the phenomenon, which does not require highly selective binding to A-T or G-C. Further support of this is seen in the decrease in G-C dependence of the melting temperature (T_m) at acid pH (29) and by direct pmr observation in a double helical hexanucleotide at high salt concentrations (30).

3. Early melting of G-C regions should be demonstrable under proper conditions with certain cations. There are some spectacular examples of this, again, with no exceptions of preferential A-T melting in aqueous systems. G-C melting with simultaneous A-T stabilization has been observed with cupric ion (19), which has been shown to bind to G(N-7)(20). Tetraalkylammonium ions can reduce the melting temperature of G-C below that of A-T (21). Uncharged analogs have no similar effect and the required interaction occurs in the large groove, as expected (22). Within the context of this model it is quite significant that tetraalkylammonium ions do stimulate *in vitro* transcription (23).

   Pre-melt regions are observed in DNA-polylysine mixtures, which remain somewhat enigmatic due to their very low melting temperature (< 30°)(24). The base composition of these regions is not known. A particularly relevant, but unexplained, observation was made by Lees and von Hippel in 1968 (25), who showed that the addition of polylysine to DNA resulted in the loss of a discrete number of slowly exchanging helical hydrogens into a very fast exchanging class. The relation between the size of this lost kinetic class and overall G-C composition is not known, but this type of study actually monitors the particular conformational change intrinsic to the model at temperatures far below T_m, e.g. 0°C. The extension of this prediction to proteins is exemplified by the fact that the highly cationic pancreatic ribonuclease has been shown to lower the melting temperature of regions of DNA to which it binds (26).

4. Cationic agents should be associated with the initiation of transcription and replication and DNA-specific protein associated with transcriptional control at melting sites should be cationic. The catabolite activating protein postulated by Dickerson, et al (14) to be necessary for "melting in" of transcriptase has an isoelectric point well above neutrality (15). The core enzyme associated with initiation in RNA polymerase is cationic (27). With respect to replication, spermidine is a requirement for the synthesis of ØX-174 (28) and M-13 DNA.

In summary, a model for G-C destabilization has been derived from nucleotide hydrogen exchange studies that is consistent with thermodynamic considerations of H-bond strength and DNA stability. This model contains the point of view that DNA stability may depend on the weakness of the interbase oxygen-to-amino H-bonds: If the nucleobase aminos were strong H-bond donors a stable double helix could not be formed, due to the competition of water for the donor sites. Sequence specific interactions of protein cationic sidechains, or non-specific interactions of small cations, both with the G(N-7) site accessible to solvent in the large helical groove, would decrease the interbase proton lifetime and allow a drastic increase in solvent (water) interaction of G-C-rich sequences exclusively. A central feature of the model's extension, the requirement for simultaneous interaction of several G(N-7) sites with cationic sidechains of a protein, is consistent with observations on the in vitro melting of DNA and suggests a renewal of certain of these studies with this interpretative possibility. One unrealized test of this hypothetical model is quite simply the demonstration of increased G-C interbase proton exchange in response to the binding of a cation, polycation or transcriptional control protein to a helical oligonucleotide in the absence of overt melting. It is quite possible that this observation has been made (see above). A control protein of this type would be specific for a G-C-rich sequence, would probably be cationic and would be termed a "nucleation protein" as opposed to a "melting protein." While the latter propagates strand separation by cooperative binding to single stranded DNA, the "nucleation protein" would initiate strand separation at a strategic sequence in the control region. It is likely that both functions could reside in a single protein so that the nucleation reaction would be obscured in melting experiments by the much larger propagation reaction demonstrated for many melting proteins, including cationic agents such as spermine and ribonuclease (33). Finally, there is an advantage provided the organism by the high G-C stability of a melting site. The G-C sequence will remain closed until a specific interaction at the G(N-7) sites in the large groove is made. Thus, G-C sequences have the property of a switch, either tightly closed to prevent unwanted initiation of transcription, or open in response to protein binding at a judicious time. Thermally unstable A-T regions, not to be trusted as initiation sites alone, are flanked by G-C regions (14). Several mutations may be required to render a G-C region sufficiently unstable to allow unwanted transcription.

Acknowledgment: This research was supported by National Science Foundation Grants BMS-74-24311 and PCM-77-20154. Part of the pmr results were obtained from the Bruker HXS 360 NMR Spectrometer of the Stanford Magnetic Resonance Laboratory, a national facility funded by National Science Foundation Grant GR23633 and National Institutes of Health Grant RR00711.

References

1. Zillig, W., Zechel, K., Rabussay, D., Schachner, M., Sethi, V.S., Palm, P., Heil, A. and Seifert, W. (1970). Cold Spr. Hbr. Symp. on Quant. Biol. 35, 47.

2. Alberts, B. and Frey, L. (1970). Nature (London) 227, 1313.

3. von Hippel, P.H. and McGhee, J.D. (1972). Ann. Rev. Biochem. 41, 231.

4. McConnell, B. and Seawell, P.C. (1972). Biochemistry 11, 4832.

5. McConnell, B. (1978). Biochemistry, submitted April 1977. Preprints available upon request.

6. McConnell, B. (1974). Biochemistry 13, 4516.

7. Crooks, J.E. (1975) in Proton Transfer Reactions, E.F. Caldin and V. Gold, eds. Halsted Press, John Wiley and Sons, New York, p. 153.

8. Teitlebaum, H. and Englander, S.W. (1975). J. Mol. Biol. 92, 55 and 79.

9. McConnell, B. and von Hippel, P.H. (1970). J. Mol. Biol. 50, 297 and 317 (see references therein).

10. Jencks, W.P. (1969). Catalysis in Chemistry and Enzymology. McGraw-Hill Inc., New York, p. 415.

11. Bloomfield, V.H., Crothers, D.M. and Tinoco, I. Jr. (1974). The Physical Chemistry of Nucleic Acids. Harper and Row, New York, pp. 349-351.

12. Hoo, D. and McConnell, B. (1978). In preparation.

13. Szybalski, W., Kubinski, H. and Sheldrick, P. (1966). Cold Spr. Hbr. Symp. on Quant. Biol. 31, 123.

14. Dickson, R.C., Abelson, J., Barnes, W.M. and Reznikoff, W.S. (1975). Science 187, 27.

15. Anderson, W.B., Schneider, A.B., Emmer, M., Perlman, R.L. and Pastan, I. (1971). J. Biol. Chem. 246, 5929.

16. Gruenwedel, D.W., Hsu, C.-H. and Lu, D.S. (1971). Biopolymers 10, 47.

17. Mandel, M. (1962). J. Mol. Biol. 5, 435.

18. Olins, D.E., Olins, A.L. and von Hippel, P.H. (1967). J. Mol. Biol. 24, 157.

19. Hiai, S. (1965). J. Mol. Biol. 11, 672.

20. Eichorn, G. (1962). Nature 194, 474.

21. Melchior, W.B. Jr. and von Hippel, P.H. (1973). Proc. Nat. Acad. Sci. 70, 298.

22. Strauss, V.P., Helfgott, C. and Puck, H. (1967). J. Phys. Chem. 71, 2250.

23. Malcolm, A.D.B., Mitchell, G.J. and Wasylyk, B. (1974). Biochem. Soc. T. 2, 863.

24. Ong, E.C. and Fasman, G.D. (1976). Biochemistry 15, 477.

25. Lees, C.W. and von Hippel, P.H. (1968). Biochemistry 7, 2480.

26. Felsenfeld, G., Sandeen, G. and von Hippel, P.H. (1963). Proc. Nat. Acad. Sci. U.S.A. 50, 644.

27. Richardson, J.P. (1969). Prog. Nuc. Acid Res. and Mol. Biol. 9, 75.

28. Wickner, W. and Kornberg, A. (1974). J. Biol. Chem. 249, 6244.

29. Woodbury, C.P. and Record, M.T. (1975). Biophys. J. 15, A-92.

30. Borer, P.N., Kan, L.S. and Ts'o, P.O.P. (1975). Biochemistry 14, 4847.

31. Eigen, M. (1964). Angew. Chemie. Int. Ed. 1, 1.

32. Allen, L.C. (1975). J. Am. Chem. Soc. 97, 6921.

33. Frank-Kamenetskii, M.D. and Lazurkin, Yu. S. (1974). Ann. Rev. Biophys. and Bioengr. 3, 127.

DISCUSSION

R.L. Somorjan: Why would you expect that the greater relative stability of G-C pairs with respect to temperature variations is necessarily paralled in the relative G-C vs A-T stabilities with respect to changes in other external variables (e.g. ionic strength or pH) ?

McConnell: I have no preconceived expectations, but wish to point out that the well known high stability character of G-C would tend to obscure consideration of G-C as a possible melting site by workers attempting to interpret stability data. I would cite as an example the statement of "preferential A-T stabilization" quoted by many workers to account for the effects of cations and polycations on DNA melts.

D.W. Jones: You have referred to the importance in solution of the effect of protonation in another part of the molecule on hydrogen-bonding ability at the NH_2 group. Useful support for this approach would be provided if one could cite examples from the solid state of pairs of relevant small molecular entities differing in this way which show a change in hydrogen-bonding ability as demonstrated by infrared spectroscopy or, preferably, X-ray or neutron diffraction. Are you aware of any such examples ?

McConnell: No, but I know of an example where the cytosine amino protons are reluctant to form H-bonds in the crystalline state. Only one of the two NH_2 protons of 1', 3'-eCmP appears to the H-bonded in the crystal structure of Coulter (1973) J. Am. Chem. Soc. 95, 570.

ON THE CONFORMATIONAL DEPENDENCE OF THE PROTON CHEMICAL SHIFTS IN NUCLEOSIDES AND NUCLEOTIDES.IV. PROTON CHEMICAL SHIFTS IN 3'-NUCLEOTIDES AS A FUNCTION OF DIFFERENT CONFORMATIONAL PARAMETERS.

CLAUDE GIESSNER-PRETTRE and BERNARD PULLMAN
Institut de Biologie Physico-Chimique, Laboratoire de Biochimie Théorique, associé au C.N.R.S., 13, rue P. et M. Curie, Paris 75005, France.

INTRODUCTION

The knowledge of the conformational dependence of the chemical shifts of protons of the nucleic acid constituents is of increasing interest for the development of the study of the conformation of polynucleotides in solution [1]-[6] (and references therein) by nuclear magnetic resonance spectroscopy since a number of experimental evidences obtained principally from spin-spin coupling constants show that conformational changes of the constituting units of the polymer occur upon polymerization [1][7]-[18]. The experimental values of the polymerization shifts are the sum of the chemical shift variation due to the conformational change undergone by the nucleotidyl unit carrying the studied nucleus and of the contributions of all the other residues of the polynucleotide chain (and eventually of its complementary strand) to the chemical shift of that nucleus. In order to determine from measured polymerization shifts the conformation of the polymer one needs to determine as precisely as possible the chemical shift variation which is due solely to the other nucleotides of the chain. This quantity can be obtained from the measured variation if the conformational dependence of the chemical shifts is known and if the conformational difference between the isolated monomer and the nucleotide inside the chain can be estimated from the variation of the coupling constants.

In solution, nucleotides (and polynucleotides) exist as an equilibrium between several conformers and it is possible to modify this equilibrium by changing the physico-chemical properties of the solution [14][19]. The variation of the chemical shifts which are measured are then function of the variations of all the different conformational parameters of the nucleotide and it was not possible up to now to determine from experiment the role of a given conformational parameter in the value of the chemical shift of a given nucleus. The theoretical determination of the conformational dependence of chemical shifts is for the moment, to our knowledge, the only tool which can give information on the chemical shift of each individual proton as a function of each individual conformational parameter of the molecule. We have

B. Pullman (ed.), Nuclear Magnetic Resonance Spectroscopy in Molecular Biology, 161-181.

therefore undertaken a systematic study of the variation of the chemical shifts of the protons of nucleosides and nucleotides as a function of the rotational angles which determine the conformations of these molecules. In the present study we are extending to 3'-nucleotides the calculations that we have previously carried out for nucleosides [20] [21] and 5'-nucleotides [6]. The results that we shall report and discuss here may be significant for the interpretation of the polymerization shifts of the protons of the nucleotidyl unit which is located at the 3' end of a polymer.

COMPUTATIONAL DETAILS

The method of computation is identical to that utilized for the nucleosides and 5'-nucleotides and is fully described in the first paper of this series [20]. The chemical shifts are calculated as the sum of three contributions : the ring current effect, the local magnetic anisotropy term and the polarization or electric field contribution. For a given conformation taken as reference we suppose all the chemical shifts equal to zero and we report only the variation ($\Delta\delta$) of these quantitities as a function of the torsional angle studied, positive variations corresponding to upfield shifts.

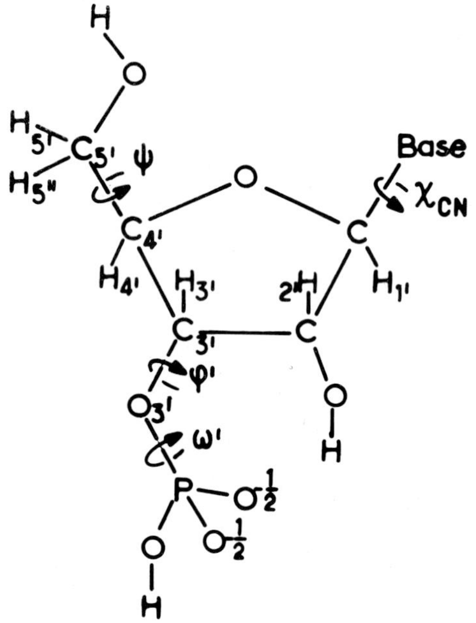

Figure 1.- Numbering of the hydrogens and designation of the rotation angles.

We have varied four angles, χ_{CN}, Ψ , φ' and ω' which are shown on figure 1. These angles are defined according to Sundaralingam's notations [22] . These notations which in every case refer to heavy atoms for the definition of the zero value of the rotation angles are, for angle φ' different from the one used by some NMR experimentalists [1] [7][8][13][28] . These authors refer to values 180° and 300° of φ' as g^- and g^+ while in the notation utilized here these conformations are called t and g^- respectively. Similarly the conformation with φ' = 60° which is called t in these PMR works corresponds to conformation g^+ in the present notation.

The conformation taken as reference has the following characteristics : χ_{CN} = 0°, Ψ = 60°, φ'= 210° and ω' = 290° which correspond to what seems from X-ray [23] -[27] and nuclear magnetic resonance studies [1][7][12][14][15][28] a set of values actually encountered in 3'-nucleotides.

The nucleotides studied are adenosine 3'-phosphate and uridine 3'-phosphate. We have examined only one purine and one pyrimidine for reasons indicated previously [6] . For each of these nucleotides we have considered the C2'-endo and the C3'-endo puckerings of the furanose ring. For a given puckering we have taken the same geometry of the ribose phosphate part of the molecule for 3'-UMP and 3'-AMP. Similarly the geometry of the base is kept constant when the puckering of the furanose ring is changed. The geometry of the ribose and of uracil comes from the study on ApU of Shefter and al. [29] , while the geometry of adenine is taken from the work of Lai and March [30] .

RESULTS

Rotation about the glycosyl bond (χ_{CN}).

The calculated chemical shifts variations for the protons of the bases as a function of χ_{CN}, reported in figure 2, show for both nucleotides that the curves are very different for the two puckerings of the ribose. The shifts of H6 for 3'-UMP and of H8 for 3'-AMP undergo variations which are similar to but smaller in magnitude than those obtained for 5'-nucleotides if the ribose is C3'-endo. For the C2'-endo nucleotides we find that these two protons are shifted upfield for a narrow range of values of χ_{CN} between 120 and 180°. Another clear-cut difference between the two puckerings concerns H2 of 3'-AMP which is strongly shifted downfield for syn conformations if the furanose ring is C3'-endo while its position is almost constant for C2'-endo ribose.

Figure 2 shows also that H5 and N3H of 3'-UMP have chemical shifts which are not sensitive to the value of the rotation angle about the glycosyl bond, in distinction to 5'-nucleotides [6] .

As in the case of 5'-nucleotides the calculated $\Delta\delta$'s for the protons of the ribose as a function of χ_{CN} are similar to those obtained for these protons in nucleosides (although numerically slightly different), so we shall not report them here.

Concerning the rotation about the glycosyl bond the main new feature that we have found for 3'-nucleotides concerns the $\Delta\delta$ of the pro-

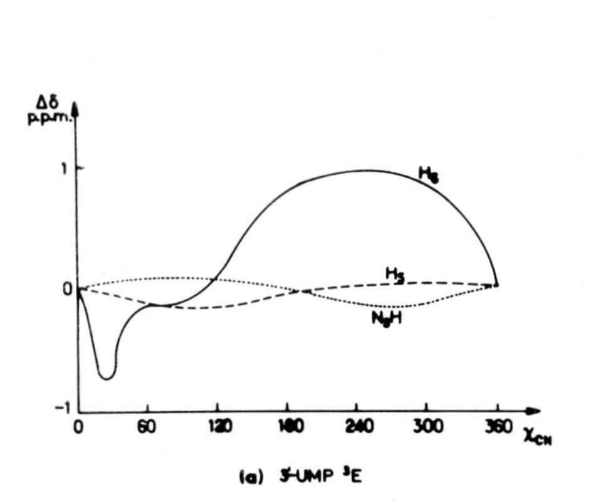

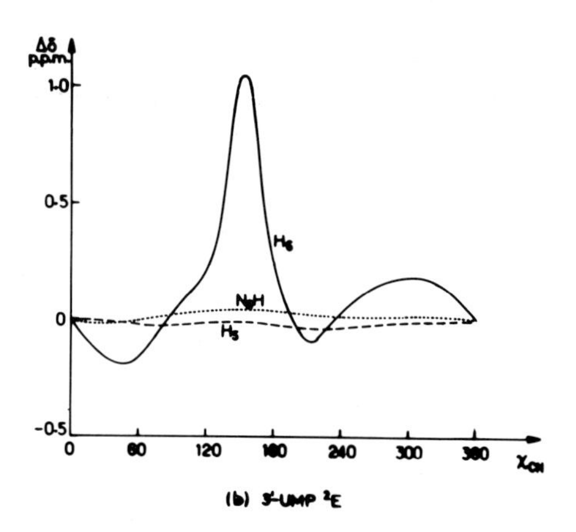

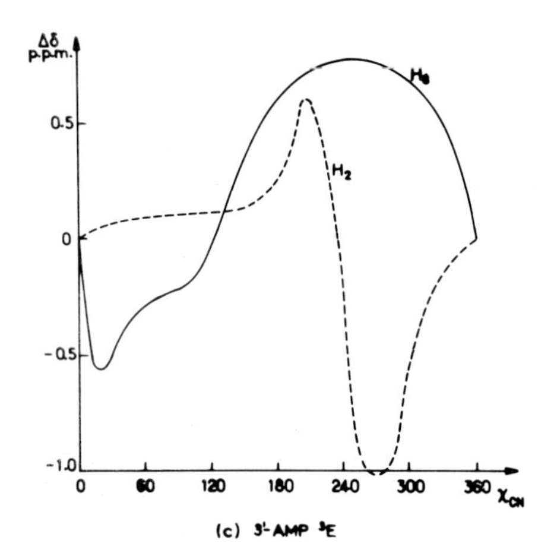

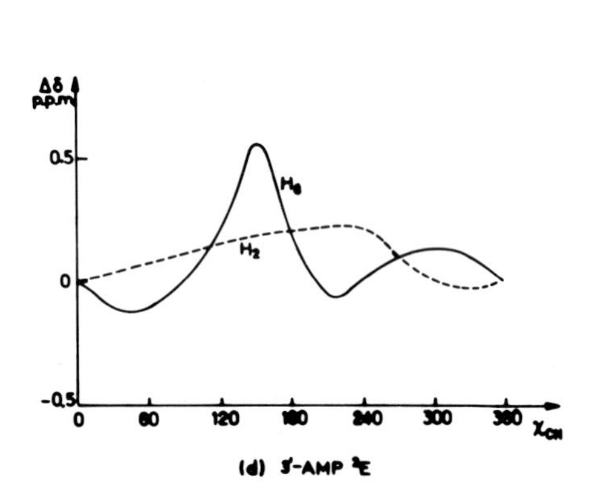

Figure 2.- Variation of the chemical shifts of the protons of the base as a function of the rotation angle about the glycosyl bond (χ_{CN}).

tons of the base in the C2'-endo conformers. In the 3'-nucleotides with a C3'-endo ribose our calculations indicate that going from an anti to a syn conformation will produce an upfield shift for the anomeric proton and for H6 of pyrimidines or H8 of purines while in those with a C2'-endo ribose the shifts of these protons of the bases have for the same conformational change a smaller chance to have a variation similar to the one of H1', because of the narrow range of values for which we calculate a large upfield $\Delta\delta$ (Figure 2b,2d) for these nuclei.

Since proton magnetic resonance spectroscopy has shown that for 3'-nucleotides in solution the C2'-endo $\rightleftharpoons$ C3'-endo equilibrium [1][7] [8][28][31][32] is often in favor of the C2'-endo puckering our results imply that a modification of the syn$\rightleftharpoons$anti equilibrium in these compounds will produce for H6 of 3'-UMP and H8 of 3'-AMP values of $\Delta\delta$ which are different from those produced by the same phenomenon in 5'-nucleotides.

If there are a number of data [28][31][33] which indicate that in solution 3'-nucleotides (principally of the purines) are to some extend syn, the X-ray experimental data show that in the crystal state they are anti in all cases studied (34 and references herein); in addition NMR data on polynucleotides in solution are in favor of an anti conformation for the nucleotide located at the 3' end of the chain [1][8][7] [13][14]. This set of results has led us to chose an anti arrangement about the glycosyl bond (χ_{CN} = 30°) for the variation of the other rotational angles that we have varied since such conformations seem on the average the most populated ones even for nucleotides having the phosphate group at the 3' position.

Rotation about the $C_{4'}$-$C_{5'}$ bond (Ψ).

The comparison of the curves on Figure 3 with the analogous ones for 5' nucleotides [6] shows that for the protons of the bases the calculated $\Delta\delta$'s as a function of Ψ are qualitatively the same for both types of nucleotide. The differences are only numerical, the variations being smaller for 3'-nucleotides than for the 5' ones but in both cases the gg conformation (Ψ = 60°) corresponds to the minimum of the curves for H6 in pyrimidines and H8 in purines. Thus for any type of nucleotides a decrease in the population of the gg conformation about the C4'-C5' bond should produce an upfield shift of either H6 or H8, according to the type of the base. The chemical shift of the other protons of the bases namely H5, N3H or H2 are not influenced by the value of Ψ as it was the case for 5'-nucleotides. We report on Figure 3 only the data for the C3'-endo ribose since the results are the same for the C2'-endo puckering with the sole difference that the calculated variations have absolute values about half of those for the C3'-endo puckering.

For the protons of the ribose, Figure 4 shows that the largest chemical shift variations are calculated for H5' and H5" (for these protons we have followed here as in the case of the 5'-nucleotides the assignment made by Remin and Shugar [35]) and that the $\Delta\delta$'s for these protons are much larger if the ribose is C3'-endo than if it is C2'-

endo. For these two protons the $\Delta\delta$ varies in opposite directions as a function of Ψ, the difference between the two curves being at its maximum for the gt conformation (Ψ=180°). This result has, as a consequence, the fact that $\Delta_{H5'H5''}$ should be sensitive to the conformation about the $C_{4'}$- $C_{5'}$ bond, especially if the 3'-nucleotide concerned is C3'-endo. The chemical shifts of the other protons of the ribose appear to have smaller variations as the value of Ψ changes. Nearly all the curves of Fig. 4 are qualitatively very different from the homologous ones for 5'-nucleotides [6]. In these previous calculations the largest $\Delta\delta$'s were calculated for H3' in all cases and for H2' if the ribose was C2'-endo. The difference between the results obtained for the two types of nucleotide is due to the fact that as $\Psi_{C_{4'}-C_{5'}}$ varies the location of the 5'-phosphate group remains constant with respect to the H5' and H5" protons but varies with respect to the other protons of the ribose in 5'-nucleotides while in the 3' nucleotides it is the location of the hydrogens H5' and H5" with respect to the 3'-phosphate group which varies with this angle. Since a detailed analysis of the calculated values shows that for both types of nucleotide the largest contribution to the chemical shift variations is due to the electric field effect of the charged phosphate, the qualitative difference found between the two types of nucleotides with the variation of Ψ is easily understood.

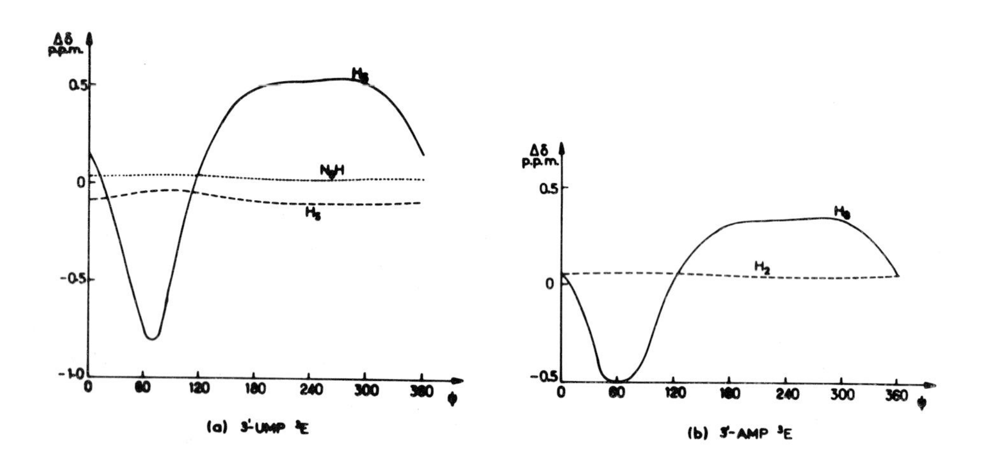

Figure 3.- Variation of the chemical shifts of the protons of the base as a function of the rotation angle about the $C_{4'}$- $C_{5'}$ bond (Ψ).

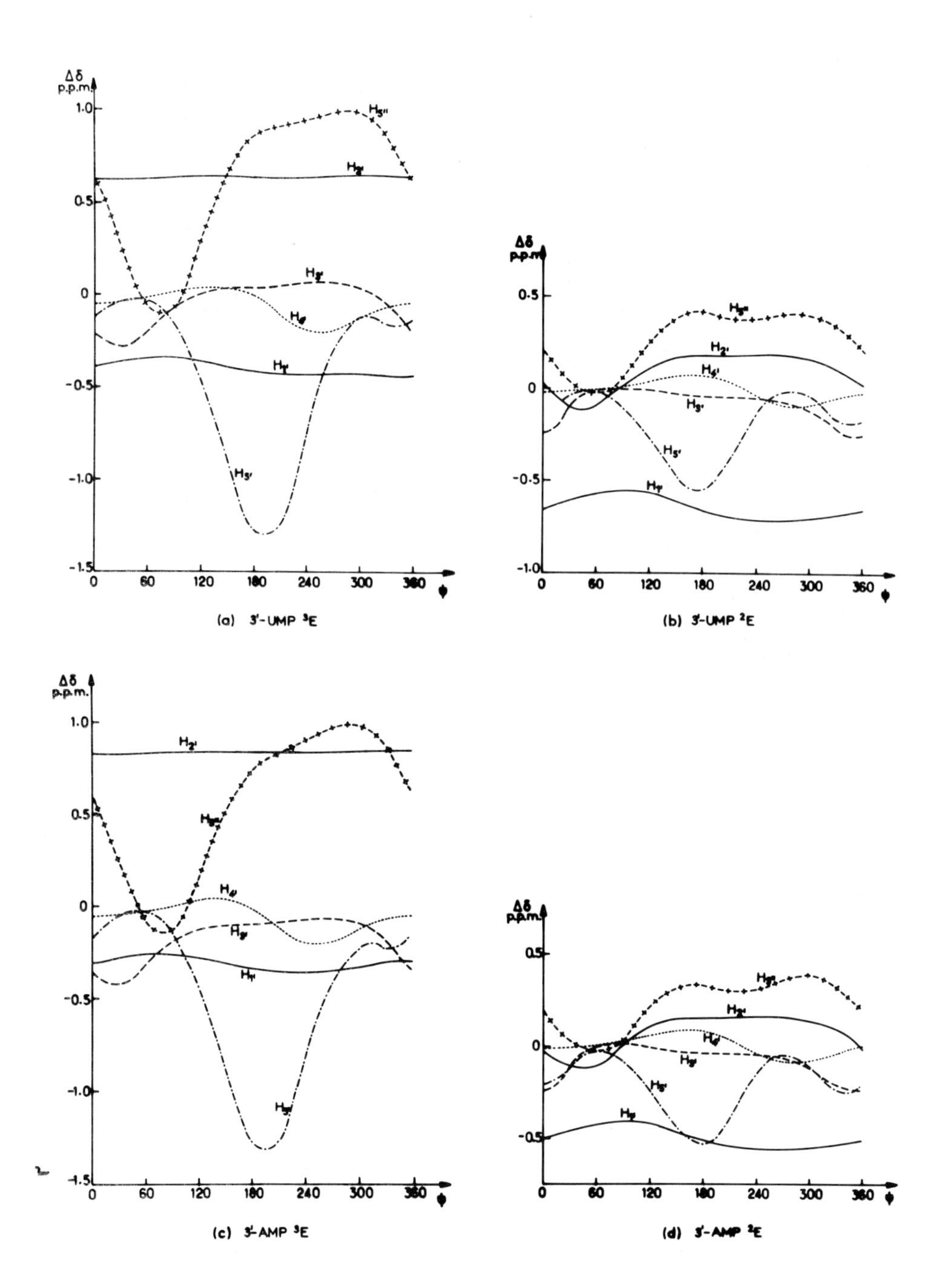

Figure 4.- Variation of the chemical shift of the non-exchangeable protons of the ribose as a function of the rotation angle about the $C_{4'}$- $C_{5'}$ bond (Ψ).

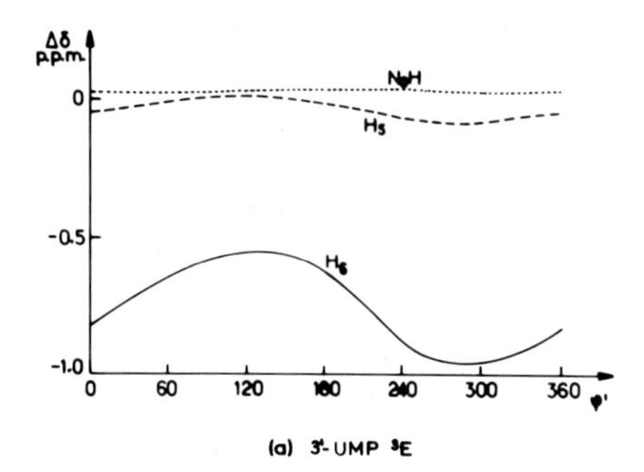

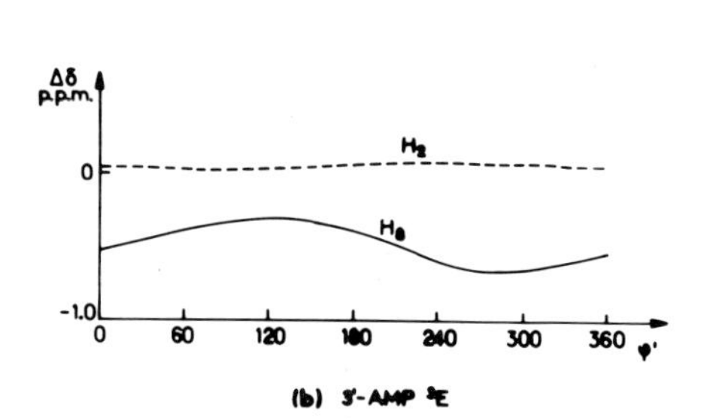

Figure 5.- Variation of the chemical shifts of the protons of the base as a function of the rotation angle about the $C_{3'}- O_{3'}$ bond (φ').

Rotation about the $C_{3'}- O_{3'}$ bond (φ').
The variation of the chemical shifts of the protons of the base are reported on Figure 5 for the C3'-endo nucleotides; the calculated $\Delta\delta$'s for the C2'-endo puckering are negligeable. Our results show that H6 of 3'-UMP and H8 of 3'-AMP undergo similar variations as a function of φ' and that the conformation g^- (φ'= 270-300°) corresponds to the lowest shift for these protons. These calculated $\Delta\delta$'s are larger for the pyrimidine than for the purine and are negligeable for the other protons of the base.

The curves of Figure 6 exhibit for most of the protons of the ribose large $\Delta\delta$ as φ' varies. They are generally quite different for the two puckerings of the furanose ring. For both puckerings H4' is strongly shifted downfield for the g^- conformation. In the C2'-endo nucleotides we calculate a very large downfield shift for H1' when $\varphi' \simeq 60°$, a conformation which has not been observed in crystal and is not predicted by theory([34] and references therein) but which cannot be eliminated from the analysis of the sole NMR data [15][28][36], while this proton does not undergo any significant $\Delta\delta$ if the ribose is C3' endo. For the C3'-endo puckering our results indicate that the shift of H5" varies like the shift of H4' but the minima and maxima of $\Delta\delta$ occur for values of φ' differing by 60° for the two protons. The calculated $\Delta\delta$ for H3' are appreciably larger if the ribose if C3'-endo than if it is C2'-endo. For this proton the shift is towards high field for the g^+ ($\varphi' \simeq 60°$) conformation in the four cases examined.

Since most of the NMR data seem to indicate that there is a rapid rotation about bond $C_{3'}- O_{3'}$ in 3'-nucleotides [28][36][37][16], any modification of the equilibrium between the different interconverting

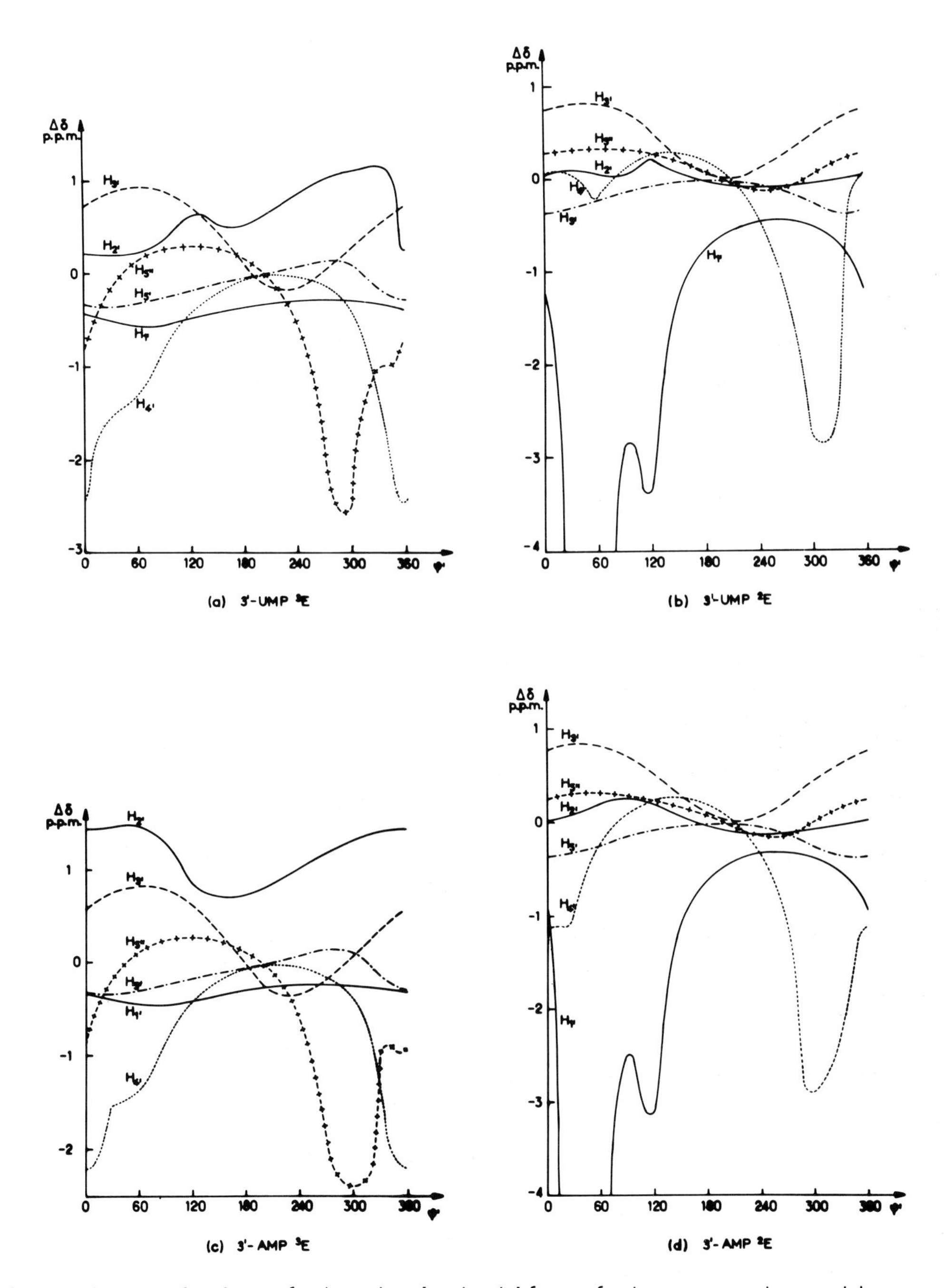

Figure 6.- Variation of the chemical shifts of the non exchangeable protons of the ribose as a function of the rotation angle about the $C_{3'}-O_{3'}$ bond (φ').

rotamers, as it seems to occur upon polymerization [1][7][8][11][16], would produce according to Figure 6 non negligeable $\Delta\delta$'s for H4', H5", H1' and to a smaller extend H3'.

If we compare Figures 5 and 6 with Figures 6 and 7 of the study on 5'-nucleotides [6] we see that the calculated $\Delta\delta$'s as a function of φ' are very different from those calculated as a function of φ. The differences are due to the fact that the variation of the location of the phosphate group with respect to the different protons of the molecule is very different in the 3' and 5' nucleotides. In 3' nucleotides we do not calculate the extremely large $\Delta\delta$ that we have found for some values of φ in 5'-nucleotides for some protons if χ and Ψ have classical values for a gg-anti conformation.

Rotation about the $O_{3'}$- P bond (ω').

When the angle about the $O_{3'}$- P is varied the calculated chemical shift variations for the protons of the base are completely negligeable since the maximum $\Delta\delta$ is of the order of 0.05 p.p.m. between completely locked conformations.

From Figure 7, which shows the variations of the chemical shifts of the protons of the ribose, we infer that H3' is the only proton which undergoes some measurable $\Delta\delta$ as a function of ω'. The shift of H2' is also slightly sensitive to the conformational arrangement about the $O_{3'}$- P bond if the ribose has the C3'-endo puckering.

The experimental value of ω' in solution cannot be obtained from NMR data but Cheng and Sarma [1][17] have interpreted some of their measurements on the variation of the shift of H3' upon temperature or polymerization by a variation of the value of ω'. They assume that in 3'-nucleotides the PO_4^- group rotates freely about the $O_{3'}$- P bond and that the restriction of this rotation brought about by polymerization produces an upfield shift on H3', since an increase of temperature which can increase the free rotation about the $O_{3'}$- P bond in dinucleoside monophosphates does produce a downfield shift for this proton. We cannot say that the calculated $\Delta\delta$ for H3' fully support this interpretation since, if our results are significant, they predict for this proton an upfield $\Delta\delta$, between a conformation with a freely rotating PO_4^- and one with a definite value of ω', which has little chance to be of some importance. We shall come back to this point later in the discussion.

Since there are a number of evidences from $J_{C2'P}$ and $J_{H2'P}$ coupling constants that in 3' nucleotides the conformation with $\varphi' \simeq 270°$ is largely populated especially if the ribose is C2'-endo [1][8][15][28], we have calculated the $\Delta\delta$ as a function of ω' for this value of φ' for 3'-AMP. The results reported on Figure 8 show that in this case the largest $\Delta\delta$'s are calculated for H5" if the ribose if C3'-endo and for H4' if it is C2'-endo. For these protons the calculated curves are qualitatively similar to those obtained for H3' but the numerical $\Delta\delta$'s are larger.

After this survey of the calculated $\Delta\delta$'s for the different conformational parameters that we have examined we see that in 3'-nucleotides the protons of the bases are less sensitive to the conformation of the

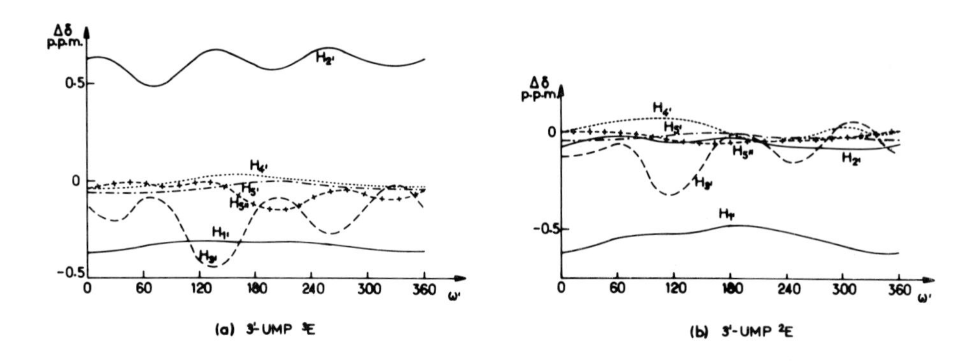

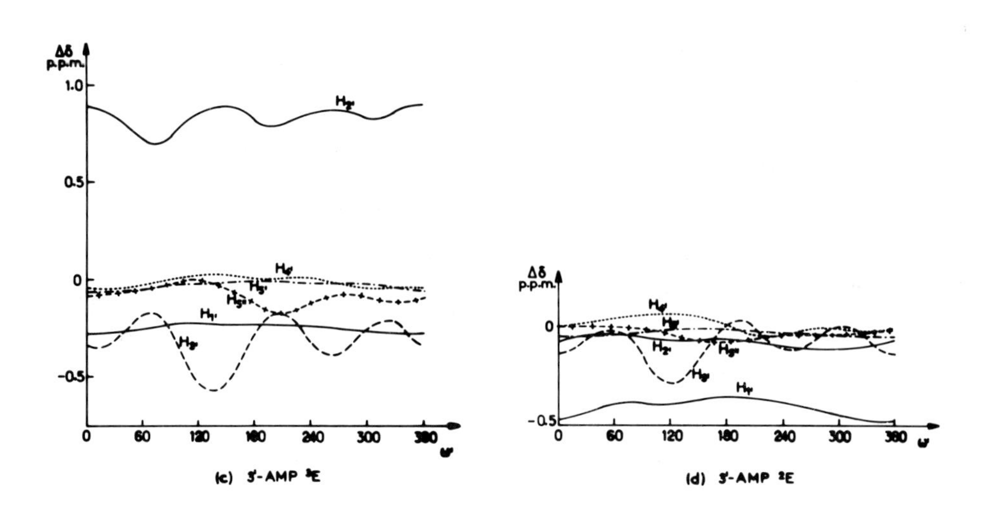

Figure 7.- Variation of the chemical shift of the non exchangeable protons of the ribose as a function of the rotation angle about the $O_{3'}$- P bond (ω') for $\varphi' = 210°$.

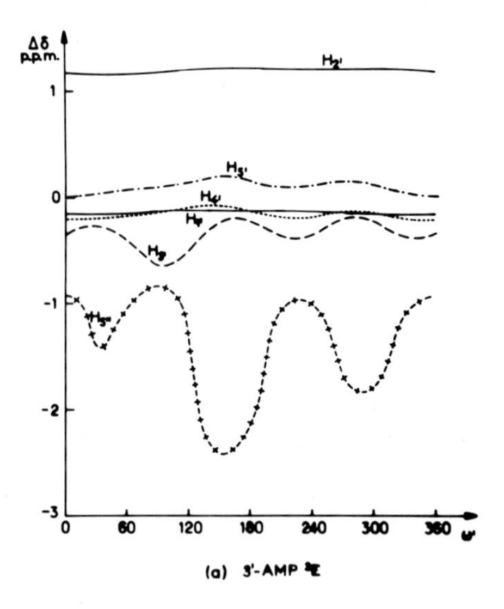

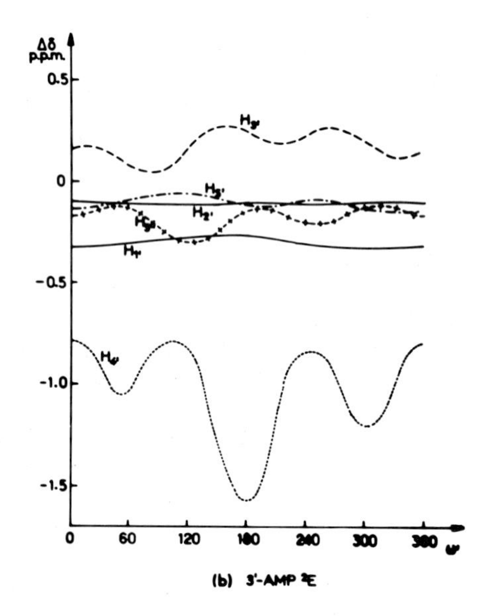

Figure 8. - Variation of the chemical shift of the non exchangeable protons of the ribose as a function of the rotation angle about the $O_{3'}$-P bond (ω') for φ' = 270°.

molecules than in 5'-nucleotides. For the protons of the ribose the situation is different according to the proton studied; in 3'-nucleotides H1', H4' and H5" undergo the largest $\Delta\delta$ as a function of the conformation of the exocyclic phosphate, while in 5'-nucleotides it is for H3' and H2' that we calculate important chemical shift variations.

We must underline that in both types of nucleotide the main contribution to the calculated $\Delta\delta$'s comes from the polarization effect of the charged phosphate group, with the exception of the rotation about the glycosyl bond.

DISCUSSION

Prior to the discussion we want to recall the limitations concerning the numerical values of the chemical shifts variations that we calculate by the semiempirical method utilized in this series of studies [6] [20][21]. This type of computations can give valuable informations on the sign and the relative importance of the $\Delta\delta$'s calculated for different protons as a function of a conformational parameter and the relative importance of the $\Delta\delta$'s calculated for one proton as a function of the different conformational parameters. On the other hand, the numerical values of the very large $\Delta\delta$'s that we calculate for some protons for some conformations cannot and should not be trusted.

The validity of our results is not tested easily since 3'-nucleotides exist in solution as a blend of several conformations but the measured variations of the proton shifts with the temperature can give an indication on the trends followed by the chemical shifts with the variation of the conformational equilibrium which is estimated from the measured values of coupling constants.

The comparison of the values reported in Table I with the curves of Figure 2 shows that our calculations do not predict for any conformational rearrangement a detectable downfield shift for H5 of pyrimidines as is measured by Lee and Tinoco [14]. But if we compare the upfield shift measured for H6 in pyrimidines and H8 in purines with our results we are led to interpret the experimental data by an increase of the average value of χ_{CN} and a decrease of the population of the gg conformation about the $C_{4'}-C_{5'}$ bond. The second part of the above statement is supported by the variations of the H4'H5' and H4'H5" coupling constants [14]. According to Figure 4 we see that a decrease in the percentage of the gg conformer is accompanied by an upfield shift of H5" and by a variation of the chemical shift difference between H5' and H5" ($\Delta_{H5'H5''}$) in agreement with experimental data (Table I). The supposed increase of the average value of χ_{CN} with the temperature is supported by the measured upfield $\Delta\delta$ for H1' and the downfield $\Delta\delta$ for H2 in 3'-AMP.

The NMR data on 3'-nucleotides tend to show that the phosphate group is rapidly reorienting itself about the $C_{3'}-O_{3'}$ bond but the equilibria deduced from experiment are quite different whether it is assumed that this phenomenon occurs between two or three conformations [15][28][36]. If we assume that only the t ($\varphi' \simeq 200°$) and g^- ($\varphi' \simeq 270°$)

TABLE I
Experimental chemical shift variations (in p.p.m.) in 3'-nucleotides due to an increase of temperature.

	3'-UMP		3'-CMP	3'AMP		3'-GMP
	(14)	(32)	(14)	(14)	(2)	(14)
H1'	0.045	0.021	0.016	0.024	0.023	0.030
H2'	0.010	-0.011	0.003	0.025		0.032
H3'	-0.018	-0.045	-0.032	-0.016		-0.018
H4'	0.018	-0.001	0.002	0.017		0.040
H5'	0.010	-0.007	0.009	0.006		-0.003
H5"	0.015	-0.003	0.011	0.016		0.016
H5 or H2	-0.015	-0.007	-0.026	-0.028	-0.030	
H6 or H8	0.121	0.064	0.120	0.019	0.022	0.041
Δgg^a	16	14	11	7		15
$\Delta\Delta$H5'H5"[b]	0.002	0.004	0.002	0.010		0.019

a) Δgg = % gg (low temperature) - % gg (high temperature)
b) $\Delta\Delta$H5'H53 = H5'H5" (high temperature) - H5'H5" (low temperature) with H5'H5" defined as in (7).

conformations are present in solution, the $\Delta\delta$'s that we calculate imply that the largest variations should be measured for H4' and H5" and that they are in the same direction. Therefore a shift from a t to a g^- conformation with increasing temperature, as deduced from vicinal PC [15] and long range PH [14] coupling constants in dinucleoside monophosphates, should, according to our results, most probably produces a downfield shift for H4' and H5". But if we assume that the three rotamers about the $C_{3'}-O_{3'}$ bond are present, then, according to our calculations, a decrease of the population of the g^+ rotamer ($\varphi' \simeq 60°$) would produce an upfield shift for the H1' proton and a downfield one for H3' and H5" of the C2'-endo molecules in agreement with the measured quantity for H1' and H3'. The downfield shift of H5" produced by this conformatinal change should be substracted from the upfield shift due to the decrease of the population of the gg conformer about the $C_{4'}-C_{5'}$ bond. The present hypothesis seems to receive some support in the case of dinucleoside monophosphates from the values measured by Lee and Tinoco [14] for $J_{H3'P}$ and $J_{H2'P}$ as a function of temperature. These authors have observed that the value of $J_{H3'P}$ decreases with increasing temperature, in agreement with a diminution of thepopulation of the trans arrangement of H3'C3'O3'P, and that the value of $J_{H2'P}$ increases in the same conditions which is in favor of an indrease of the population of the g^- conformer. Unfortunately the comparable data are not available for the 3'-mononucleotides with the exception of 3'-UMP for which Schleich et al. [32] have measured a small decrease of $J_{H3'P}$ when the temperature is varied from 23 to 88°C.

For φ' = 210°, a largely populated conformation, our results show that the three classical rotamers about the $O_{3'}$- P bond correspond to three equivalent maxima for the $\Delta\delta$ of H3' and to three equivalent minima for the $\Delta\delta$ of H2', so that a modification of the population of the different rotamers should not produce any measurable variations of the shifts of these protons. Thus if the interpretation of Cheng and Sarma [1][17] of the downfield shift observed for H3' with temperature and the calculated curves of Figure 7 are both qualitatively correct we must suppose that the conformational rearrangement taking place around the $O_{3'}$- P bond involves values of ω' such as 120-150° for example, quite different from the generaly encountered staggered arrangements. For the g^- conformation ($\varphi' \approx 270°$) about $C_{3'}$- $O_{3'}$ which seems to accompany the C2'-endo puckering of the ribose [28][17][1][13], it does not appear from our results that any systematic change in the value of ω' could produce a negative $\Delta\delta$ for H3', but in this case a modification of the equilibrium about the $O_{3'}$- P bond between the classical rotamers would produce variations of the chemical shift of either H4' or H5" since the minima occuring on the curves for the gauche forms have a quite different depth from the one occuring for the trans conformation (Figure 8).

After this discussion on the downfield shift observed for H3' but not observed in 5'-nucleotides [14][6] with increase in temperature we must recall that a variation in the value of χ_{CN} can also produce a downfield $\Delta\delta$ for this proton in C3'-endo nucleotides [20][21]. The difference between the measured $\Delta\delta$ for H3' in 3' and 5'-nucleotides has then to be interpreted as signifying that the variation in the value of χ_{CN} occurs between different values of this angle in the two types of nucleotides the presence of the phosphate group at the 5' position impeding some values of χ_{CN} which are allowed for 3'-nucleotides.

Even if we have been able to provide interpretations for the observed trends of the shift variations of most protons we must recall that none of our results is able to interpret the measured downfield shift for H5 in 3'-UMP and 3'-CMP when the temperature is increased. In view of the failure of the theory this experimental data which has been observed by Lee and Tinoco [14] for 3'- but not for 5'-nucleotides can perhaps be attributed to some solvent effects or to a modification of the $^3E \rightleftharpoons {}^2E$ equilibrium. In this series of works we have calculated the $\Delta\delta$'s due to the modification of the value of rotation angles about exocyclic bonds for C2'-endo and C3'-endo ribose but we have no results for the $\Delta\delta$'s due to a modification of the puckering of the furanose ring, and there is no reason that the chemical shifts should be independent of this degree of freedom since many interatomic distances are different in C2'-endo and C3'-endo nucleosides even if χ, Ψ, φ and ω have equal values in both conformers. For the conformational parameters that we have studied, bond lengths and bond angles can be supposed to be constant at first approximation but this is no

TABLE II
Dimerization shifts $\delta_{(monomer)} - \delta_{(dimer)}$ in p.p.m. for ApA (for the

	Exp.						conf.1	
	(14)	(7)	(13)	A-RNA	B-DNA	a	b	c
Ap unit								
H1'	0.293	0.260	0.195	0.001	0.680	0.043	0.214	0.171
H2'	0.175	0.200	0.118	-0.135	-0.265	-0.065	-0.276	-0.587
H3'	0.121	0.124	0.077	-0.365	-0.178	-0.223	-0.027	0.146
H4'	0.108	0.140	0.114	-0.049	0.028	-0.015	-0.015	0.135
H5'	-0.025	0.058	0.064	-0.089	0.004	-0.012	-0.064	-0.123
H5''	0.008	0.067	0.095	-0.041	-0.019	-0.007	0.006	0.938
H2	0.355	0.295	0.236	1.269	1.373	0.438	0.418	0.418
H8	0.180	0.156	0.073	-0.049	0.027	0.061	0.133	0.297
pA unit						d	e	f
H1'	0.205	0.180	0.118	0.047	0.121	0.240	0.459	0.456
H2'	0.355	0.170	0.268	0.165	-0.125	0.304	0.393	0.417
H3'	0.072	0.	-0.015	0.107	-0.079	0.226	0.402	0.396
H4'	0.095	0.015	0.018	0.037	-0.046	-0.013	-0.009	-0.012
H5'	-0.264	-0.245	-0.255	-0.254	-0.348	-0.169	-0.189	0.120
H5''	0.010	-0.051	-0.068	0.103	-0.028	-0.024	-0.005	-0.143
H2	0.133	0.201	0.068	0.088	0.159	0.110	0.103	0.096
H8	0.277	0.250	0.195	0.840	-0.343	1.014	0.864	0.801

In conformation 1 to 4 and a to f the riboses have a C3'-endo pucker
Conformation parameters of the dimer

conf.1 $\chi_1=20°$, $\Psi_1=60°$, $\varphi'_1=210°$, $\omega'_1=290°$, $\chi_2=20°$, $\Psi_2=60°$, $\varphi_2=180°$, $\omega_2=$

conf.2 $\chi_1=40°$, $\Psi_1=60°$, $\varphi'_1=210°$, $\omega'_1=290°$, $\chi_2=20°$, $\Psi_2=60°$, $\varphi_2=180°$, $\omega_2=$

conf.3 $\chi_1=20°$, $\Psi_1=60°$, $\varphi'_1=210°$, $\omega'_1=290°$, $\chi_2=40°$, $\Psi_2=60°$, $\varphi_2=180°$, $\omega_2=$

conf.4 $\chi_1=20°$, $\Psi_1=60°$, $\varphi'_1=210°$, $\omega'_1=80°$, $\chi_2=20°$, $\Psi_2=60°$, $\varphi_2=180°$, $\omega_2=$

Conformational parameters of the monomer taken as reference.

conf.a $\chi= 20°$, $\Psi= 60°$, $\varphi'=210°$, $\omega'=290°$

conf.b $\chi= 40°$, $\Psi= 60°$, $\varphi'=210°$, $\omega'=290°$

conf.c $\chi= 40°$, $\Psi= 60°$, $\varphi'=270°$, $\omega'=80°$

conf.d $\chi= 20°$, $\Psi= 60°$, $\varphi=180°$, $\omega=290°$

conf.e $\chi= 40°$, $\Psi= 60°$, $\varphi=180°$, $\omega=290°$

conf.f $\chi= 40°$, $\Psi= 60°$, $\varphi=180°$, $\omega=80°$

numbering of angle χ, Ψ, φ' and ω' see [1][7][8][13])

Calc.								
Conf.2			Conf.3			Conf.4		
a	b	c	a	b	c	a	b	c
-0.131	0.040	-0.003	0.147	0.318	0.275	-0.017	0.154	0.111
0.146	-0.060	-0.374	0.137	-0.072	-0.383	-0.245	-0.454	-0.707
-0.319	-0.223	-0.050	-0.034	0.162	0.335	0.171	0.367	0.540
-0.016	-0.016	0.134	0.046	0.046	0.196	0.008	0.008	0.142
0.012	-0.040	-0.099	-0.036	0.016	-0.075	-0.107	-0.159	-0.218
-0.019	-0.006	0.926	0.048	0.061	0.992	0.706	0.719	1.651
0.294	0.274	0.274	0.436	0.416	0.416	-0.024	-0.044	-0.044
-0.026	0.046	0.210	0.208	0.280	0.444	0.230	0.302	0.466
d	e	f	d	e	f	d	e	f
0.332	0.551	0.548	0.020	0.239	0.236	-0.033	0.186	0.183
0.359	0.448	0.472	0.216	0.305	0.329	-0.053	0.036	0.060
0.188	0.364	0.358	0.049	0.225	0.219	-0.046	0.130	0.124
0.009	0.013	0.010	0.017	0.013	-0.016	-0.045	-0.041	0.044
-0.181	-0.201	0.108	-0.148	-0.168	0.128	0.311	0.331	-0.022
-0.046	-0.027	-0.165	-0.034	0.024	-0.162	0.238	0.257	0.119
0.163	0.156	0.149	0.085	0.078	0.071	-0.007	-0.014	-0.021
0.807	0.657	0.694	1.075	0.676	0.613	0.188	0.038	0.025

290°

290°

290°

80°

longer the case for the equilibrium $^2E \rightleftharpoons {}^3E$ and our semiempirical method of calculation does not take into account fully in its present state the contribution that the variation of bond lengths and bond angles makes to the value of the chemical shifts.

In order to illustrate the role of the effect of the conformational rearrangement on the value of the polymerization shifts we have taken the example of ApA which is experimentally one of the most studied compound. In Table II we report together with the experimental values, the calculated dimerization shifts for six different conformations of this dinucleoside monophosphate.

For the two conformations called A-RNA and B-DNA the calculations are done for the dimer with the geometrical data given by Arnott for A-RNA [40] and B-DNA [41] with the assumption that the isolated monomers as well as the dimers have in solution the rigid conformation exhibited by the crystals of RNA or DNA [38][39]. Since it is well known that, in solutions, mononucleotides as well as dinucleoside monophosphates exist as a mixture of several conformers it is not unexpected that the results reported in Table II under the headings A-RNA and B-DNA do not compare favorably with the experimental values.

For the conformations of ApA numbered 1 to 4 we have calculated the dimerization shift by the following procedure : 1) for one conformation the chemical shift of a given proton is calculated as for the mononucleotide; 2) from this value we substract the value of the shift calculated for this nucleus when it belongs to the corresponding mononucleotide in a given conformation, a to c for the 3'-unit and d to f for the 5'-unit.

The main difference between the results reported for conformations 1 to 4 and those reported under the heading A-RNA and B-DNA is that in the former case the calculated value includes the "neighbouring effect", that is the contribution to the computed shift of the other nucleotidyl unit of the dimer as well as the chemical shift variation due to the "rearrangement effect" (that means the $\Delta\delta$ due to the conformational difference between the isolated mononucleotide taken as reference and the corresponding nucleotidyl unit in the dimer), while in the second case, only the "neighbouring effect" is taken into account.

Although conformations a and b of 3'-AMP and d and e of 5'-AMP differ only by the value of χ_{CN} (difference of 20°) for the four conformations of ApA that we consider in Table II, our results show that for most protons the calculated values of the dimerization shifts are quite different according to the conformation chosen as reference state of the monomer. Some experimental results on dinucleoside monophosphates require for their interpretation the hypothesis that some "rearrangement effect" contributes to the measured dimerization shifts [1][7][8][13], but the feature illustrated by the results of Table II is that the exact conformation of the monomer might be as important as the conformation of the dimer in the determination of the dimerization shift. As an example, for H1' of the Ap unit we see that for conformation 1 we calculate a dimerization shift differing by 0.171 p.p.m. according to the choice a or b for the conformation of 3'-AMP. For this same proton the dimerization shifts for conformations 1 and 2 of the dimer differ 0.174 p.p.m. if 3'-AMP is supposed to have conformation b. The above results cannot be due to an unreasonable "rearrangement effect" since the conformations that we have retained for both monomers are, on experimental grounds, among the most populated ones [28][42].

The comparison of the different calculated values with the measured ones reported in Table II shows as a general tendency that the agreement between theory and experiment is poorer for the Ap unit than for the pA one. We are tempted to interpret this disagreement either by a great flexibility of 3'-AMP [28][32] or by a large contribution to the "rearrangement effect" of the shift of the equilibrium ^{2}E ^{3}E which occurs upon dimerization [7][13][14], a contribution which we cannot evaluate for the moment.

The calculated values of Table II confirm our previous statement [6] that H2 of adenine and the imino protons of uracil (or thymine) and guanine are those for which the dimerization shift is, to a rather large extend, not influenced by the "rearrangement effect" but unfortunately Table II shows also that the shifts of these protons are not the most sensitive to the exact conformation of the polynucleotide chain.

This series of calculations [20][21][6] on the conformational dependence of chemical shifts illustrates the need of a continuously more precise determination of the conformation of the mononucleotides and of the contribution of the "rearrangement effect" to the polymerization shift, if one wants to be able to fully interpret the NMR measurements of polymers in solution.

Acknowledgments.- We wish to thank Drs. Lee and Tinoco for communicating to us the microfilm with the values of the chemical shifts and coupling constants and Dr. R.H. Sarma for the corrected ones for ApA.

REFERENCES

[1] Cheng, D.M. and Sarma, R.H. : J. Amer. Chem. Soc. 99, 7333 (1977)
[2] Borer, P.N., Kan, L-S. and Ts'o P.O.P. : Biochemistry, 14, 4847 (1975).
[3] Kroon, P.A., Kreishman, G.P., Nelson, J.H. and Chan, S.I. : Biopolymers, 13, 2571 (1974).
[4] Patel, D.J. and Canuel, L. : Proc. Nat. Acad. Sci. U.S.A., 73, 674 (1976).
[5] Altona, C., Van Boom, J.H. and Haasnoot, C.A.G. : Eur. J. Biochem. 71, 557 (1976).
[6] Ribas Prado, F., Giessner-Prettre, C. and B. Pullman, J. Theoret. Biol., in press.
[7] Lee, C.H., Ezra, F.S., Kondo, N.S., Sarma, R.H. and Danyluk, S.S.: J. Am. Chem. Soc. 98, 3627 (1976).
[8] Ezra, F.S., Lee, C.H., Kondo, N.S., Danyluk, S.S. and Sarma, R.H.: J. Am. Chem. Soc. 99, 1977 (1977).
[9] Ts'o, P.O.P., Kondo, N.S., Schweizer, M.P. and Hollis, D.P. : Biochemistry, 8, 997 (1969).
[10] Kondo, N.S., Ezra, F. and Danyluk, S.S.: F.E.B.S. Letters, 53, 213 (1975).
[11] Evans, F.E. and Sarma, R.H.: Nature, 263, 567 (1976).
[12] Sarma, R.H. and Danyluk, S.S. : Int. J. Quantum Chem. Quantum Biol. Symp. 4, 269 (1977).
[13] Kondo, N.S. and Danyluk, S.S. : Biochemistry, 15, 756 (1976).
[14] Lee, C-H. and Tinoco, I. : Biochemistry, 16, 5403 (1977).
[15] Alderfer, J.L. and Ts'o, P.O.P.: Biochemistry, 16, 2410.
[16] Lee, C-H., Evans, F.E. and Sarma, R.H. : F.E.B.S. Letters, 51, 73 (1975).
[17] Cheng, D.M. and Sarma, R.H. : Biopolymers, 16, 1687 (1977).
[18] Singh, H., Herbut, M.H., Lee, C-H. and Sarma, R.H. : Biopolymers, 15, 2167 (1976).
[19] Prestegard, J.H. and Chan, S.I. : J. Am. Chem. Soc. 91, 2843 (1969).
[20] Giessner-Prettre, C. and Pullman, B. : J. Theoret. Biol. 65, 171 (1977).
[21] Giessner-Prettre, C. and Pullman, B. : J. Theoret. Biol. 65, 189 (1977).
[22] Sundaralingam, M. : Biopolymers, 7, 821 (1969).
[23] Sundaralingam, M. and Jensen, L.H. : J. Mol. Biol. 13, 914 (1965).
[24] Hecht, J. and Sundaralingam, M. : J. Am. Chem. Soc. 94, 4314 (1972).
[25] Sundaralingam, M. : Acta Cryst. 21, 495 (1966).
[26] Eckstein, F., Saenger, W. and Suck, D. : Biochem. Biophys. Res. Com. 964 (1972).
[27] Saenger, W., Suck, D. and Eckstein, F. : Eur. J. Biochem. 46, 559 (1974).
[28] Davies, D.B. and Danyluk, S.S. : Biochemistry, 14, 543 (1975).
[29] Shefter, E., Barlow, M., Sparks, R.A. and Trueblood, K.N. : Acta Cryst. B25, 895 (1969).
[30] Lai, T.F. and Marsh, R.E. : Acta Cryst. B28, 1982 (1972).
[31] Tran-Dinh Son and Chachaty, C. : Biochim. Biophys. Acta, 335, 1 (1974).

[32] Schleich, T., Blackburn, B.J., Lapper, R.D. and Smith, I.C.P. : Biochemistry, 11, 137 (1972).
[33] Tran-Dinh Son, Chachaty, C. and Gueron, M. : J. Am. Chem. Soc. 94, 7903 (1973).
[34] Pullman, B. and Saran, A. : Progress in Nucleic Acid Research and Molecular Biology, 18, 215 (1976).
[35] Remin, M. and Shugar, D. : Biochem. Biophys. Res. Com. 48, 636 (1972).
[36] Smith, I.C.P., Mantsch, H.H., Lapper, R.D., Deslauriers, R. and Schleich, T. : Conformation of Biological Molecules and Polymers, Vth Jerusalem Symposium, E.D. Bergmann and B. Pullman Eds. Academic Press, New York, 381 (1973).
[37] Evans, F.E., Lee, C-H., and Sarma, R.H. : Biochem. Biophys. Res. Com. 63, 103 (1975).
[38] Kan, L.S., Kast, J.R., Ts'o, D.Y. and Ts'o, P.O.P., in press.
[39] Kan, L.S., Ts'o, D.Y., Ts'o, P.O.P., Ribas Prado, F., Giessner-Prettre, C. and Pullman, B., unpublished results.
[40] Arnott, S., Hukins, D.W.L., Dover, S.D., Fuller, W. and Hodgson, A.B. : J. Mol. Biol. 81, 107 (1973).
[41] Arnott, S. and Hukins, D.W.L. : J. Mol. Biol. 81, 93 (1973).
[42] Davies, D.B. and Danyluk, S.S. : Biochemistry, 13, 4417 (1974).

DISCUSSION

Gorenstein: The very large shifts must reflect severe steric distortion and should properly be discounted so long as some degrees of conformational freedom remain in molecules. It would be very useful to compare your calculated shift changes with observed experimental data such as the cyclonucleoside studies of Dr. Danyluk where the conformational angle χ is locked.

Giessner-Prettre: In most cases the large shifts which are calculated for some values of a rotational angle are effectively greatly reduced or even totally leveled off if a second rotational angle is varied or if the puckering of the furanose ring is changed. It seems that it might be hasardous to compare our results with experimental data on cyclonucleosides. The chemical structure of cyclo and non cyclonucleosides are different and, if the conformation of the molecule is one parameter which determines the value of the chemical shifts of its different nuclei, the exact chemical structure of the molecule studied is the second. In addition similar compounds like 2,2'-anhydro-1-(β-D-arabinofuranosyl)-uracil and 2,3'-anhydro-1-(β-D-arabinofuranosyl)-uracil do not have the same puckering, so the value of the rotation angle about the glycosyl bond is not the only conformational parameter which is different between these two molecules.

SOME COMMENTS AND COMPARISONS CONCERNING THE USE OF "RING-CURRENT" CALCULATIONS IN ELUCIDATING MOLECULAR CONFORMATIONS

R.B. MALLION
The King's School, Canterbury, United Kingdom

INTRODUCTION

Use of "ring-current" ideas in the elucidation of molecular conformations by ^{1}H nmr-spectroscopy continues unabated, twenty years after their foundation (Pople 1956). Such applications almost invariably make use of one (or both) of two basic methods, one entirely classical, and the other quantum mechanical (though still very much empirical). These seem to have become known, respectively, as the 'Johnson-Bovey' and the 'Haigh-Mallion' approaches; both are misnomers: the Johnson-Bovey (1958) tables (Emsley *et al.* 1965; Bovey 1969) were computed by appeal to a (corrected) version of a classical equation due to Waugh and Fessenden (1957) (see also Waugh 1958; Mallion 1969 and 1971; and Marynick & Onak 1969), while the Haigh-Mallion tables (Mallion 1969; Haigh & Mallion 1972) are founded on a trivial extension (Haigh & Mallion 1971) of McWeeny's (1958) quantum-mechanical "ring-current" theory.

In the period 1974-1977 alone, these two sets of tables were together cited more than 200 times in the chemical and biochemical literature, the longer-established Johnson-Bovey tables being favoured over the quantum-mechanical ones in a ratio of about 4 to 1. Furthermore, many recent applications of these techniques have been in the elucidation of the conformation, in solution, of molecules of biological interest; (*e.g.*, recent references, randomly selected, are Robillard *et al.* 1976; Giessner-Prettre & Pullman 1976 and 1977; Robillard 1977; Gettins & Dwek 1977; Dwek *et al.* 1977; and Perkins *et al.* 1977.) These methods have also been used in the study of dilution effects (Bartle *et al.* 1974; Abraham *et al.* 1977). It seems timely, therefore, as well as appropriate to the present audience, to compare and contrast the numerical predictions of these two approaches, to dwell somewhat on their respective shortcomings, and to consider in general the overall propriety of applying "ring-current" criteria in conformational studies.

NUMERICAL COMPARISONS BETWEEN THE PREDICTIONS OF THE TWO THEORIES

B. Pullman (ed.), Nuclear Magnetic Resonance Spectroscopy in Molecular Biology, 183-191.

The first point to recall is that both sets of tables are based on methods which, from a theoretician's point of view, are extremely crude and that, consequently, *neither* can be expected to give realistic predictions in *all* regions of space. The Waugh-Fessenden-Johnson-Bovey treatment regards the π-electrons as being in two classical current-'loops', above and below the plane of a benzenoid ring, while the McWeeny-Haigh-Mallion approach is essentially founded on London's (1937) theory of diamagnetic "ring-currents", modified to take account of the presence of a 'test'-dipole which is introduced in order to probe the secondary magnetic-field that these "ring currents" are assumed to produce. The latter method is usually, though not necessarily, based on a simple Hückel molecular-orbital and, for the evaluation of certain integrals encountered, it frequently invokes an approximation which has since become known as the 'London approximation'; (for discussion on this see Haigh & Mallion 1971; Amos & Roberts 1971; Roberts 1971 and 1974; and Coulson *et al.* 1975). In both sets of tables, the effects of these various simplifications and approximations on predicted shieldings are by no means the same for nuclei situated at all points in the vicinity of a benzenoid ring (assumed to bear unit — *i.e.* 'benzene' — "ring-current"). We now therefore make a purely numerical comparison of the predictions of the Johnson-Bovey (1958) and Haigh-Mallion (1972) tables i) at points near the benzenoid ring in question, and ii) at large distances from that ring; we shall then examine possible reasons for the differences that will be found.

i) At Points Near the Ring

In this comparative discussion, all lengths and coordinates will be specified in units of benzene carbon-carbon bond-length (1.397 Å), denoted by the symbol a. All points will be specified in terms of their cylindrical polar-coordinates, (ρ, z, ϕ), referred to the benzene-ring centre as origin (see Figure 1 of Haigh & Mallion (1972))

Figure 1 shows predictions (from the two sets of tables) of shielding, $\Delta\sigma$, at points in the ring plane (*i.e.* with $z = 0$) as a function of ρ, with $\phi = 0$. Until about 1.5 units from the ring centre, the breakdown of the assumptions of both theories at points so near the ring prevents both curves from being well-behaved in this region — the two curves do, in fact, approach realistic values of $\Delta\sigma$ from infinity, but from *opposite* directions! For values of ρ greater than about 1.8, the "ring-current" deshielding predicted by the classical theory is greater than that predicted by the molecular-orbital method. This is consistent (a) with previous observations (Mallion 1971) that, as far as in-plane calculations on planar, benzenoid hydrocarbons are concerned, the Johnson-Bovey tables tend to overestimate downfield shifts, and (b) with findings (Haigh *et al.* 1970) that, *with appropriate parameterisation*, the McWeeny molecular-orbital method accounts quantitatively for the observed downfield-shifts in planar, condensed benzenoid hydrocarbons.

In Figure 4 of Haigh & Mallion (1972), the predicted "ring-current" shielding (from the classical and quantum-mechanical theories) is shown as a function of vertical distance above the ring centre (*i.e.* as a function of z, with $\rho = 0$). These plots indicate that while the curves arising from both theories are well-behaved, for all values of z, shielding predictions from the Johnson-Bovey tables are consistently, and significantly, greater than those of the Haigh-Mallion tables. This is certainly in accord with subsequent observation and experience (*e.g.* Haigh & Mallion 1971; Rose 1973; Crespi *et al.* 1973; Mallion 1975).

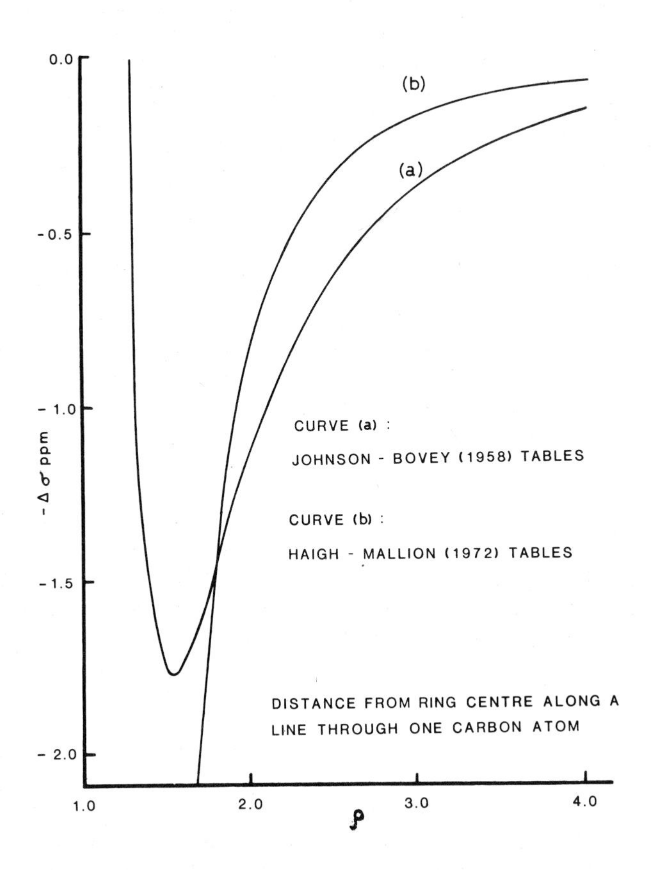

Figure 1: Comparison between predictions of the Johnson-Bovey (1958) and Haigh-Mallion (1972) tables, for points *in* the plane of a benzene ring.

ii) <u>At Large Distances from the Ring</u>

To compare predictions of the two theories at large distances from the ring, it is sufficient to confine attention to calculations concerning points that are *in* the plane of the benzenoid ring ($z = 0$) and that lie along a line through the ring centre and one carbon-atom ($\phi = 0$). Figure 2 indicates that the *ratio* of the deshielding predicted by the two theories varies with ρ giving a smooth curve which appears to level off to a constant value (of a little over 2.6) at between 5 and 6 carbon-carbon bond-lengths from the ring centre. Furthermore, the value of this ratio is maintained at least as far as *fifteen* carbon-carbon bond-lengths from the centre of the benzenoid ring in question; (this is the largest distance at which I have calculated it — see Mallion (1969)).

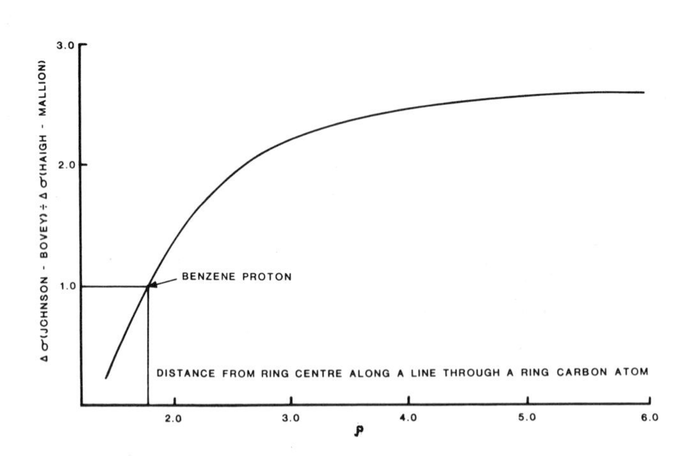

Figure 2. Comparison between predictions of the Waugh-Fessenden (1957) -Johnson-Bovey (1958), and McWeeny (1958) - Haigh-Mallion (1971) methods.

We now investigate these discrepancies by a more-detailed consideration of how both theories are calibrated by appeal to the common assumption that a benzene proton resonates at *ca.* 1.5 ppm to lower field than does an 'otherwise analogous' olefinic proton. What emerges from calculations based on the McWeeny (1958) and Haigh-Mallion (1971, 1972) theories is the *ratio* $\Delta\sigma/\Delta\sigma_{benzene}$, where $\Delta\sigma$ is the "ring-current" shielding at the proton in question and $\Delta\sigma_{benzene}$ is the corresponding shielding calculated, by the same method, to be experienced by a standard benzene-proton, because of the "ring current" in benzene; this ratio is then converted to a shift (in ppm) downfield of the 'olefinic' reference simply by *multiplying* by the factor of 1.5 ppm (or, in the case of the Haigh-Mallion (1972) tables, by a factor of 1.56 ppm (see Haigh *et al.* 1970)). It is clear, therefore, than *any* absolute downfield-shift calculated by the McWeeny-Haigh-Mallion approach — whether it be for a point close to, or distant from, a benzenoid ring — always depends *linearly* on the multiplicative factor $\left(\frac{1.5(6)}{\Delta\sigma_{benzene}}\right)$. At points more-distant from the ring, $\Delta\sigma$ does of course become very small anyway so that, in an absolute sense, the actual numerical value of the multiplicative factor becomes less significant; nevertheless, even at very large distances from the ring, the downfield shift predicted by the semi-empirical quantum-mechanical methods should in principle be linearly dependent upon

1) the value assumed for the "ring-current" contribution to the down-field shift of a benzene proton; (this is probably not an important source of error);
2) the reciprocal of the quantity $\Delta\sigma_{benzene}$, which is proportional to what is sometimes termed the calculated "geometric-factor" for a benzene proton (Haigh *et al.* 1970; Mallion 1973). Because a benzene proton is only *ca.* 1.78 units from the ring centre, the London integral-approximation and the McWeeny 'test-dipole' approximation are likely to be drastic and hence the calculated geometric-factor for a benzene proton may be very seriously in error (Haigh & Mallion 1971; Mallion 1973; Roberts 1971 and 1974). This latter claim is well substantiated by Table 1 of Mallion (1973) in which geometric factors calculated by the McWeeny (1958) method are compared with those calculated *via* an approximation based on the Biot-Savart Law; this comparison shows that, for points distant from the ring, geometric factors calculated by the two methods differ only in the fourth decimal-place but that the two geometric factors calculated for a benzene proton (-1.0436 and -0.6583) differ by *ca.* 50%.

The dependence of the predictions of the Waugh-Fessenden-Johnson-Bovey method on the value assumed for the "ring-current" contribution to the downfield shift of a benzene proton is rather more subtle (Mallion 1971); the notional position of the 'lobes' of the precessing electron-cloud is varied until the 'lobe separation' is such that for $\rho = 1.777$ in the plane of the ring (the position of a benzene proton) the calculation predicts a shielding value of -1.5 ppm. This having been done, the value of the downfield shift of benzene is not used again, but the 'lobe separation' found by the above process is retained in all future calculations. Thus, the influence of the value assumed for the "ring-current" shift of benzene is only an *implicit* one in the remaining calculations and very rapidly ceases to have much effect on the predicted shieldings as distance from the ring is increased. This can be seen by examining the equation arising from the Waugh-Fessenden (1957)-Johnson-Bovey (1958) method:

$$\Delta\sigma = \frac{ne^2}{6\pi mc^2 a} \frac{1}{[(1+\rho)^2+z^2]^{\frac{1}{2}}} \left[K(k) + \frac{1-\rho^2-z^2}{(1-\rho)^2+z^2} E(k)\right] \qquad (1)$$

This is the shielding provided by n electrons circulating in a loop of radius a, at a point with cylindrical coordinates (ρ, z) with respect to *the centre of the current loop*; k is the modulus of the Complete Elliptic Integrals K and E, details of which are given in Mallion (1971). When a point *in* the ring plane is being considered (as here), $z=p$, where the π-electron 'lobe'-separation is $2p$.

Now it will be evident from equation (1) that for points at large distances from the ring, in the ring plane — *i.e.* when

$\rho >>> z - z^2(=p^2)$ becomes negligible compared with ρ^2, $(1+\rho)^2$ and $(1-\rho)^2$. Hence, as $\rho \to \infty$ for points in the ring plane, predictions of the Johnson-Bovey (1958) tables are effectively independent of the 'lobe separation', p (*i.e.* are independent of any particular, physically-reasonable value assumed for the "ring-current' downfield-shift of a benzene proton and used to determine the 'lobe-separation' in the first place.) Furthermore, since no *ratios* of calculated shifts are taken, such long-distance predictions of the Johnson-Bovey tables, unlike those of the Haigh-Mallion tables, are *not* bedogged by the vestige of any *theoretical* value for the "ring-current" downfield shift of a benzene proton which has been estimated by invoking approximations that are possibly invalid for a point so close to the ring centre.

It is also of interest to compare the predictions of the classical point-dipole method of Pople (1956) with those of the McWeeny method, at large distances from the ring. On the point-dipole theory, the shielding at a point distance R from the ring centre, in the plane of the ring, is

$$\Delta\sigma = \frac{e^2a^2}{2mc^2} \times \frac{1}{R^3} \tag{2}$$

and, on the McWeeny theory, since the "ring current" for benzene is -1/9, and the geometric factor can be written as $\frac{\mu}{R^3}$ where (McWeeny 1958)

$$\mu = 1 + \frac{9}{8}\,\frac{1}{R^3} + \frac{843}{128}\,\frac{1}{R^4} \cdots , \tag{3}$$

the shielding at a point distance R from the ring centre, on a line through the ring centre and a ring carbon-atom, is (McWeeny 1958):

$$\Delta\sigma = \frac{2\beta}{3}\left(\frac{2\pi e}{hc}\right)^2 \frac{S^2}{a^3}\left[-\frac{1}{9}\right]\left(\frac{\mu}{R^3}\right) \tag{4}$$

where, at very large distances from the ring, $\mu \to 1$, by virtue of equation (3). (In equation (4), β is the standard Hückel carbon-carbon resonance-integral, and S is the area of a benzene ring). Now, for the McWeeny theory to give the same $\Delta\sigma$-prediction as the Pople point-dipole approach at some *very large distance* R from the ring centre, we must have

$$\frac{e^2a^2}{2mc^2} = +\frac{2\beta}{3}\left(\frac{2\pi e}{hc}\right)^2 \frac{S^2}{a^3}\left[-\frac{1}{9}\right] \tag{5}$$

This comparison yields a value for β, the Hückel resonance-integral; the required value is *ca.* 381 KJ mol^{-1}. The value previously found to be appropriate (Haigh *et al.* 1970) when predictions of the McWeeny (1958) theory were compared with experiment was *ca.* 146 KJ mol^{-1}, and these two figures are *again in the ratio 2.6*. It thus seems that, in order to give shielding predictions in agreement with those of the Pople (1956) point-dipole method at *large* distances from a benzenoid ring, the McWeeny (1958) theory would have to use a β-value about 2.6 times the size of that found to be appropriate in an empirical investigation of the secondary fields at points *near* the ring. Furthermore, a semi-classical approach based on the Biot-Savart Law (Mallion 1973) — the predictions of which are certainly *algebraically* coincident with those of the McWeeny theory at points *distant* from ring centres — requires *yet another* β-value in order to reproduce experimental shieldings at points *near* benzenoid rings!

CONCLUSIONS

I should like, therefore, to emphasise the following:
1) The 'point-dipole' method of Pople (1956) and the Johnson-Bovey (1958) tables give the same deshielding-predictions for points at large distances from the ring, *in* the ring plane, as would be expected (see Abraham *et al.* 1977, for a recent numerical confirmation of this), but the Haigh-Mallion (1972) tables predict corresponding deshielding-values which are smaller by a factor of *ca.* 2.6. This is explained by the remarks in 3), below.
2) At points above the ring plane, the Johnson-Bovey tables always predict greater shielding than the Haigh-Mallion tables; it is for points in just this region of space that errors arising from the London integral-approximation — on which the Haigh-Mallion tables are based — are most severe (see Haigh & Mallion 1971, especially pp. 966-968). Even the Johnson-Bovy tables, however, underestimate such observed shieldings (Haigh & Mallion 1971; Crespi *et al.* 1973).
3) Empirical parameterisations of "ring-current" theories — *particularly those which involve taking ratios to benzene* — must be treated with extreme caution (Mallion 1973). Because the approximations of these theories are generally least valid for points *near* a benzenoid ring, those parameterisations based on a calibration with experimental data for points *close to* conjugated rings (*e.g.* Haigh *et al.* 1970; Mallion 1973) will not necessarily be appropriate for deriving absolute shieldings from relative shifts predicted, by the theory, at more-distant points — where, paradoxically, the actual *approximations* of the theory become *more* tenable than they are for points in close proximity to ring centres! This is because those features of such empirical parameterisations which are especially designed to compensate for the particularly gross approximations that arise when "ring-current" theories are concerned with points *near* benzenoid rings, will, if such parameterisations are injudiciously applied, outlive their purpose and *persist* — quite out of context — in the estimation of shieldings at the more-distant points. This reservation should be borne firmly in mind when interpreting calculations based on the Haigh-Mallion tables; for reasons outlined in this discussion, however, the long-distance predictions of the Johnson-Bovey tables are exempt from this particular criticism — although these tables *are* liable to many other criticisms because they derive from an entirely classical theory.

These considerations indicate that, despite recent theoretical advances, the "ring-current" idea (except when applied to the specially-parameterised case of the protons in planar, condensed, benzenoid hydrocarbons) essentially remains only a *semi*-quantitative conceptual aid; therefore, until a set of tables is available that are based on a quantum-mechanical method in which no appeal is made to the London integral-approximation, "ring-current" tables should be used only to provide evidence of a corroboratory nature in cases where a molecular conformation has been almost unequivocally established by other physical means. Those investigations in which numerical predictions of presently available "ring-current" tables have been used as the corner-stone of a conformational study would appear to be based on a disquietingly insecure foundation.

ACKNOWLEDGEMENTS

I am very grateful to the Israel Academy of Sciences and Humanities, the Hebrew University of Jerusalem, and the *Fondation Edmond de Rothschild* of Paris for their kind hospitality, and to the Royal Society of London for their generous award of a travel grant.

REFERENCES

Abraham, R.J., Fell, S.C.M. & Smith, K.M.: 1977, *Org.Magn.Resonance* 9 367.
Amos, A.T. & Roberts, H.G.Ff.: 1971, *Molec.Phys.* 20, 1089.
Bartle, K.D., Mallion, R.B., Jones, D.W. & Pickles, C.K.: 1974, *J.Phys. Chem.* 78, 1330.
Bovey, F.A.: 1969, *Nuclear Magnetic Resonance Spectroscopy*, Academic Press, London and New York, Appendix.
Coulson, C.A., Gomes, J.A.N.F. & Mallion, R.B.: 1975, *Molec.Phys.* 30, 713.
Crespi, H.L., Norvis, J.R., Boys, J.P. & Katz, J.J.: 1973, *Ann. New York Acad.Sci.* 222, 800.
Dwek, R.A., Wain-Hobson, S., Dower, S., Gettins, P., Sutton, B., Perkins, S.J. & Givol, D. : 1977, *Nature* 266, 31.
Emsley, J.W., Feeney, J. & Sutcliffe, L.H.: 1965, *High Resolution Nuclear Magnetic Resonance Spectroscopy*, Pergamon Press, Oxford, Vol. 1, Appendix B.
Gettins, P. & Dwek, R.A.: 1977, in R.A. Dwek, I.D. Campbell, R.E. Richards, R.J.P. Williams (ed.), *NMR in Biology*, Academic Press, London, New York and San Francisco, p.125.
Giessner-Prettre, C. & Pullman, B.: 1976, *Biochem.Biophys.Res.Commun.* 70, 578.
Giessner-Prettre, C. & Pullman, B.: 1977, *J.Theor.Biol.* 65, 171.
Haigh, C.W. & Mallion, R.B.: 1971, *Molec.Phys.* 22, 955.
Haigh, C.W. & Mallion, R.B.: 1972, *Org.Magn.Resonance* 4,203.
Haigh, C.W., Mallion, R.B. & Armour, E.A.G.: 1970, *Molec. Phys.* 18, 751.

Johnson, C.E. & Bovey, F.A.: 1958, *J.Chem.Phys.* 29, 1012.
London, F.: 1937, *J.Physique Radium* 8, 397.
Mallion, R.B.: 1969 'Nuclear Magnetic Resonance: A Theoretical and Experimental Study of the Spectra of Condensed, Benzenoid Hydrocarbons', University of Wales (Ph.D. Thesis).
Mallion, R.B.: 1971, *J.Chem.Soc. (B)*, p.681. (See also Addendum on the last page (unnumbered) of the 'Index Issue' (No. 13), 1971).
Mallion, R.B.: 1973, *Molec.Phys.* 25, 1415.
Mallion, R.B.: 1975, in R.K. Harris (ed.), *Specialist Periodical Report; Nuclear Magnetic Resonance Vol. IV*, The Chemical Society, London, ch.1, esp.pp.50-51.
Marynick, D. & Onak, T.: 1969, *J.Chem.Soc. (A)*, p. 1797.
McWeeny, R.: 1958, *Molec.Phys.* 1, 311.
Perkins, S.J., Dower, S.K., Gettins, P., Wain-Hobson, S. & Dwek, R.A.: 1977, *Biochem.J.* 165, 223.
Perkins, S.J., Radda, G.K. & Richards, R.E.: 1977, *Europ.J.Biochem.*, 'A Hydrogen-Deuterium Exchange Study of the Amide Protons of Polymyxin B by Nuclear Magnetic Resonance', (in press).
Pople, J.A.: 1956, *J.Chem.Phys.* 24, 1111.
Roberts, H.G.Ff.: 1971, *Chem.Phys.Letters* 11, 259.
Roberts, H.G.Ff.: 1974, *Molec.Phys.* 27, 843.
Robillard, G.T.: 1977, in R.A. Dwek, I.D. Campbell, R.E. Richards & R.J.P. Williams (ed.), *NMR in Biology*, Academic Press, London, New York and San Francisco, p.201.
Robillard, G.T., Tarr, C.E., Vosman, F. & Berendsen, H.J.C.: 1976, *Nature* 262, 363.
Rose, P.I.: 1973, *Org.Magn.Resonance* 5, 187.
Waugh, J.S.: 1958, *J.Amer.Chem.Soc.* 80, 6697.
Waugh, J.S. & Fessenden, R.W.: 1957, *J.Amer.Chem.Soc.* 79, 846.

CARBON-13 NMR RESULTS IN THE FIELD OF NITROGEN HETEROCYCLES
THE DETECTION OF PROTONATION SITES

Harald Günther*, Angela Gronenborn, Ulrich Ewers
und Horst Seel
Institute of Organic Chemistry, University of
Cologne, Greinstr. 4, D-5000 Köln, Germany

Compared to hydrocarbons, the nitrogen free electron pair adds an extra dimension to the chemical behaviour of nitrogen heterocycles. Especially in the biochemistry of these systems coordination and second order bonding are important areas where their Lewis base properties play a predominant role. Methods that allow to measure pK_a-values and that enable the determination of protonation sequences in unsaturated polyaza-heterocycles deserve, therefore, considerable attention.

Recently, we have established the usefulness of ^{13}C-NMR parameters to detect protonation sites in pteridines (1,2) and an example of this work is given with the structure elucidation of the pterin ions (Fig. 1). Besides chemical shifts

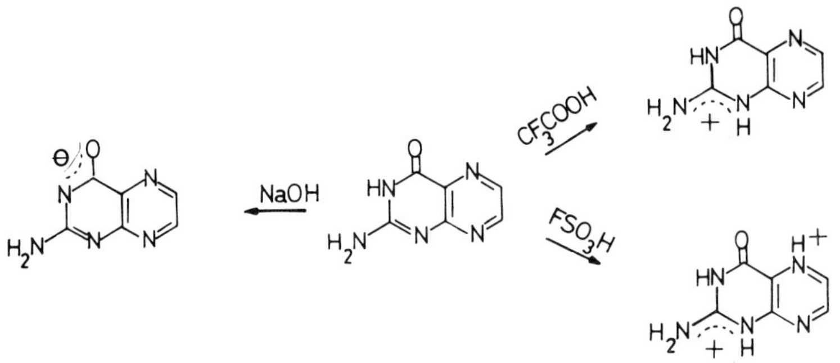

Figure 1. The structure of pterin ions.

characteristic changes in $^{13}C,^{1}H$ coupling constants were recognized that may serve as an analytical tool for the detection of protonation sequences in nitrogen heterocycles. From the experimental observations made for pteridines (1, 3) as well as for other nitrogen heterocycles (2,4) the following predictions can be made for changes of $^{13}C,^{1}H$ coupling

B. Pullman (ed.), Nuclear Magnetic Resonance Spectroscopy in Molecular Biology, 193-200.

constants over one, two, and three bonds, respectively:

(1) N-protonation will increase the $^{1}J(^{13}C,^{1}H)$ data in the protonated ring.

(2) N-protonation will decrease geminal $^{13}C,^{1}H$ coupling in fragment (a) and increase geminal $^{13}C,^{1}H$ coupling in fragment (b). The first observation can be rationalized with arguments provided by the POPLE-BOTHNER-BY MO theory of geminal H,H coupling (1,5).

(3) N-protonation will decrease vicinal $^{13}C,^{1}H$ coupling in fragment (c).

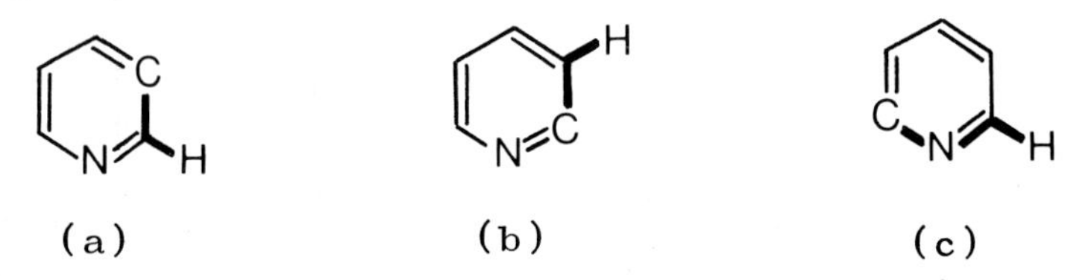

Fig. 2 summarizes some typical results for quinolines and protonated quinolines.

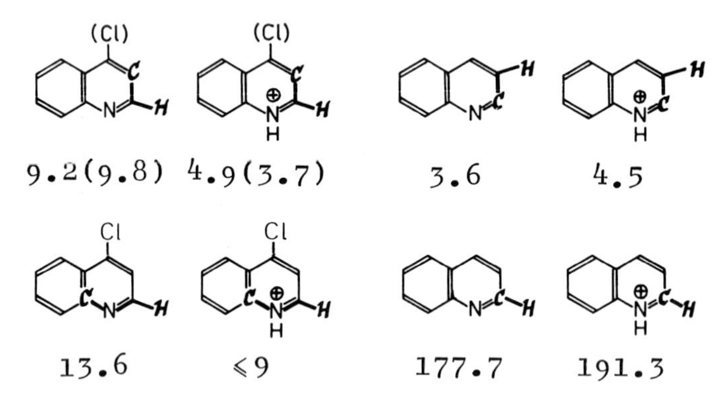

Figure 2. Experimental results for selected $^{13}C,^{1}H$ coupling constants (Hz) of unprotonated (solvent $CDCl_3$) and protonated (solvent 5N H_2SO_4) quinoline and 4-chloro-quinoline.

In order to support these findings further it was desirable to measure the protonation effects more precisely for the parent system pyridine itself, where a complete analysis of the ^{13}C-NMR spectrum was available (6), but data for the hydrochloride remained unknown. We therefore decided to study this system using our newly developed technique for the exact determination of $^{13}C,^{1}H$ coupling constants that was recently successfully applied to pyridine (7) and other compounds (8). In the present case, the three isomeric pyridinium ions 1 - 3 were prepared and their ^{13}C-NMR-spectra recorded using deuterium decoupling. Line broadening

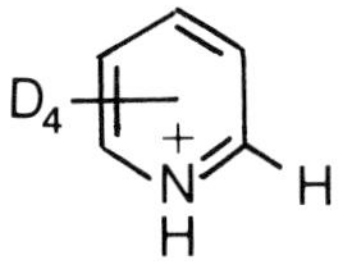

1

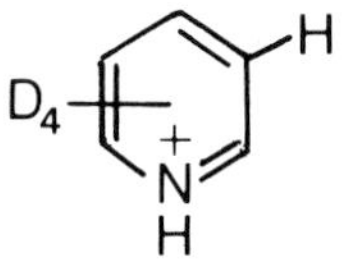

2

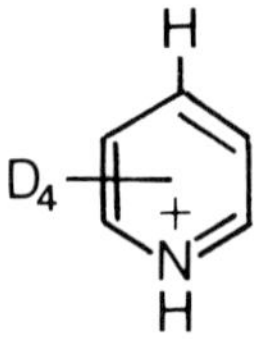

3

originating from unresolved $^{14}N,^{13}C$ coupling was detected mainly for the α-carbons C-2,6, as apparent from the spectrum of isomer 1 (Fig. 3). Only slight broadening was present for the lines of the γ-carbon C-4, whereas the ß-carbons C-3,5 showed no such effect. This is in accord with the findings $^1J(^{13}C_\alpha\text{-}N) > {^3J}(^{13}C_\gamma\text{-}N) > J(^{13}C_\beta\text{-}N)$ made for ^{15}N-pyridine hydrochloride (9). Since triple resonance experiments of the type $^{13}C\{^2H,^{14}N\}$ were unsuccessful, the smaller splittings (J < 0.5 Hz) could not be resolved. The complete results are collected in Table 1 and 2. Those changes relevant to our discussion above are illustrated by the following diagrams:

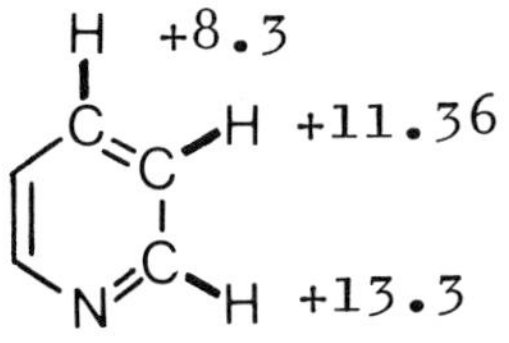

$\Delta^1J(^{13}C,^1H)$

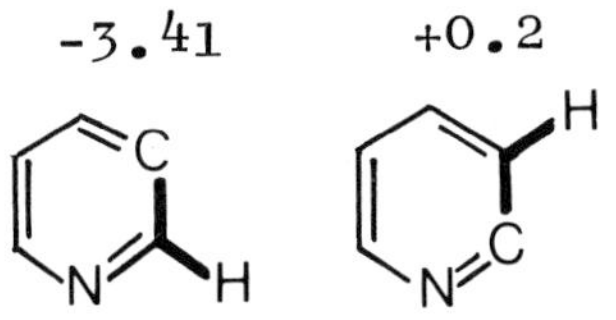

$\Delta^2J(^{13}C,^1H)$

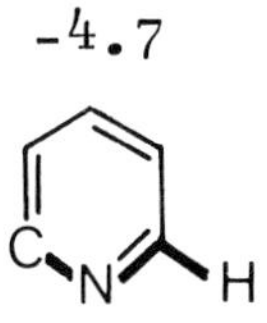

$\Delta^3J(^{13}C,^1H)$

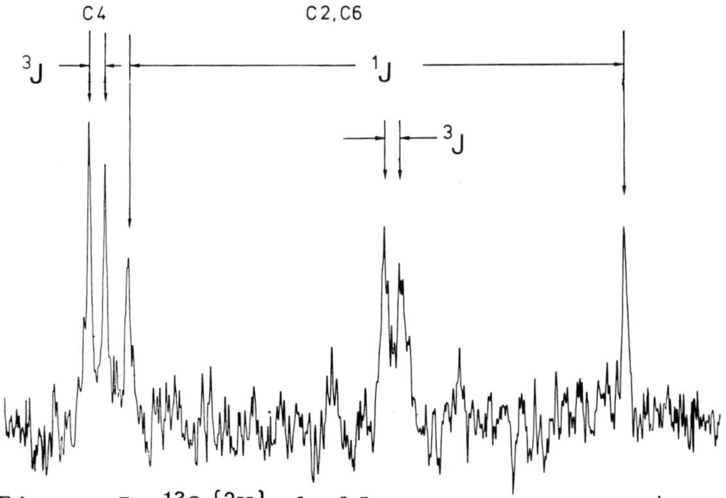

Figure 3. $^{13}C\{^2H\}$-double resonance spectrum of isomer 1 (C-2, C-6, and C-4 resonances only).

Table 1. $^{13}C,^{1}H$ Coupling Constants For Pyridine Hydrochloride (Hz)

	H_{α}	H_{β}	H_{γ}	$H_{\beta'}$	$H_{\alpha'}$
C-2	190.7	3.3	6.9	a)	6.4
C-3	5.13	173.95	a)	7.20	a)
C-4	6.10	a)	169.43	a)	6.10

a) not resolved

Table 2. The Effect of Protonation on the $^{13}C,^{1}H$ Coupling Constants in Pyridine (Hz)

	H_{α}	H_{β}	H_{γ}	$H_{\beta'}$	$H_{\alpha'}$
C-2	+13.3	+0.2	+0.1	---	-4.7
C-3	-3.41	+11.36	---	+0.61	---
C-4	-0.25	---	+8.3	---	-0.25

These values now yield a firm basis for further discussions of experimentally observed trends and a number of examples from investigations carried out in our laboratory are treated in the following section.

Selected Applications of ^{13}C-NMR Measurements

A molecule of particular interest in the light of the protonation scheme developed for pterin (see above) is isopterin, where the two functional groups in the pyrimidine ring have interchanged their position. If the second protonation is governed by the principle of maximum charge separation, it should occur at N-8 instead of N-5. That this is indeed the case is established from the $^{13}C,^{1}H$ coupling constants observed for the mono- and dication:

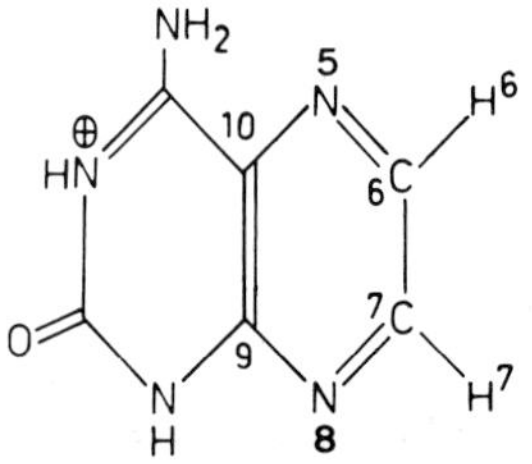

	i,j	+	++
$^{2}J(^{13}C,^{1}H)$:	6,7	10.7 (10)	5.8 (15)
	7,6	11.6 (10)	13.4 (5)
$^{3}J(^{13}C,^{1}H)$:	10,6	11.0 (10)	~10 (5)
	9,7	10.4 (12)	7.3 (12)

The values in paranthesis are those of pterin (3) for comparison.

These conclusions are fully supported by the chemical shifts measured for the anion (in 5N KOH), the mono-cation (in CF_3COOH), and the dication (in FSO_3H). As Fig. 4 demonstrates, the typical α-carbon upfield shifts upon NH^+-formation in the second protonation step are observed for C-9 and C-7 and not for C-10 and C-6 as in the case of pterin.

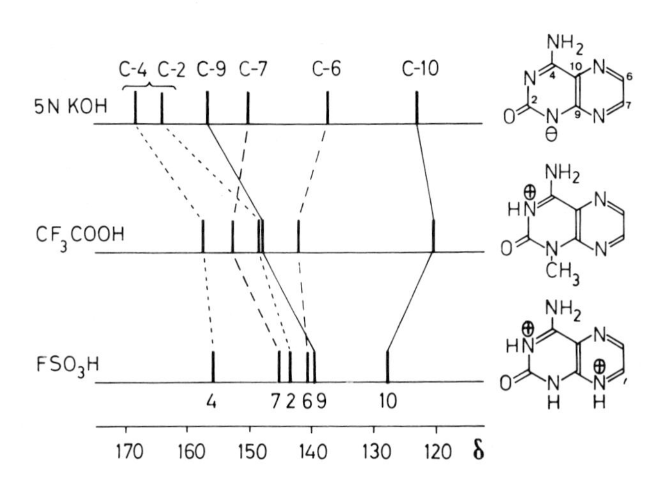

Figure 4. Protonation shifts for the ^{13}C-resonances of isopterin.

The results of an investigation of 2,4-dimethoxy-pyrimidine (4), measured in $CDCl_3$ and in 5N H_2SO_4, are summarized in Table 3. From the chemical shifts and the changes of the coupling constants N-1 protonation is indicated. We find upfield shifts for the resonances of C-2 and C-6, an increase for the geminal coupling constant between C-6 and H-5, and decreases for the geminal coupling constant between C-5 and H-6 and the vicinal coupling constant between C-2 and H-6.

A similar study was performed on adenosin monophosphate (5), where spectral results obtained in neutral and basic medium are compared with those found in acidic medium in Table 5. The criteria developed above for the location of the protonation site lead to N-1 protonation, in accord with other findings (10). Upfield shifts result for C-2 and C-6, both in ortho-position to N-1. The strong increase observed for ^1J(C-2,H-2) indicates protonation in the six-membered ring. The only vicinal coupling that shows a decrease is ^3J(C-6,H-2).

Another system of biological interest is folic acid (6).

Table 3. Chemical Shifts (ppm) and $^{13}C,^{1}H$ Coupling Constants (Hz) of Neutral (o) and Protonated (+) 2,4-Dimethoxy-pyrimidine

	C-2	C-4	C-5	C-6
δ(o)	166.4	172.4	102.9	159.2
δ(+)	160.9	175.7	104.0	148.2
Δδ	-5.5	+3.3	+1.1	-11.0
$^{1}J(^{13}C,^{1}H)$ (o)	---	---	170.8	179.3
(+)	---	---	181.9	191.5
ΔJ			+11.1	+12.2
$^{2}J(^{13}C,^{1}H)$ (o)	---	a)	8.5	2.4
(+)	---	a)	3.7	3.7
ΔJ			-4.8	+1.3
$^{3}J(^{13}C,^{1}H)$ (o)	13.4	a)	---	---
(+)	8.5	a)	---	---
ΔJ	-4.9			

a) Only one splitting of 8.5 (o) and 9.8 (+) Hz, respectively, was measured. It can be assigned tentatively to the vicinal coupling ^{3}J(C-4,H-6).

Table 4. Chemical Shifts (ppm) and $^{13}C,^{1}H$ Coupling Constants of Adenosin-monophosphate in Neutral (o) and Acidic (+) Solution

	C-2	C-4	C-5	C-6	C-8
δ(o) a)	153.6	149.6	119.2	156.2	141.0
δ(+) b)	145.9	149.2	119.4	150.9	143.4
Δδ	-7.7	-0.4	+0.2	-5.3	+2.4
$^{1}J(^{13}C,^{1}H)$ (o) c)	202.5	---	---	---	215.3
(+) d)	216.0	---	---	---	218.4
ΔJ	+13.5				+3.1
$^{3}J(^{13}C,^{1}H)$ (o) c)	---	11.0, 5.0	11.0	11.0	---
(+) d)	---	14.0, 5.5	11.0	7.3	---
ΔJ		+3.0,+0.5	0.0	-2.7	

a) H_2O/D_2O, pH 7.1; b) 1N H_2SO_4; c) H_2O/D_2O, pH 11.0; d) 2N H_2SO_4.

4 **5** **6**

Here, due to the reduced number of protons present in the pteridine ring, the arguments must be based predominantly on chemical shifts. These are illustrated in Fig. 6 for measurements made for the anion (2N NaOH), the uncharged species (DMSO), the dication (10N H_2SO_4) and the trication (FSO_3H). In the first step N-3 is reprotonated in going from the anion to the neutral compound. Accordingly, δ(2) and δ(4) move to higher field. In the second step, using 10N H_2SO_4 as solvent, two protons are added. One protonates N-1, as is seen from the upfield shifts for the resonances of C-2 and C-9. The second proton adds to N-ß of the side chain, as indicated by the shielding observed for C-6 and several carbons of the side chain, as C-α and C-1' of the benzene nucleus. Finally, a third proton is added to folic acid in FSO_3H, this time again in the pteridine ring. Since C-9 as well as C-7 are strongly shielded, the protonated nitrogen must be N-8. For this final step further evidence comes from the $^{13}C,^{1}H$ coupling constants. The one bond coup-

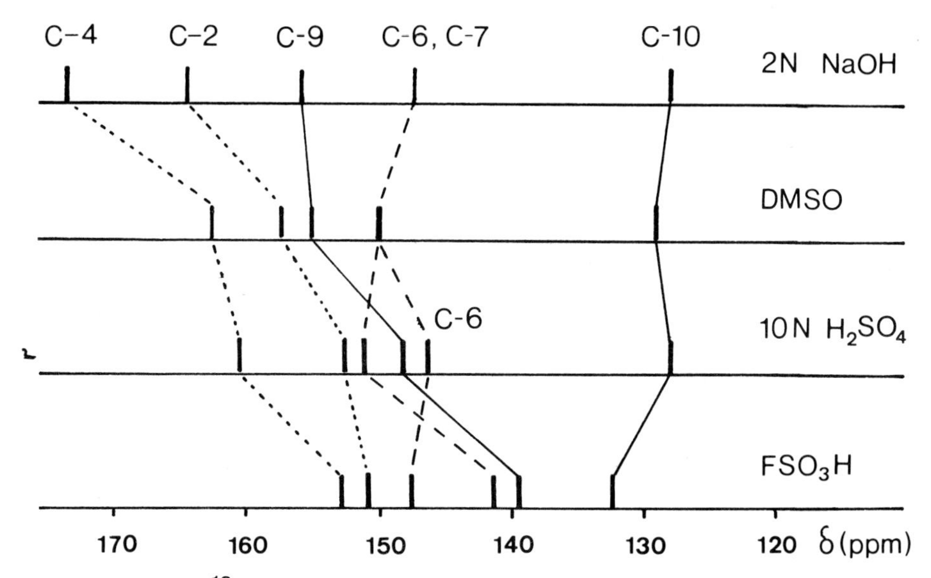

Figure 6. ^{13}C-resonances of the pteridine part of folic acid in different solvents.

ling ^{1}J(C-7,H-7) increases from 191.5 to 203.7 Hz, whereas the vicinal coupling ^{3}J(C-9,H-7) decreases from 11.0 to 7.3 Hz. Thus, the protonation sequence of folic acid is N-1, N-ß, N-8, different from that expected on the basis of the pterin results that suggested N-5 as the third protonation site. These findings underline that caution must be used if protonation sequences are derived from observations made for model compounds, even if their structure closely resembles that of the compound of interest.

We are indebted to Prof. Dr. W. Pfleiderer, University of Konstanz, for a generous gift of isopterin. The support of this work by the Deutsche Forschungsgemeinschaft and the Fonds der Chemischen Industrie is gratefully acknowledged.

References

* New adress: Gesamthochschule Siegen, Fachbereich 8, Adolf-Reichwein-Str., D-5900 Siegen 21, Germany

1. U. Ewers, H. Günther, and L. Jaenicke, Chem. Ber. 107 3275 (1974).

2. U. Ewers, A. Gronenborn, H. Günther, and L. Jaenicke, in "Chemistry and Biology of Pteridines" W. de Gruyter, Berlin 1976, p. 687.

3. G. Müller and W. v. Philipsborn, Helv. Chim. Acta 56, 2680 (1973).

4. P. Birdsall, J. Feeney, and P. Partington, J.C.S. Perkin II, 2145, 1973.

5, J. A. Pople and A. A. Bothner-By, J. Chem. Phys. 42, 1339 (1965).

6. M. Hansen and H. J. Jakobsen, J. Magn. Resonance 10, 74 (1973).

7. H. Günther, H. Seel, and H. Schmickler, ibid. 28, 145 (1977).

8. H. Günther, H. Seel, and M.-E. Günther, Org. Magn. Resonance 11, in print.

9. a) T. Bundgaard and H. J. Jakobsen, Tetrahedron Lett. 1976, 1621; b) R. Lichter and J. D. Roberts, J. Amer. Chem. Soc. 93, 5218 (1971).

10. J. Elguero, C. Marzin, A. R. Katritzky, and P. Linda, Advanc. Heterocyclic Chem., Suppl. 1, Academic Press, New York 1976.

HG 78

NMR STUDIES OF THE MOLECULAR DYNAMICS OF PEPTIDES AND PROTEINS

Kurt Wüthrich, Gerhard Wagner and Arno Bundi
Institut für Molekularbiologie und Biophysik
Eidgenössische Technische Hochschule
8093 Zürich-Hönggerberg, Switzerland

ABSTRACT

High resolution NMR techniques were used to characterize the molecular structures contained in the dynamic ensembles of species which constitute the solution conformations of polypeptide chains. For a group of globular proteins, measurement of the denaturation temperature, the rate of interconversion between globular and denatured conformation at the denaturation temperature, the exchange kinetics of internal amide protons and the intramolecular mobility of aromatic rings showed that the molecular structures contained in the native conformation do not include species of the structure type obtained by thermal denaturation. Evidence is described for the occurrence of hydrophobic stability domains in globular proteins, which might also play a role as nucleation sites in the folding process. For a series of synthetic linear oligopeptides, pH titration shifts of amide proton resonances were used to study intramolecular hydrogen bonding between carboxyl groups and backbone amide groups. With this technique, small populations of structured species could be demonstrated even in peptides which would from other spectral parameters be classified as flexible "random coils". With regard to studies of protein folding pathways, it is of particular interest that local spatial structures thus observed were also seen in the molecular conformations of globular proteins, both in crystals and in solution.

1. INTRODUCTION

Among the techniques used for studies of polypeptide conformations in solution, high resolution nuclear magnetic resonance (NMR) techniques are quite unique in that they can provide a many-parameter characterization of the molecular structures (Wüthrich, 1976). A particularly attractive use of NMR is for complementation of structural data obtained

B. Pullman (ed.), Nuclear Magnetic Resonance Spectroscopy in Molecular Biology, 201-210.

by other methods, mainly X-ray studies in single crystals, with information on the molecular dynamics. Molecular rate processes on a time scale from years down to picoseconds may be manifested in parameters which characterize static and transient states of the nuclear spin systems (Wüthrich, 1976). In this lecture two investigations of the dynamic ensembles of structural species which constitute the solution conformations of polypeptide chains are described, one dealing with a group of small globular proteins, the other with a selection of synthetic linear oligopeptides.

2. A DYNAMIC MODEL OF GLOBULAR PROTEIN CONFORMATIONS

2.1. Proteins used for the NMR studies

The experimental data were obtained from a group of proteins related to the basic pancreatic trypsin inhibitor (BPTI). BPTI is a small globular protein of molecular weight 6500. It consists of one polypeptide chain with 58 amino acid residues, including 3 disulfide bonds, and 4 phenylalanines and 4 tyrosines as the only aromatics (Kassell and Laskowski, 1965). The crystal structure at 1.5 Å resolution was extensively refined (Deisenhofer and Steigemann, 1975). In the proton and carbon-13 NMR spectra of BPTI, a large number of resolved resonance lines were assigned to particular residues in the amino acid sequence (Snyder et al., 1975; De Marco et al., 1977; Richarz and Wüthrich, 1977, 1978; Wüthrich et al., 1977, 1978) and related to particular aspects of the molecular conformation (Masson and Wüthrich, 1973; Karplus et al., 1973; Wüthrich and Wagner, 1975; Wagner et al., 1976; Brown et al., 1976, 1978). Hence, NMR spectral similarities and differences between BPTI and related proteins could on the basis of these earlier studies be interpreted in terms of structural features in well defined locations of the protein molecule. Furthermore, BPTI was also a suitable parent compound for the present project since its globular form is outstandingly stable in aqueous solution (table 1) (Vincent et al., 1971; Masson and Wüthrich, 1973; Karplus et al., 1973), which leaves a large experimentally accessible range for studies of modified BPTI species with reduced stability. The chemical modifications of BPTI and two homologous proteins used in this project are characterized in the footnotes to table 1.

2.2. Structural data from high resolution NMR experiments

With the highly refined X-ray structure (Deisenhofer and Steigemann, 1975) and the well resolved NMR spectra it could be demonstrated that the average molecular conformations of BPTI in the crystal and in solution are, with the exception of localized regions on the protein surface (Brown et al., 1976, 1978), very similar (Wüthrich et al., 1978) and the structure type seen in BPTI crystals prevailed also in aqueous

Table 1 Stability and dynamic properties of the globular conformation of BPTI and six related proteins.

Protein [a]	T_{den} (^oC) [b]	k_{ex} (min^{-1}) [c]	$\Delta G^{\ddagger}_{rot}$ (kcal mol^{-1}) [d]
BPTI	> 95	$2 \cdot 10^{-7}$	14.2
CTI	70	$6 \cdot 10^{-5}$	(13.0)
HPI	60	$5 \cdot 10^{-3}$	13.3
R-BPTI	76	$1 \cdot 10^{-4}$	14.2
BPTI*	88	$5 \cdot 10^{-5}$	13.9
Des(A,R)BPTI	> 65	$7 \cdot 10^{-5}$	13.9
TRAM-BPTI	> 95 [e]	--- [e]	14.2

a BPTI, basic pancreatic trypsin inhibitor; HPI, isoinhibitor K from Helix pomatia (Tschesche and Dietl, 1975); CTI, cow colostrum trypsin inhibitor (Čechová et al., 1971); R-BPTI, modified BPTI obtained by reduction of the disulfide bond 14-38, with the cysteinyl residues protected by carboxaminomethylation (Vincent et al., 1971); BPTI*, modified BPTI obtained by cleavage of the peptide bond Lys 15-Ala 16 (Jering and Tschesche, 1976); Des(A,R)BPTI, modified BPTI obtained by cleavage of the peptide bond Lys 15-Ala 16 and removal of Ala 16 and Arg 17 (Jering and Tschesche, 1976); TRAM-BPTI, modified BPTI obtained by transamination of the α-amino group of Arg 1 (Brown et al., 1978).

b T_{den} is the denaturation temperature at pD 5.0.

c k_{ex} is the rate constant for the exchange of the most stable amide proton at pD 4.5 and 36°.

d $\Delta G^{\ddagger}_{rot}$ is the activation energy for 180° flips of the aromatic ring of Phe 45 about the C^{β}-C^{γ} bond at 4°.

e BPTI and TRAM-BPTI were compared under different conditions. In 1 M guanidium chloride solution at pH 6.0, T_{den} was 75° for BPTI and 68° for TRAM-BPTI. At pH 6.0 and 55°, k_{ex} was approximately 8 times larger for TRAM-BPTI than for native BPTI.

solutions of all the proteins in table 1 (Wüthrich et al., 1976; Wagner, 1977; Wagner et al., 1978). That the conformation type seen for BPTI in single crystals was preserved also in a group of related proteins provided the quite unique opportunity to investigate the influence of strictly local variations of the polypeptide structure on the molecular conformation, and thus to obtain new insights into basic aspects of protein dynamics. NMR was used to measure the denaturation temperature, the amide proton exchange rates with the solvent in D_2O solution and the intramolecular mobility of the aromatic rings in the proteins of table 1.

The denaturation temperatures (table 1) were determined from the transition of the ^{1}H NMR spectrum of the globular proteins to a random coil spectrum (Wüthrich, 1976), where at T_{den} equal concentrations of globular and denatured protein were observed. At the denaturation temperature, the exchange between globular and denatured protein was slow on the NMR time scale (Wüthrich et al., 1978).

The amide proton exchange rates were obtained for numerous individual protons by measurements of the decrease of the resonance intensities with time after dissolving the protein in D_2O. Approximately 25 interior amide protons could thus readily be studied (Masson and Wüthrich, 1973; Karplus et al., 1973; Wagner, 1977). Table 1 contains the rate constant for the most slowly exchanging amide proton in each of the different proteins.

The high resolution ^{1}H NMR spectra provide an unambiguous criterion to determine rotational motions of the aromatic rings of phenylalanine and tyrosine about the C^{β}-C^{γ} single bond (Wüthrich, 1976). In BPTI, three aromatic rings were found to undergo 180° flip motions (Hetzel et al., 1976) with frequencies of the order of 50 to 1500 sec^{-1} at 40° and activation energies $\Delta G^{\ddagger}$ of the order of 15 kcal mol^{-1}, one ring was immobilized over the entire accessible temperature range, with $\Delta G^{\ddagger}$ $\approx$ 20 kcal·mol^{-1} at 80°, and four rings were flipping too rapidly to be studied quantitatively (Wüthrich and Wagner, 1975; Wagner et al., 1976). In table 1 the activation energy for the rotational flips of a selected ring is listed for the different proteins.

With regard to the discussions in the following sections, two features may be singled out from the data in table 1. One is that the amide proton exchange rates are correlated with the thermal stability of the globular conformation, i.e. in proteins with lower denaturation temperature the amide proton exchange is faster. Secondly, in contrast, the rotational motions of the aromatic rings are not correlated with the thermal stability or the amide proton exchange, i.e. for corresponding rings essentially identical flip frequencies and activation energies for the flipping motions prevail in all the proteins listed in table 1, unless, of course, the immediate ring environment was affected by the protein modification.

2.3. Description of the solution conformation of globular proteins by a dynamic ensemble of rapidly interchanging molecular structures

Earlier investigations of amide proton exchange in globular proteins have resulted in the suggestion that the exchange occurs via admixture of "open" structures, O(H), to the "closed" globular form of the molecules, C(H) (Hvidt and Nielsen, 1966).

$$C(H) \underset{k_2}{\overset{k_1}{\rightleftharpoons}} O(H) \xrightarrow[D_2O]{k_3} O(D) \underset{k_1}{\overset{k_2}{\rightleftharpoons}} C(D) \qquad (1)$$

In this model exchange of interior amide protons with deuterions of the solvent, D_2O, occurs from the open structures O(H) with the rate k_3. Depending on the relative rates for the closing of the protein, k_2, and the acid/base catalyzed exchange of exposed amide protons in the open forms, k_3, the overall proton exchange rate k_{ex} may be governed either by an EX_1 or an EX_2 mechanism, i.e. one has that $k_{ex} = k_1$ or $k_{ex} = (k_1/k_2)k_3$, respectively (Hvidt and Nielsen, 1966; Englander et al., 1972).

For BPTI, an EX_2 mechanism was indicated by the pH dependence of k_{ex} in the range from 0.5 to 11.0 (Wagner, 1977). The correlation between denaturation temperature and amide proton exchange rates for the different proteins in table 1 further implies that the equilibrium $C(H) \rightleftharpoons O(H)$ involves global variations of the protein structure. This is further supported by the observations that, with few exceptions, the order of the relative exchange rates for the individual amide protons was the same in all the proteins studied (Wagner, 1977; Wagner et al., 1978).

In earlier interpretations of the amide proton exchange scheme (1) it was concluded that the open structures O(H) which mediate exchange of the most stable amide protons, are identical with the molecular structures obtained by thermal denaturation (Tanford, 1970; Englander et al., 1972). This conclusion was based to a large extent on the near equality for several proteins of ΔG^o between the structures C(H) and O(H) in the reaction scheme (1), and ΔG^o between the globular conformation and the denatured protein (Tanford, 1970). The data in table 1 now imply that this equality is coincidential. The lack of a correlation between the frequencies of the 180° flips of the aromatic rings and the denaturation temperature or the amide proton exchange rates clearly shows that the open forms O(H) which mediate the exchange of interior amide protons (equation (1)) cannot be in the same class as the thermally denatured proteins. It is implied that the local environment of the aromatic rings in the closed form C(H) of the proteins is essentially preserved also in the open forms O(H). Independently, admixture among the open species O(H) of molecular structures corresponding to the thermally denatured protein was excluded on the grounds that the transition from denatured to globular BPTI was slow on the NMR time scale even at the denaturation temperature (Wüthrich et al., 1978), and that a slow closing rate k_2 would not be compatible with the observed kinetics of the amide proton exchange (Wagner, 1977).

Overall, the above considerations imply that the "globular solution conformations" of the proteins in table 1, and probably of proteins in general, are best described as dynamic ensembles of all the structural species C(H) and O(H) involved in the amide proton exchange, where the Boltzmann law favors population of species closely related to C(H) over the population of more extremely opened forms. The observed correlation between amide proton exchange rates and denaturation temperature (table 1) may be rationalized by the assumption that the energy barrier $\Delta G^{\ddagger}$ separating the globular conformation from the denatured forms of the protein can be overcome only by cooperative structural fluctuations resulting from overlap of different open structures O(H). With the increase of the population of states O(H) when approaching the denaturation temperature, the probability of such cooperative processes is greatly enhanced, so that denaturation sets in.

2.4. Hydrophobic stability domains in globular protein structures - possible nucleation sites in the folding process

That the hydrophobic pockets which enclose the aromatic rings in the proteins of table 1 are apparently preserved in the open structures O(H) is compatible with the assumption that the molecular fluctuations promoting the amide proton exchange consist primarily of intramolecular translational and rotational motions of intact hydrophobic domains relative to each other. Such fluctuations would primarily open the hydrogen bonded secondary structures which link the different hydrophobic pockets (Deisenhofer and Steigemann, 1975), while decomposition of the hydrophobic domains would occur only as a consequence of the cooperative fluctuations leading to protein denaturation.

Hydrophobic stability domains as evidenced here in one class of globular proteins might quite generally play an essential role in the architecture of protein molecules, a role which may so far have been assigned too one-sidedly to the regular secondary structures adopted by the polypeptide backbone (Dickerson and Geis, 1969). In particular, that the preservation of the hydrophobic pockets surrounding the aromatic rings is not correlated with the overall stability of the globular protein conformation manifested in the denaturation temperature, indicates that such hydrophobic domains might be formed at an early stage of polypeptide folding, possibly even as primary nucleation sites for the folding process.

3. MOLECULAR CONFORMATIONS IN LINEAR OLIGOPEPTIDES

3.1. Amide proton titration shifts in polypeptide chains

Investigations of synthetic model peptides H-Gly-Gly-X-Ala-OH, where X stands for one or several of the common amino acid residues, showed that only a very limited number of amide protons in a polypeptide chain show sizeable intrinsic pH titration shifts, i.e. shifts arising from through-bond interactions with ionizeable groups (Bundi and Wüthrich, 1977, 1978). These are the amide proton of the C-terminal amino acid residue, the amide protons of Asp residues and the residues following Asp in the amino acid sequence, and possibly the amide proton of the residue next to the N-terminus. Intrinsic titration shifts are upfield with increasing pH. In addition, downfield amide proton titration shifts were observed in globular proteins (Lauterwein et al., 1977; Wagner, 1977) and in synthetic oligopeptides (Bundi and Wüthrich, 1978). From comparative proton NMR studies of homologous peptides with and without free carboxylic acid groups, downfield amide proton titration shifts in the pH range from 2 to 6 could be related to conformation dependent hydrogen bonding interactions with carboxylate groups (Bundi and Wüthrich, 1978).

The following are some typical experimental amide proton titration shifts between pH 2 and 6. In the tetrapeptide H-Gly(1)-Gly(2)-Glu(3)--Ala(4)-O(H), downfield shifts of 0.062 ppm and 0.172 ppm were observed for the amide protons of the residues 2 and 3, respectively. The corresponding titration shifts in H-Gly-Gly-Glu-Ala-OCH_3 were 0.077 ppm and 0.228 ppm, in H-Gly-Gly-Gln-Ala-OH 0.021 ppm and -0.006 ppm, in H-Gly-Gly-Met-Ala-OH 0.028 ppm and 0.020 ppm. From comparison of the pK_a values obtained from the amide proton titrations with those for the carboxylic acid groups measured independently by carbon-13 NMR, the large downfield shifts in the Glu peptides were shown to arise from hydrogen bonding with the Glu side chain carboxylate, while the titration shifts in the other peptides could be related to through space interactions with the C-terminus (Bundi and Wüthrich, 1978).

3.2. Determination of small populations of hydrogen bonded structures in flexible linear peptides

On the basis of the experimental observations and the assumptions described in the following, amide proton titration shifts can be used to estimate the population of specific hydrogen bonded structures in the ensemble of species constituting the solution conformation of polypeptide chains.

In the crystal structure of BPTI, which was obtained from crystals grown at pH 10.5, the side chain of Glu 49 is hydrogen bonded to its amide proton (Deisenhofer and Steigemann, 1975). The pK_a value for Glu 49 in BPTI was measured by ^{13}C NMR to be in the range from 3.6 to 3.8 (Richarz and Wüthrich, 1978). In H_2O solution of BPTI, an amide proton titrated 0.408 ppm downfield with a pK_a value of 3.85, and no other amide proton resonance showed a titration shift with a pK_a between 3.5 and 7.9 (Wagner, 1977). These data combined with the above mentioned observations with Glu containing oligopeptides imply that the titration shift of 0.408 ppm has to be assigned to Glu 49. Assuming that the hydrogen bond linking the side chain carboxylate with the amide proton of Glu 49 is nearly 100 % populated in the globular protein and that all the hydrogen bonds in question have similar geometries (or else that the titration shift is not markedly dependent on the hydrogen bond geometry), a 10 % population of a carboxylate-amide proton hydrogen bond in a dynamic ensemble of molecular species is estimated to give rise to a downfield titration shift of 0.04 ppm. Using this number, one has that among the molecular structures contained in the solution conformation of the tetrapeptide H-Gly-Gly-Glu-Ala-OH, approximately 20 % include a hydrogen bond between the amide proton of Gly 2 and the Glu side chain carboxylate, and approximately 50 % a hydrogen bond between the amide proton and the side chain of Glu 3. In the peptides H-Gly-Gly-X-Ala-OH with inert side chains of X, e.g. Met or Gly, the amide proton titration shifts of the residues 2 and 3 indicate that the solution conformation contains approximately 5 % of the molecules in a γ-turn type conformation (Némethy and Printz, 1972). The populations of hydrogen bonded species thus estimated for aqueous solutions of these peptides are compatible with the other NMR parameters. While the preferred spatial arrangement of the Glu side chain was also manifested in the chemical shifts of the γ-methylene protons and the temperature coefficients of the amide proton chemical shifts (Bundi and Wüthrich, 1978), it could only be characterized in detail from the amide proton titration studies. The small admixtures of hydrogen bonded species to the predominantly extended flexible arrangement of the peptide backbone was, however, not measureably evidenced in the conventional NMR parameters.

3.3. Structured species in "random coil" conformations of peptide chains

The molecular flexibility of a predominantly extended peptide chain is considerably less restricted than that of a densely packed globular protein. Sizeable fluctuations of the torsion angles about individual single bonds are hence expected among the species which constitute the solution conformation of such a molecule. It is then quite interesting that amide proton titration shifts indicate that the solution conformations of peptides which would from other evidence, e.g. the vicinal spin-spin coupling constants J(HNHα), be classified as "random coils", include surprisingly high populations of particular structured species.

These observations appear of interest also with regard to possible folding pathways for polypeptide chains (Némethy and Scheraga, 1977; Baldwin, 1978). It is particularly intriguing that the predominant spatial arrangement of the Glu side chain in a small linear peptide is also seen for Glu in the globular protein BPTI (Deisenhofer and Steigemann, 1975). Interestingly, pronounced non-random spatial arrangements were also observed for aromatic side chains in linear oligopeptides (Bundi et al., 1975; Wüthrich and De Marco, 1976), and a theoretical investigation indicated that the orientations of bulky amino acid side chains expected from the nearest neighbor interactions in short peptide fragments prevailed also in the crystal structure of BPTI (Gelin and Karplus, 1975).

ACKNOWLEDGEMENT

Financial support by the Swiss National Science Foundation (project 3.0040.76) is gratefully acknowledged.

REFERENCES

Baldwin, R.L.: 1978, Trends in Biochemical Sciences 3, 66.
Brown, L.R., De Marco, A., Wagner, G. and Wüthrich, K.: 1976, Eur. J. Biochem. 62, 103.
Brown, L.R., De Marco, A., Richarz, R., Wagner, G. and Wüthrich, K.: 1978, Eur. J. Biochem. (submitted).
Bundi, A., Grathwohl, Ch., Hochmann, J., Keller, R., Wagner, G. and Wüthrich, K.: 1975, J. Magn. Reson. 18, 191.
Bundi, A. and Wüthrich, K.: 1977, FEBS Lett. 77, 11.
Bundi, A. and Wüthrich, K.: 1978, Biopolymers (submitted).
Čechová, D., Jonáková, V. and Šorm, F.: 1971, in H. Fritz and H. Tschesche (eds.), Proc. Int. Conf. on Proteinase Inhibitors, W. de Gruyter, Berlin, p. 105.
Deisenhofer, J. and Steigemann, W.: 1975, Acta Cryst. B31, 238.
De Marco, A., Tschesche, H., Wagner, G. and Wüthrich, K.: 1977, Biophys. Struct. Mechanism 3, 303.
Dickerson, R.E. and Geis, I.: 1969, The Structure and Action of Proteins, Harper and Row, New York.
Englander, S.W., Downer, N.W. and Teitelbaum, H.: 1972, Ann. Rev. Biochem. 41, 903.
Gelin, B.R. and Karplus, M.: 1975, Proc. Nat. Acad. Sci. US 72, 2002.
Hetzel, R., Wüthrich, K., Deisenhofer, J. and Huber, R.: 1976, Biophys. Struct. Mechanism 2, 159.
Hvidt, A. and Nielsen, S.O.: 1966, Advan. Protein Chem. 21, 287.
Jering, H. and Tschesche, H.: 1976, Eur. J. Biochem. 61, 443.
Karplus, S., Snyder, G.H. and Sykes, B.D.: 1973, Biochemistry 12, 1323.
Kassell, B. and Laskowski Sr., M.: 1965, Biochem. Biophys. Res. Commun. 20, 463.
Lauterwein, J., Wüthrich, K., Schweitz, H., Vincent, J.P. and Lazdunski, M.: 1977, Communication at the 11th FEBS meeting, Copenhagen, Abstract B5-1 363.
Masson, A. and Wüthrich, K.: 1973, FEBS Lett. 31, 114.
Némethy, G. and Printz, M.P.: 1972, Macromolecules 5, 755.
Némethy, G. and Scheraga, H.A.: 1977, Quart. Rev. Biophysics 10, 239.
Richarz, R. and Wüthrich, K.: 1977, FEBS Lett. 79, 64.

Richarz, R. and Wüthrich, K.: 1978, Biochemistry (in press).
Snyder, G.H., Rowan III, R., Karplus, S. and Sykes, B.D.: 1975, Biochemistry 14, 3765.
Tanford, Ch.: 1970, Adv. Protein Chem. 24, 1.
Tschesche, H. and Dietl, T.: 1975, Eur. J. Biochem. 58, 439.
Vincent, J.P., Chicheportiche, R. and Lazdunski, M.: 1971, Eur. J. Biochem. 23, 401.
Wagner, G.: 1977, Konformation und Dynamik von Protease-Inhibitoren: ^{1}H NMR Studien, ETH Zürich (Ph.D. Thesis Nr. 5992).
Wagner, G., De Marco, A. and Wüthrich, K.: 1976, Biophys. Struct. Mechanism 2, 139.
Wagner, G., Wüthrich, K. and Tschesche, H.: 1978, Eur. J. Biochem. (in press).
Wüthrich, K.: 1976, NMR in Biological Research: Peptides and Proteins, North Holland, Amsterdam.
Wüthrich, K. and De Marco, A.: 1976, Helv. Chim. Acta 59, 2228.
Wüthrich, K. and Wagner, G.: 1975, FEBS Lett. 50, 265.
Wüthrich, K., Wagner, G. and Tschesche, H.: 1976, in H. Peeters (ed.), Proc. XXIII Coll.-Proteids of the Biological Fluids, Pergamon Press, London, p. 201.
Wüthrich, K., Wagner, G., Richarz, R. and De Marco, A.: 1977, in R.A. Dwek, I.D. Campbell, R.E. Richards and R.J.P. Williams (eds.), NMR in Biology, Academic Press, London, p. 51.
Wüthrich, K., Wagner, G., Richarz, R. and Perkins, S.J.: 1978, Biochemistry (in press).

STRUCTURAL INFORMATION FROM PHOTO-CIDNP IN PROTEINS

R. Kaptein
Department of Physical Chemistry, University of Groningen, Groningen, The Netherlands

1. INTRODUCTION

Recently we have developed a 360 MHz laser photo-CIDNP method for the study of proteins in solution (Kaptein et al , 1978 a,b). The method rests on a reversible photo-reaction of a dye with the amino acid residues tyrosine, histidine and tryptophan. When these residues are accessible to the photo-excited dye, nuclear spin polarization can be generated in their side chains resulting in selective enhancements of the corresponding nmr lines. Subtraction of "light" and "dark" spectra leads to greatly simplified spectra. These difference spectra contain structural information, because the CIDNP intensities are related to the extent of exposure of the polarizable groups. The method aids in making nmr peak assignments and allows monitoring of the polarized groups to study a variety of interactions in which the protein may be involved.

In this paper an outline of the method will be given. Both technical and chemical aspects of the generation and detection of the polarization effects will be discussed. The CIDNP behaviour of the aromatic amino acids provides the basic information necessary for applying the method to protein systems. The proteins that will be considered are the bovine pancreatic trypsin inhibitor (BPTI), HEW lysozyme, and ribonuclease A and S. These proteins have been extensively studied by nmr and X-ray crystallographic methods and may therefore suitably serve as model systems. Since the active sites of enzymes are necessarily accessible (at least to substrates!) there is a good chance of observing CIDNP in active site residues. Examples of this will be given in the cases of both ribonuclease and lysozyme.

2. THE LASER PHOTO-CIDNP METHOD

CIDNP originates from magnetic interactions in pairs of free radicals and is observed in their reaction products (Kaptein, 1977). In spite of numerous applications of this effect in organic chemistry it has not thus far been observed in proteins or other biological

B. Pullman (ed.), Nuclear Magnetic Resonance Spectroscopy in Molecular Biology, 211-229.

macromolecules. In order to be useful in these systems the polarization should be generated in the protein itself and preferably not in a variety of reaction products as is the case in most radical reactions. We have accomplished this by employing a cyclic photoreaction of a dye added in low concentration ($\sim 2 \times 10^{-4}$ M) to a solution of the substrate. For tyrosine (TyrH) the photoreaction with the dye (D) can be written as follows.

$$D \xrightarrow{h\nu} {}^{1}D \longrightarrow {}^{3}D \qquad (1)$$

$$^{3}D + TyrH \longrightarrow DH\cdot + Tyr\cdot \qquad (2)$$

$$DH\cdot + Tyr\cdot \longrightarrow D + TyrH \qquad (3)$$

In most cases the triplet dye (^{3}D) abstracts the phenolic hydrogen atom of tyrosine. Nuclear polarization arises from the radical pair recombination step (3). For model compounds such as N-acetyl tyrosine many dyes react according to reactions (1) - (3) with concomitant generation of CIDNP (Kaptein et al., 1978 a)†. However, in most of our work we have employed one particular class of dyes, the flavins. Their photochemistry has been studied in great detail (McCormick, 1977). They can be excited by visible light and quite efficiently induce polarization in three amino acid side chains (Tyr, His, and Trp) in cyclic reactions with no other polarized products. The results presented in this paper have been obtained with 3-N-carboxymethyl lumiflavin (flavin I).

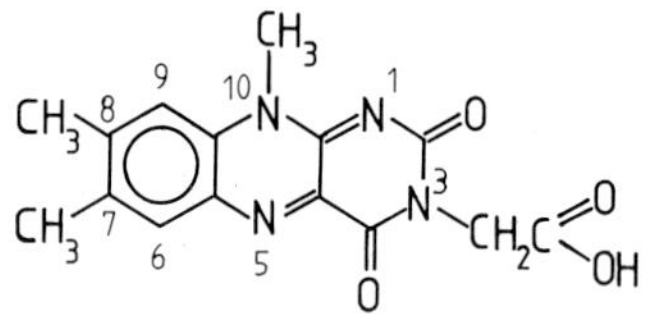

flavin I

Detection of CIDNP in the complex nmr spectra of proteins requires the use of advanced instrumentation such as a spectrometer operating at superconducting fields and a high power argon laser for light excitation. As a light source we have employed a Spectra Physics model 171 argon ion laser. The sample is irradiated in the probe of the Bruker HX-360 spectrometer operating at 360 MHz in the pulse Fourier transform mode. The laser beam was directed via a computer controlled shutter and a mirror through the bottom of the probe. Temperature control of the

---

†After we had communicated our preliminary results on tyrosine CIDNP to Dr. K.A. Muszkat at the Nato Advanced Study Institute on "Chemically Induced Magnetic Polarization" (Urbino, Italy, April 1977) Dr. Muszkat succeeded in observing this effect at 90 MHz (cf. J.C.S. Chem. Commun., 1977, p. 872).

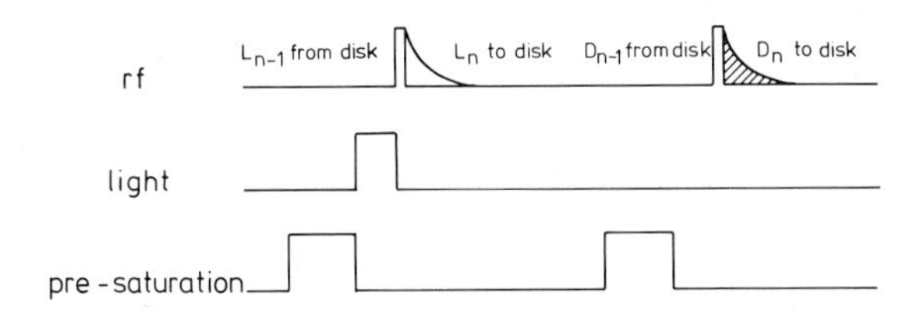

Figure 1. Pulse sequence for the photo-CIDNP difference method. L and D refer to light and dark FID's respectively. Gating of light and rf pulses is controlled by a BNC-12 computer and a NIC-293 pulse controller.

sample is achieved by passing preheated (or cooled) nitrogen through a dewar equipped with an antireflection coated window. Detection of polarization in an envelope of strongly overlapping lines can best be accomplished by a combination of the pre-saturation and difference methods used by Schäublin et al.(1977). The pulse sequence is shown in Figure 1. Alternatingly light (L) and dark (D) free induction decays (FID's) are collected, which can be subtracted to yield the pure CIDNP spectrum. For time averaging the appropriate accumulated FID (L_{n-1} or D_{n-1}) is loaded from disk into core memory and stored back on disk after co-adding the next FID. In this way the effects of slow drift or changing homogeneity due to temperature variations are minimized. The pre-saturation pulse can either be broadband rf irradiation in order to saturate the complete spectrum or, alternatively it can be used in the coherent mode to saturate a strong solvent (e.g. HDO) line. Typical operating conditions are as follows:
1 sec pre-saturation pulse, 0.4 sec light pulse (5 or 7 Watt, multiline), 7 sec delay between 90° pulses. The 14 sec repetition time for the whole cycle is then sufficiently long so that no serious heating occurs. By contrast, in continuous light irradiation experiments even at much lower power levels the thermal load due to the absorption of light may present a serious problem.

3. GROUND RULES: CIDNP IN THE AMINO ACIDS TYROSINE, HISTIDINE AND TRYPTOPHAN

3.1 N-Acetyl tyrosine

The pulse sequence discussed above has been applied to a solution of 10^{-2} M N-acetyl tyrosine and 2×10^{-4} flavin I (Kaptein et al, 1978 a). Figure 2a, b and c show the light spectrum, the dark spectrum and their difference respectively. Strong emission is observed for the 3,5 (ortho) protons of the tyrosine ring and enhanced absorption for the 2,6 ring protons and the β-CH_2 group. The sign of the polarization is determined by the chemical reactions (1) to (3), the hyperfine coupling constants and g-factor of the tyrosyl radical (g=2.0041; Tomkiewicz et al. 1972)

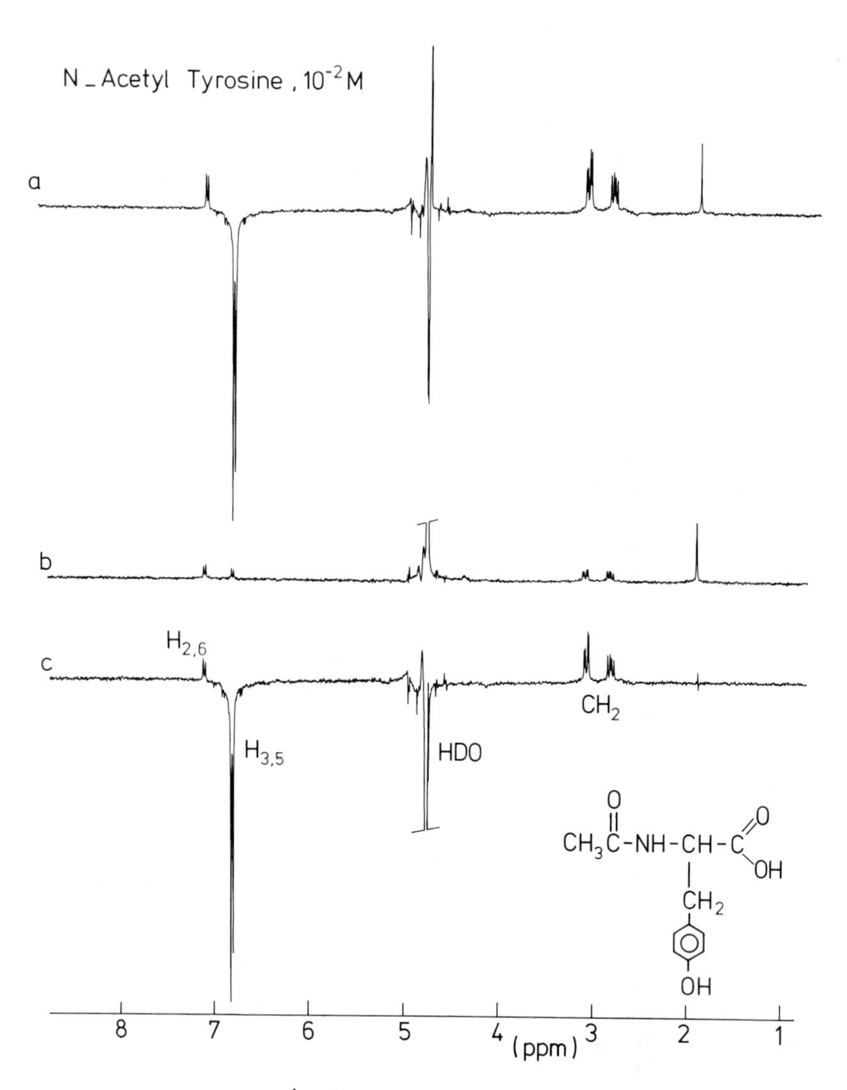

Figure 2. 360 MHz ^{1}H FT nmr spectra (10 pulses) of 10^{-2} M N-acetyl tyrosine and 2 x 10^{-4} M flavin I taken with the pulse sequence of Figure 1, (a) light spectrum (0.4 sec laser irradiation, 5 W, (b) dark spectrum, (c) difference a-b.

and the g-factor of the flavosemiquinone radical (g=2.0030; Ehrenberg and Eriksson,1964). It can be easily predicted by the CIDNP sign rules (Kaptein, 1971). For instance, for the 3,5 ring protons the net effect rule would read

$$\Gamma(3.5) = \mu \cdot \varepsilon \cdot \Delta g \cdot A(3.5) = +++- = - \text{ (emission)}.$$

The polarization is fairly constant over a wide pH range (1-10) with a steep decrease above pH=10 corresponding to the pK_a for ionization of the phenolic OH.

It turns out that the positive enhancement of the 2,6 protons is only for a small part due to direct polarization. It arises mainly from polarization transfer by cross-relaxation from the 3.5 protons. For small molecules this mechanism yields polarization opposite to that of the primary polarization (Closs and Czeropski, 1977). Dipolar cross-relaxation is also responsible for the nuclear Overhauser effect (NOE), which changes sign for slowly tumbling macromolecules. It will be shown that similarly to the NOE the sign of the polarization transferred by cross-relaxation also changes in proteins (see sections 4 and 5).

3.2 N-Acetyl histidine

Figure 3 shows the results of a photo-CIDNP experiment for a solution of 2.5 x 10^{-2} M N-acetyl histidine and 2 x 10^{-4} M flavin I in D_2O at pH = 7.5. The C-2 and C-4 protons of the imidazole ring show strong positive polarization, whereas the β-CH_2 group shows emission in this case. This is in accordance with an intermediate histidyl radical in which the unpaired electron spin is delocalized over the ring. The cyclic reaction scheme is probably very similar to that for tyrosine with the triplet flavin as the reactive species. The g-factor of the histidyl radical is then lower than that of the flavosemiquinone ($g < 2.0030$). The pH dependence of the histidine CIDNP effect shows a gradual increase going from low to high pH and levels out at pH = 7. The protonated imidazole ring seems to be somewhat less polarizable than the non-protonated form.

3.3 N-Acetyl tryptophan

The CIDNP effects observed in the case of N-acetyl tryptophan (5 x 10^{-3} M, flavin I 2 x 10^{-4} M) are shown in Figure 4. Again positive enhancements are found for the ring protons and emission for the β-CH_2 group. The aromatic region of the nmr spectrum of Trp consists of two doublets at 7.7 and 7.5 ppm (C-4 H and C-7 H), two triplets at 7.3 and 7.2 ppm (C-6 H and C-6 H) and a single line at 7.3 ppm of the C-2 proton. Only the C-4 H doublet, the C-6 H triplet, the C-2 H line and the β-CH_2 multiplet are found to be enhanced. This polarization pattern corresponds with a delocalized tryptophyl radical with appreciable spin density on the 2, 3, 4 and 6 carbons. The pH dependence shows an increase around pH = 4.3, a plateau between pH 5 and 10 and a steep increase at pH = 11. A study of the methylated amino acids (Kaptein, to be published) suggests that for Tyr and His hydrogen abstraction is the primary step, whereas for Trp electron transfer yielding the tryptophyl cation radical is the dominant pathway.

With the flavin dye no polarization has been observed for the other common amino acids with the possible exception of a small positive effect for the methyl group of methionine, which is not likely to be of practical importance. Some amino acids (most notably cysteine) do react, however, with the photo-excited flavin as is evidenced by bleaching. Phenylalanine is the only aromatic residue that is not polarizable.

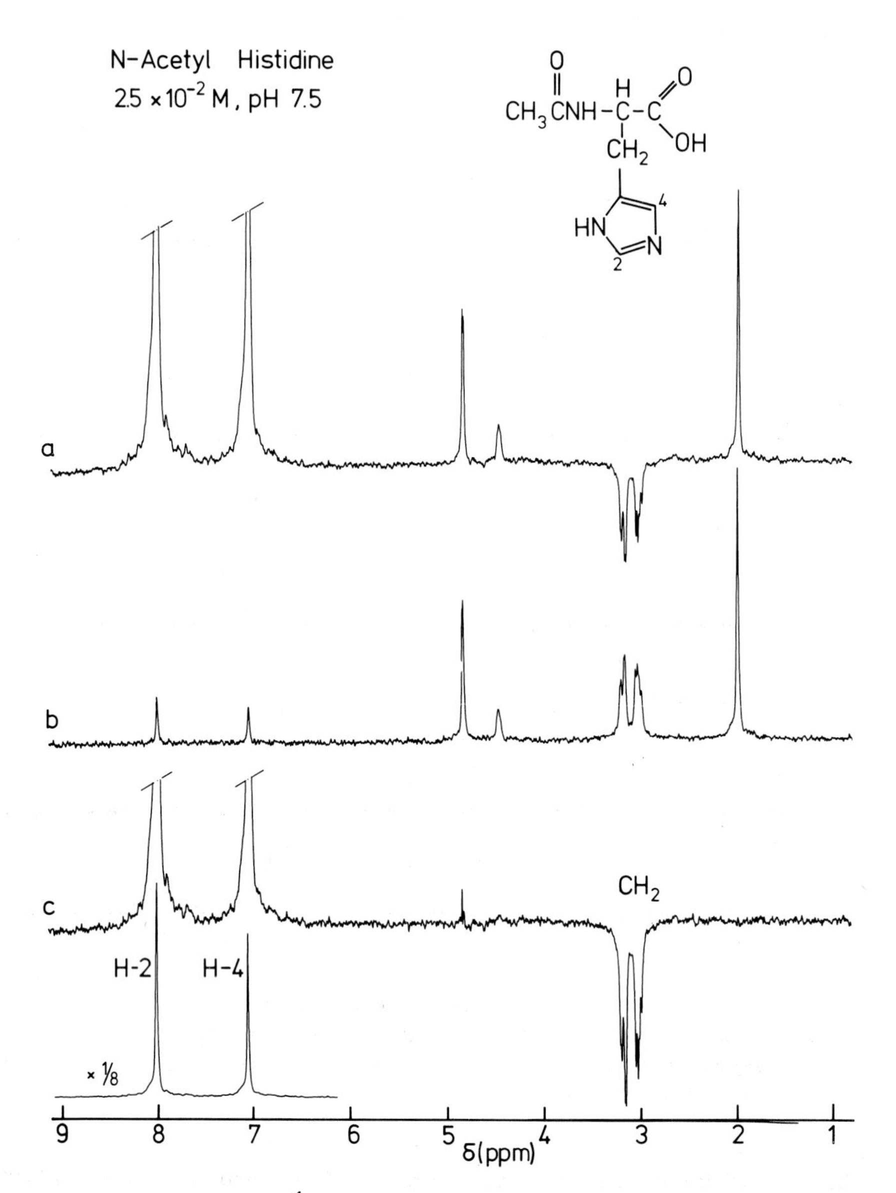

Figure 3. 360 MHz ^{1}H FT nmr spectra (10 pulses) of 2.5 x 10^{-2} M N-acetyl histidine and 2 x 10^{-4} M flavin I at pH = 7.5 taken with the pulse sequence of Figure 1, (a) light spectrum, (b) dark spectrum, (c) difference a-b. The lower trace is a part of spectrum c with the vertical scale reduced by a factor 8.

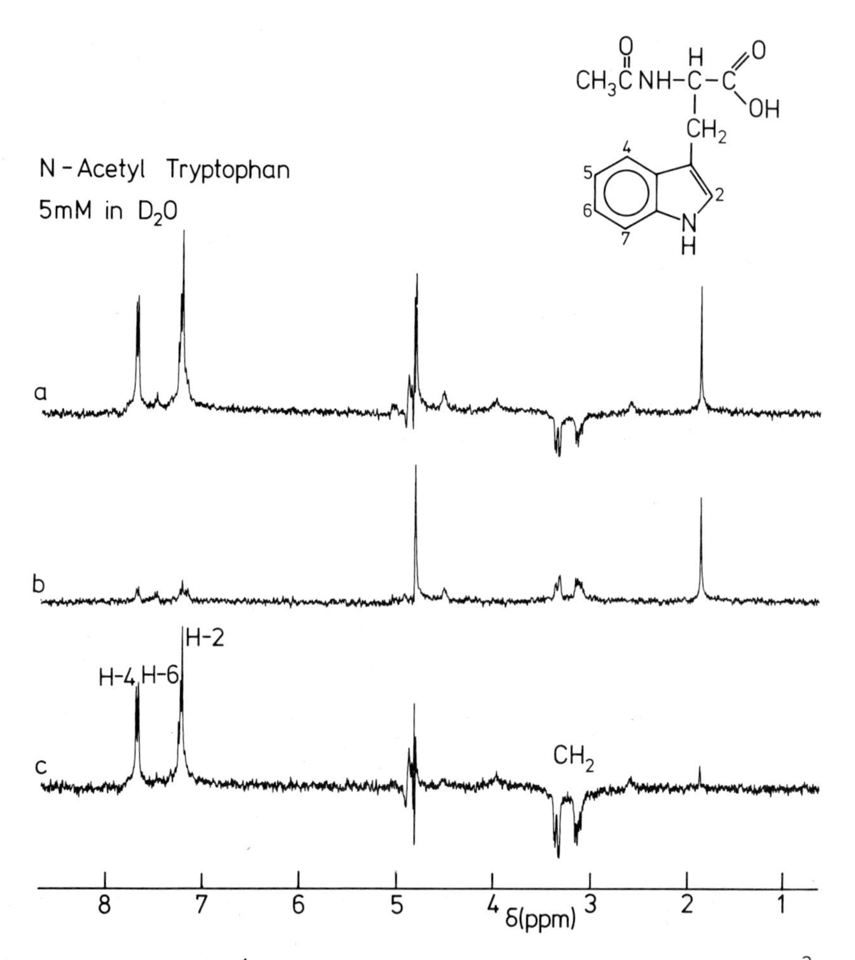

Figure 4. MHz ^1H FT nmr spectra (10 pulses) of 5 x 10^{-3} M N-acetyl tryptophan and 2 x 10^{-4} M flavin I in D_2O, pulse sequence of Figure 1, (a) light spectrum, (b) dark spectrum, (c) difference a-b.

It should be noted that the factors that determine the relative polarization intensities (spin density distribution and g-factors) are not expected to change very much when the radicals are part of a protein. Therefore, apart from relaxation effects and chemical shift differences the polarization patterns of Figures 2, 3 and 4 are expected to occur also for protein bound residues. It is quite convenient that nature has placed the g-factor of the flavosemiquinone in between those of the Tyr and Trp radicals, because this results in polarization of opposite sign for these residues. The enhanced absorption of the Trp ring protons can be easily distinguished from that of the His C-2 and C-4 protons on the basis of the pH dependent chemical shifts of the latter.

Finally we should consider the possibility of specific binding of the flavin dye to proteins. It is known that flavins form complexes with tryptophan and somewhat weaker complexes with tyrosine (Draper and Ingraham, 1970). The way in which this will affect the CIDNP intensities will strongly depend on the strength of the complex. Weak binding to a polarizable residue should enhance the CIDNP effect, since there is a larger probability of reaction. However, the radical pair mechanism requires separation of the radical pair before the spin-selective reaction takes place. Thus, strong complexation i.e. a low off-rate of the flavosemiquinone should not be favourable for large polarization effects. As we shall see in section 6 there are indications for specific binding of flavin I to lysozyme.

4. BOVINE PANCREATIC TRYPSIN INHIBITOR (BPTI)

BPTI is a small globular protein of molecular weight 6500 containing four tyrosines and four phenylalanines as the only aromatic residues. Its crystal structure is accurately known (Deisenhofer and Steigemann, 1975). Detailed nmr studies of the protein have been made by Wüthrich et al. (1977). The nmr lines of all four tyrosines have been assigned (Snyder et al., 1976). Thus, BPTI serves as an excellent test case for the photo-CIDNP method. Figure 5 shows the laser irradiated, dark, and difference spectra obtained for a solution of 5 mM BPTI and 2 x 10^{-4} M flavin I (Kaptein et al., 1978 b). The most prominent features of the difference spectrum (Figure 5c) are the two emission lines at 7.04 and 6.74 ppm. On the basis of chemical modification studies Snyder et al. (1976) assigned these lines to Tyr 10 and Tyr 21. The other tyrosines, Tyr 23 and Tyr 35, do not show any effect. Inspection of the X-ray structure (Deisenhofer and Steigemann, 1975) reveals that Tyr 10 and Tyr 21 lie at the surface of the protein, whereas Tyr 23 and Tyr 35 are more or less buried in the interior. For a compact rigid protein such as BPTI it is expected that these features still hold for the structure in solution. Thus, we conclude that in this case the photo-CIDNP method discriminates between exposed and buried tyrosine side chains.

It is interesting to compare some other properties of the tyrosine residues of BPTI in the light of the present results (see Table 1). The pK_a's of the phenolic hydroxyl groups of Tyr 10 and Tyr 21 are closer to the value of the free amino acid (pK_a = 10.1), than those of the residues 23 and 35. The higher pK's probably reflect a more hydrophobic environment. The results of nitration with tetranitromethane are also in agreement with a more easy access to Tyr 10 and Tyr 21. However, extensive iodination results in reaction of Tyr 35 as well. In general chemical modification studies do not always seem to be reliable for the distinction between exposed and buried groups and have the possible disadvangtage of introducing structural perturbations.

Some minor features of Figure 5c are also worth noting. First, the small emission at 7.30 ppm belongs to the 2,6 ring protons of Tyr 10

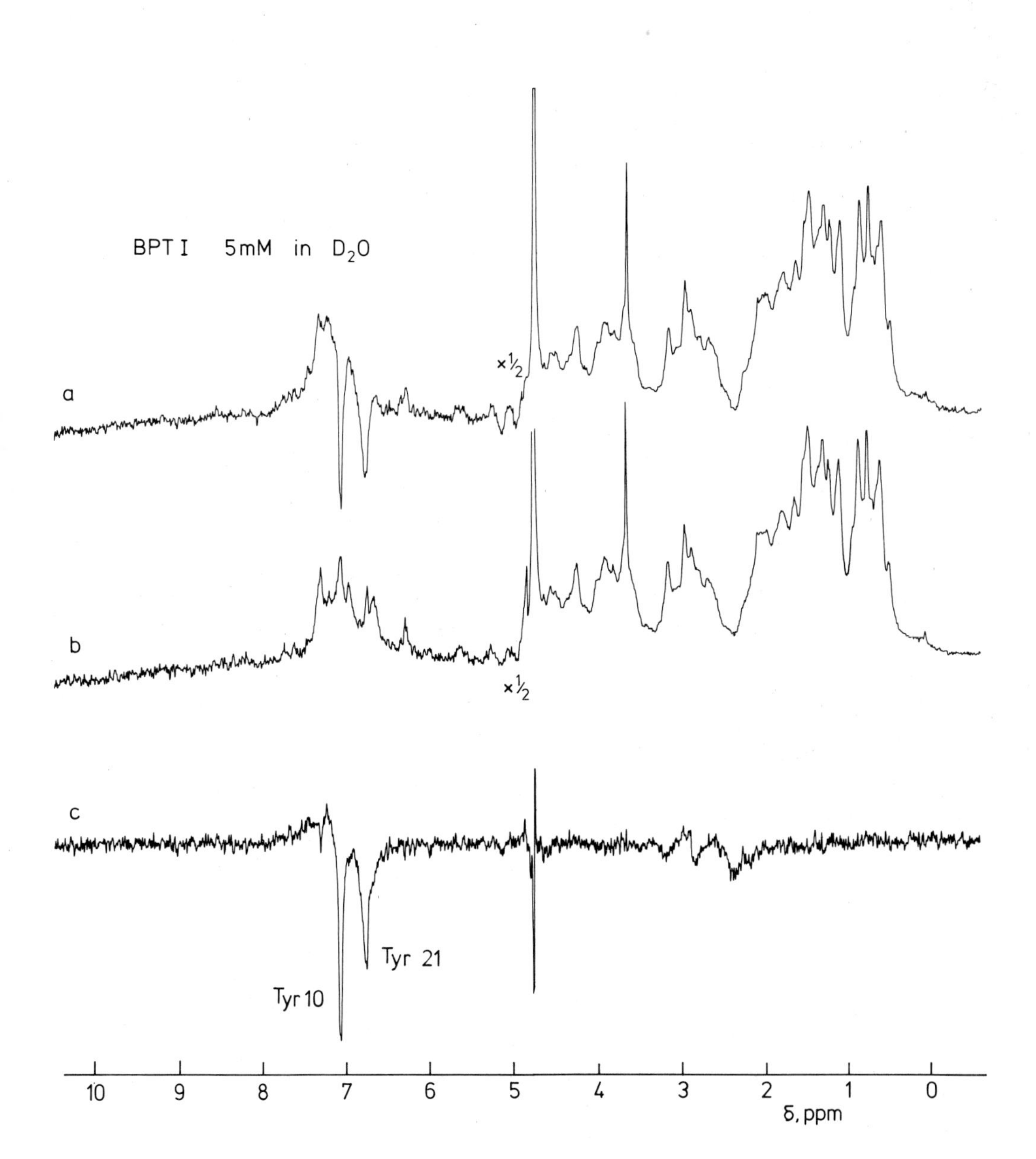

Figure 5. 360 MHz ^{1}H FT nmr spectra (40 pulses) of 5×10^{-3} M BPTI and 2×10^{-4} M flavin I in D_2O at pH = 5.6, temperature 21°C, (a) light spectrum (laser pulse 0.6 sec, 5 W), (b) dark spectrum, (c) difference a-b.

Table 1. Some properties of the tyrosine residues of BPTI

Tyrosine residue	10	21	23	35
chemical shift of 3,5 protons, in ppm (DSS) (Snyder et al., 1976)	7.04*	6.74*	6.30	6.76,6.80
pK_a (Wagner et al., 1976)	9.9	10.4	11.5	11.1
nitration (Meloun et al., 1968)	$3\text{-}NO_2$	$3\text{-}NO_2$	-	-
iodination (Sherman and Kassell, 1968)	3-I	3-I	-	$3,5\text{-}I_2$

*emission observed in photo-CIDNP experiment.

and is an example of the cross-relaxation effect mentioned in section 3.1. In the slower tumbling protein polarization transfer from the 3,5 protons results in polarization opposite to that found in the free amino acid (cf. Figure 2). There is also polarization observable in the region around 3 ppm, which on account of the quality of the subtraction must be considered real. The positive components are undoubtedly due to $\beta\text{-}CH_2$ protons of Tyr 10 and Tyr 21. Interestingly, the small emission at 2.91 ppm most probably belongs to the $\varepsilon\text{-}CH_2$ group of Lys 41, which has been assigned by Brown et al.(1976) by observing a titration behaviour coupled to that of Tyr 10. This shows that Tyr 10 and Lys 41 are very close in space, just as is required by the cross-relaxation mechanism. Polarization transfer to surrounding shells of nuclei could afford valuable structural information similar to that from NOE measurements.

5. RIBONUCLEASE

Most nmr work on ribonuclease has concentrated on the four histidine C-2 proton resonances that are shifted out of the aromatic envelope (Meadows et al., 1967; Markley, 1975a). The nmr assignments of the C-2 and C-4 protons of all four histidines are now firmly established (Markley, 1975a). Not much is known, however, about the ^1H resonances of the other aromatic residues (six Tyr and three Phe), although Markley (1975b) has made an attempt at assigning Tyr 25. In section 5.1 we shall compare the photo-CIDNP results for RNase A and RNase S with X-ray data (Richards and Wyckoff, 1971). RNase S is a complex of the S-peptide and S-protein obtained from RNase A by subtilisin cleavage of the bond between residues 20 and 21. It has the same enzymatic activity as RNase A (Richards and Vithayathil, 1959). The effect on the photo-CIDNP spectrum of inhibitor binding is discussed in section 5.2.

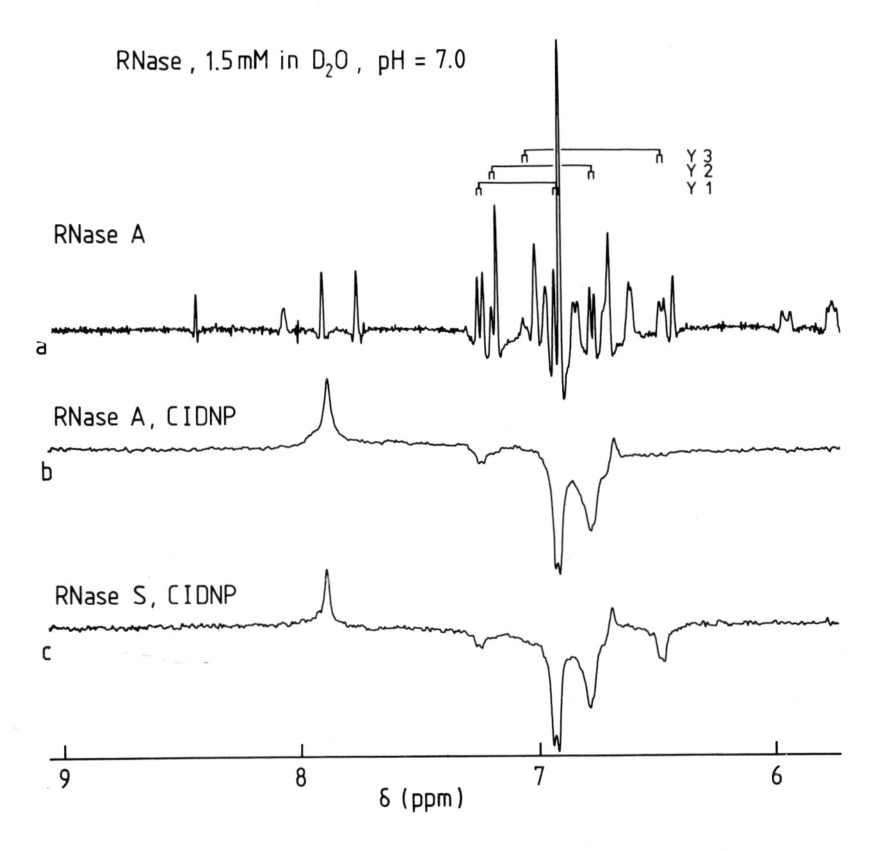

Figure 6. (a) Aromatic region of the 360 MHz ^1H nmr spectrum of 1.5 x 10^{-3} M ribonuclease A in D_2O , pH = 7.0, 38°C, 2000 pulses. The resolution is increased by multiplication of the FID by a sine bell, (b) photo-CIDNP difference spectrum of RNase A (25 pulses), (c) photo-CIDNP difference spectrum of RNase S (same conditions). Pairs of doublets indicated by Y1, Y2 and Y3 belong to tyrosines.

5.1 Photo-CIDNP in RNase A and RNase S

Figure 6a shows the aromatic part of the 360 MHz spectrum of bovine pancreatic ribonuclease A (1.5 mM in D_2O, pH = 7.0, 38°C). The spectral resolution was artificially enhanced by multiplication of the FID by a sine bell (Wüthrich et al., 1977). The spectrum of RNase S is very similar. The doublet resonances of three of the six tyrosine residues indicated by Y1, Y2, and Y3 have been indentified by double resonance methodes (Lenstra et al., 1978). The photo-CIDNP difference spectra of RNase A and RNase S taken under the same conditions are shown in Figure 6b and 6c. Positively enhanced lines at 7.92 ppm and 6.71 ppm belong to the C-2 and C-4 protons of His 119. This active site residue is also the most exposed of the four histidines as judged from the X-ray structure (Richards and Wyckoff, 1971). For RNase A (figure 6b) strong emissions are observed at 6.94 and 6.78 ppm due to the 3,5 protons of the tyrosines Y1 and Y2. The corresponding 2,6 protons are polarized by cross-relaxation and show up as weak emissions at 7.26 and 7.19 ppm respectively. These tyrosines have not yet been assigned to specific residues. However, inspection of the X-ray structure suggests that Tyr 76, 92 and 115 are the most exposed tyrosines, so that these would seem likely candidates for Y1 and Y2. Further work on this assignment problem is in progress.

Comparison of Figure 6b with Figure 6c reveals a remarkable difference between RNase A and RNase S. In RNase S an extra tyrosine (Y3) is polarized, while the remainder of the spectrum is virtually identical. The only differences in the crystal structures of RNase A and S occur in the region of the residues 16-23. Thus, we conclude that Y3 must be identified with Tyr 25, which is close to this region. Apparently the opening of the subtilisin loop renders Tyr 25 accessible to the photo-excited flavin. Further evidence for this assignment comes from the pH dependence of the chemical shift of the Y3 protons (Lenstra et al., 1978). These protons show a titration with the pK_a of His 48, which is adjacent to Tyr 25. We note that the present assignment differs from that of Markley (1975b).

5.2 Binding of the inhibitor 2'-CMP

We attributed the positively enhanced lines in Figure 6b to the active site His 119 on the basis of its known chemical shifts (Markley, 1975a). This is nicely confirmed by a photo-CIDNP experiment in the presence of the competitive inhibitor cytidine 2'-monophosphate, which is known to bind with its phosphate group to His 119. In figure 7 the normal photo-CIDNP difference spectrum of 1.5 mM RNase A is compared with that in the presence of 8 mM 2'-CMP. It can be seen in Figure 7b that the positive lines have disappeared from the spectrum. Thus, the inhibitor blocks accessto His 119 or otherwise inactivates this residue towards the photoreaction. Since histidines play an important role in the catalytic mechanism of many enzymes, the enhancement of histidine resonances by CIDNP is of particular interest.

6. HEW LYSOZYME

Hen egg white lysozyme catalyzes the hydrolysis of cell wall polysaccharides. The crystal structures of the enzyme and of some of its oligosaccharide complexes have been studied by X-ray methods (Blake et al., 1967). Its aromatic residues (three Tyr, one His, six Trp, and three Phe) give rise to a complex nmr spectrum in the region of 6 to 8 ppm and cause large upfield shifts for a number of methyl groups. Extensive nmr studies on lysozyme (McDonald and Phillips, 1970; Campbell et al., 1975) have led to the assignments of many lines to specific residues. The nmr work of the Oxford group has recently been reviewed by Dobson (1977). Here we shall examine the flavin sensitized photo-CIDNP spectrum of lysozyme and discuss the effects of binding of the

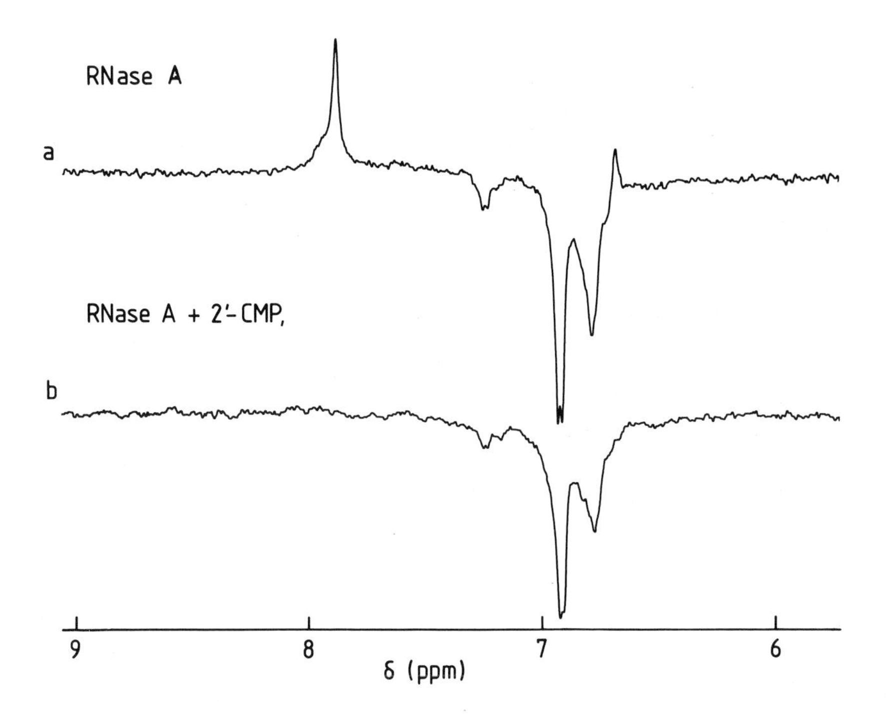

Figure 7. (a) Aromatic region of the 360 MHz photo-CIDNP spectrum of 1.5 x 10^{-3} M RNase A and 2 x 10^{-4} M flavin I in D_2O, pH = 7.0, 38°C; (b) idem in the presence of 8 x 10^{-3}M 2'-CMP.

competitive inhibitor N-acetylglucosamine (NAG).

6.1 Tryptophan photo-CIDNP in lysozyme

The standard pulse sequence of Figure a has been applied to a 2 mM solution of HEW lysozyme with 4 x 10^{-4} M flavin I in D_2O at pH = 5.4. At the temperature of the experiment (55°C) the protein is still in its folded conformation.

Figure 8a, b and c show the,light, dark and difference spectra respectively. The polarized lines indicated by F6, F8, and F10 belong to the C-6H, 8-methyl and 10-methyl groups of flavin I. The remainder of the spectrum of Figure 8c can be analysed in terms of two polarized tryptophan residues indicated by Trp I and Trp II. Tryptophans are found at positions 28, 62, 63, 108, 111 and 123 in the chain. Identification of the polarized residues relies on the crystal structure and on the nmr assignments of all tryptophan C-2 protons made by the Oxford group (R.J.P.Williams, 1978). Estimates by Glickson et al. (1971) of solvent exposure of the tryptophan indole rings and our own estimates of dye accessibility from inspection

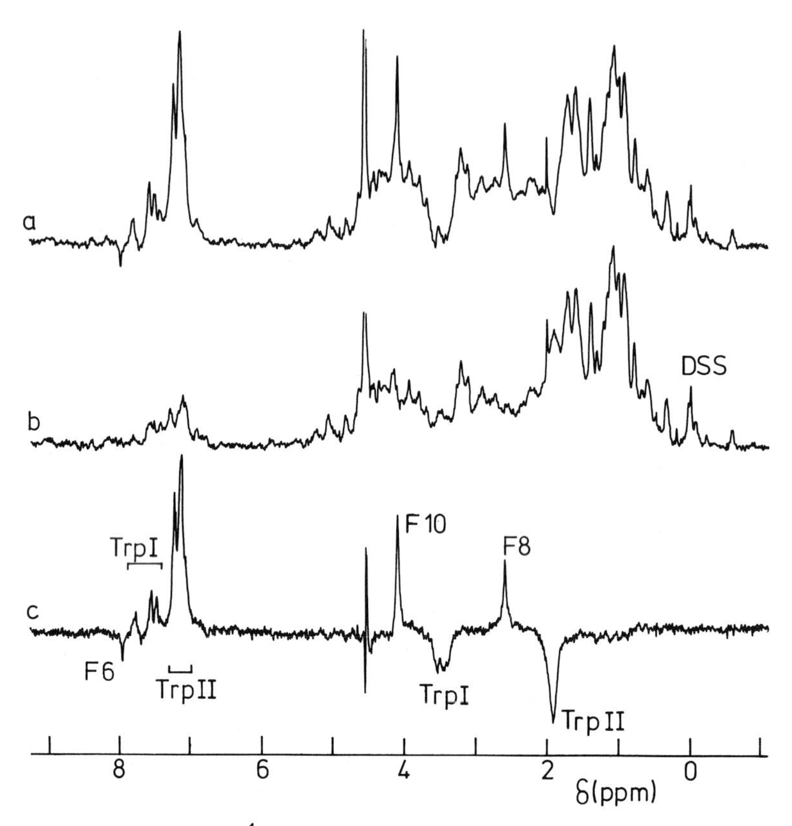

Figure 8. 360 MHz ^{1}H nmr spectra of 2 x 10^{-3}M HEW lysozyme and 4 x 10^{-4}M flavin I in D_2O (pH = 5.4, temp. 55°C) taken with the pulse sequence of Figure 1 (a) light spectrum, (b) dark spectrum, (c) difference a-b. Polarized flavin lines are indicated by F6, F8 and F10. Trp I and Trp II denote lines belonging to two tryptophan residues (see text).

of a molecular model based on the crystal coordinates indicate the following. Trp 28 is completely buried in the interior. The approach of the flavin dye to Trp 108 and Trp 111 should be very difficult, where as access to Trp 62, 63 and 123 seems possible. The indole ring of Trp 62 is by far the most exposed and therefore we attribute the strongly polarized lines (Trp II) to this residue. This assignment is corroborated by the fact that the single C-2 H line of Trp 62 was found to resonate at 7.10 ppm (Williams, 1978). Moreover, the large upfield shift of the β-CH_2 emission at 1.89 ppm (shifted from its normal position of 3.3 ppm) is probably due to a ring current effect as would be expected for the β-CH_2 of Trp 62 considering its proximity to the indole ring of Trp 63.

The assignment of the other polarized tryptophan (Trp I) is less certain at present. However, it seems most likely that Trp I should be identified with Trp 63. The broad C-2 H line found at 7.69 ppm by Williams (1978) for this residue should then correspond with the low field absorption at 7.75 in Figure 8c. Assignment to Trp 123 (C-2 H at 7.58 ppm, Williams, 1978) would be in conflict with the recent assignment of the low field doublet of aromatic region to Trp 28 (Perkins et al., 1977). Table 2 summarizes these results.

Table 2. Assignments of the polarized tryptophan residues of lysozyme (cf. Figure 8.).

	C-2 H	C-4 H	C-6 H	β-CH_2
Trp I (63)[a]	7.75[b]	7.53	7.45	3.53, 3.42
Trp II(62)	7.10	7.20	7.05	1.89
free tryptophan	7.3	7.7	7.3	3.3

(a) probable assignment.
(b) chemical shifts in ppm.

It should be noted that the side chains of both Trp 62 and Trp 63 are situated in the active site cleft of lysozyme. Their indole NH atoms are involved in binding to sugar oxygen atoms of the polysaccharide substrates at binding site C (Blake et al., 1967). The inhibitor N-acetylglucosamine (NAG) also binds at this site. Therefore, it seemed interesting to study the photo-CIDNP effects of lysozyme in the presence of NAG.

6.2 Binding of N-acetylglucosamine

The photo-CIDNP difference spectrum of 2 mM lysozyme in D_2O in the presence of 50 mM NAG is presented in Figure 9a. The spectrum of Figure 9b was taken under exactly the same conditions but in the absence of NAG. Remarkably, an *enhancement* of the CIDNP intensities occurs in the presence of NAG. Furthermore, the polarized lines are shifted. This can be clearly seen in Figure 9c, which is the difference of spectrum (a) (at half the vertical scale) and spectrum (b). The most pronounced shifts occur for the lines of Trp 62: an upfield shift for the C-2 H at 7.1 ppm, a splitting of the β-CH_2 protons at 1.9 ppm and upfield shift of the α-CH at 4.5 ppm. The emission of this latter proton is also present in Figure 8c, but at the higher temperature it is obscured by the HDO line. The α-CH polarization is still another example of polarization transfer by cross-relaxation (in this case from the β-CH_2 group). The splitting of the β-CH_2 protons by 0.2 ppm corresponds with a conformational change upon NAG binding, whereby one proton moves towards the plane of Trp 63 by approximately 0.1 Å, whereas the other proton moves away from this plane over the same distance. Smaller shifts occur for the lines of the other tryptophan. Interestingly the flavin 10-methyl line at 4.0 ppm

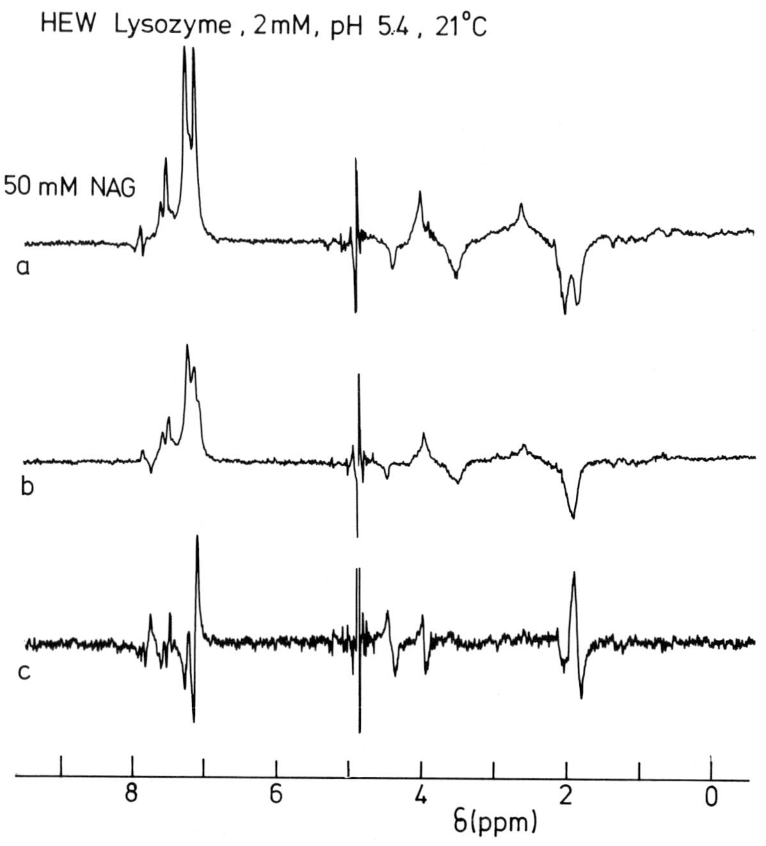

Figure 9. (a) Photo-CIDNP difference spectrum of 2 x 10^{-3}M lysozyme with 4 x 10^{-4} M flavin I in D_2O (pH = 4.6, temp. 22°C) in the presence of 5 x 10^{-2} M N-acetylglucosamine, (b) idem without inhibitor, (c) difference ½ a - b.

also shifts in the presence of NAG. Similar shifts have been observed in the normal nmr spectra of both flavin and lysozyme upon mixing. This indicates that flavin I (weakly) binds to lysozyme. The fact that this binding is affected by formation of the lysozyme/NAG complex suggests that the flavin binds in the region of Trp 62 and Trp 63, close to the sugar binding site C.

Finally we shall speculate on the observed increase of the CIDNP intensities upon NAG binding. The solvent exposure of the indole rings of Trp 62 and Trp 63 is only slightly reduced in the lysozyme/NAG complex (Glickson et al., 1971), so that electron transfer reactions should still be possible (as opposed to NH abstraction which should be impeded in the complex). The CIDNP increase may be related either to the primary photochemical reaction or to the efficiency of the polarization process (or both). In the first case electron donation of the sugar oxygen atoms bound to the ring NH's of Trp 62 and Trp 63 may increase the reactivity of the rings towards electron transfer to the triplet flavin. In the second case loosening of the flavin/ lysozyme complex by NAG binding would promote the separation of the radical pair and hence increase the CIDNP efficiency as we discussed in section 3.3. The origin of the polarization increase will be further investigated. Nevertheless, in spite of these uncertainties it is clear that the selective CIDNP enhancements of active site tryptophan residues permit detailed studies of the binding of various inhibitors to lysozyme.

7. CONCLUSIONS

Some features of the laser photo-CIDNP method will now be summarized. We have shown that by employing a dye as sensitizer nuclear spin polarization can be generated in native proteins. When flavins are used as dyes nmr lines of Tyr, His, and Trp residues can be selectively enhanced when these residues are accessible to the photo-excited dye. Thus, the method constitutes a "surface probe" with the very high resolution of the nmr technique, which is capable of resolving lines due to individual residues. Large enhancements, often of the order of a magnitude or more, are translated in considerable time savings. Combined with a difference technique the method results in substantial simplifications of the complex protein nmr spectra and in an effective increase in the spectral resolution for the polarized groups. This permits detailed studies of various interactions of proteins, such as the enzyme-inhibitor interactions discussed in sections 5.2 and 6.2 for RNase A and lysozyme. For both enzymes the photo-CIDNP spectrum proved to be very sensitive to the presence of inhibitors, albeit in completely different ways ! Examples of applications to the study of protein-nucleic acid interactions will be treated by Hilbers et al. elsewhere in this volume. We believe that in spite of its brief existence the photo-CIDNP method is potentially a powerful tool for the study of protein structure in solution.

Acknowledgements. This work was supported by the Netherlands Foundation for Chemical Research (SON) with financial aid from the Netherlands Organisation for the Advancement of Pure Research (ZWO). I am indebted to Dr.R.J.P.Williams (Oxford) for communicating the assignments of the tryptophan C-2 H resonances of lysozyme prior to publication and to Dr.F.Müller (Wageningen) for a gift of flavin I. I thank my coworkers K.Dijkstra and K.Nicolay for their skilful experimental assistance. Stimulating discussions with Dr.H.J.C.Berendsen, Dr.G.T.Robillard and Dr.L.J.Berliner are gratefully acknowledged.

8. REFERENCES

Blake, C.C.F., Johnson, L.N., Mair, G.A., North, A.C.T., Phillips, D.C. and Sarma, V.R.: 1967, *Proc.Royal Soc.* B167, pp.378-388.

Brown, L.R., De Marco, A.Wagner, G. and Wüthrich, K.: 1976,

Eur. J.Biochem. 62, pp. 103-107.
Campbell, I.D., Dobson, C.M., and Williams, R.J.P. : 1975, *Proc.Royal Soc.* A345, pp. 41-59.
Closs, G.L. and Czeropski, M.S. : 1977, *Chem.Phys.Letters* 45, pp.115-116.
Deisenhofer, J. and Steigemann, W. : 1975, *Acta Cryst.B* 31, pp.238-250.
Dobson, C.M.: 1977, in R.A.Dwek, I.D. Campbell, R.E. Richards, and R.J.P. Williams (Eds), *NMR in Biology*, Academic Press, London,pp.63-94.
Draper, R.D. and Ingraham, L.L.: 1970, *Arch. Biochem. Biophys.* 139, pp. 265-268.
Ehrenberg, A. and Eriksson, L.E.G.: 1964, *Arch. Biochem. Biophys.* 105, pp. 453- 454.
Glickson, J.D., Phillips, W.D., and Rupley, J.A.: 1971, *J.Amer. Chem. Soc.* 93, pp. 4031-4038.
Kaptein, R. : 1971, *Chem.Commun.* pp.732-733.
Kaptein, R. : 1977, in L.T.Muus, P.W.Atkins, K.A.Mclauchlan, and J.B.Pedersen (Eds), *Chemically Induced Magnetic Polarization*, Ch. 1, pp. 1-16.
Kaptein, R. Dijkstra, K., Müller F., van Schagen, C.G. and Visser, A.J.W.G.: 1978a, *J.Magn. Res.* (in press).
Kaptein, R. , Dijkstra, K. and Nicolay, K.: 1978b (submitted for publication).
Lenstra, L., Bolscher, B., and Kaptein, R.: 1978 (to be published).
Markley, J.L.: 1975a, *Accounts Chem.Res.*, 8, pp. 70-80.
Markley, J.L.: 1975b, *Biochemistry*, 14, pp. 3554-3561.
McCormick, D.B., : 1977, *Photochem.Photobiol.*, 26, pp. 169-182.
McDonald, C.C., and Phillips, W.D.: 1970 in G.D.Fasman and S.N.Timasheff (Eds), *Fine Structure of Proteins and Nucleic Acids*, Marcel Dekker, New York, Ch.1 pp. 1-48.
Meadows, D.H., Markley, J.L., Cohen, J.S., and Jardetzky, O.: 1967, *Proc. Natl. Acad. Sci. U.S.A.*, 58, pp. 1307-1313.
Meloun, B., Fric., I., and Sorm, F.: 1968, *Eur.J.Biochem.* , 4, pp. 112-117.
Perkins, S.J., Johnson, L.N., Phillips, D.C., and Dwek, R.A.: 1977, *FEBS Letters* 82, pp. 17-22.
Richards, F.M., and Vithayathil, P.J. : 1959, *J.Biol.Chem.*, 234, pp. 1459-1465.
Richards, F.M., and Wyckoff, H.W.: 1971, in P.D.Boyer (Ed.) , *The Enzymes*, Academic Press, New York, Vol IV, Ch.24,pp.647-806.
Schäublin, S., Wokaun, A. and Ernst, R.R.: 1977, *J.Magn. Res.* ,27, pp. 273-302.
Sherman, M.P. and Kassell, B.: 1968, *Biochemistry*, 7 pp. 3634-3641.
Snyder, G.H., Rowan III, R., Karplus, S., and Sykes, B.D.: 1976, *Biochemistry*, 14, pp. 3765-3777.
Tomkiewicz, M., McAlpine, R.D., and Cocivera, M.: 1972, *Can.J.Chem.*, 50, pp. 3849-3856.
Wagner, G., De Marco, A., and Wüthrich, K.: 1976, *Biophys. Struct. Mech.*, 2 pp.139-158.
Williams, R.J.P.: 1978 (personal communication).
Wüthrich, K., Wagner, G., Richarz, R., and De Marco, A.: 1977, in R.A. Dwek, I.D.Campbell, R.E.Richards, and R.J.P.Williams (Eds.), *NMR in Biology*, Academic Press, London, pp. 51-62.

DISCUSSION

Rüterjans: Did you observe also enhancements of ring resonances of aromatic amino acids of the interior of proteins ? If so your beautiful method would also be a probe for the dynamic behaviour of proteins indicating that the fluctuations of protein structures may be as large as to allow the flavin molecule to penetrate.

Apparently not all tyrosine or histidine resonances of the protein surface are enhanced (His 105 of Rnase A is supposed to be on the surface). Does this result reflect the low association constant of the association of the flavin to these amino acid residues ?

Is the enhancement of Tyr 25 or Rnase S dependent on the presence or absence of nucleotides ? Can you derive interactions of Tyr 25 with Hop 14 or His 48 from the pH dependence of the enhanced resonance ?

Kaptein: We have not observed CIDNP effects of residues in the interior of proteins. However, in the case of lysozyme at higher temperatures we did observe polarization due to tyrosines, which is not present in the low temperature spectra. Apparently, these tyrosines become accessible to the photo-excited flavin at the higher temperature.

As to your question regarding RNase I do not know why we do not see His 105 which should be only slightly less accessible than His 119. There should not be any association of the flavin with histidine residues. We have not yet studied the behaviour of the Tyr 25 resonances in sufficient detail to allow conclusions about nucleotide binding or interaction with Asp.

NITROGEN 15 NUCLEAR MAGNETIC RESONANCE INVESTIGATIONS ON AMINO ACIDS

F.Blomberg and H.Rüterjans
Institut für Physikalische Chemie der Universität
D-44oo Münster, Germany

Although nitrogen 15 nuclear magnetic resonance spectroscopy appears to be very useful for investigating structures of nitrogen containing molecules, the low natural abundance and unfavourable sensitivity of the nitrogen 15 nucleus have limited its application. These disadvantages can be overcome by NMR pulse techniques, large volume probes or enrichment of the nitrogen 15 isotope. Especially for biological molecules enrichment of the isotope seems to be necessary, since in most cases there is not much material available.
In many cases the nitrogen containing groups of amino acids, i.e. the imidazole ring of histidine, the guanidinium group of arginine or the amide group of asparagine and glutamine play a crucial role in the catalytic function of enzymes (1). In order to investigate the mechanisms of such enzymes by nitrogen 15 NMR it is necessary to understand the behaviour of the nitrogen resonances of simple molecules. Hence, the purpose of this paper is to report results of nitrogen 15 and carbon 13 NMR work on several amino acids, i.e. chemical shifts and coupling constants and to derive from these data some features about the structure and conformation of these compounds as well as informations about their interactions with the solvent water.

EXPERIMENTAL

Nitrogen 15 enriched (95%) glycine, glutamine (amide nitrogen enriched), lysine (α-amino nitrogen enriched), proline, arginine (terminal nitrogens of the guanidinium group enriched) and histidine (imidazole nitrogens enriched) were purchased from Rohstoff-Einfuhr GmbH, Düsseldorf,Germany. N-acetyl-histidine (imidazole nitrogens enriched) was prepared according to standard procedures (2).
NMR measurements were performed using 2 ml of an o.2 M solution. All nitrogen 15 NMR measurements were carried out on a Bruker HFX 9o at 9.12 MHz, the carbon 13 NMR experiments

B. Pullman (ed.), Nuclear Magnetic Resonance Spectroscopy in Molecular Biology, 231-246.

were run with a Bruker WH 27o at 67.89 MHz. In all cases the deuterium lock signal was provided by deuteriumoxide in a coaxial capillary inside the 1o mm sample tube (Wilmad Glass Co., Buena, N.J., USA).
pH values were measured directly in the NMR sample tube using a special combined electrode (Ingold, Frankfurt, Germany) and a Radiometer pH-meter (Model PHM 26).
To remove traces of paramagnetic impurities which may broaden or obscure nitrogen 15 NMR resonances due to their influence on the relaxation times and the NOE we have used a purification procedure described elsewhere (3).
Titration curves were calculated and fitted to the pH-dependence of the chemical shifts and coupling constants using the Henderson-Hasselbalch equation (3).

RESULTS AND DISCUSSION

Chemical shifts

The chemical shift values of the different ionizised species of the amino acids are listed in Table 1. The chemical shifts are given in parts per million referenced to the external standard ammoniumnitrate (ammonium nitrogen enriched, 4 M in 2 M nitric acid).

Table 1 Nitrogen 15 chemical shift values of some amino acids.

Chemical shifts, ppm [a]	Cation	Amphion	Anion
Glycine	- 1o.2	- 11.8	- 2.5
Lysine (α NH_2 enr.)	- 19.3	- 21.1	- 14.4
Proline	- 34.1	- 36.1	
Glutamine (Amide N enr.)	- 92.7	- 92.7	- 92.7
Arginine (Guanid. NH_2 enr.)	- 52.4 [ß]	- 52.5	- 52.1 [+]
Histidine $N\pi$	- 155.1 [ß]	- 211.o	- 197.1
Histidine $N\tau$	- 152.7 [ß]	- 157.7	- 173.4
N-Acetyl-Histidine $N\pi$	- 155.5	- 155.5	- 2oo.8
N-Acetyl-Histidine $N\tau$	- 152.1	- 152.1	- 169.o

a negative sign means downfield from external 4 M $^{15}NH_4NO_3$ in 2 M HNO_3

ß Cation represents the dicationic species as well as the monocationic species

\+ Guanidinium group deprotonated at pH values >12 shows a downfield shift to values < -63 ppm

The chemical shift or the screening constant σ is usually discussed with respect to the contributions from a diamagnetic and a paramagnetic term (4). The diamagnetic term is supposed to remain roughly constant upon smaller changes of the chemical environment and hence should produce only small changes of the chemical shift (5). Since it is very difficult to evaluate exact values of nitrogen chemical shifts at the moment, it seems to be reasonable to rationalize changes in nitrogen chemical shifts by empirical substituent rules (6) or qualitative interpretation using theoretical approximations of the screening constants (7).

On deprotonation of an adjacent carboxylate group the nitrogen resonances of amino groups shift downfield by about 2 ppm. This shift in signal position is by a factor of about o.5 smaller compared to the shifts observed for peptide nitrogen resonances in small peptides during deprotonation of the neighbouring carboxylate group (6). Since the diamagnetic contribution to the screening constant is supposed to be dominating in the case of amino groups, and a larger paramagnetic term has to be considered for nitrogens in peptide bonds this difference seems to be reasonable. An upfield shift of about 8 ppm is observed on deprotonation of the amino group itself. Very similar shifts of amino nitrogens are observed on deprotonation of amino groups of peptides (8). The formation of the lone electron pair is certainly responsible for this shift to higher field.

The amide nitrogen resonance of glutamine remains unaffected by protonation and deprotonation processes at either the α amino or the carboxylate group. Also the nitrogen resonances of the guanidinium group of arginine do not change upon changes at the α amino or carboxylate group. Obviously changes of substituents far away from the observed nitrogen (more than 3 bonds) do not influence the chemical shift of the considered nucleus as far as there is no direct interaction between the nitrogen and the ionizing group.

The deprotonation of the guanidinium group of arginine which occurs at pH values above pH=12 results in a drastic downfield shift of that nitrogen resonance to values of about -65 ppm. This is probably due to an increase of the paramagnetic part of the shielding constant, since a lone electron pair occurs at one of the sp^2 hybridized nitrogens. Since a rapid tautomeric equilibrium exists between both terminal amino nitrogens of the guanidinium group and since the paramagnetic term has a negative sign the average signal of both nitrogens is shifting downfield.

The pH dependence of the chemical shifts of both imidazole nitrogens of histidine is shown in Fig.1. The two nitrogen resonances could be assigned to the π and the τ nitrogens of the imidazole ring due to the couplings of the nitrogen 15 nuclei to the neighbouring protons. The chemical shift values of both nitrogens are very similar at low pH values

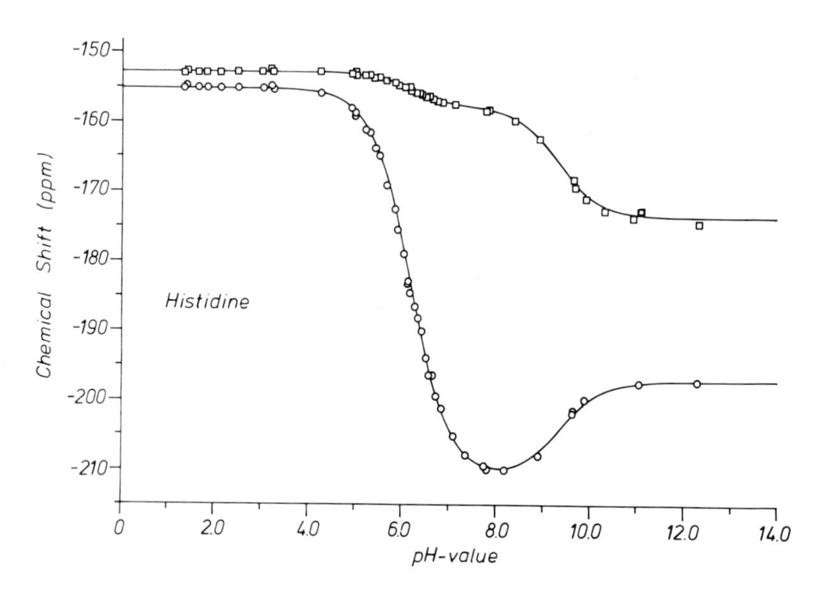

Fig. 1 pH dependence of the nitrogen 15 chemical shift of both imidazole nitrogens of histidine (N π : O , N τ : □). The solid lines represent the calculated titration curves.

(-152.7 ppm and -155.1 ppm for N τ and N π respectively). On deprotonation of the carboxyl group no change of the resonance positions is observed. This is reasonable since the carboxyl group is far away from the observed nuclei. With further increase of the pH value both resonances shift to lower field. However, the shift of the N π resonance is about 56 ppm whereas the shift of the N τ resonance is only 5 ppm. This shift in signal position is clearly due to the deprotonation of the imidazole ring and the titration curve with its pK value of 6.15 for N π (6.19 for N τ) which has been fitted to the chemical shift data is in good agreement with literature data (9).
The deprotonation of the imidazole ring decreases the π electron density at the N π, diminishes the polarization of the σ framework and the occurence of the lone electron pair seems to lower the average excitation energy of the paramagnetic shielding constant (1o). All these effects are expected to enlarge the paramagnetic part of the screening constant and to produce such a remarkable shift of the N π resonance to lower field. However, the small shift of the τ nitrogen resonance can only be explained in terms of the tautomeric equilibrium depicted in Fig.2. Since we are observing weighted averages of the two states of either of the two nitrogens, the slight shift of the τ nitrogen resonance to lower field on deprotonation of the imidazole ring

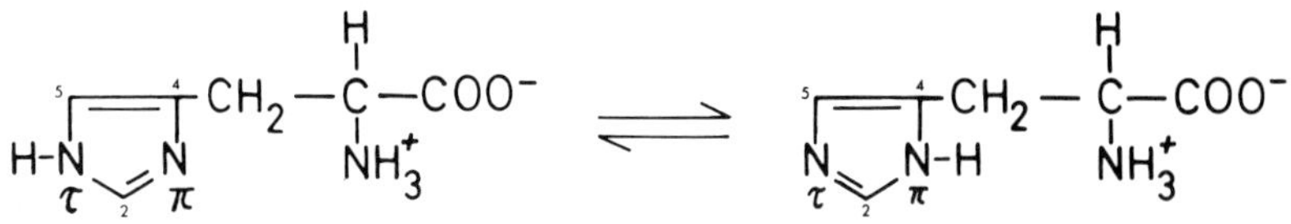

Fig. 2 Tautomeric equilibrium of the imidazole ring of histidine.

indicates that this nitrogen is also transformed - to a much lesser extent than the π nitrogen - to a nitrogen being deprotonated in the equilibrium. However, the downfield shift produced by the tautomeric equilibrium is nearly cancelled by the upfield shift owing to the electronic changes which occur as a consequence of the deprotonation process at the π nitrogen. Hence, only the small shift of about 5 ppm is observed.
Above pH=8 both resonances shift in opposite directions obviously due to the deprotonation of the α amino group. The N π signal is shifting upfield whereas the N τ signal is shifting downfield. Since the place of deprotonation is more than three bonds apart from the nitrogens of interest we have to assume a direct interaction between the α amino group and one of the nitrogens. Because of sterical reasons such an interaction is only possible via the π nitrogen. However, this interaction of the protonated α amino group and the lone electron pair of the π nitrogen seems to shift the tautomeric equilibrium depicted in Fig.2 to the species with the π nitrogen in the deprotonated state. The deprotonation of the α amino group causes the breakdown of that interaction, a considerable amount of species with the τ nitrogen bearing a lone electron pair is formed, which results in the downfield shift of the τ nitrogen resonance position. On the other hand the π nitrogen resonance is shifting upfield owing to the partial conversion of this nucleus to a nitrogen with a hydrogen attached to it. Because of the tautomeric equilibrium the same amount of τ and π nitrogen is converted during the deprotonation of the α amino group, hence the shift in signal position of the τ and π nitrogen is the same however of inverted sign.

In N-acetyl-histidine such an interaction between the positively charged α amino group and the lone pair at the N π position is not possible. The deprotonation of the imidazole ring of that compound should result in shifts of its N π and N τ resonances which are similar to those observed for histidine after deprotonation of the α amino group. Indeed we obtained values of the chemical shifts (-2oo.8 ppm and -169.o ppm) for the N π and N τ resonances of N-acetyl-histidine which agree quite well with the values determined for histidine (-197.1 ppm and -173.4 ppm for N π and Nτ ,Table 1). However, the slight difference between the shifts of both compounds - compared to the shifts of the histidine nitrogens above pH=9.3 the N π resonance of N-acetyl-histidine is shifted about 4 ppm downfield whereas the N τ resonance is shifted upfield by the same amount - indicates that for N-acetyl-histidine the tautomeric equilibrium is shifted a little bit more to the species with N π bearing a lone electron pair.

Coupling constants

Usually coupling constants are interpreted as being governed by three terms: the Fermi contact term (FC term), the term of interaction between the angular momentum of the electron orbitals and the nuclear spin, and the term representing the interaction between the dipoles of the electrons and the nucleus. In many cases the Fermi contact term is the predominant one and there has been good success in the interpretation of coupling constants using only this term (11). Based on this assumption a relationship between the s-character of the hybrid orbitals of the bound nuclei and the coupling constant can be derived.

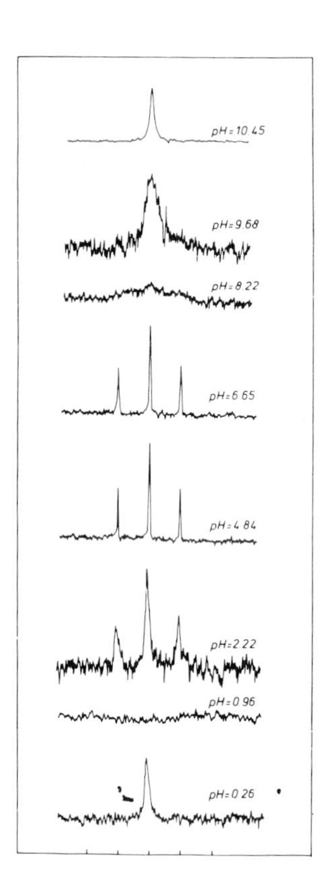

Fig. 3 Undecoupled nitrogen 15 spectra of L-glutamine enriched with nitrogen 15 in amide position as a function of pH.

One-bond nitrogen 15-hydrogen coupling constants are observed for several amino acids. The measured values (-74.7 Hz for glycine, -73.7 Hz for the α amino group of lysine, -89.7 Hz for the amide group of glutamine, and -91.5 Hz for the terminal nitrogens of the guanidinium group of arginine) essentially follow the semiempirical relation of Bourn and Randall (12) which was derived using the theoretical dependence of the FC term on s-electron densities to predict the one-bond coupling constants from the s-character of the bonds at the nitrogen.
Chemical exchange of the protons attached to the nitrogens of amino or amide groups or even of heterocyclic compounds with water protons

does occur at just the proper rate to affect the nitrogen-hydrogen scalar interaction. These effects are demonstrated in Fig.3 for the amide nitrogen of glutamine, where the dependence of the undecoupled nitrogen 15 spectra on pH is shown. Broadening of the triplet occurs as the rate of chemical exchange gets into the order of magnitude of the coupling constant. Further decrease or increase of the pH results in a sharp resonance when the exchange rates are such that:

$$\tau_e^2 \cdot J^2 >> 1 \qquad (1)$$

Values of the exchange rates are obtained using a line shape analysis according to the procedure of Gutowsky et al.(13, 14). The results are depicted in Fig.4. For α amino groups base catalysis apparently accelerates the exchange of protons when the pH is increased. For these nitrogens the multiplet due to the scalar interaction with protons is only observed at very acid pH values. An increase of the exchange rates of protons bound to the amide structure of glutamine or to the terminal nitrogens of the guanidinium group of arginine at alkaline pH values indicates that apparently also for these groups base catalysis of the exchange is occuring. However, below pH=2 the exchange rates of the protons bound to the amide or guanidinium group increase again in contrast to the exchange rates of the protons bound to α amino groups. It seems that also in the acid pH region

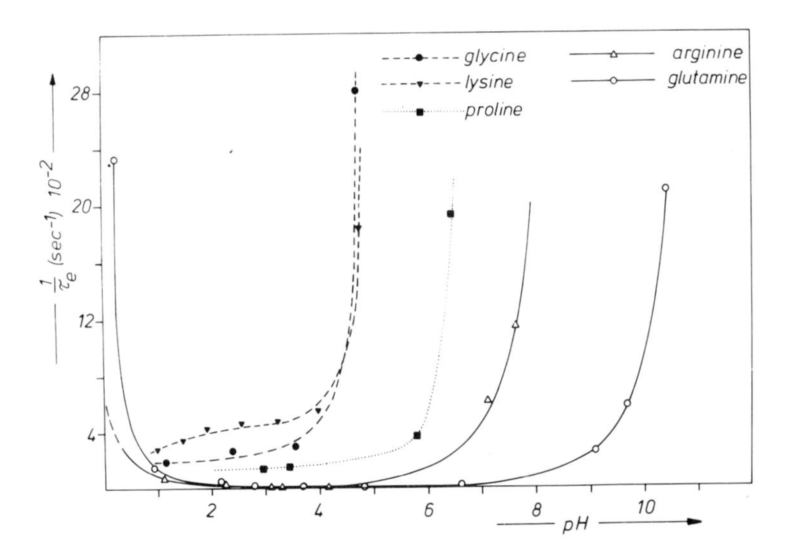

Fig. 4 pH dependence of the exchange rates of protons attached to the nitrogen 15 in several amino acids.

an interaction of the exchangeable protons with a base is necessary in as much as an acceptor group is needed. Both groups contain either an oxygen or a second amino group which may act as an auxiliary base. The exchange rate may be accelerated in acid medium because a water molecule is formed as an acceptor for a proton bound to the nitrogen just by a transfer of a proton from a hydronium ion to the auxiliary base.

For the whole investigated pH range the exchange of the protons attached to the nitrogens of the imidazole ring of histidine is so fast, that no direct coupling is observed. The line width of the nitrogen 15 signals with and without proton broad band decoupling is almost identical. If a value of 80-90 Hz (as indicated by INDO calculations) for this coupling is assumed, the averaging out of the coupling indicates that the exchange rate of these protons must be larger than 1000 sec^{-1}.

The determination of the one-bond carbon 13-nitrogen 15 coupling constants of the imidazole ring of histidine and N-acetyl-histidine (Table 2) has been carried out using the assignment of the carbon resonances to the corresponding carbon atoms (15,16). The pH dependence of these couplings reflects the various protonation states of histidine and N-acetyl-histidine. At pH values below pH=5 the imidazole ring is protonated. INDO calculations for a model compound (4-methyl-imidazole) indicate, that in this pH region the Fermi contact term is large and negative and is dominating the couplings of both nitrogens to the neighbouring carbons. Using the least square parameters derived by Schulman and Venanzi (17) we get a good agreement between the measured and calculated coupling constants (Tab.2). With deprotonation of the imidazole ring the formation of the lone electron pair reduces the contribution of the Fermi contact term which results in a less negative value of the coupling constants. Due to the tautomeric equilibrium the lone electron pair is formed at both nitrogens on deprotonation of the imidazole ring to a certain extent. The contribution of the Fermi contact term to the couplings of both nitrogens is reduced according to that extent. Hence the observed different changes of the coupling constants of both nitrogens are reasonable and reflect the different extent of lone pair formation at the π and τ nitrogen.

The INDO calculations indicate that the Fermi contact term of the coupling of the nitrogen bearing the lone electron pair should be large and positive, whereas the orbital term remains relatively small and positive. We would expect that some of the nitrogen-carbon coupling constants therefore should invert the sign on deprotonation of the imidazole ring. From the pH dependence of the coupling constants experimental evidence for such a change in sign is obtained for the one-bond C4-Nπ coupling constant of histidine (Tab.2).

	Coupling constants, Hz							
	measured					calculated		
	Histidine			N-acetyl-histidine		4-methyl-imidazole		
	Cation[a]	Amphion	Anion	Amphion[b]	Anion	Cation	Nπ-H Tautomer	Nτ-H Tautomer
$^{2}J_{^{15}N\pi-^{1}H2}$	- 6.1	-10.2	-9.6	- 4.6	-9.8	- 2.5	- 4.9	- 9.8
$^{3}J_{^{15}N\pi-^{1}H5}$	- 3.0	- 1.8	-2.2	- 4.2	-2.0	- 8.7	- 6.3	- 0.3
$^{2}J_{^{15}N\tau-^{1}H2}$	- 4.6	- 8.2	-8.8	- 4.8	-7.9	- 2.8	-10.6	- 7.8
$^{2}J_{^{15}N\tau-^{1}H5}$	- 4.8	- 5.9	-6.6	- 4.9	-6.6	- 1.2	-10.6	- 2.2
$^{1}J_{^{13}C2-^{15}N\pi}$	-16.0	- 2.7	-6.9	-16.9	-6.5	-15.2	- 9.2	+ 8.4
$^{1}J_{^{13}C4-^{15}N\pi}$	- 9.9	+ 4.1	-4.7	-10.7	-4.2	- 9.4	-15.5	+ 2.4
$^{2}J_{^{13}C\beta-^{15}N\pi}$	n.d.[+]	- 4.6	-3.8	n.d.[+]	-3.9	+ 1.1	+ 0.2	- 2.2
$^{1}J_{^{13}C2-^{15}N\tau}$	-16.1	-10.1	-6.4	-16.3	-7.4	-16.1	+ 8.9	-12.5
$^{1}J_{^{13}C5-^{15}N\tau}$	-11.6	-10.4	-7.3	-10.6	-8.2	- 9.8	+ 2.4	-13.2
$^{2}J_{^{15}N\pi-^{15}N\tau}$	+ 0.9	- 0.6	-0.9	+ 1.0	n.d.[+]	+ 3.4	- 4.7	- 4.5

a Cation represents the cationic as well as the dicationic species of histidine

b Amphion represents the amphionic as well as the cationic species of N-acetyl-histidine

\+ not detected

From the pH dependence of the nitrogen 15 chemical shifts of the imidazole nitrogens of histidine and N-acetyl-histidine it has been concluded that at pH=9.3 the tautomeric equilibrium of the deprotonated imidazole ring of histidine is shifted toward a larger amount of species with the τ nitrogen being deprotonated. As a consequence of this behaviour the coupling constants of the π nitrogen become less positive, whereas the couplings of the τ nitrogen become less negative. Since in N-acetyl-histidine an interaction between the lone electron pair of the π nitrogen and the α amino group is not possible, we observe nitrogen-carbon coupling constants after deprotonation of the imidazole ring of that compound which are essentially the same as those observed for histidine after deprotonation of the α amino group. Again from the changes of the nitrogen-carbon coupling constants our assumption of an interaction between the imidazole ring and the α amino group of histidine is confirmed.

Two-bond coupling constants between the nitrogen of α amino groups and the Cα proton could not be detected. Hence we have to conclude in agreement with other authors that two-bond nitrogen-hydrogen coupling constants of sp^3 hybridized nitrogens are very small (18).
Two-bond coupling constants of sp^2 or sp hybridized nitrogens are dependent on the existence and orientation of a lone electron pair at the nitrogen.
The proximity of a lone pair should make a positive contribution to the reduced nitrogen-carbon coupling constant. The trans configuration of a lone electron pair relative to the coupled carbon produces a small coupling constant. Hence the values of the couplings of Nπ and Nτ of histidine and N-acetyl-histidine to C5 and C4 respectively could not be determined (19). The coupling of the Nπ to the Cß carbon can only be determined at pH values above pH=6.2 and pH=7.1 for histidine and N-acetyl-histidine, respectively. Because of the positive contribution of the cisoid lone pair to the reduced coupling constants we suppose this value to be negative. Further evidence for this assumption can be derived from INDO calculations on 4-methyl-imidazole, since the calculated coupling constants Nπ-C (methyl) for both possible tautomers of the deprotonated imidazole ring have a negative sign (Tab. 2).
Also the Nπ-Nτ coupling constant is dependent on the existence and orientation of a lone electron pair at either nitrogen. From our measurements we have to conclude, that this coupling inverts sign on deprotonation of the imidazole ring of histidine. The calculations on 4-methyl-imidazole suggest that this coupling is positive below pH=6.2 and negative above this pH value (Table 2).
Usually relatively large nitrogen 15-hydrogen coupling constants are observed in the case of a cisoid configuration of the lone electron pair at the nitrogen relative to the

considered hydrogen atom, whereas small absolute values are found in a transoid configuration or in compounds with the lone pair protonated (2o-22). In histidine and N-acetyl-histidine these two-bond coupling constants are supposed to be negative from theoretical considerations of some other heterocyclic compounds and from our calculations on 4-methyl-imidazole (19,23). Upon deprotonation of the imidazole ring all nitrogen 15-hydrogen coupling constants become more negative, since a lone electron pair is occuring at both nitrogens - the $N\pi$ and the $N\tau$ - to a certain extent according to the tautomeric equilibrium. Since our calculations reveal that the value of the $N\tau$-H5 coupling constant is quite unaffected by the loss of a proton on deprotonation of the imidazole ring as long as the τ nitrogen remains in the tautomeric state with $N\tau$ protonated, the change of that coupling constant is almost completely reflecting the partial conversion of $N\tau$ to a nitrogen bearing a lone electron pair. We have used the pH dependence of that coupling constant to determine the amount of the tautomeric equilibrium (Fig.2) at several pH values. Above pH=6.2 the observed coupling constant can be expressed by:

$$J_{(obs)} = \gamma_{(NH)} \cdot J_{(NH)} + \gamma_{(N:)} \cdot J_{(N:)} \qquad (2)$$

where $\gamma_{(N:)}$ and $\gamma_{(NH)}$ are the mole fractions of $N\tau$ bearing a lone electron pair or a proton, respectively; $J_{(N:)}$ is the two-bond nitrogen-hydrogen coupling constant for a five-membered heterocyclic system with a lone pair at the nitrogen. Its value is found to be -14.4 Hz for various compounds (19,22,24). $J_{(NH)}$ is the corresponding coupling constant of a heterocyclic system with its nitrogen fully protonated. From our measurements this value is determined to be -4.8 Hz. Very similar values are obtained for other fully protonated five-membered heterocyclic compounds (19,22,24-26). Using these values we obtain from eq. (2)

$$\gamma_{(N:)} = - o.1o4 \cdot J_{(obs)} - o.5 \qquad (3)$$

With eq.3 we get a value for $\gamma_{(N:)}$ of o.12 at pH=8.2. The assumed interaction between the protonated α amino group and the deprotonated imidazole ring shifts the tautomeric equilibrium as already outlined discussing the chemical shift data. A larger portion of $N\pi$ is converted to a nitrogen bearing a proton. Since the two-bond $N\tau$-H5 coupling constant is a probe for this conversion, this constant should be more negative above pH=9.3 and indeed this is observed. In our quantitative treatment according to eq.3 the $\gamma_{(N:)}$ value is shifted from o.12 to o.2. Also the other two-bond nitrogen-hydrogen coupling constants are changing upon deprotonation of the α amino group of histidine. As expected the couplings of N π become less negative, whereas the

couplings of $N\tau$ are getting more negative. The magnitude of these changes is the same namely o.6 Hz (Tab. 2).
Since the proposed interaction between the protonated amino group and the lone electron pair of the $N\pi$ is not possible in N-acetyl-histidine, as already discussed, the two-bond nitrogen-hydrogen coupling constants of this compound after deprotonation of the imidazole ring should be similar to those observed for histidine above pH=9.3, and indeed this is found. Using eq. 3 we can also calculate the value of the tautomeric equilibrium of N-acetyl-histidine ($\gamma_{(N:)}$ =o.19).

Three-bond coupling constants are known to be strongly dependent on the dihedral angle between the coupled nuclei (27-29).
In the case of lysine the three-bond coupling constant of the α amino nitrogen coupling to the Cß protons could be determined ($|2.9|$ Hz). This constant can in principle be used to determine the dihedral angle χ of the side chain or to estimate the population of the three possible rotamers, if the nitrogen 15-hydrogen coupling constants of the pure gauche and trans conformations are known. However, because of the small value of that coupling constant relatively sharp resonances are required for measuring these quantities. Hence the observed exchange broadening at several pH values prevents the determination of that constant.
In the case of histidine and N-acetyl-histidine two three-bond couplings can be observed, the coupling of the $N\pi$ to H5 and to the Cß protons of the side chain. The pH dependence of the coupling of $N\pi$ to H5 again reflects the conversion of a protonated imidazole nitrogen into one which bears a lone electron pair. Also this coupling is sensitive to the change of the tautomeric equilibrium of histidine above pH=9.3. The sign of this constant is assumed to be negative analogous to the corresponding values in pyridine and some oximes (19,2o). Further evidence for this assumption follows from the similar behaviour of these compounds on protonation and from the calculation of this coupling constant in the case of 4-methyl-imidazole (Table 2).
The second observed coupling, the coupling between $N\pi$ and the Cß protons, should allow certain insights into the conformation of the bond between C4 and Cß of histidine. The coupling constants between the π nitrogen and the ß protons can be represented by the following equations:

$$\begin{aligned} J_{(N\pi-H\beta a)} &= g1 \cdot J(I) + g2 \cdot J(II) + g3 \cdot J(I) \\ J_{(N\pi-H\beta b)} &= g1 \cdot J(II) + g2 \cdot J(I) + g3 \cdot J(I) \end{aligned} \qquad (4)$$

if rapid exchange between the possible rotamers in solution is assumed. g1, g2, and g3 are the mole fractions of these rotamers as depicted in Fig. 5. g1 and g2 correspond to the

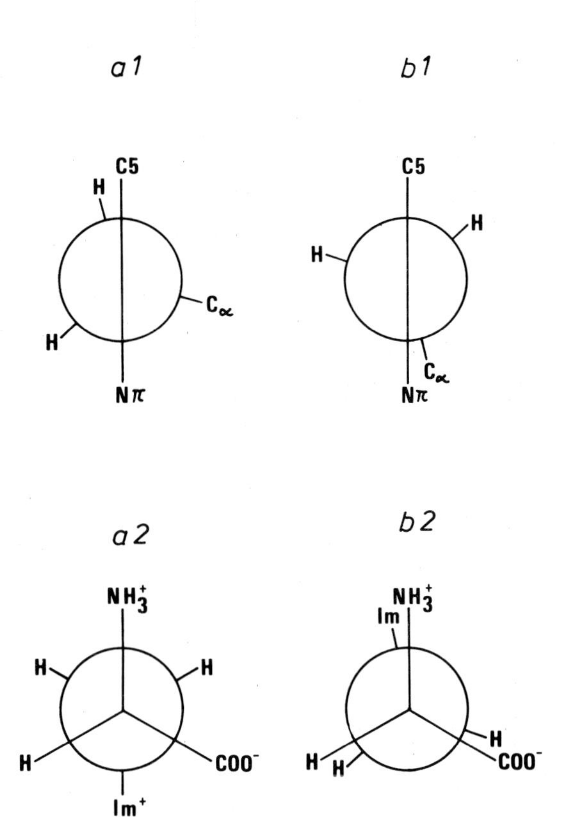

Fig. 5 Conformations of histidine in aqueous solution as proposed by nitrogen 15 and carbon 13 NMR studies (a) conformation of histidine below pH=6.2 (b) conformation of histidine between pH=6.2 and pH=9.3. Upper part: conformation of the C4-Cß bond, lower part: conformation of the Cα-Cß bond.

rotamers with either Ha or Hb in gauche position to Nπ , g3 is related to the rotamer with both protons gauche to Nπ . J(I) and J(II) are the three-bond coupling constants for the pure gauche and trans conformation of a Cß proton relative to N π , respectively. Since we observe different values for the coupling of Nπ to Ha and Hb at low pH values we have to conclude that for histidine the rotamers g1 and/or g2 have to be considered in this pH region. Model building and the charges on the imidazole ring, the carboxylate and the α amino group prompted us to propose that the carboxylate group is located above the ring plane which is positively charged. In this conformation the positive charges of the imidazole ring and the α amino group are furthest apart, and because of the repulsive coulomb forces this arrangement becomes plausible. The fact that we do not detect any change in the nitrogen 15 chemical shifts of both ring nitrogens of histidine during deprotonation of the carboxylate group supports this interpretation, such a change should be expected if there were any hydrogen bonding between the carboxylate group and the protonated imidazole ring. With deprotonation of the imidazole ring two effects are observed. The magnitudes of the three-bond Nπ-Hß coupling constants decrease and the couplings become undetectable above pH=7 which can be attributed to electronic changes in the ring system itself and to changes of the conformation of the C4-Cß bond. On the other hand the couplings of the two ß protons of histidine to N π become equivalent. The interaction between the α amino group and the imidazole ring in the pH range of 6.2 - 9.3 which is evident from the chemical shift data and from nearly all discussed couplings is only possible if both Cß-H bonds are forming an angle of about $12o^{o}$ to the C4-N π bond. In this position the protons become equivalent relative to the imidazole ring as depicted for the correspondent rotamer b1 in Fig. 5. Hence we have to assume that the population of the rotamers with either of the ß protons eclipsed relative to N π are only small.

Nitrogen 15 NMR proves to be not only an excellent tool to get information about ionization processes of biologically interesting molecules but also to be sensitive to electronic and conformational changes in those compounds. Hence this method should be a good supplement to carbon 13 and hydrogen NMR work and should even be superior to these methods in some aspects such as the detection of electronic changes in nitrogen containing systems and - in aqueous solution - measurements of exchange rates of protons directly attached to nitrogens. Furthermore the various couplings between nitrogen 15 and carbon 13 or hydrogen atoms can add useful information about the ionization groups, the hybridization of involved nuclei and in specific cases some conformationally related properties.

Acknowledgement: F.B. thanks the Studienstiftung des Deutschen Volkes for a fellowship.

REFERENCES

1. E.Reich, D.B.Rifkin, and E.Shaw, "Proteases and Biological Control", Cold Spring Harbor Conferences on Cell Proliferation,Vol.2,Cold Spring Harbor Laboratory, Cold Spring Harbor,N.Y., 1975, pp 1-64.
2. A.Neuberger, Biochem.J. 32, 1452-1456 (1938).
3. F.Blomberg, W.Maurer, and H.Rüterjans, J.Am.Chem.Soc. 99, 8149-8159 (1977).

4. N.F.Ramsey, Phys.Rev. 78, 699-7o3 (195o).
5. K.A.K.Ebraheem, G.A.Webb, and M.Witanowski, Org.Magn. Reson. 8, 317-32o (1976).
6. D.Gattegno, G.E.Hawkes, and E.W.Randall, J.Chem.Soc., Perkin Trans. 2, 1527-1531 (1976).
7. G.A.Webb and M.Witanowski, in: "Nitrogen NMR", M.Witanowski and G.A.Webb,Ed., Plenum Press, New York, N.Y., 1973, pp 1-39.
8. V.Markowski, T.B.Posner, P.Loftus, and J.D.Roberts, Proc.Natl.Acad.Sci. USA 74, 13o8-13o9 (1977).
9. A.Deutsch and P.Eggleton, Biochem.J. 32, 2o9-211 (1938).
1o. M.Witanowski, L.Stefaniak, H.Januszewski, and Z.Grabowski, Tetrahedron 28, 637-653 (1972).
11. J.N.Murrell, Prog. Nucl. Magn. Reson. Spectros. 6, 1-6o (197o).
12. A.J.R.Bourn and E.W.Randall, Mol.Phys. 8, 567-579(1964).
13. H.S.Gutowsky, D.W.McCall, and C.P.Slichter, J.Chem.Phys. 21, 279-292 (1953).
14. J.Kaplan, J.Chem.Phys. 28, 278-282 (1958).
15. W.F.Reynolds, I.R.Peat, M.H.Freedman, and I.R.Lyerla,Jr., J.Am.Chem.Soc. 95, 328-331 (1973).
16. W.Horsley, H.Sternlicht, and J.S.Cohen, J.Am.Chem.Soc. 92, 68o-686 (197o).
17. (a) J.M.Schulman and T.Venanzi, J.Am.Chem.Soc. 98, 47o1-47o5 (1976); (b) ibid. 98, 6739-6741 (1976).
18. R.L.Lichter and J.D.Roberts, Spectrochim.Acta 261, 1813-1814 (197o).
19. R.L.Lichter and J.D.Roberts, J.Am.Chem.Soc. 93, 5218-5224 (1971).
2o. (a) D.Crépaux, J.M.Lehn, and R.R.Dean, Mol.Phys. 16, 225-239 (1969); (b) D.Crépaux, J.M.Lehn, and R.R.Dean, ibid. 14, 547-555 (1968).
21. D.Crépaux and J.M.Lehn, Org.Magn.Reson. 7, 524-526(1975).
22. J.P.Kintzinger and J.M.Lehn, J.Chem.Soc.,Chem.Commun. 66o-661 (1967).
23. J.P.Jacobson, O.Snerlin, E.J.Pedersen, J.T.Nielsen, and K.Schaumburg, J.Magn.Reson. 1o, 13o-138 (1973).
24. J.P.Kintzinger and J.M.Lehn, Mol.Phys. 14, 133-145(1968).
25. M.M.King, H.J.C.Yeh, and G.O.Dudek, Org.Magn.Reson. 8, 2o8-212 (1976).
26. J.M.Briggs, E.Rahkamaa, and E.W.Randall, J.Magn.Reson. 11, 416-42o (1973).
27. M.Karplus, J.Chem.Phys. 3o, 11-15 (1959).
28. V.F.Bystrov, Yu.D.Gavrilov, and V.N.Solkan, J.Magn.Reson. 19, 123-129 (1975).
29. M.Barfield and H.L.Gearhart, Mol.Phys. 27, 899-9o2 (1974).

DISCUSSION

Navon: A similar phenomenon of π minimum near the PK value was observed by Jack Cohen (J. Am. Chem. Soc...) for carbon-13 NMR of acetic acid. However he has found that this effect was due in his case to impurities of paramagnetic ions. Is it possible that the same explanation is true in your results on histidine ?

Blomberg: Unfortunately the data concerning your question are not included in the paper due to the limitation of space. However, we have several evidences that the maximum of the relaxation rate of N and the deviation from the expected "titration-curve"-behavior of the relaxation rate of N τ are due to an association phenomenon.

1) We have purified the NMR- sample very carefully using a procedure described in Ref. 3 of this paper.

2) The pH-dependences of the relaxation-rates of both imidazole nitrogens of histidine can be fitted with a model, which assumes a dimerisation of a positively charged and a neutral imidazole ring. The association constants calculated using this model independently for both nitrogens are very similar (about 2 kcal/mole).

3) If paramagnetic impureties are responsible for the observed effects, the N_π resonance should be influenced much more than the N_τ resonance, since N_π is certainly the nucleus at which complexation takes place as determined by several other authors. However we observe just the opposite behavior.

4) We observe a maximal relaxation time for N_π at about pH = 7 ($\simeq$ 25 sec). At this pH value minima of relaxation times are observed in as purified samples.

5) The NOB of both nitrogens (tautomeric equilibrium) is deduced on titration of the imidazole ring of histidine due to the loss of a proton. However no bell-shape curve of the NOE values of both imidazole nitrogens in dependence of pH could be observed.

However, such a behavior should be expected if the binding of paramagnetic impureties is maximal at the pH of the imidazole ring.

All those evidences exclude the possibility that the pH dependence of the relaxation rates of both imidazole nitrogens of histidine is produced by paramagnetic impureties to a larger extent - at least at the pH of the imidazole ring.

Hence we have to consider a stacking association of a positively charged and a neutral imidazole ring of histidine. We obtain very similar results in the case of N-acetyl-histidine also, which provides further evidence for this type of interaction.

ONE-BOND C,H COUPLING CONSTANTS, A NEW PARAMETER IN CONFORMATIONAL STUDIES OF PEPTIDES? THE CYCLO-SARCOSYLS.*)

by H. Egli, U. Vögeli and W. von Philipsborn
Institute of Organic Chemistry
University of Zurich, Switzerland

SUMMARY

Carbon-hydrogen spin coupling constants over one bond, $^1J(C,H)$, were determined for a number of simple open-chain model peptides as well as for *cyclo*-disarcosyl, *cyclo*-tetrasarcosyl and *cyclo*-octasarcosyl. The dependence of $^1J(C\alpha,H)$ on the ionization state of the amino and carboxyl ends of an open-chain peptide and the relative orientation of C-H bond and p_z-orbital of the amide nitrogen is demonstrated. The solution conformation of the *cyclo*-sarcosyl peptides is discussed as a test case for the new parameter. In peptide systems containing N-methyl groups, the conformational information obtained from $^1J(C,H)$ can be corroborated by the vicinal coupling constant $^3J(C,H)$ between N-methyl carbon and α-CH proton.

1. INTRODUCTION

In recent years high-resolution NMR spectroscopy and, in particular, spin-spin coupling constants have played a major role in conformational analysis of peptide systems [2]. Since proton-NMR has dominated such studies for a long time, there exists a large amount of data on vicinal H,H coupling constants, i.e., J[HN-C(α)H] (angle Φ) and J[HC(α)-C(β)H] (angle χ), whereby in both cases a Karplus-Bystrov-type relationship between $^3J(H,H)$ and the torsional angle is utilized to obtain conformational information. Further, geminal H,H coupling constants have proven useful in the study of

*) ^{13}C-NMR spectroscopy, Part XX, for Part XIX see [1].

B. Pullman (ed.), Nuclear Magnetic Resonance Spectroscopy in Molecular Biology, 247-260.

cyclic oligopeptides containing glycine and sarcosine [3,4]. In addition, isotopic labelling, for example, by ^{15}N has made possible investigations of heteronuclear coupling constants such as $^3J(^{15}N,H)$ and $^2J(^{15}N,H)$ [2]. With the advent of ^{13}C NMR and with subsequent sensitivity improvements by the use of large sample-diameters and high fields from superconducting magnets, the observation of C,H coupling constants in natural ^{13}C-isotope abundance becomes feasible, at least for smaller peptides. In general, fully proton-coupled ^{13}C spectra of peptides containing different amino acids will be very complicated and difficult to analyze, at least with respect to vicinal and geminal C,H coupling. The one-bond $^{13}C,^1H$ coupling constants ($^1J(C,H)$), however, are large and can even be extracted from off-resonance decoupled spectra, with subsequent corrections for H_2-field perturbation. To date, very few data are available on the $^1J(C,H)$ parameter in amino acids and peptides, and the structural dependence including effects of different ionization states has not yet been investigated at all. On the basis of some earlier calculations and experimental results [5,6,7,8], it can be expected that $^1J(C\alpha,H)$ will also show conformational dependence due to lone-pair effects from the amide nitrogen. Furthermore, the value of this coupling constant may be affected by the orientation of the Cα-H bond relative to the π-system of the amide carbonyl. Such effects may become important in peptides with preferred conformations.

In search of suitable model compounds for the conformational effects, we have chosen the <u>cyclo</u>-sarcosyls c-Sar_2, c-Sar_4 and c-Sar_8 [9]. For these three peptides X-ray crystal analyses [10], as well as conformational studies in solution by proton-NMR [3,4] were already published. On the other hand, there is no severe sensitivity problem in obtaining proton-coupled ^{13}C spectra due to sufficient solubility and C_2 or C_i symmetry of the three molecules. Finally, the presence of N-methyl groups allows the determination of the vicinal coupling constant $^3J(\underline{C}H_3,C\alpha\underline{H})$ which will be shown to corroborate the conformational information deduced from $^1J(C,H)$.

2. EXPERIMENTAL

The ^{13}C spectra were measured on a VARIAN XL-100-12 spectrometer in the FT mode using V-4410 and V-4418 probe heads with 10 and 18 mm inserts, respectively. The instrument is equipped with an SL-124 computer station including a disk system. Proton-coupled spectra were obtained under gated decoupling conditions with spectral widths of 5000 Hz and 2 sec. acquisition time corresponding to a digital resolution of 0.5 Hz. The reproducibility of the $^1J(C,H)$ coupling constants measured under these conditions is better than ± 0.5 Hz. The amino acids and linear peptides were measured in D_2O and the pD values adjusted by addition of NaOD or DCl. The pH meter readings were then corrected using the relation pD = pH + 0.4. For the cyclic sarcosyl peptides $CDCl_3$ was used as a solvent with the addition of small amounts (< 5%) of TFA to increase solubility. Concentrations were as follows: c-Sar_2 (1 m), c-Sar_4 (0.2 m) and c-Sar_8 (0.05 m). Chemical shifts are referred to TMS as an internal standard.

3. RESULTS

There are only very few data in the literature about $^1J(C,H)$ in amino acids and peptides [2]. We have, therefore, decided to first examine some non-cyclic oligopeptides together with the corresponding amino acids in order to obtain reference data for conformationally flexible peptides and to study ionization and solvent effects.

The $^1J(C,H)$ coupling constants for Gly_2, Gly_3, Gly_4, as well as glycine, sarcosine and alanine in the respective cationic, zwitterionic and anionic forms are summarized in Tables 1a-c.

The total range for α-CH_2 coupling constants extends from 136 to 146 Hz whereby CH_2 groups at the amino end (a) and at the carboxyl end (b) can be clearly distinguished from the central CH_2 groups (c) on the basis of the titration curves for the chemical shifts and coupling constants. The values of $^1J(C,H)$ for a CH_2 group of type (c) in glycyl peptides are found in the very narrow range from 140 to 141 Hz and they exhibit only a very small pH dependence. N-methylation does not lead to a significant change in $^1J(C,H)$ as shown by the data for glycine and sarcosine (as well as the corresponding peptides). Cα-alkylation results in a small increase of $^1J(C,H)$ as seen for the alanine data. Amino acids with different side chains are not expected to yield values which are very different from these reference data. Solvent effects on $^1J(C,H)$ are within 1-2 Hz for the three

Table 1a. $^{1}J(C\alpha,H)$ [Hz, D_2O] in Linear Glycyl Peptides

pD	Gly(a)-Gly(b)		Gly(a)-Gly(c)-Gly(b)		
1	144.7	141.0	144.6	140.9	140.8
3	144.5	140.6	144.5	140.9	140.8
6	144.4	139.4	144.4	140.7	139.5
8	143.5	139.3	143.9	140.7	139.5
13	136.9	138.9	137.2	140.1	139.2

pD	Gly(a)-[Gly(c)]$_2$-Gly(b)		
3	144.8	141.0	141.0

Table 1b. Range of $^{1}J(C\alpha,H)$ [Hz] for Glycyl Peptides

	NH_2-$C_\alpha H$	-$C_\alpha H$-COOH	-NH-$C_\alpha H$-CO-
cation	145	141	141
zwitterion	144.5	139.5	140.5
anion	137	139	140

Table 1c. Substituent Effects on $^{1}J(C\alpha,H)$ [Hz]

	NH_2-$\underline{CH_2}$-COOH	CH_3NH-$\underline{CH_2}$-COOH	NH_2-$\underline{CH}CH_3$-COOH
cation	145.2	145.4	146.3
zwitterion	143.7	143.7	145.0
anion	135.7	135.0	138.3

Table 1d. Solvent Dependence of $^{1}J(C\alpha,H)$ [Hz]

	AlaOEt-HCl	SarOEt-HCl	c-Sar$_2$
D_2O	147.6	146.0	143.9
CD_3OD	146.4	145.7	143.0
$CDCl_3$	146.3	144.6	142.2 142.5 a)

a) Value obtained with addition of TFA (< 5%)

solvents D_2O, CD_3OD and $CDCl_3$ (Table 1d). Trifluoroacetic acid added in small amounts (< 5%) to a $CDCl_3$ solution does not cause significant effects on the Cα,H coupling constants in the sarcosyl peptides. Thus, we have shown that the one-bond C,H coupling constants of N-terminal and C-terminal Cα,H groups are sensitive to ionization effects, whereas central C,H coupling constants in peptides appear in a very narrow and constant range (140-141 Hz).

For the cyclic peptides of sarcosine, however, $^1J(C\alpha,H)$ varies over a much wider range from 137 to 150 Hz which is due to specific conformational effects as shown in Section 4. The data for c-Sar_2, c-Sar_4 and c-Sar_8 are listed in Table 2.

The simplest cyclic sarcosyl peptide in which rotation about the N-C(α) and N-C(O)-bonds is restricted, is the diketopiperazine c-Sar_2. Only one $^1J(C\alpha,H)$ coupling constant (142.5 Hz) is observed which is slightly higher than the value for open-chain peptides discussed above. The value of 142.5 Hz is considered to be an average between two coupling constants of the methylene carbon with the diastereotopic protons of a non-planar sixmembered ring (see Sect. 4).

Cyclo-tetrasarcosyl (c-Sar_4), for which X-ray analysis as well as proton NMR in solution have demonstrated the presence of a center of symmetry, gives two ^{13}C chemical shifts for the methylene carbons and four ^1H chemical shifts indicating the presence of four types of diastereotopic methylene protons. Hence, four different $^1J(C,H)$ coupling constants can be expected. In fact, the methylene resonance at 52.0 ppm exhibits two coupling constants of the same magnitude (138.2 Hz), whereas the other methylene resonance at 49.9 ppm yields two very different coupling constants (137.5 and 150.0 Hz). The correlation of the carbon chemical shifts and one-bond coupling constants with the four proton shifts and with the two geminal $^2J(H,H)$ coupling constants (18 Hz and 14 Hz) was achieved by a series of off-resonance decoupling experiments. Based upon the assignment of the proton shifts given by Dale and Titlestad [3], the ^{13}C signal at 52.0 ppm must be assigned to C(5) and C(11), the signal at 49.9 ppm to C(2) and C(8) (Fig. 1). Further, the diastereotopic methylene protons H^i, H^o (Fig. 2) at C(2),

Table 2. ^{13}C Shifts [ppm], $^{1}J(C,H)$ and $^{3}J(C,H)$ Coupling Constants [Hz] in Cyclic Sarcosyl Peptides

	c-Sar_2	c-Sar_4		c-Sar_8			
δ[C(α)]	51.5	52.0[a]	49.9[a]	51.6	50.8	50.8	50.3
δ[C(Me)]	33.2	37.2	33.6	36.5	36.3	36.1	35.7
δ[CO]	163.2	171.3	167.3	169.6	169.1	168.8	167.3
$^{1}J(C\alpha H)$	142.5	138.2	137.5; 150.0	140.0	137.5	137.5	142.5
$^{1}J(CH_3)$	139.6	139.8	140.0	139	138.5	140	139.5
$^{3}J(\underline{C}H_3,C\alpha H)$	⩽ 0.5	4.0	6.3; 1.6	-	-	-	-
$^{3}J(\underline{C}\alpha H,C\underline{H}_3)$	3.3	3.0	3.0	-	-	-	-
$^{2}J(H,H)$	-	-18	-14	-17	-18	-18	-16

a) assignment see text

C(8) and C(5), C(11) respectively, can be correlated with the one-bond C,H coupling constants via the proton shifts. This assignment of the $^{1}J(C,H)$ coupling constants will be confirmed, as shown in Sect. 4, by vicinal C,H coupling constants between the N-methyl carbon and the diastereotopic methylene protons. When the observed $^{1}J(C,H)$ values are compared with the reference data for the non-cyclic glycyl peptides it is evident that the rather large value of 150.0 Hz deserves special attention.

Cyclo-octasarcosyl (c-Sar_8) was previously shown [4] to exhibit in the ^{1}H NMR spectrum ($CDCl_3$) two N-methyl signals ($CDCl_3$/benzene: 4 signals) and four AB systems in the methylene region indicating the presence of eight types of diastereotopic methylene protons. Also in the solid state, the molecule possesses an approximate two-fold symmetry [10c,d]. The ^{13}C-spectrum shows four N-methyl signals, four amide carbonyl resonances, but only three chemical shifts can be resolved for the four types of non-equivalent methylene carbons expected. The coincidence of two lines could not be resolved even at a field of 8.4 Tesla. Assuming C_2 symmetry also for the structure of the molecule in solution, eight different $^{1}J(C\alpha,H)$ coupling constants can in principle be expected. The three ^{13}C lines at 50.8 ppm (4 carbons), 51.6 ppm (2 carbons) and 50.3 ppm (2 carbons) under proton-coupled conditions are split into triplets with one-bond C,H coupling constants of 137.5 ± 1, 140 ± 1 and 142.5 ± 1 Hz, respectively. With the same procedure as outlined for c-Sar_4, the carbon chemical shifts and $^{1}J(C,H)$ coupling constants were correlated with the proton NMR data. Although an assignment of the carbon and proton chemical shifts has not been achieved for c-Sar_8, it is clear that each of the methylene resonances exhibits only one $^{1}J(C,H)$ coupling constant (within experimental errors). The value of 137.5 Hz for the four methylene carbons at 50.8 ppm is at the lower end of the range for $^{1}J(C,H)$, whereas the values for the two other methylene signals (140.0, 142.5 Hz) are close to the conformationally averaged data in open-chain peptides (140-141 Hz) and in c-Sar_2 (142.5 Hz).

As an independent parameter $^{3}J(C,H)$ can be used to confirm the conformational conclusions which will be drawn from the one-bond coupling constants. The sarcosyl peptides are unique in this respect since a vicinal coupling between the N-methyl carbon and the two diastereotopic $C\alpha$,H protons can be obtained from the non-decoupled spectrum of the N-methyl group. This C,H coupling constant will depend on the torsional angle ϕ. The corresponding data for c-Sar_4 and c-Sar_2 are given in Table 2. Information about the torsional angle ϕ can also be obtained from the fragment ^{13}CO-N-Cα-H, as previously shown by Bystrov [11]. This coupling constant, however, is more difficult to extract from the spectrum since the carbonyl carbon in the peptide chain is coupled to the vicinal CH_2 protons on the one side and to the geminal CH_2 protons on the other side.

The $^{13}CH_3$-N-C(α)-H coupling constant is smaller than 0.5 Hz in c-Sar_2, whereas in c-Sar_4 values of 6.3 Hz for $^{13}CH_3$-N-C(2,8)-H^i, 1.6 Hz for $^{13}CH_3$-N-C(2,8)-H^o, and 4.0 Hz for $^{13}CH_3$-N-C(5,11)-$H^{i,o}$ have been obtained. It can be seen that the methylene groups (C(2,8)) which show two distinctly different $^1J(C,H)$ values (137.5; 150.0 Hz) do also give two different $^3J(C,H)$ coupling constants. On the other hand, the C(5) and C(11) methylene groups exhibit only one $^1J(C,H)$ value (138.2 Hz) and one $^3J(C,H)$ value (4.0 Hz).

In the conformational analysis of the peptide fragments presented in Sect. 4, the geminal H,H coupling between the diastereotopic methylene protons will also be used as a criterion. Barfield and Grant have shown how this parameter depends on the relative orientation of the C-H bond and the plane of the π-electrons of the carbonyl group [12]. Corresponding data for c-Sar_4 and c-Sar_8 already reported by Dale and Titlestad are included in Table 2.

4. DISCUSSION

We will first discuss the rather rigid cyclic tetrapeptide c-Sar_4 for which the largest variation in the magnitude of the $^1J(C,H)$ parameter is observed. Later, we shall proceed to the more flexible and conformationally averaged structures. For c-Sar_4 proton NMR studies in $CDCl_3$ solution [3,9] have revealed the existence of a single and rigid centrosymmetric conformation in which the configuration of the amide units is cis, trans, cis, trans. This conformation was in fact determined by X-ray crystal analysis [10b]. It is shown in Fig. 1 and will be used as a basis for the interpretation of the $^1J(C,H)$ data. The space-filling molecular model illustrates that the structure is very compact and not capable of facile ring inversion.

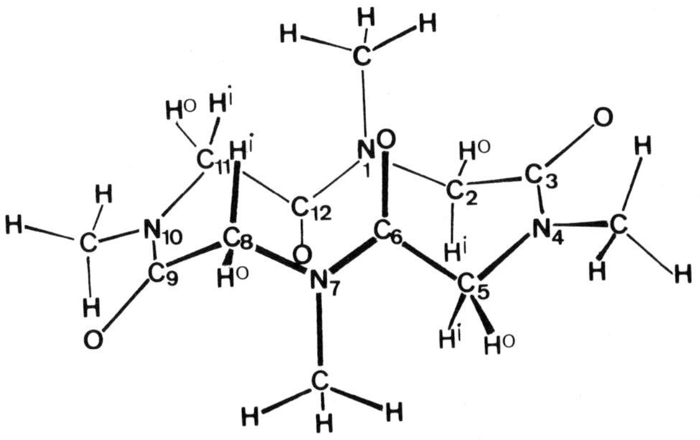

Figure 1. Cyclo-tetrasarcosyl

The steric situation for the two types of methylene groups can be illustrated by Newman-projections along the N-CH_2 axis (angle ϕ) and the CH_2-CO axis (angle ψ), as shown in Fig. 2a for the C(2), C(8) methylene groups and in Fig. 2b for the C(5), C(11) methylene groups. The torsional angles ϕ and ψ were taken from the X-ray work.

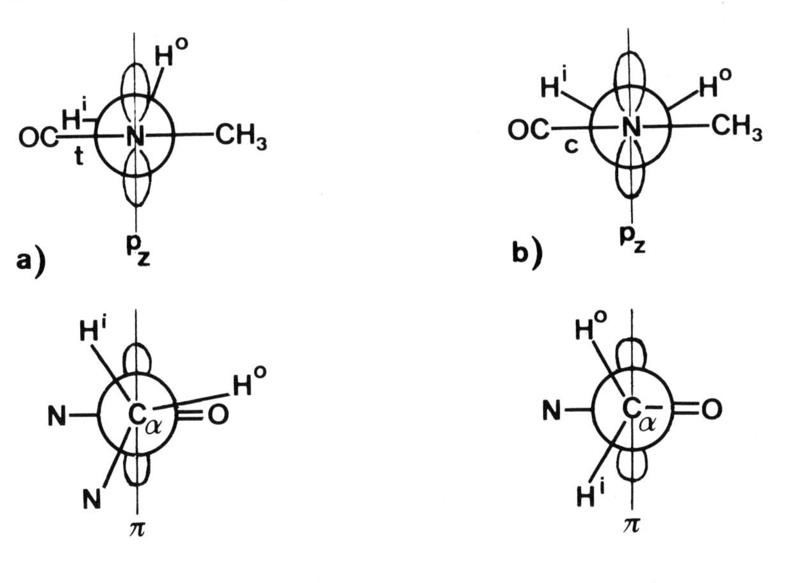

$^1J(C\alpha,H^o)$ = 150.0 Hz $\quad$ $^1J(C\alpha,H^o)$ = 138.2 Hz

$^1J(C\alpha,H^i)$ = 137.5 $\quad$ $^1J(C\alpha,H^i)$ = 138.2

Figure 2. Segmental Conformations of Cyclo-tetrasarcosyl

The different values for $^1J(C,H^i)$ and $^1J(C,H^o)$ (Fig. 2a) can be rationalized if it is assumed that the p_z-orbital of the amide nitrogen gives a positive contribution to $^1J(C,H)$ for small torsional angles between the C,H bond and the axis of the p_z-orbital. These torsional angles are 30° for H^o and 90° for H^i. The large difference between the angles is confirmed for the conformation of the peptide in solution by the vicinal C,H coupling constants between the N-methyl carbon and H^i ($^3J(C,H)$ = 6.3 Hz) on the one hand and between the same carbon and H^o (1.6 Hz) on the other hand. The correlation of the large torsional angle (180°) with the value of 6.3 Hz and of the 60° angle with the value of 1.6 Hz agrees qualitatively with the known Karplus-Bystrov-type angular dependence of vicinal C,H coupling. Thus, there is no reason to believe that the solution conformation of c-Sar_4 deviates much from the crystal conformation.

The conformation of the analogous fragment depicted in Fig. 2b is in full accord with the above interpretation. Thus, the very similar orientations of the two C,H bonds relative to the p_z-orbital (approx. 60°) are reflected in the same $^1J(C,H)$ coupling constants (138.2 Hz) and in two $^3J(C,H)$ coupling constants of very similar magnitude (~4 Hz) for the torsional angles of 30° and 150°, respectively.

Fig. 2 which also illustrates the orientation of the diastereotopic methylene protons with respect to the π-orbital of the carbonyl group is useful to rationalize the difference in the $^2J(H,H)$ coupling constants for the two fragments. The value of 18 Hz (large negative contribution from π-system) is according to Barfield and Grant [12] in good agreement with the conformation for the fragment containing the C(5) and C(11) methylene group.

The question must now be asked to what extent the orientation of the π-system of the carbonyl group also contributes to the observed $^1J(C,H)$ coupling constant. The projection formulae in Fig. 2 illustrate that small torsional angles (30-35°) between the axis of the π-orbital and the CH bond axis are connected with small $^1J(C,H)$ coupling constants (137-138 Hz) whereas the C-H^o bond axis in the C(2,8)-methylene groups combines a large torsional angle (85°) with the largest $^1J(C,H)$ coupling constant of 150.0 Hz. If the π-orbital would act in an analogous manner as a p_z-orbital on nitrogen leading to a positive contribution to $^1J(C,H)$, an opposite trend in the coupling constants would be expected. Further evidence for the indication that the nitrogen p_z-orbital yields a positive contribution to $^1J(C,H)$ comes from the corresponding coupling constants in the N-methyl groups. The average value for the three _cyclo_-sarcosyl peptides is 139.5 ± 1.0 Hz and thus significantly larger than the smallest experimental value for $^1J(C,H)$ of the Cα-H bonds. Therefore, it appears that hyperconjugative effects leading to an increase in the $^1J(C,H)$ coupling constants are more effective between the C(α)-H bond and the p_z-orbital on nitrogen than with the p_z-orbital on carbon.

The 24-membered ring of c-Sar_8 appears to retain a preferred conformation in solution since Titlestad has shown that for the eight methylene groups four AB-systems are observed in the proton spectrum which do not show exchange broadening below 40° [4]. Hence, there is no 'ring inversion' at room temperature on the proton NMR time scale. The proton and carbon NMR spectra are consistent with a two-fold symmetry axis for the ring conformation in solution, and from the X-ray crystal analysis [10c] the amide configuration sequence is _cis_,_cis_, _trans_,_trans_, _cis_,_cis_, _trans_,_trans_. On the basis of variable temperature ^{1}H-NMR studies between -80°C and room temperature it was concluded that the crystal conformation is retained in solution.

The $^1J(C,H)$ coupling constants observed indicate that at least certain segments of the peptide show a considerable flexibility, since the values of 140.0 and 142.5 Hz are in the range of the data for conformationally averaged systems. The third coupling constant (137.5 Hz) can be discussed together with the large geminal coupling of 18 Hz for the diastereotopic protons of the same methylene group. This situation is reminiscent of a segmental conformation already discussed for c-Sar_4,cf. Fig. 2b. It is noteworthy that four methylene groups of c-Sar_8 exhibit the same ^{13}C shift, only one value for the $^1J(C,H)$ coupling constant and a common value for the geminal H,H coupling (18 Hz). This result cannot be interpreted on the basis of the crystal conformation. We conclude that the orientation of the two diastereotopic methylene protons relative to the nitrogen p_z-orbital must be very similar in these four methylene groups and of the type illustrated in Fig. 3a. The other four methylene groups form two pairs with different ^{13}C shifts, $^1J(C,H)$ and $^2J(H,H)$ coupling constants, respectively. The observed $^1J(C,H)$ data (140.0 ± 1 and 142.5 ± 1 Hz) suggest that these segments constitute the more flexible part of the structure (Fig. 3b). A space-filling molecular model of the octapeptide confirms these conformational conclusions and allows an assignment of the rigid segments to the CH_2-N bond at the _cis_-amides, whereas there is considerable torsional mobility for the CH_2-N bond at the trans-amide systems.

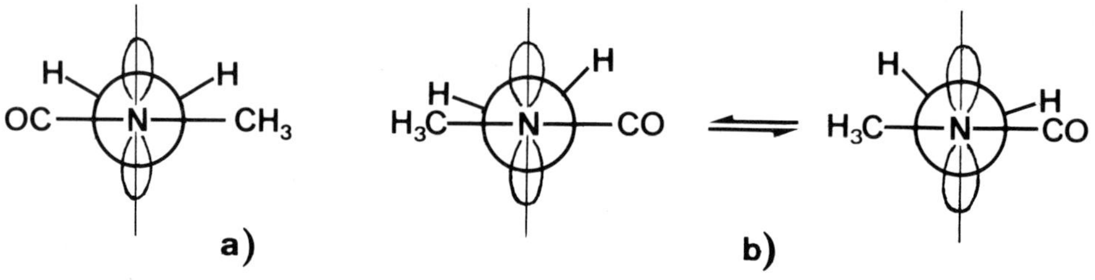

cis-amide

$^1J(C\alpha,H)$ = 137.5 Hz

trans-amide

$^1\bar{J}(C\alpha,H)$ = 140.0 Hz
142.5

Figure 3. Segmental Conformations of Cyclo-octasarcosyl

In c-Sar_2, as mentioned in Sect. 2, only one value for the $^1J(C,H)$ parameter is observed. This fact, together with the magnitude of this coupling constant (142.5 Hz), indicates conformational averaging between diastereotopic protons of a non-planar sixmembered ring. For the crystal structure a flattened chair-form was determined by X-ray analysis [10a]. In solution, this structure is expected to undergo rapid ring inversion resulting in an averaged value for the two otherwise non-equivalent $^1J(C,H)$ coupling constants. Since the molecule exhibits a dipole moment of 1.13 D in benzene solution [4], chair conformations as well as a planar six-membered ring can be ruled out. A flexible boat or twist-boat is the most likely conformation.

The $^1J(C\alpha,H)$ coupling constants of open-chain peptides with amino acids other than glycine and sarcosine may in principle contain conformational information since variations in rotamer population (angle Φ) are expected to affect the conformationally averaged $^1J(C,H)$ value. Since such data are not yet available the values for glycyl and sarcosyl peptides reported in this paper can only serve as the first reference data for peptides in general. In the present series, it appears that cyclic peptides of restricted torsional mobility (cf. Fig. 3b) give averaged coupling constants in a wider range (140-143 Hz) than open-chain peptides (140-141 Hz).

Lone-pair contributions to one-bond C,H coupling constants were already previously observed and calculated for a variety of systems. For example, the effect of a nitrogen lone-pair was detected in N-methyl-aziridine, in which for the C,H bonds syn and anti to the lone-pair $^{1}J(C,H)$ values of 171 and 161 Hz respectively were obtained [7]. In the acetaldoxime stereoisomers the corresponding data for the C,H coupling constants are 177 and 163 Hz. Calculations in the frame work of the Pople-Santry theory have shown that the stereospecificity is due to hyperconjugative electron delocalization from the lone-pair into the C,H antibonding orbitals [5,6,8]. A synperiplanar arrangement of lone-pair and C-H bond leads to a positive contribution to the coupling constant, whereas an antiperiplanar arrangement should give a (smaller) negative contribution. Both from experimental and theoretical considerations, maximum ΔJ values of 15-20 Hz can be expected for the extreme conformations. The stereochemical dependence of $^{1}J(C,H)$ was also calculated for the CH_3-N and CH_3-CO groups in N-methylacetamide and the results were compared with rotationally averaged coupling constants in some simple amides [13].

Lone-pair effects from oxygen atoms are well-known in oxygen heterocycles. They have served as a criterion to differentiate between α- and β-anomeric stereoisomers of carbohydrates such as hexopyranoses [14] and pentopyranoses [15].

The peptide case constitutes a special situation in so far as the 'lone-pair' of the nitrogen atom is in fact a p_z-orbital which is part of the molecular orbital of the amide system. Our data seem to indicate that the variation in $^{1}J(C,H)$ coupling constants which must be attributed to 'lone-pair' effects can be as high as 12 Hz. For very favorable relative orientations of the C-H bond and the nitrogen 'lone-pair' even larger effects may be expected.

ACKNOWLEDGEMENTS

The authors would like to thank Prof. J. Dale and Dr. K. Titlestad for the samples of c-Sar_4 and C-Sar_8. This work has been supported by the Swiss National Science Foundation.

REFERENCES

[1] U. Vögeli, W. von Philipsborn, K. Nagarajan and M.D. Nair, Helv. Chim. Acta 61, 607 (1978)

[2] cf. V.F. Bystrov in: Progress in Nuclear Magnetic Resonance Spectroscopy, J.W. Emsley, J. Feeney and L.H. Sutcliffe, eds., Vol. 10, Part 2, p. 41, Pergamon Press Oxford, 1976

[3] J. Dale and K. Titlestad, Chem. Commun. 1970, 1403

[4] K. Titlestad, Acta Chem. Scand. B 29, 153 (1975)

[5] V.M.S. Gil and J.J.C. Teixeira-Dias, Mol. Phys. 15, 47 (1968)

[6] G.E. Maciel, J.W. McIver, Jr., N.S. Ostlund and J.A. Pople, J. Amer. Chem. Soc. 92, 1 (1970)

[7] T. Yonezawa and I. Morishima, J. Mol. Spectry. 27, 210 (1968)

[8] T. Yonezawa, I. Morishima, K. Fukuta and Y. Ohmori, J. Mol. Spectry. 31, 341 (1969)

[9] J. Dale and K. Titlestad, Chem. Commun. 1969, 656

[10] a) P. Groth, Acta Chem. Scand. 23, 3155 (1969); b) ibid. 24, 780 (1970); c) ibid. 27, 3217 (1973); d) K. Titlestad, P. Groth, J. Dale and M.Y. Ali, Chem. Commun. 1973, 346

[11] a) V.F. Bystrov, Y.D. Gavrilov and V.N. Solkan, J. Magn. Reson. 19, 123 (1975); b) V.F. Bystrov and V.N. Solkan, Bull. Acad. Sci. USSR, Div. Chem. Sci. 23, 1232 (1974)

[12] M. Barfield and D.M. Grant, J. Amer. Chem. Soc. 85, 1899 (1963)

[13] Y.D. Gavrilov, V.N. Solkan and V.F. Bystrov, Bull. Acad. Sci. USSR, Div. Chem. Sci. 24, 2368 (1975)

[14] K. Bock and C. Pedersen, J. Chem. Soc. Perkin II, 1974, 293

[15] K. Bock and C. Pedersen, Acta Chem. Scand. B 29, 258 (1975)

NMR SATURATION TRANSFER STUDIES OF THE CATALYSIS OF THE REVERSIBLE HYDRATION OF ACETALDEHYDE BY CARBONIC ANHYDRASE

Dalia Cheshnovsky and Gil Navon
Department of Chemistry, Tel-Aviv University, Tel Aviv, Israel

The enzyme carbonic anhydrase catalyses the interconversion of carbon dioxide and bicarbonate at a rate which is one of the fastest known for enzyme catalysis[1].

$$CO_2 + H_2O \rightleftharpoons HCO_3^- + H^+ \qquad (1)$$

The enzyme is capable of catalysing other reactions such as hydrolysis of some carboxylic, sulfonic and carbonic esters (2-4) and the reversible hydration of aldehydes (5-7) and pyruvic acid (8), although with much smaller rates as compared with the hydration of CO_2.

Since the process of obtaining NMR spectra is relatively slow, NMR investigations of the binding of small molecules to enzymes are limited mostly to inhibitors, activators, or very slowly reacting substrates (9-10). However, the main strength of the NMR method in investigating reaction mechanisms is for exchange reactions under conditions of equilibrium. Thus the NMR method seemed to us very suitable for the investigation of the reversible hydration of acetaldehyde (Eq. 2) and its catalysis by carbonic anhydrase.

$$CH_3CH{=}O + H_2O \rightleftharpoons CH_3CH(OH)_2 \qquad (2)$$

The spectrum of acetaldehyde in D_2O solution is given in Fig. 1a. It is seen that the spectrum is composed of two well separated spectra of the aldehyde and its hydrate with almost equal intensity.

The kinetics of the exchange between the aldehyde and its hydrate can be investigated by the line broadening technique. Since the lines of acetaldehyde are well separated from those of the hydrate the limit of slow exchange applies, and the line broadening due to the exchange should be equal to $1/\pi\tau$ where τ is the exchange lifetime. Thus, if the line broadenings of the aldehyde and the hydrate are determined by exchange only, their ratio should be equal to K_{eq}=[hydrate]/[aldehyde]. However, this was not found to be the case when carbonic anhydrase was added to an aqueous solution of acetaldehyde. The peaks of the acet-

B. Pullman (ed.), Nuclear Magnetic Resonance Spectroscopy in Molecular Biology, 261-271.

aldehyde were broadened to a greater extent than those of the hydrate, much more than the value expected from the equilibrium constant K_{eq}. Thus it seemed to us that some other factors such as binding to the enzyme contributed to the linewidth and the exchange lifetime could not be evaluated from the line broadening. Therefore the saturation transfer method of Forsen and Hoffman (11-13) and Gupta and Redfield (14) was used for the investigation of the enzyme catalyzed exchange between acetaldehyde and its hydrate.

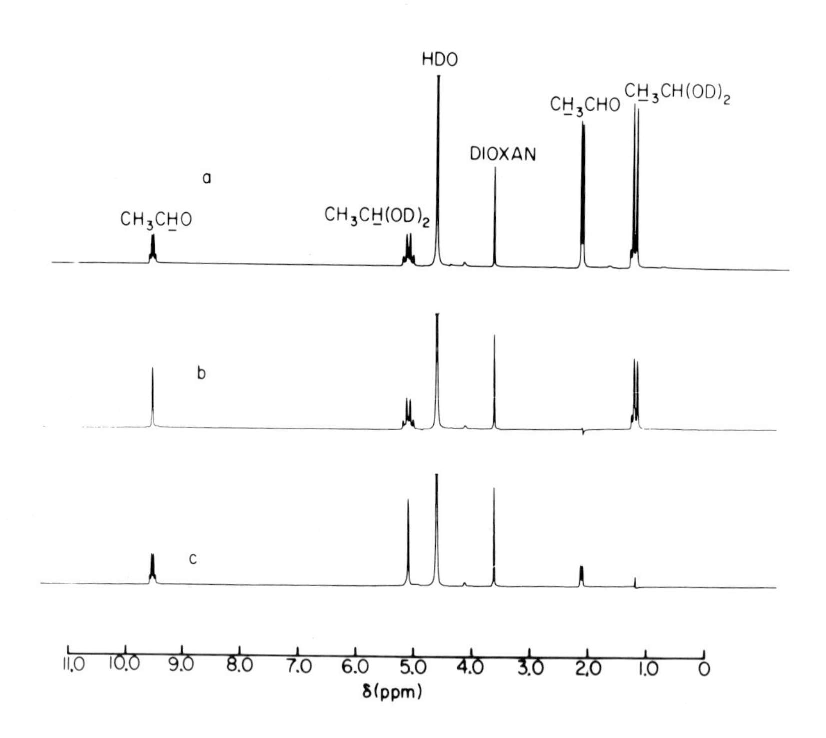

Figure 1. NMR spectra of D_2O solution of acetaldehyde in the presence of 1.3×10^{-4} M bovine carbonic anhydrase B in 0.02 M phosphate buffer, pH meter reading 7.5. The total concentration of the acetaldehyde and its hydrate was 0.46 M. The temperature was 28°C.
a. No double irradiation. b. Double irradiation at the position of the acetaldehyde methyl protons. c. Double irradiation at the position of acetaldehyde hydrate methyl protons.

The saturation transfer method

When exchange occurs between species A and B, Bloch equation for the magnetization parallel to the magnetic field, M_z, can be written as:

$$\frac{dM_z^A}{dt} = -\frac{M_z^A - M_o^A}{T_1^A} - \frac{M_z^A}{\tau_A} + \frac{M_z^B}{\tau_B} \quad (3)$$

where M_o^A and M_o^B are the equilibrium magnetizations of species A and B, $\tau_A$and τ_B are their exchange lifetimes and T_1^A is the longitudinal relaxation time of A in the absence of exchange. Under steady state conditions, $dM_z^A/dt=0$, and by using the equality $\tau_A/\tau_B=M_o^A/M_o^B$, the following relation is obtained:

$$\frac{\tau_A}{T_1^A} = \frac{M_z^A/M_o^A - M_z^B/M_o^B}{1 - M_z^A/M_o^A} \quad (4)$$

In the saturation transfer method double irradiation is applied on a resonance line of one of the species, say species B, and the intensity of the corresponding group in the other species, A, is recorded. This is shown in Figs. 1b and 1c. It is seen (Fig. 1b) that when the aldehyde methyl protons are irradiated, in addition to the decoupling which occurs in the quartet of the aldehyde proton, a reduction in the size of the signal of the hydrate methyl protons is occuring. This is due to a transfer of saturation caused by the chemical exchange. Similar phenomena occur when the doublet due to the hydrate is being irradiated (Fig. 1c). Since the intensity of the line is proportional to M_z the ratios M_z^A/M_o^A and M_z^B/M_o^B can be obtained by integration of the lines of groups A and B respectively, with and without irradiation.

Typical curves of M_z^A/M_o under increasing irradiation intensities on group B are shown in Fig. 2. It is seen that after an initial rapid diminuation of M_z^A/M_o^A which corresponds to saturation of group B there is a further decrease in the intensity following the increase of H_2 rf power. This is due to a direct saturation of line A caused by the tail of the spectral density of H_2 which is centered at the resonance frequency of line B. This effect was taken into account by extrapolation of the slow portion of the decrease of line A to $H_2=0$ as shown in Fig. 2.

Using the WH-90 Fourier transform NMR spectrometer where the decoupling of rf field H_2 is introduced by the time sharing method, i.e. H_2 is on when the receiver is off, it is possible to measure the residual intensity of line B which is being irradiated. This is shown in Fig. 3. Thus it is possible to measure both M_z^A/M_o^A and M_z^B/M_o^B under low H_2 intensities where the effect of direct saturation can be neglected. By this method the exchange lifetime τ_A can be measured using only one experiment, with and without double irradiation. The applicability of Eq. 4 is demonstrated in Table 1 where τ_A/T_1^A was measured with various intensities of H_2. Also, the value obtained by increasing H_2, so that signal B is completely saturated, and extrapolating the values of M_z^A/M_o^o to zero H_2 intensity is given for comparison. It is seen that the value of τ_A/T_1^A was independent of the irradiation intensity indicating the validity of the method.

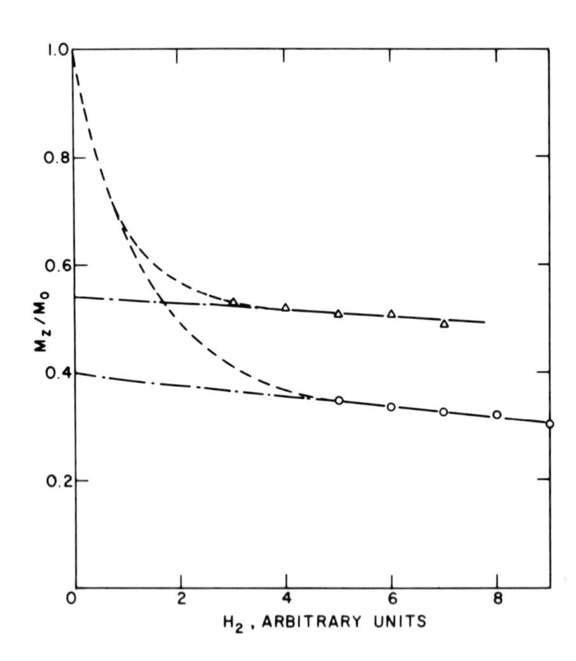

Figure 2. An illustration of the determination of the extrapolated M_z/M_o. (Δ) relative intensities of the acetaldehyde hydrate methyl protons with double irradiation at the acetalde hyde methyl protons resonance. The solution contained $1.25x10^{-4}$M carbonic anhydrase, acetaldehyde and its hydrate total concentration 0.46M, 0.02M phosphate buffer, pH meter reading 7.5. The temperature was 27°C. (o) The same as the previous curve but the methyl doublet of the acetaldehyde was observed while that of the acetaldehyde hydrate was being irradiated, and the enzyme concentration was $5.2x10^{-5}$M.

A comparison between the saturation-transfer method with exchange broadening is possible in a system where no causes other than chemical exchange are expected to contribute to the line-broadening of the exchanging species. Such a system is the acid catalyzed exchange reaction between acetaldehyde and its hydrate. Indeed, as can be seen from Table 2, the specific rate constants calculated from the two methods are the same within the experimental error, nnd furthermore the ratio of the forwards and reverse rate constants agrees in the two methods with $K_{eq.}$=1.15 which was measured independently from the ratio of the areas under the peaks of the two species.

For the calculation of exchange lifetimes according to Eq. 4 the longitudinal relaxation times in the absence of the exchange should be known. Since some exchange is induced by the buffer, the T_1 relaxation times of the various lines in the spectrum of acetaldehyde and its

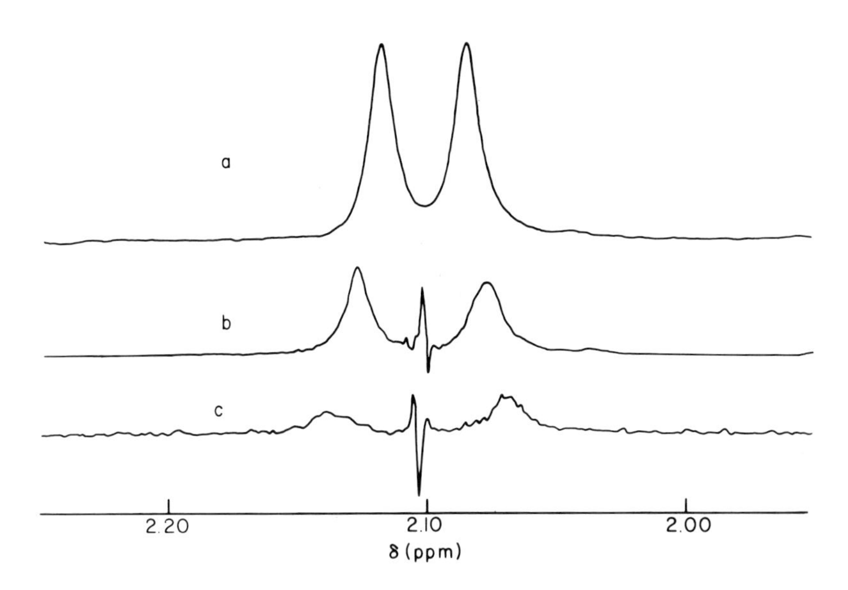

Figure 3. NMR spectra on an expanded scale of the acetaldehyde methyl protons without double irradiation (spectrum a) and with couble irradiation (spectrum b and c). The average intensity of irradiation in spectrum c is twice as much as in spectrum b.
Number of scans was: 10, 50 and 10 in spectra a, b and c respectively. The solution contained 5.0×10^{-4}M carbonic anhydrase, acetaldehyde and its hydrate total concentration 0.46M, 0.02M phosphate buffer, pH meter reading 7.5. The temperature was 19°C.

Table 1. Relative magnetizations of methyl protons of acetaldehyde and its hydrate at low levels of double irradiation.[a]

Double irradiation pulse width μ sec	M_z^A/M_o^A	M_z^B/M_o^B	τ_A/T_1^A
0	1.0	1.0	
0.8	0.69	0.45	0.74
1.0	0.58	0.28	0.72
1.5	0.54	0.20	0.74
2.0	0.47	0.10	0.68
extrapolated	0.415	0.0	0.71

a. Acetaldehyde and its hydrate total concentration was 0.46M at a phosphate buffer concentration of 0.02M and pH meter reading 7.5. Enzyme concentration was 5×10^{-4}M and the temperature was 19°C.

Table 2. Rate constant of the acid catalysed reversible hydration of acetaldehyde.

pD[a]	2.31	2.50	3.62
method	LB	LB	ST
$1/\tau$(hyd), sec^{-1}	4.54	2.95	0.223
$1/\tau$ (dehyd), sec^{-1}	3.97	2.64	0.185
k_{D^+}(hyd)[c], $M^{-1}sec^{-1}$	927	933	930
k_{D^+}(dehyd)[c], $M^{-1}sec^{-1}$	811	835	771
k_{D^+}(hyd)/k_{D^+}(dehyd)	1.14	1.12	1.20

[a]Calculated by adding 0.41 to the pH-meter reading.

[b]LB = line broadening, $1/\tau$ is calculated according to the relation $1/\tau=\pi(\Delta\nu-\Delta\nu(o))$ where $\Delta\nu(o)$ is the linewidth at neutral pH. ST = saturation transfer method as explained in the text.

[c]k_{D^+} is calculated by dividing $1/\tau$ by D^+ concentration.

hydrate were measured as a function of buffer concentration and were extrapolated to zero buffer concentration. Using the values of T_1 at the various temperatures $1/\tau$ (dehydration) as a function of temperature was found. At every temperature the rate of the reverse reaction, $1/\tau$ (hydration), was calculated using $1/\tau$ (dehydration) and the value of K_{eq} at the same temperature.

The enzymatic catalysis of the reversible hydration reaction.

The net exchange rate caused by the enzyme $1/\tau_E$ is given in Eq. 5

$$1/\tau_E = 1/\tau - 1/\tau_o \qquad (5)$$

where $1/\tau$ and $1/\tau_o$ are the measured exchange rates in the presence and the absence of the enzyme respectively.

Phenomenologically we can write

$$1/\tau_E = k_{enz} \quad [E] \qquad (6)$$

where [E] is the molar concentration of the enzyme.

An Arrhenius plot of k_{enz} for the dehydration reaction, k_{enz}^{dehyd}, gave a good straight line (see Fig. 4), corresponding to an activation energy of 11.8 ± 1.8 kcal/mole. Our results at 0°C, k_{enz}^{dehyd}=19,400 M^{-1} min^{-1} can be compared with the result obtained by pocker and Meany (5). The value given in their paper at 0°C and pH 7.91, k_{enz}=48,000 M^{-1} min^{-1} is the sum of k_{enz}^{hyd} and k_{enz}^{dehyd}, so that using our equilibrium constant K_{eq}=3.0 at 0°C the value of k_{enz}^{dehyd}=12,000 M^{-1} min^{-1} is calculated from their results. The reason for the discrepancy can be linked to the fact that bovine carbonic anhydrase with a specific activity of 1600 units/mg, according to the method of Wilbur and Anderson, (15) was used in their work as compared to 3000 units/mg in the present work.

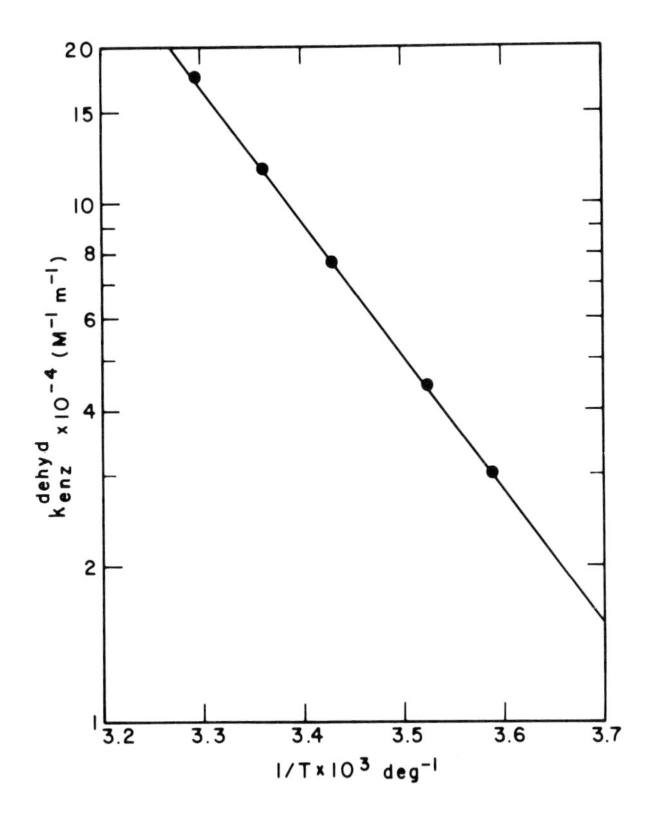

Figure 4. Arrhenius plot of k_{enz}^{dehyd}. Phosphate buffer concentration was 0.02M, pH meter reading 7.5. Acetaldehyde and its hydrate total concentration 0.45M. Concentrations of carbonic anhydrase were in the range of $1.0x10^{-4}$ - $1.4x10^{-3}$M.

In order to establish whether the catalysis is specific to carbonic anhydrase and is not due to a general catalysis by the protein we first measured the possible effects of bovine serum albumin on the reversible hydration reaction of acetaldehyde. No effect was detected at albumin concentrations as high as $5.5x10^{-4}$M at a temperature of 31°C and pH 6.73.

As a further indication for the participation of the active site of carbonic anhydrase in the catalysis, we measured the effect of two carbonic anhydrase specific inhibitors: p-toluensulfonamide and azide, on the enzyme-catalysed exchange reaction. Both inhibitors were found to inhibit the exchange reaction. In the case of the strongly bound inhibitor, p-toluensulfonamide, the exchange rate drops to a value equal to that in the absence of the enzyme at inhibitor to enzyme ratio of slightly above unity. For the azide ion a more gradual inhibition is observed as is shown in Fig. 5. Thus the data was plotted using the linear plot of

$$I_o / (1-V/V_o) = E_o - K_1 V_o/V \quad (7)$$

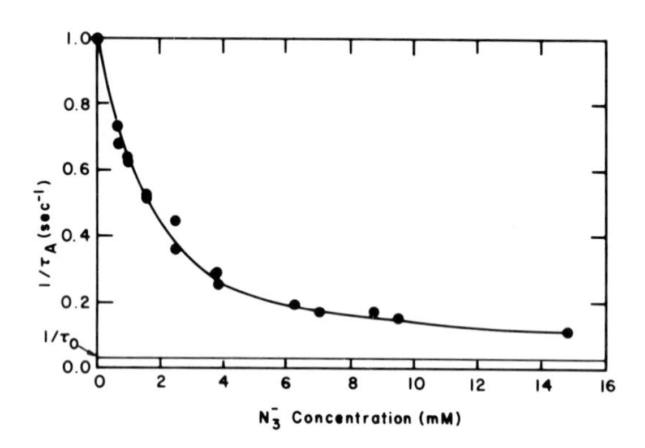

Figure 5. Dehydration rate constant of acetaldehyde hydrate as a function of azide ions concentration. Enzyme concentration was 3.4×10^{-4}M. Acetaldehyde and its hydrate total concentration 0.4M in 0.02M phosphate buffer, pH meter reading 7.5. The temperature was 32°C. $1/\tau_0$ is the rate constant in the absence of the enzyme.

The absence of a consistent deviation from the linear plot indicates a simple one-site inhibition pattern. The binding constant of the azide ion obtained from this plot $K_I = (1.3 \pm 0.2) \times 10^{-3}$M at 32°C and pH 7.5 compares favourably with the value of $K_I = 5.9 \times 10^{-4}$M obtained by Pocker and Stone (16) at 25°C and pH 7.55 by the inhibition of p-nitrophenylacetate hydrolysis.

The effect of binding on the relaxation rates

Values of the observed broadening per mole added enzyme, are given in Table 3. The contribution of the exchange to the molar line broadening is given by the rate constant k_{enz} calculated according to Eq. 6.

Table 3. The contributions of exchange and binding to the line broadening of the methyl protons of acetaldehyde and its hydrate caused by carbonic anhydrase[a]

	Aldehyde	Hydrate
$\pi\Delta\nu_p/[E]$	3.8 ± 0.2	2.3 ± 0.2
k_{enz}	2.3 ± 0.3	1.9 ± 0.2
$1/(T_{2b}[E])$	1.5 ± 0.5	0.4 ± 0.4
$\pi\Delta\nu_p/[E]$ [b]	0.9 ± 0.2	0.2 ± 0.2

[a] Data is given in units of $10^3 sec^{-1}M^{-1}$. Enzyme concentration was $1.1x10^{-3}$M. Total concentration of acetaldehyde and its hydrate was 0.46M, phosphate buffer 0.02M and pH meter reading 7.5. The temperature was 25°C.

[b] In the presence of excess p-toluenesulfonamide.

The broadening remained after substracting the effect of exchange (Eq.8) is attributed to binding to the enzyme.

$$1/T_{2b}[E] = \pi\Delta\nu_p/[E] - k_{enz} \tag{8}$$

The line broadenings caused by the enzyme were considerably reduced upon the addition of the inhibitor p-toluenesulfonamide, but a resudual broadening was observed even in the presence of excess inhibitor. This is shown in Fig. 6. Similar values of residual broadening were also observed in the presence of excess azide ions. In addition to the inhibition of the enzyme catalyzed exchange reaction, the inhibitors are expected to abolish the line broadening due to the binding to the active site of the enzyme. Thus, the residual broadenings in the presence of excess p-toluenesulfonamide given in the last line of Table 3, are attributed to binding to sites other than the active site. It is seen from Table 3 that the binding to the active site as well as to other sites on the enzyme affect the linewidth of the acetaldehyde to a greater extent than the linewidth of the hydrate.

The effect of the binding to the enzyme on T_1 could not be measured for the aldehyde and the hydrate separately since the exchange between them which is catalyzed by the enzyme, tends to equalize them. The result in the limit of fast exchange is a weighted average relaxation rate given (17) by the expression:

$$\overline{1/T_1} = f_A/T_{1A} + f_B/T_{1B} \tag{9}$$

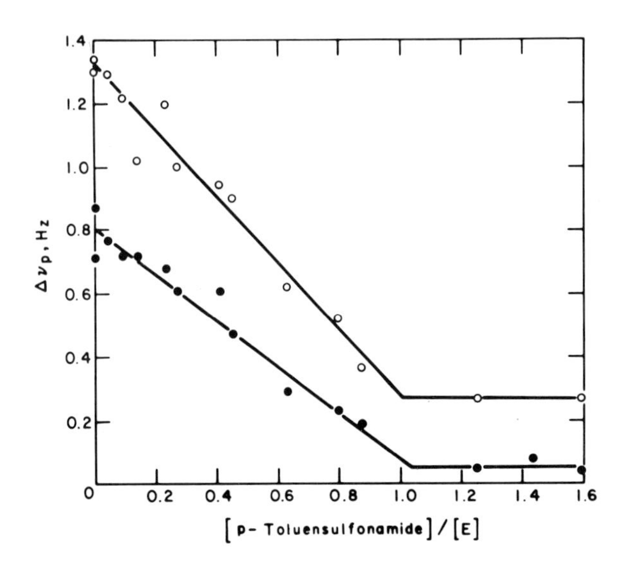

Fig. 6. Line broadenings of the methyl protons of acetaldehyde (o) and its hydrate (•) by carbonic anhydrase as a function of p-toluenesulfonamide concentration. Enzyme concentration was 1.1×10^{-3}M. Acetaldehyde and its hydrate total concentration 0.42M, 0.02M phosphate buffer. pH meter reading 7.5. The temperature was 25°C.

where f_A and f_B are the molar fractions of species A and B and T_{1A} and T_{1B} are their longitudinal relaxation times in the absence of exchange. This limit was obtained at high enzyme concentrations where the relaxation rates of the aldehyde and the hydrate were found to be equal. The contribution of the binding to the average relaxation rate, $\overline{1/T_{1b}}$, was calculated according to Eq. 9.

$$\overline{1/T_{1b}} = \overline{1/T_{1E}} - \overline{1/T_1}(0) \qquad (10)$$

where $\overline{1/T_{1E}}$ is the average relaxation rate obtained in the presence of high enzyme concentrations and $\overline{1/T_1}(0)$ was calculated according to Eq. 9 using the T_1 values in the absence of exchange. $1/T_{1b}$[E] was found to be $17M^{-1}$ sec^{-1} at 25°C. This value is a weighted average of the effect of binding on $1/T_1$ of the aldehyde and the hydrate peaks. Since the binding induced broadening of the aldehyde peaks is much larger than that of the hydrate it is expected that most of the binding effect on T_1 is due to the aldehyde. In the calculation of the exchange lifetimes from the saturation transfer data we used the T_1 values in the absence of added enzyme. It is seen from the present result that the effect of binding to the enzyme is indeed negligible. Even if we count the whole effect on $\overline{1/T_{1b}}$ as due to the hydrate, the effect on the calculated exchange lifetime would be smaller than 0.2 percent.

Conclusions

We can conclude that the line broadening of acetaldehyde induced by carbonic anhydrase is a combination of three contributions: catalysis of the exchange reaction between the aldehyde and its hydrate, binding to the active site of the enzyme and binding to other sites on the enzyme. By the use of the saturation transfer method it was possible to measure the exchange kinetics without interference from the effects of binding. This, together with the investigation of the effect of inhibitors allowed us to obtain the relative magnitude of the different contributions to the line broadening.

References

1. For a recent review see: P. Wyeth and R.H. Prince, Inorg. persp. Biol. Med. 1, 37 (1977) and references therein.
2. J.M. Armstrong, D.V. Myers, J.A. Verpoorte and J.T. Edsall, J. Biol. Chem. 241, 5137 (1966).
3. Y. Pocker and J.T. Stone, Biochemistry 7, 3021 (1968).
4. Y. Pocker and L.J. Guilbert, Biochemistry 13, 70 (1974).
5. Y. Pocker and J.E. Meany, Biochemistry 4, 2535 (1965).
6. Y. Pocker and J.E. Meany, Biochemistry 6, 239 (1967).
7. Y. Pocker and J.E. Meany, J. Chem. Soc. 89, 631 (1967).
8. Y. Pocker and J.E. Meany, J. Phys. Chem. 74, 1486 (1970).
9. R.A. Dwek, "Nuclear Magnetic Resonance in Biochemistry: Application to enzyme systems". Clarendon press, Oxford, 1973.
10. T. Kushnir and G. Navon, in H. Sund and G. Blauer (Eds) "Protein Ligand Interactions", Walter de Gruyter, Berlin, 1975. pp. 174-190.
11. S. Forsen and R.A. Hoffman, Acta Chem. Scand. 17, 1787 (1963).
12. S. Forsen and R.A. Hoffman, J. Chem. Phys. 39, 2892 (1963).
13. S. Forsen and R.A. Hoffman, J. Chem. Phys. 40, 1189 (1964).
14. R.K. Gupta and A.G. Redfield, Biochem. Biophys. Res. Commun. 41, 273 (1970).
15. K.M. Wilbur and N.G. Anderson, J. Biol. Chem. 176, 147 (1948).
16. Y. Pocker and J.T. Stone, Biochemistry 7, 2936 (1968).
17. H. Strehlow and J. Frahm, Ber. Bunsen. Ges. 79, 57 (1975).

CONFORMATION AND INTERACTION WITH SMALL MOLECULES OF CYCLIC HEXAPEPTIDES AS STUDIED BY NUCLEAR MAGNETIC RESONANCE SPECTROSCOPY AND CIRCULAR DICHROISM

YUKIO IMANISHI and SHUNSAKU KIMURA
Department of Polymer Chemistry, Kyoto University,
Kyoto, Japan

1. INTRODUCTION

Enzymes and other biopolymers exhibit many exquisite functions, resulting from the cooperation of several elemental factors. The most important factor must be the selection and the recognition of a substrate by proteins. In enzyme reactions, this is important as the pre-rate-determining step in the catalysis as well as this is related to the allosteric control of the action. It is also important in the hormone — receptor interaction. Therefore, the substrate-recognition could be the basis of all life phenomena. To investigate the substrate-recognition, cyclic peptides are very useful protein models. Cyclic peptides have been shown to bind metal cations and ammonium compounds by the intramolecular cooperation of carbonyl groups(1). Furthermore, they are so suitable for the detailed analysis of conformation by spectroscopy(2) that we can discuss the function of cyclic peptides in relation to their conformational properties.

We have synthesized and investigated a series of cyclic hexapeptides containing N-substituted amino acids, which are cyclo$(Sar)_6$, cyclo$(Sar-Sar-Gly)_2$, cyclo$(L-Pro-Sar-Gly)_2$, and cyclo$(Sar-Gly-Gly)_2$ (3,4,5). It has been reported that the numbers of different conformations due to the cis/trans isomerism of the N-substituted peptide bonds decreased in the above order, and that the first two cyclic hexapeptides, which are very flexible, bound alkali metal cations effectively. The proline residue made the cyclic hexapeptide more soluble in organic solvents. These observations point to cyclo$(L-Pro-Sar-Sar)_2$ which could be a synthetic ionophore transporting selectively metal ions through a hydrophobic membrane.

In the present investigation, we synthesized cyclo$(L-Pro-Sar-Sar)_2$ and investigated its specific interaction with various small molecules using nuclear magnetic resonance (nmr) and circular dichroism(cd).

2. SYNTHESIS OF CYCLO$(L-PRO-SAR-SAR)_2$

B. Pullman (ed.), Nuclear Magnetic Resonance Spectroscopy in Molecular Biology, 273-283.

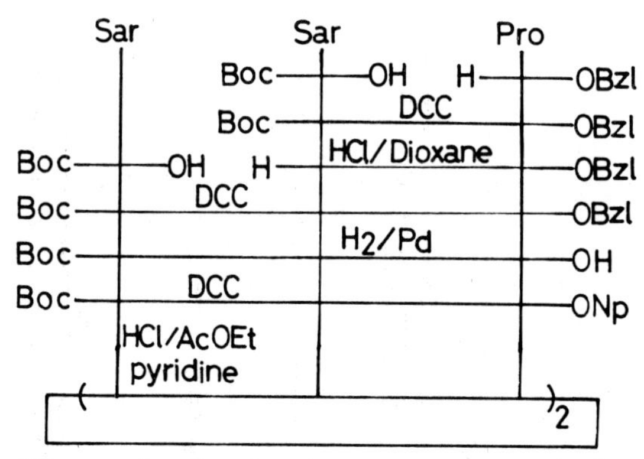

Figure 1. Synthetic route to cyclo(L-Pro-Sar-Sar)$_2$

Cyclo(L-Pro-Sar-Sar)$_2$ was synthesized in the way illustrated in Figure 1. The linear tripeptide was obtained by the stepwise elongation of the chain, and its active ester hydrochloride was cyclo-dimerized in the excess of pyridine to yield the cyclic hexapeptide. Cyclo(L-Pro-Sar-Sar)$_2$ was identified by the elemental analysis, mass spectrum, nmr, infrared (ir) spectrum, and thin layer chromatography.

Cyclo(L-Pro-Sar-Sar)$_2$ was soluble either in water or in organic solvents such as methanol, ethanol, chloroform, and dioxane as expected.

3. CONFORMATION OF CYCLO(L-PRO-SAR-SAR)$_2$

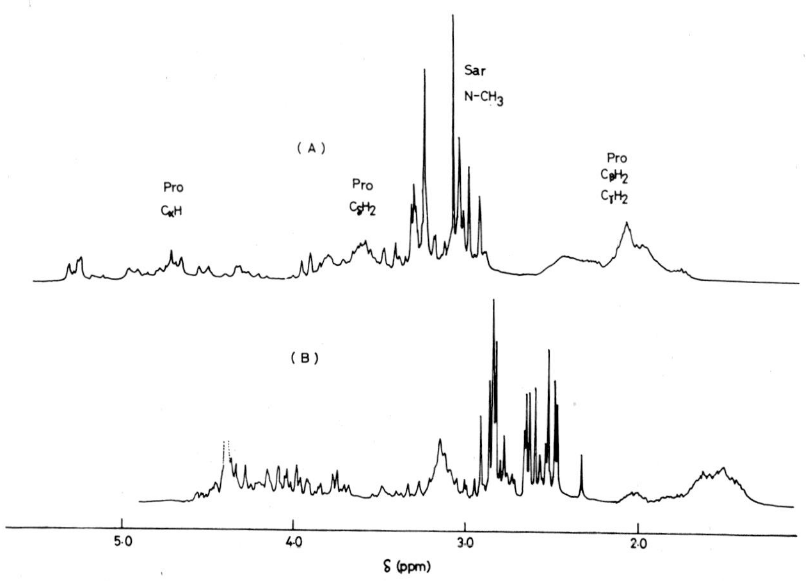

Figure 2. 270MHz nmr spectra of cyclo(L-Pro-Sar-Sar)$_2$: (A), in $CDCl_3$, 29.4mM; (B), in D_2O, 26.0 mM.

270MHz nmr spectra of cyclo(L-Pro-Sar-Sar)$_2$ in $CDCl_3$ and D_2O are shown in Figure 2. Many resonance signals appear for individual protons; for example, the N-CH_3 protons of sarcosyl residues give more signals than ten. In case of a C_2-symmetric conformation the N-CH_3 protons of four sarcosyl residues in cyclo(L-Pro-Sar-Sar)$_2$ should give two resonance signals, while in case of an asymmetric conformation four resonance signals. Therefore, cyclo(L-Pro-Sar-Sar)$_2$ must take many different conformations. This situation is undoubtedly related to the cis/trans isomerization of the N-substituted peptide bonds. Apparently, the N-CH_3 proton signals are separated into two groups. When C_6D_6 was added to the $CDCl_3$ solution of cyclo(L-Pro-Sar-Sar)$_2$, all resonance signals shifted to upfield, but the upfield shift of the signals included in the low-field group was more significant than that of the high-field group. The effect of benzene, together with the literature information(6,7,8), indicate that the low-field signals are ascribed to N-CH_3 protons involved in a trans-peptide bond, and that the high-field signals are ascribed to those involved in a cis-peptide bond.

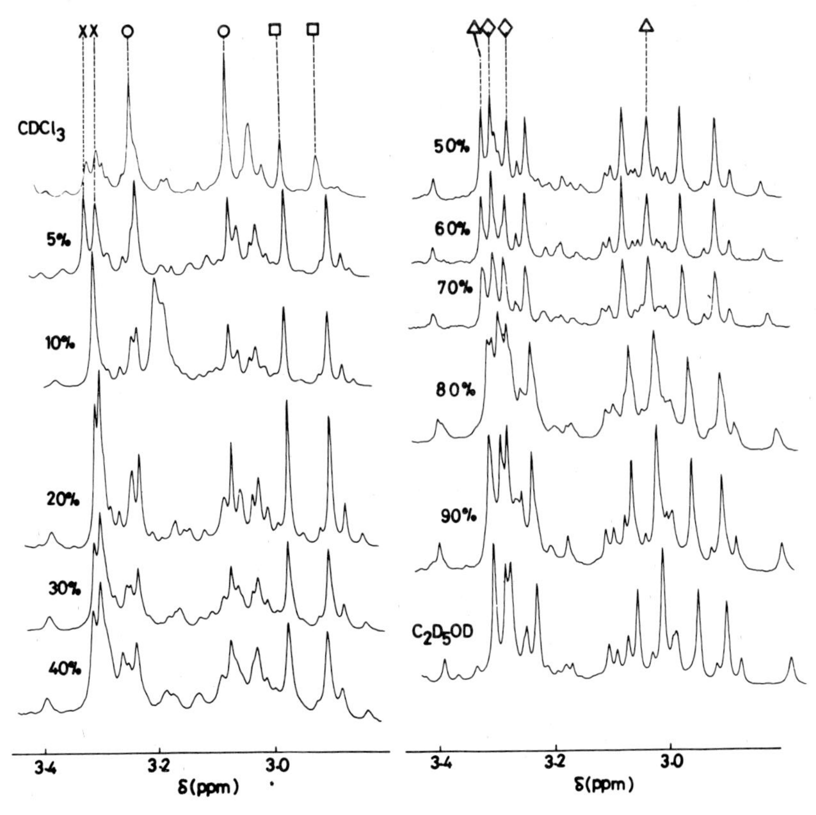

Figure 3. N-CH_3 proton signals of 270MHz nmr spectra of cyclo-(L-Pro-Sar-Sar)$_2$ in $CDCl_3$/C_2D_5OD mixture: a pair of signals is indicated by the same mark; C_2D_5OD vol% is given in the figure.

Figure 3 shows the 270MHz nmr spectra of the sarcosyl N-CH_3 protons of cyclo(L-Pro-Sar-Sar)$_2$ in $CDCl_3/C_2D_5OD$ mixture having various compositions. It is seen in Figures 2 and 3 that the nature of solvent affects profoundly the conformational equilibrium of cyclo(L-Pro-Sar-Sar)$_2$ and that more conformational isomers appear in more polar solvent. Figure 3 also shows that the addition of small amount of C_2D_5OD to $CDCl_3$ solution of cyclo(L-Pro-Sar-Sar)$_2$ causes a significant change of conformational equilibrium, in which the hydrogen bonding of C_2D_5OD with cyclo(L-Pro-Sar-Sar)$_2$ must be deeply involved. Looking for possible C_2-symmetric conformations, the resonance signals having equal intensities and receiving a similar solvent effect were paired. As summarized in Table 1, cyclo(L-Pro-Sar-Sar)$_2$ in $CDCl_3/C_2D_5OD$ mixture consists of seven different C_2-symmetric conformers. There must be some asymmetric conformers which are involved in the conformational equilibrium less extensively.

Table 1. Characteristics of C_2-symmetric conformation in $CDCl_3$ — C_2D_5OD mixture.

Mark[a)]	α, β[b)]	Note
○	C, T (or T, C)	Major conformation (36%) in $CDCl_3$
△	T, C (or C, T)	Major conformation (20%) in C_2D_5OD
□	C, C	Present in any solvent composition
×	T, T	Prevails in 5 — 40 vol % C_2D_5OD
◇	T, T	Prevails in 30 — 100 vol % C_2D_5OD
	C, T	Minor conformation
	T, C	Minor conformation

a) $N\text{-}CH_3$ proton signals marked in Figure 3.
b) Cis(C) or trans (T) with α and β peptide bonds of cyclo(Pro $\overset{\alpha}{-}$ Sar $\overset{\beta}{-}$ Sar)$_2$.

A comparison of cyclo(L-Pro-Sar-Sar)$_2$ with cyclo(L-Pro-Sar-Gly)$_2$ is interesting. For the latter, only two conformers were present in $CDCl_3$, CD_3OD, and $(CD_3)_2SO$ (4). The conformational difference between two cyclic hexapeptides must be ascribed to the different properties of sarcosyl and glycyl residues. The peptide bond involving a glycyl residue usually takes a trans conformation only and tends to form an intramolecular hydrogen bond. These properties reduce greatly the conformational freedom of cyclo(L-Pro-Sar-Gly)$_2$. Therefore, cyclo(L-Pro-Sar-Sar)$_2$ is a flexible molecule in a static sense.

3.2. Conformational calculation using a hard-sphere model

Using suitable values for individual bond lengths and bond angles, linear cyclic hexapeptide chains shown in Figure 4 were generated in a computer (9). Then the first and the 13th carbon atoms were brought into a distance closer than 2.5 Å (cyclization), and the self-intersecting chains were avoided. In the calculation, only C_2-symmetric conformations were taken into account, and therefore the independent confor-

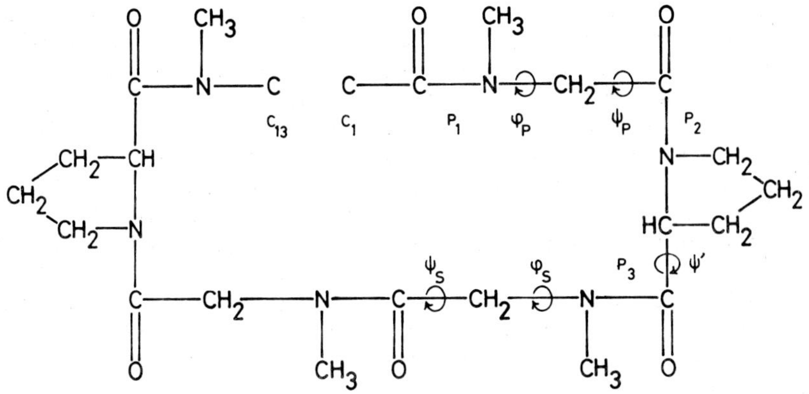

Figure 4. Conformational variables in a linear hexapeptide for the computer calculation.

mational variables were the peptide bonds P_1, P_2, and P_3, and the dihedral angles φ_p, ψ_p, φ_s, ψ_s, and ψ'. The dihedral angle ψ' has been known to take a few discrete values which depend on the conformation of the peptide bond P_3. $\varphi_s - \psi_s$ dipeptide maps were calculated for four possible isomeric dipeptide bonds P_3-P_1. $\varphi_p - \psi_p$ dipeptide maps were calculated for four possible isomeric dipeptide bonds P_1-P_2. Combining these fragment peptides, the numbers of cyclic, non-self-intersecting chains were counted and summarized in Table 2.

Table 2. Nature of peptide bonds and allowed cyclic chains.

P_1, P_2, P_3 a)	Number of cyclic chains without atomic overlaps
T , T , T	30
T , T , C	11
T , C , T	0
T , C , C	0
C , T , T	8
C , T , C	0
C , C , T	4
C , C , C	0

a) Cis(C) or trans(T) with P_1, P_2, and P_3 peptide bonds.

With regard to the conformation of peptide bonds, four types (TTT, TTC, CTT, and CCT) were proved to occur by calculation. By nmr investigation, CTC or CCC conformation was detected as well, while neither of them was allowed to exist by calculation. In these conformations the peptide bonds keep a C_2-symmetry, but other atoms must occupy asymmetric positions and move faster than the nmr time scale (10). Therefore, cyclo(L-Pro-Sar-Sar)$_2$ was proved to be flexible in a dynamic sense as well.

4. INTERACTION OF CYCLO(L-PRO-SAR-SAR)$_2$ WITH SMALL MOLECULES

4.1. Ion — dipole interaction of cyclo(L-Pro-Sar-Sar)$_2$ with metal cations

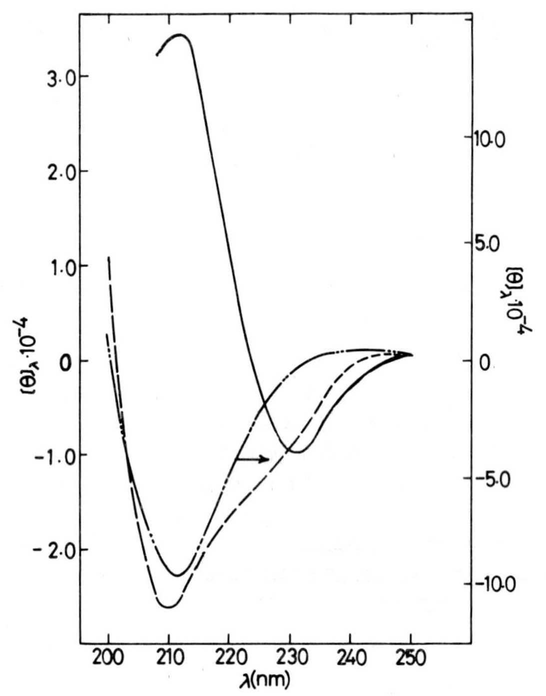

Figure 5. Effect of $LiClO_4$ on the cd spectrum of cyclo(L-Pro-Sar-Sar)$_2$ in ethanol: concentration of cyclo(L-Pro-Sar-Sar)$_2$, 9.5 x 10^{-4} M: molar ratio of $LiClO_4$ to cyclo(L-Pro-Sar-Sar)$_2$; (——), 0; (- - -), 3; (— ·· —), 37.

Cyclo(L-Pro-Sar-Sar)$_2$ and $LiClO_4$ had an ion — dipole interaction in ethanol to form an 1:1-complex and consequently the cd spectrum of cyclo(L-Pro-Sar-Sar)$_2$ was changed as shown in Figure 5. In accompany with the decrease of the strength of the Cotton effect at 212 nm, the sarcosyl N-CH_3 region of the 270MHz nmr spectrum became much simpler as shown in Figure 6. The nmr spectra indicate that in the presence of excess $LiClO_4$ cyclo(L-Pro-Sar-Sar)$_2$ converged into a single asymmetric conformation, in which four sarcosyl peptide bonds assume trans, trans, trans, and cis conformations. This asymmetric conformation does not prevail in ethanol solution without $LiClO_4$, but it is very stabilized by acquiring energy through the interaction of Li^+ with carbonyl groups (ion — dipole interaction). From the change of cd spectrum and the assumption of 1:1-complex formation, the association constant was determined to be 2.3 x $10^2 M^{-1}$. KCl and $Ba(ClO_4)_2$ interacted with cyclo-(L-Pro-Sar-Sar)$_2$ in ethanol, and caused a similar conformational change of the cyclic hexapeptide. The Ba^{2+} complex was ethanol-insoluble. Its ir spectrum showed better-resolved carbonyl absorptions than free peptide.

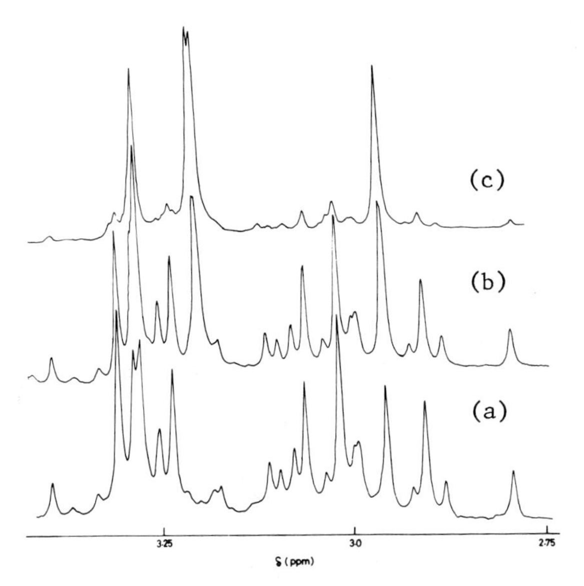

Figure 6. Effect of $LiClO_4$ on the nmr spectrum of cyclo(L-Pro-Sar-Sar)$_2$ in C_2D_5OD: molar ratio of $LiClO_4$ to cyclo(L-Pro-Sar-Sar)$_2$; (a), 0; (b), 0.43; (c), 1.9.

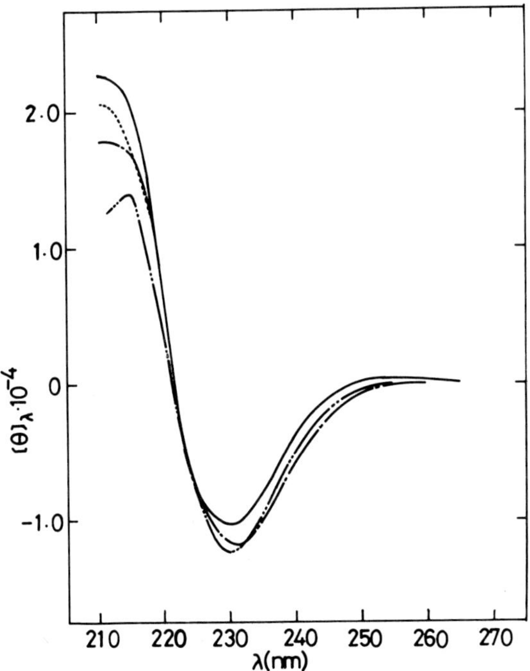

Figure 7. Effect of ammonium salts on the cd spectrum of cyclo-(L-Pro-Sar-Sar)$_2$ in dioxane/ethanol (4:5 v/v) mixture: (——), free peptide; (····), DL-Val-OEt·HCl, 4.9 times; (— · —), DL-Phe-OEt·HCl, 3.4 times; (— ·· —), DL-Phe-OEt·HCl, 5.4 times.

On the other hand, $Cu(ClO_4)_2$ and cyclo(L-Pro-Sar-Sar)$_2$ formed different types of complexes. The complex formation caused the increase of the rotational strength at 212 nm and a strong induced circular dichroism at the charge-transfer transition of Cu^{2+}. In this system 1:1- and 1:2-complexes were formed.

4.2. Hydrogen bonding of cyclo(L-Pro-Sar-Sar)$_2$ with ammonium ion

Amino acid ester hydrochlorides were added to cyclo(L-Pro-Sar-Sar)$_2$ in organic solvents. Very little change of the cd spectrum of cyclo(L-Pro-Sar-Sar)$_2$ was observed when an excess amount of DL-valine ethyl ester hydrochloride was added in dioxane/ethanol(4:5 v/v) mixture. When DL-phenylalanine ethyl ester hydrochloride, which possesses either an ammonium group or a phenyl group, was added, a conformational change of cyclo(L-Pro-Sar-Sar)$_2$ took place to some extent, which was evidenced by the decrease of the strength of Cotton effect at 210 nm (see Figure 7). Taking into account the observations about the additive effect of other related substances, it was concluded that an ammonium group and a phenyl group make an action to cyclo(L-Pro-Sar-Sar)$_2$ cooperatively to alter the conformational distribution.

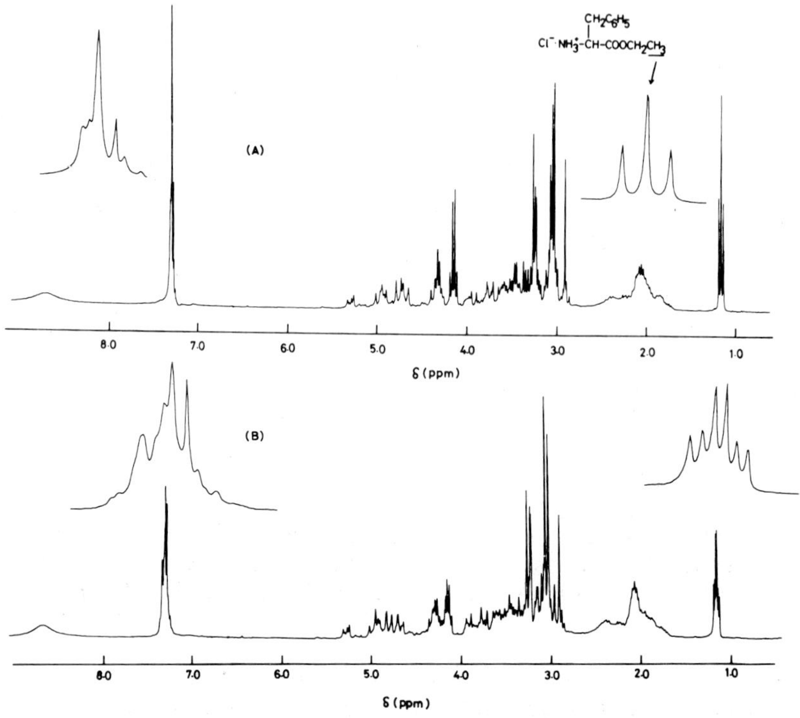

Figure 8. 270 MHz nmr spectra of cyclo(L-Pro-Sar-Sar)$_2$ in $CDCl_3$ with ammonium salts: (A), L-Phe-OEt·HCl, 1.1 times; (B), DL-Phe-OEt·HCl, equimolar.

The triplet signal of methyl protons in L-phenylalanine ethyl ester hydrochloride suffered only a chemical shift change when it was mixed with cyclo(L-Pro-Sar-Sar)$_2$, as shown in Figure 8. Free ammonium salts and complexed ones must be exchanging more rapidly than the nmr time scale. It is seen also from Figure 8 that two sets of triplet signals having equal intensities appear with DL-phenylalanine ethyl ester hydrochloride when the latter was mixed with cyclo(L-Pro-Sar-Sar)$_2$. This phenomenon could be explained in two different ways (11): the association constant with cyclo(L-Pro-Sar-Sar)$_2$ is different between the optical enantiomers; or the chiral cyclic hexapeptide and D- or L-ammonium salts formed two types of diastereomeric complexes having different structures.

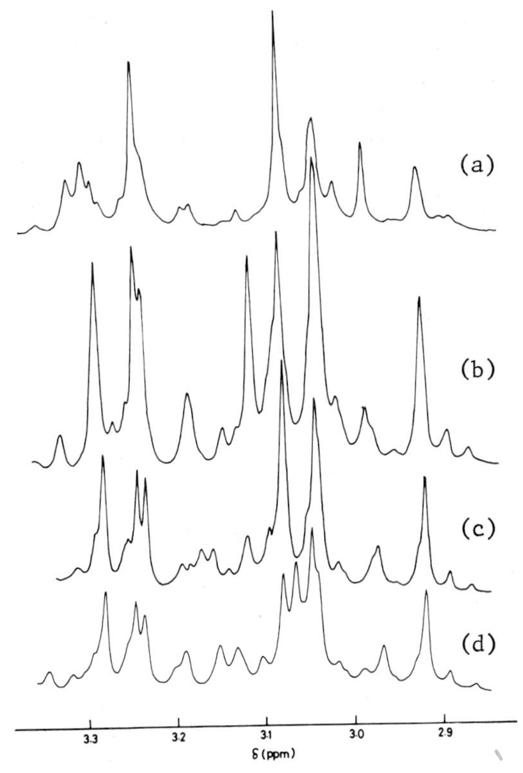

Figure 9. N-CH$_3$ proton signals of 270 MHz nmr spectra of cyclo(L-Pro-Sar-Sar)$_2$ in $CDCl_3$ with ammonium salts: molar ratio of DL-Phe-OEt·HCl to cyclo(L-Pro-Sar-Sar)$_2$; (a), 0; (b), 0.31; (c), 1.0; (d), 1.5.

As seen from Figure 9, four new signals of N-CH$_3$ protons of cyclo-(L-Pro-Sar-Sar)$_2$ appeared with increasing amount of DL-phenylalanine ethyl ester hydrochloride added. The nmr change suggests the conformational change of cyclo(L-Pro-Sar-Sar)$_2$ induced by the hydrogen bonding plus the aromatic-amide interaction. But, the cyclic hexapeptide never converged into a single conformation under the presence of excess ammonium salts, which contrasts sharply to the addition of metal ions.

When DL-valine benzyl ester hydrochloride was added to the $CDCl_3$ solution of cyclo(L-Pro-Sar-Sar)$_2$, two sets of doublet signals having different intensities were observed with the pendant CH_3 groups of the ammonium salt (see Figure 10). The exchange between free ammonium salts and complexed ones must be slower than the nmr time scale in the case of DL-valine benzyl ester hydrochloride. This observation is different from that in the case of DL-phenylalanine ethyl ester hydrochloride, although both molecules possess an ammonium group as well as a phenyl group. Apparently, the exchange rate depends strongly on the spatial arrangement of an ammonium group and a phenyl group.

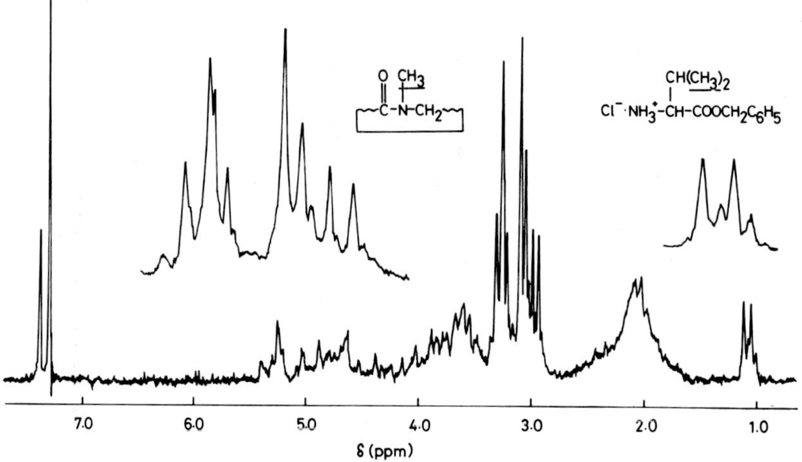

Figure 10. 100 MHz nmr spectrum of cyclo(L-Pro-Sar-Sar)$_2$ in $CDCl_3$ with DL-Val-OEt·HCl: molar ratio of DL-Val-OEt·HCl to cyclo(L-Pro-Sar-Sar)$_2$, 0.32.

5. CONCLUSION

Chiral, N-alkylated cyclic hexapeptide, cyclo(L-Pro-Sar-Sar)$_2$ was synthesized and it was found to be soluble in a wide range of solvents. Owing to the cis/trans isomerization of six N-alkylated peptide bonds and the absence of intramolecular hydrogen bonding, cyclo(L-Pro-Sar-Sar)$_2$ is very flexible in a thermodynamic sense as well as in a kinetic sense.

Cyclo(L-Pro-Sar-Sar)$_2$ interacted with various metal cations. In particular, the interaction with Li^+ in ethanol was so strong that the conformational mixture comprising at least seven C_2-symmetric conformers turned into a single asymmetric conformer. Cyclo(L-Pro-Sar-Sar)$_2$ interacted also with ammonium salts, and the interaction was strengthened by the cooperation of a phenyl group. With cyclo(L-Pro-Sar-Sar)$_2$ — aromatic ammonium salt complexes, the stability or the structure of complex was influenced by the chirality of the ammonium salt, and the exchange between free and bound ammonium salts was influenced by the spatial arrangement of an ammonium group and a phenyl group.

REFERENCES

1. Y. Imanishi, Adv. Polym. Sci., 20, 1 (1976).
2. D.W. Urry and M. Ohnishi, "Spectroscopic Approaches to Biomolecular Conformation," Amer. Med. Ass., Chicago, Ill., Ch. VII, pp. 263 – 303.
3. T. Sugihara, Y. Imanishi, and T. Higashimura, Biopolymers, 14, 723 (1975).
4. T. Sugihara, Y. Imanishi, and T. Higashimura, Biopolymers, 14, 733 (1975).
5. T. Sugihara, Y. Imanishi, and T. Higashimura, Biopolymers, 15, 1529 (1976).
6. J.V. Hatton and R.E. Richards, Mol. Phys., 3, 253 (1960).
7. J.V. Hatton and R.E. Richards, Mol. Phys., 5, 139 (1962).
8. M. Sisido, Y. Imanishi, and T. Higashimura, Biopolymers, 11, 399 (1972).
9. G. Nemethy, Biopolymers, 3, 155 (1965).
10. V. Madison, C.M. Deber, and E.R. Blout, J. Am. Chem. Soc., 99, 4788 (1977).
11. C.M. Deber and E.R. Blout, J. Am. Chem. Soc., 96, 7566 (1974).

^{13}NMR STUDIES ON OPIOID PEPTIDES; ^{13}C-ENRICHED METHIONINE-ENKEPHALIN AND α-ENDORPHIN

Roxanne Deslauriers*, Pierre Tancrède*, W. Herbert McGregor***, Dimitrios Sarantakis***, Rajmund L. Somorjai**, and Ian C.P. Smith*.
*Divisions of Biological Sciences and **Chemistry, National Research Council of Canada, Ottawa, Ontario, Canada K1A 0R6; ***Wyeth Laboratories, Philadelphia, Pa. 19101, U.S.A.

^{13}C NMR spectroscopy is becoming increasingly useful in the study of dynamical aspects of peptide conformation (Deslauriers and Smith, 1976; Bleich *et al.* 1976) as well as in the study of interactions between peptide hormones and macromolecules (Blumenstein and Hruby, 1977; Convert *et al.* 1977). We report here studies of the relative mobilities of moieties within the recently-discovered opiate agonists, the enkephalins (Hughes *et al.* 1975) and endorphins (Li and Chung, 1976a; and Lazarus *et al.* 1976), the first peptides reported to have morphine-like activity. The enkephalins are pentapeptides (Tyr-Gly-Gly-Phe-Met or Tyr-Gly-Gly-Phe-Leu) which can be extracted from brain tissue; the larger endorphins are thought to be enzyme-cleavage products of β-lipotropin, a 91-residue peptide synthesized in the pituitary (Li and Chung, 1976b). α-Endorphin comprises residues [61-76] of β-lipotropin, and the first five amino acids of α-endorphin are identical in sequence to methionine-enkephalin. It was thought that comparison of the dynamical properties of enkephalin and α-endorphin might provide insight into the structure-activity relations of these peptides (Ronái *et al.* 1977).

Morphine is known to interact sterospecifically with lipids such as cerebroside sulfate and phosphatidyl serine (Abood *et al.* 1977). It was of interest to investigate whether the enkephalins also interact with lipid dispersions and if so, the conformational aspects of the interaction. We have used enkephalins enriched specifically to 90% in ^{13}C at the α-carbons of the glycine residues to probe the interactions with lecithin and phosphatidyl serine.

MATERIALS AND METHODS

[2-[2-^{13}C]glycine]methionine-enkephalin and [3-[2-^{13}C]glycine] methionine-enkephalin, 90% enriched in ^{13}C were synthesized according to the procedure described in Tancrède *et al.* (1978). α-Endorphin (Tyr-Gly-Gly-Phe-Met-Thr-Ser-Glu-Lys-Ser-Gln-Thr-Pro-Leu-Val-Thr) was synthesized following the method described in Tancrède *et al* (1978).

B. Pullman (ed.), Nuclear Magnetic Resonance Spectroscopy in Molecular Biology, 285-296.

TABLE 1. Chemical shifts of α-endorphin and methionine-enkephalin in D_2O.[a]

Residue	α-Endorphin[(b)] δ (ppm)	Enkephalin[(c)] δ (ppm)	α-Endorphin Residue	δ (ppm)	α-Endorphin Residue	δ (ppm)
Tyr αCH	54.97	55.72	Thr αCH	60.15	Thr αCH	60.56
βCH_2	37.49	37.11	βCH_2	68.08	βCH_2	68.08
γC	126.55	126.58	γCH_3	19.76	γCH_3	19.76
δCH	131.96	132.00	C=O	171.04	C=O	172.00
εCH	117.02	117.07				
ξC	156.37	156.41	Ser αCH	56.68	Pro αCH	61.60
C=O	173.78	173.78	βCH_2	62.15	βCH_2	30.40
			C=O	172.57	γCH_2	25.78
Gly αCH_2	43.44	43.60			δCH_2	49.60
C=O	172.80	172.20	Glu αCH	54.96	C=O	174.79
			βCH_2	28.04		
Gly αCH_2	43.44	43.39	γCH_2	33.60	Leu αCH	53.80
C=O	172.80	171.83	δC=O	182.50	βCH_2	40.81
			C=O	174.34	γCH_2	25.42
Phe αCH	55.65	55.99			δCH_3	23.14
βCH_2	38.06	38.18	Lys αCH	54.73	δCH_3	22.16
γC	137.22	137.37	βCH_2	28.04	C=O	174.79
δCH	129.91	130.41	γCH_2	23.14		
εCH	130.31	129.89	δCH_2	31.37	Val αCH	58.15
ξCH	128.35	128.34	εCH_2	40.34	βCH	30.40
C=O	174.79	171.01	C=O	174.06	γCH_3	19.63
					γCH_3	18.83
Met αCH	53.66	53.34	Ser αCH	56.04	C=O	174.34
βCH_2	31.37	31.34	βCH_2	62.15		
γCH_2	30.40	30.50	C=O	172.18	Thr αCH	61.21
$S-CH_3$	15.30	15.25			βCH_2	69.03
C=O	176.50	176.34	Gln αCH	54.14	γCH_3	19.91
			βCH_2	27.31	C=O	172.18
			γCH_2	32.21		
			δC=O	178.84		
			C=O	174.06		

a) Assignments for the carbonyls are tentative. Chemical shifts are given with respect to TMS, external reference, and are accurate to 0.05 ppm. b) c = 100 mg/ 1.7 ml D_2O, pH = 5.0. c) c = 100 mg/ml, pH = 3.6. The residues in α-endorphin are listed in columns according to their position in the peptide beginning at the N-terminus.

Egg lecithin and phosphatidyl serine (beef brain) were purchased from Lipid Products and Serdary Research, respectively.

^{13}C NMR spectra were obtained in D_2O on Varian CFT-20, XL-100 and Bruker HX-270 spectrometers operating at 20, 25 and 68 MHz, respectively, using methods described in Tancrède *et al.* (1978). Viscosity measurements were performed on the samples used for NMR studies with an Ostwald (flow) microviscometer (Tanford, 1967).

Optimal diffusion constants for overall anisotropic rigid body reorientation of enkephalin were computed by least-squares fitting of the observed ^{13}C spin-lattice relaxation times of proton-bearing carbons (Somorjai and Deslauriers, 1976). Correlation times for internal motion about individual C-C bonds in side chains were estimated according to methods described by Deslauriers and Somorjai (1976).

RESULTS AND DISCUSSION

The ^{13}C NMR spectrum of methionine-enkephalin is shown in Figure 1.

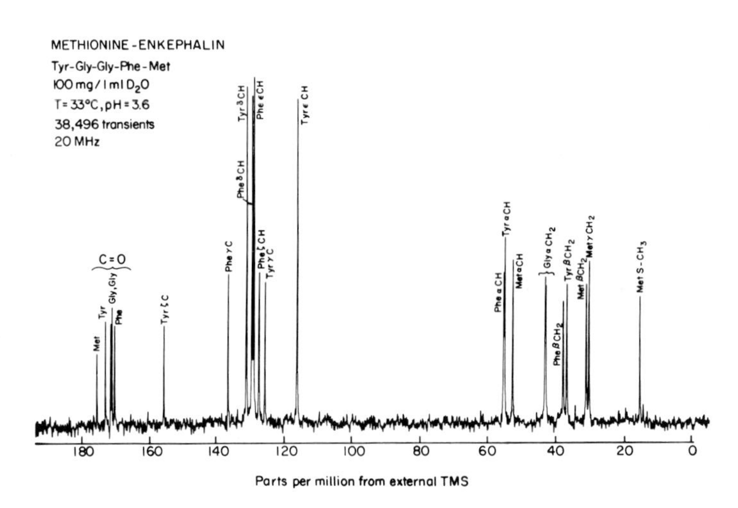

Figure 1. ^{13}C NMR spectrum of methionine-enkephalin.

The ^{13}C chemical shifts are presented in Table 1. Unambiguous assignment of the glycyl residues is provided by the specifically ^{13}C-enriched enkephalins. The tyrosyl and phenylalanyl residues are readily assigned by pH effects on these residues (Christl and Roberts, 1972).

CORRELATION TIMES FOR INDIVIDUAL CARBONS IN ENKEPHALIN

Figure 2 shows the ^{13}C spin-lattice relaxation times (T_1) of individual proton bearing carbons in methionine-enkephalin at a concentration of 100 mg/ml in D_2O. The T_1 values multiplied by N, the number of protons directly bonded to the carbon under study, are in good agreement with those reported by Combrisson *et al.* (1976).

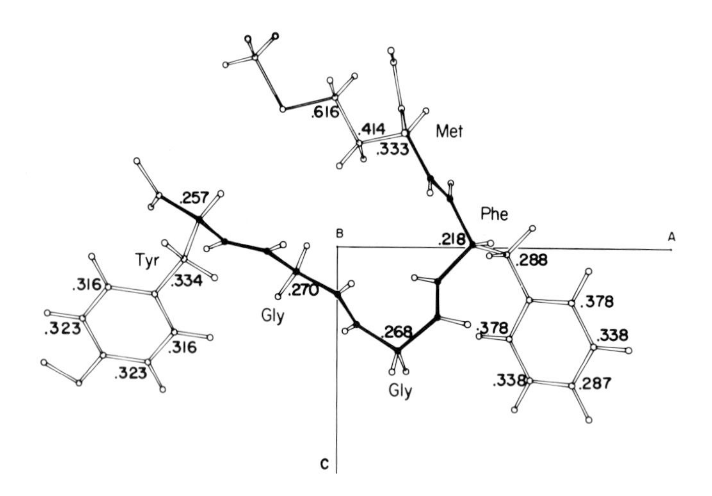

Figure 2. ^{13}C NT_1 values (in seconds) for methionine-enkephalin.

Leucine-enkephalin has recently been found to have a folded conformation in the solid state (Smith and Griffin, 1978). The structure presented in Figure 2 is one of those compatible with the ^1H NMR spin-spin coupling data of Jones *et al.* (1976).

The optimal correlation times for overall molecular reorientation (τ_{mol}) of methionine-enkephalin were determined for rotational diffusion about the principal axes (A, B, C in Figure 2) of the moment of inertia tensor; $\tau_x = 3.2 \times 10^{-10}$, $\tau_y = 1.7 \times 10^{-10}$ and $\tau_z = 9.9 \times 10^{-11}$ sec rad^{-1}. They are independent of any physically reasonable starting values for τ_x, τ_y or τ_z, and insensitive to the molecular conformation chosen for the least-squares fitting (Tancrède *et al.* 1978). The optimal values of the correlation times for internal motion (τ_{int}) which will reproduce the observed T_1 values of the side chains are shown in Table 2. The least-squares fit was performed using a folded conformation of enkephalin with optimal correlation times for overall molecular motion of $\tau_x = \tau_y = 2.4 \times 10^{-10}$, $\tau_z = 9.9 \times 10^{-11}$ sec rad^{-1}. The former two values are averaged because the computer program used in these calculations assumes that the overall motion of the molecule has at least cylindrical symmetry (Deslauriers and Somorjai, 1976).

Table 2. Correlation times for internal motion in the side-chains of enkephalin

Carbon	τ_{int} (sec.)	T_1obs. (sec.)	T_1calc. (sec.)
Tyr $C_\beta H_2$	3.1×10^{-10}	0.167	0.168
Phe $C_\beta H_2$	8.5×10^{-10}	0.144	0.147
Phe $C_\Theta H$	5.5×10^{-11}	0.287	0.289
Met $C_\beta H_2$	1.2×10^{-10}	0.207	0.208
Met $C_\gamma H_2$	7.7×10^{-12}	0.308	0.307

MOBILITY OF GLYCYL RESIDUES

One particularly interesting structural feature of enkephalin is the presence of two non-terminal glycine residues. Glycine residues have been reported to undergo increased internal motions when compared with other, optically active, residues in the same position of a peptide sequence (Deslauriers *et al.* 1977b). Figure 2 shows that the NT_1 values of the two glycine residues in enkephalin are essentially equal at peptide concentrations of 100 mg/ml. There does not seem to be any large difference between the NT_1 values of the α-carbon of the terminal tyrosine residue and the two glycine residues. Overlap of the ^{13}C resonances of the two glycine α-carbons at 20 MHz prevent measurement of accurate T_1 values. We therefore enriched enkephalin in ^{13}C at each of the glycine α-carbons; this also allowed us to measure the concentration dependence of the T_1 values for the enriched carbons at concentrations from 100 mg/ml to 2 mg/ml. The T_1 values increased linearly with decreasing peptide concentration, the increase in T_1 values being more pronounced for the [2-glycine] residue. At peptide concentrations of 2 mg/ml the NT_1 value of the [3-glycine] residue was 0.230 sec and that of the [2-glycine] residue 0.260 sec. As the peptide concentration decreases, the correlation time for overall molecular motion decreases; the correlation time for the internal motion (or the segmental motion in this case) of the [2-glycine] residues decreases also. The observed T_1 values were pH-independent.

MOLECULAR AGGREGATION

Enkephalin has been reported to aggregate under various conditions (Khaled *et al.* 1977). The continuity of the T_1 *vs* concentration plot indicates no discontinuous change in the rotational correlation time (effective molecular weight) of enkephalin under our experimental conditions. Rotational correlation times of particles in solution can sometimes be correlated with macroscopic sample viscosity. We measured T_1 values and macroscopic viscosities for both [2-[2-^{13}C]glycine] methionine enkephalin and [3-[2-^{13}C]glycine] methionine enkephalin at concentrations varying between 2 mg/ml and 50 mg/ml and at temperatures varying between 5 and 60°C. If a particle is undergoing Brownian rotational diffusion in solution, the relation between the rotational correlation time of the particle (τ_{mol}), the viscosity (η), and the temperature (T) is

$$\tau_{mol} = \frac{V_m \eta}{kT}$$

where V_m is the molecular volume and k is the Boltzmann factor (Tanford, 1967). The volume obtained by plotting $\tau_{eff}T$ versus η, where τ_{eff} is the observed effective correlation time for the [3-glycine] residue, is 586 Å^3. This can be compared with the value of 543 Å^3 calculated for enkephalin by the method of atomic increments (Edwards, 1970), which gives an estimate of the minimum molecular volume by summing the Van der Waals radii of the constitutent atoms. The close correspondence between the calculated molecular volume and that obtained from our experimental data suggests that enkephalin is monomeric in aqueous solution at concentrations of 2-50 mg/ml, and 5-60°C.

SEGMENTAL MOTION OF [2-GLYCINE]

Knowing the correlation time for overall molecular motion, it should be possible to determine the correlation time for internal motion within the peptide backbone of enkephalin by methods similar to those employed for calculating the correlation times of internal motion in the side chains (Deslauriers *et al.* 1977a). We take the correlation time of the [3-glycine] residue as representative of overall molecular motion; from this we calculate the correlation time for internal motion of the [2-glycine] residue. This is at best an approximation because the [3-glycine] residue may also undergo internal motion. The correlation time for internal motion (or segmental motion) of the [2-glycine] residue increases with increasing peptide concentration, going from 6.8×10^{-11} to 7.8×10^{-11} sec rad^{-1} for a change in peptide concentration from 2 mg/ml to 15 mg/ml. The overall molecular motion was characterized by $\tau_z = 9.9 \times 10^{-11}$ and $\tau_x = \tau_y = 2.4 \times 10^{-10}$ sec rad^{-1}. Thus although the peptide is monomeric under our experimental conditions, the correlation time for internal motion of the [2-glycine] residue increases with increasing peptide concentration. This is probably a result of the increase in macroscopic viscosity of the solution.

ACTIVATION ENERGY FOR MOLECULAR REORIENTATION

The temperature dependence of our τ_{eff} values can be used to extract a value for the activation energy for molecular reorientation using the Arrhenius expression

$$\tau = \tau_0 \cdot \exp\ (E_a/RT) \tag{2}$$

where E_a is the activation energy, R the gas constant and T the absolute temperature. Using equation (2) and the temperature dependence of τ for the [2-glycine] and [3-glycine] residues at various concentrations, we have found the activation energy for rotation to vary between 20-26 kJ/mole, with an estimated error of ±25% due to a ±10% error in T_1 values.

STUDIES ON α-ENDORPHIN

α-Endorphin has been shown to have 60% of the activity of enkephalin (Ronái *et al.* 1977). The 68 MHz ^{13}C spectrum of α-endorphin is shown in Figure 3. The narrowness of the resonance indicates that the peptide is not highly aggregated and probably exists in a wide variety of rapidly interconverting conformations. The ^{13}C chemical shifts for the peptides are given in Table 1 and NT_1 values in Figure 4.

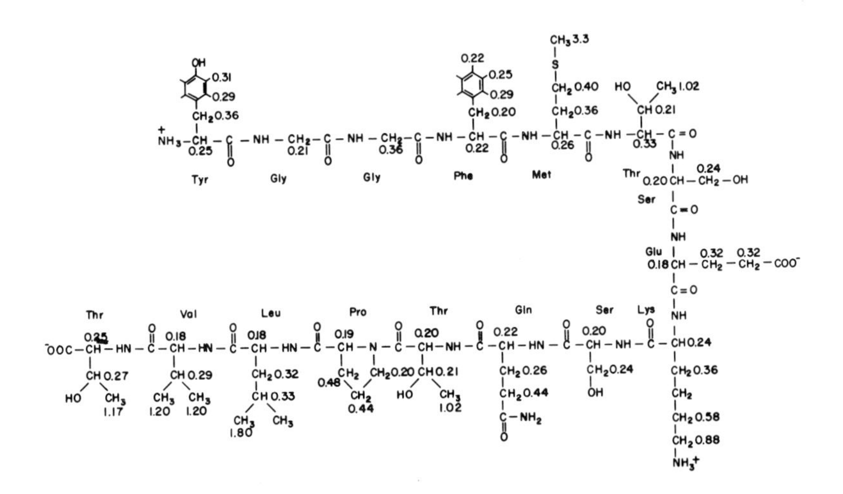

Figure 4. NT_1 values of carbons in α-endorphin (100 mg/1.7 ml).

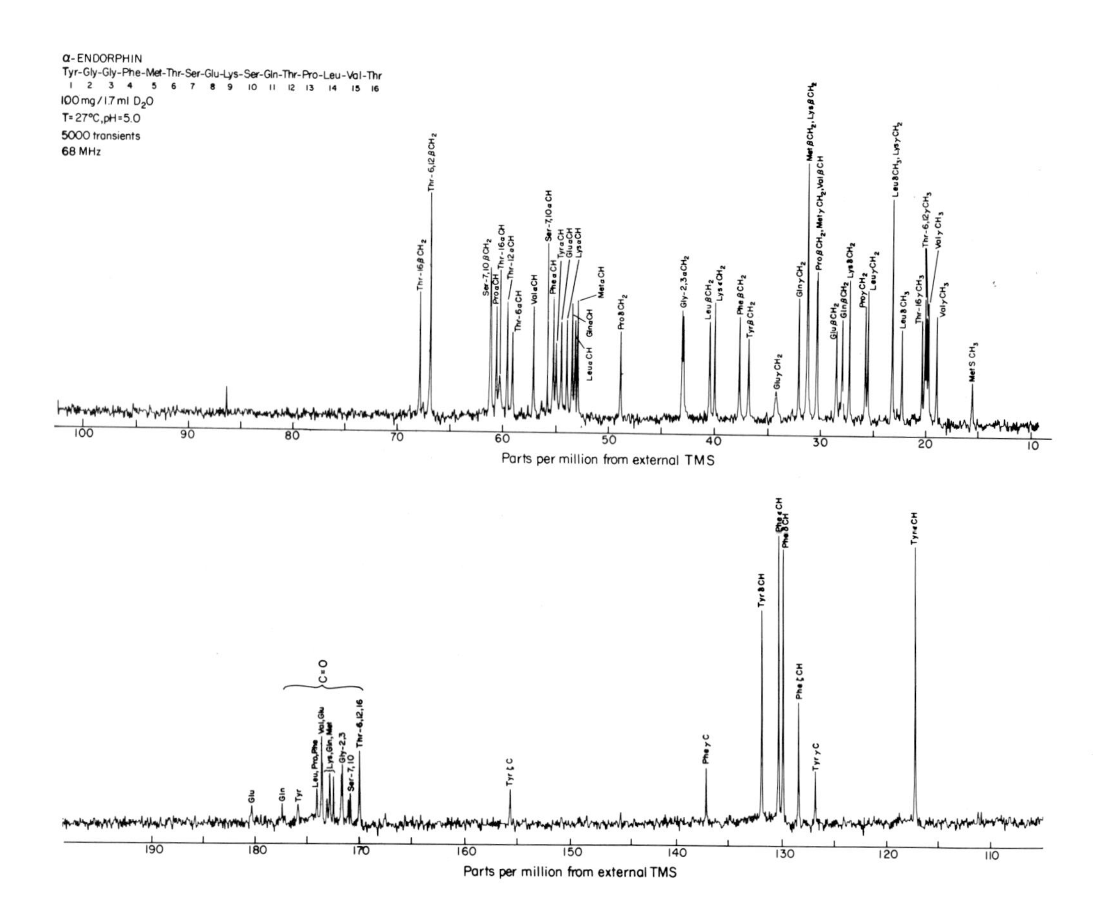

Figure 3. ^{13}C NMR spectrum (68 MHz) of α-endorphin.

Comparison of these data with those obtained for enkephalin shows that incorporation of the pentapeptide into enkephalin causes a greater decrease in T_1 values (40%) for the backbone carbons than for the side chains (25%). This may reflect the fact that the side-chains undergo considerable internal motion, which is less affected by a change in the overall molecular correlation time than motions within the peptide backbone (Deslauriers *et al.* 1977b). Furthermore, the NT_1 values of the enkephalin portion of the α-endorphin sequence differ little from those of the remainder of the peptide. Thus, it seems that in α-endorphin, the enkephalin sequence is not buried or otherwise restricted by the remainder of the peptide sequence. It would be of interest to study β-endorphin (residues [61-91] of β-lipotropin), which is eight times more active than enkephalin, shows prolonged biological activity (Loh *et al.* 1976), and may have a preferred conformation in solution (Hollosi *et al.* 1977; Austen *et al.* 1977).

ENKEPHALIN-LIPID INTERACTIONS

Morphine has been shown to interact stereospecifically with such lipids as cerebroside sulfate and phosphatidyl serine (Abood *et al.* 1977). By analogy, the enkephalins might also show an interaction with lipid systems, since they are competitive with morphine *in vivo*. Table 3 shows the results of preliminary studies using lecithins and phosphatidyl serine.

These results show that enkephalin does interact with phosphatidyl serine. The interaction is pH-dependent; around neutrality a strong interaction is evident in the decreased T_1 value. This interaction can be eliminated under basic pH-conditions. We have verified that the decrease in T_1 was not due to paramagnetic impurities by bubbling H_2S through the samples. The observed effect at pH is reversible. Increasing the temperature under basic conditions, yields T_1 values characteristic of the free species. An increase in temperature under neutral or acidic pH conditions causes an increase of T_1 values; however, the T_1 values remain shorter than those observed for the free species. Further studies remain to be performed to elucidate the exact nature of the interaction, namely as a function of pH concentration of added peptide, and position of label.

REFERENCES

Abood, L.G., Salem, Jr., N., MacNeil, M., Bloom, L., and Abood, M.E.: 1977, *Biochim. Biophys. Acta* 468, 51.
Austen, B.M. and Smyth, D.G.: 1977, *Biochem. Biophys. Res. Commun.* 77, 86.
Bleich, H.E., Cutnell, J.D., and Glasel, J.A.: 1976, *J. Amer. Chem. Soc.* 15, 2455.
Blumenstein, M. and Hruby, V.J.: 1977, *Biochemistry* 16, 5169.

Table 3. Interactions between enkaphalin and lipids

Concentration* Lipid mg/ml	Concentration* Enkaphalin mg/ml	pH	T°C	T_1 (sec)	δ (ppm)	$\Delta\nu$ (Hz)
	2.8	1.5-10.0	32	0.23	43.47	6
			59	0.39		
EL 28	2.8	3.5	32	0.23	N.O.	N.O.
56	2.8	3.5	32	0.24		
PS 12	3.0	6.8	32	0.20	43.47	6
		11.8	32	0.25	43.45	6
PS 33	3.0	7.4	32	0.22	43.46	6
		12.0	32	0.23	43.45	6
PS 64	0.75	6.8	32	0.22	43.42	6
		11.9	32	0.24	43.44	6
PS 91	1.1	4.5	35	0.14	N.O.	10
		10.0	35	0.21	N.O.	8
PS 91	3.3	4.8	35	0.14	N.O.	14
			57	0.22	N.O.	10
		9.4	57	0.44	N.O.	

*Concentrations are given in mg/ml D_2O.

EL = egg lecithin.

PS = phosphatidyl serine.

Enkephalin = [2-[2-^{13}C glycine] methionine enkephalin.

N.O. = not observed.

BIBLIOGRAPHY

Bradbury, F.A., Smith, D.G., Snell, C.R., Birdsall, N.J.M., and Hulme, E.C.: 1976, *Nature* 260, 793.
Christl, M. and Roberts, J.D.: 1972, *J. Amer. Chem. Soc.* 94, 4565.
Combrisson, S., Roques, B.P., and Oberlin, R.: 1976, *Tetrahedron Lett.* 38, 3455.
Convert, O., Griffin, J.H., Di Bello, C., Nicolas, P., and Cohen, P.: 1977, *Biochemistry* 16, 5061.
Deslauriers, R. and Smith, I.C.P.: 1976, in *Topics in Carbon-13 NMR Spectroscopy*, Vol. 2. G.C. Levy, ed. John Wiley and Sons, N.Y. pp. 1-80.
Deslauriers, R. and Somorjai, R.L.: 1976, *J. Amer. Chem. Soc.* 98, 1931.
Deslauriers, R., Ralston, E., and Somorjai, R.L.: 1977a, *J. Mol. Biol.* 113, 697.
Deslauriers, R., Levy, G.C., McGregor, W.H., Sarantakis, D., and Smith, I.C.P.: 1977b, *Eur. J. Biochem.* 75, 343.
Edward, J.T.: 1970, *J. Chem. Educ.* 47, 261.
Hollósi, M., Kajtár, M., and Gráf, L.: 1977, *FEBS Letters* 74, 185.
Hughes, J., Smith, T.W., Kosterlitz, H.W., Fothergill, L.A., Morgan, B.A., and Marris, H.R.: 1975, *Nature* 258, 577.
Jones, C.R., Gibbons, W.A., and Garsky, V.: 1976, *Nature* 262, 779.
Khaled, M.A., Long, M.M., Thompson, W.D., Bradley, R.J., Brown, G.P., and Urry, D.W.: 1977, *Biochem. Biophys. Res. Commun.* 76, 224.
Lazarus, L.H., Ling, N., and Guillemin, R.: 1976, *Proc. Natl. Acad. Sci. U.S.A.* 73, 2156.
Li, C.H. and Chung, D.: 1976a, *Proc. Natl. Acad. Sci. U.S.A.* 73, 1145.
Li, C.H. and Chung, D.: 1976b, *Nature* 260, 622.
Loh, H.H., Cho, T.M., Wu, Y.C., and Way, E.L.: 1974, *Life Sci.* 14, 2231.
Loh, H.H., Tseng, L.F., Wei, E., and Li, C.H.: 1976, *Proc. Natl. Acad. Sci. U.S.A.* 73, 2895.
Ronái, A.Z., Gráf, L., Székely, J.I., Dunai-Kovacs, Z., and Bajusz, S.: 1977, *FEBS Lett.* 74, 182.
Smith, G.D. and Griffin, J.F.: 1978, *Science* (in press).
Somorjai, R.L. and Deslauriers, R.: 1976, *J. Amer. Chem. Soc.* 98, 6460.
Tancrède, P., Deslauriers, R., McGregor, W.H., Ralston, E., Sarantakis, D., Somorjai, R.L., and Smith, I.C.P.: 1978, *Biochemistry* (submitted).
Tanford, C.: 1967, *Physical Chemistry of Macromolecules*, John Wiley, N.Y.

DISCUSSION

Navon: Did you try to compare your C^{13} T_1 data of enkephalin with H' relaxation times ? These are of course much easier experiments and since there are many methylene groups, where the H-H distance is fixed, you may obtain additional information on the motion of the different groups.

Deslauriers: We have not tried to compare ^{13}C T_1 data with ^1H T_1 data. We have not measured ^1H relaxation times in the enkephalins, I am not aware that any one else has either. Although the experiments may be easier to perform, they are not necessarily easier to interpret. I think interpretation of ^1H relaxation times would be easier in cases where the peptide is rigid and the ^1H-^1H distances known. In enkephalin, there are probably many interconverting structures where the relevant ^1H-^1H distances are not easily defined. It would however be possible to analyze some of the ^1H relaxation data, where as you say, the relevant distances are fixed.

Birdsall: There appears to be a remarkable increase in the π of the α- carbon of threonine-66 of α-endorphin relative to the α-carbon atoms of neighbouring residues. Do you consider that thus increase in π and presumably in the local segmental motion of the backbone might be an explanation for the fact that the met^{65}- thr^{66} peptide bond is the site of enzymatic cleavage of longer endorphin molecules to give methionine enkephaline ?

I agree with your comment that the T_1 value of the α-carbon of this threonine residue seems anomalously high. This warried us because α-endorphin also has a C-terminal threonine residue which might be expected to have longer π values due to segmental motion of the end of the peptide. We could have mis-assigned the various threonine residues. However we have performed studies at various pH values which allowed us to assign unambiguously the terminal residues as well as the titratable side-chains. Therefore we believe our assignment to be correct and the observed π value correct for this residue. Such large differences between adjacent, optically active, residues are very unusual; only glycine residues have previously shown such effects. Cleavage of the peptide of this point could occur due to greater accessibility of cleavage enzymes at this point. However I do not believe it is known whether or not it is α-endorphin which is cleaved to yield enkephalin rather than a larger fragment of β-lipotropin.

NMR STUDIES OF SELECTIVELY DEUTERIATED AND FLUORINE LABELLED DIHYDROFOLATE REDUCTASE

J. Feeney,
Division of Molecular Pharmacology,
National Institute for Medical Research,
Mill Hill, London NW7 1AA

Dihydrofolate reductase (EC.1.5.1.3) catalyses the reduction of dihydrofolate to tetrahydrofolate by NADPH. The enzyme has an essential role in nucleotide synthesis and is of considerable pharmacological interest being the target for important 'anti-folate' drugs such as methotrexate (I) and trimethoprim (II) (1,2). As part of a wider programme aimed at understanding the factors influencing the specificity of inhibitor binding to dihydrofolate reductase, we have been using high resolution NMR techniques to study the complexes formed when substrate analogues bind to the enzyme (3-12). In such studies it is important to resolve resonances which can be assigned to nuclei from the ligand or from individual residues in the enzyme. Unfortunately, even for a small enzyme such as dihydrofolate reductase from Lactobacillus casei (M.Wt. 17,800) the ^1H spectrum at 270 MHz is very complex and only a few resonances corresponding to single amino acid residues can be detected. This is illustrated in Figure 1 which

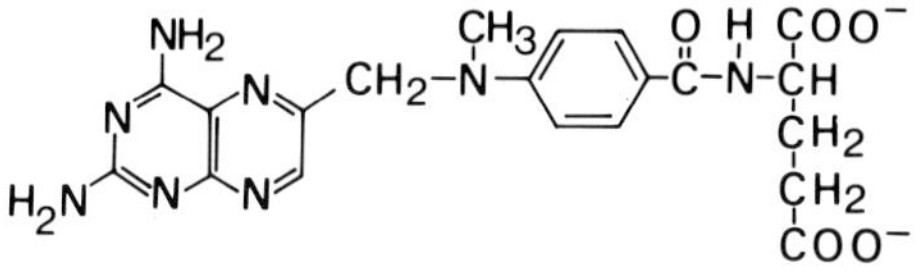

I

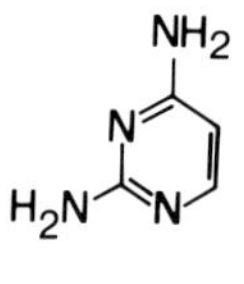

III

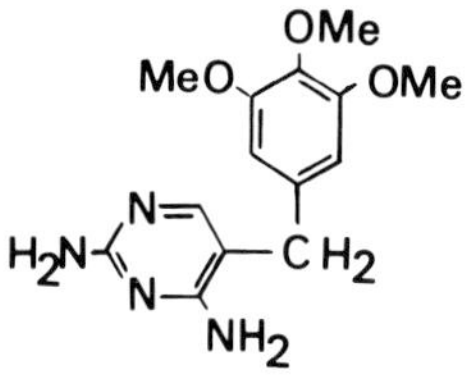

II

B. Pullman (ed.), Nuclear Magnetic Resonance Spectroscopy in Molecular Biology, 297-310.

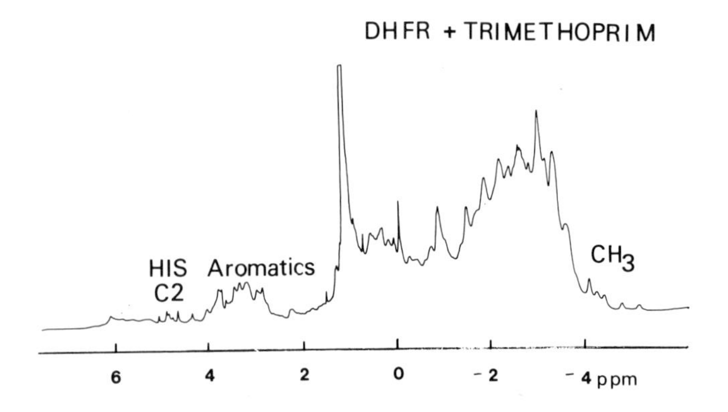

Figure 1. The 270 MHz ^{1}H nmr spectrum of a 1 mM solution of dihydrofolate reductase in the presence of 1 molar equivalent of trimethoprim.

shows the 270 MHz ^{1}H spectrum of the enzyme in the presence of one molar equivalent of trimethoprim. Most of the spectrum consists of a complex envelope containing overlapped signals from the different amino acids in the enzyme. The only signals which can be assigned to individual amino acids are the resonances from six histidine C2-H protons (4-5 ppm to low field of the dioxan reference) and the high field resonances (-4 to -5 ppm) corresponding to methyl groups in four aliphatic residues which are in close proximity to aromatic rings in the protein structure. The complex envelope of peaks between 2.7 and 4.0 ppm contains resonances from some 90 aromatic protons, none of which give clearly resolved signals. Furthermore, the signals from the aromatic protons of bound trimethoprim would also be expected to appear in this region of the spectrum.

In order to investigate the aromatic residues in detail it is necessary to resort to isotopic substitution methods to provide the required spectral simplification. For this reason we have studied enzymes containing selectively deuteriated aromatic residues and also containing ^{19}F-labelled amino acids. The deuteriation approach was pioneered some years ago by Jardetzky (13) and Katz (14) and their colleagues. More recently Sykes et al. (15) and other workers (11, 12, 16, 17, 18) have incorporated fluorine labelled amino acids into proteins and examined their NMR spectra.

The purpose of this present paper is to review the information obtained from our studies of selectively deuteriated and fluorine labelled dihydrofolate reductase (3-12). The enzyme used in these studies was isolated and purified from a methotrexate resistant strain of Lactobacillus casei as described by Dann and coworkers (19). This strain of L.casei is auxotropic for all the aromatic amino acids

(R.W. King and P. Scudder, unpublished results) and thus is a convenient organism to use for incorporating isotopically labelled amino acids, unchanged, into dihydrofolate reductase. In each case the labelled enzymes were prepared by growing L.casei on a medium containing a mixture of the pure amino acids including the appropriate isotopically labelled amino acid.

SPECTRAL SIMPLIFICATION BY USING ISOTOPIC LABELLING

Fluorine-labelled enzymes

Dihydrofolate reductases containing 3-fluorotyrosine and 6-fluorotryptophan have been prepared separately by growing L.casei on media containing the appropriate fluorine labelled amino acid (11). As expected, very simple ^{19}F spectra are obtained for these enzymes since the only resonances observed are those of the labelled amino acid. For example, the 3-fluorotyrosine labelled enzyme gives rise to five separate signals corresponding to the five tyrosine residues (see Figure 2). In the absence of ligands some of the ^{19}F signals are very broad (~30 Hz) and there is considerable overlap of signals.

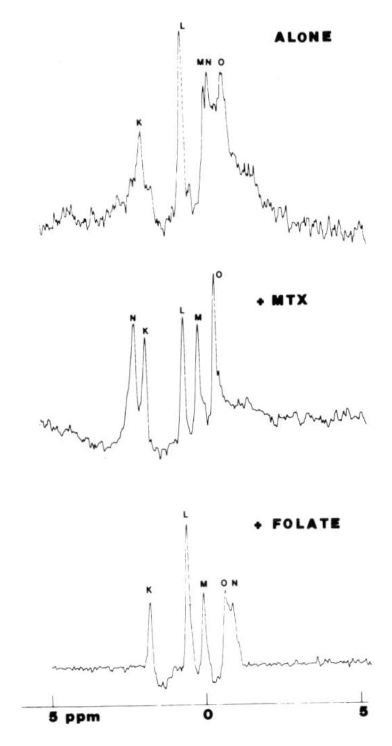

Figure 2. The 94.1 MHz ^{19}F spectra (proton noise decoupled) of 3-fluorotyrosine containing dihydrofolate reductase and its binary complexes with methotrexate and folate.

When methotrexate is added to the enzyme the signals become sharper and three of them (Y_N^F, Y_M^F and Y_O^F) show changes in chemical shifts. It will be seen later that in all cases where we have studied signals from single residues, the signals from the enzyme alone are invariably broader than those in spectra from enzyme complexes with tightly binding ligands such as methotrexate. The most likely explanation is that the enzyme in the absence of ligands exists in more than one conformational state which results in line broadening either due to the presence of species with slightly different shifts in slow exchange or to exchange broadening. The line-narrowing observed in the presence of an inhibitor such as methotrexate corresponds to the enzyme adopting a single conformational state.

In complexes with tightly binding ligands such as methotrexate, the exchange between free and bound forms of the enzyme is slow and this makes it difficult to connect signals from the complex with those from the free enzyme. To overcome this problem we have examined the enzyme in the presence of increasing amounts of fragments of methotrexate such as 2,4-diaminopyrimidine (III) which bind more weakly and are more likely to be in fast exchange. By monitoring the progressive shifts observed on increasing the ligand concentration it is possible to assign the signals being influenced by the different parts of the inhibitor. Thus, 2,4-diaminopyrimidine influences three of the five tyrosine residues such that the spectrum in the presence of excess ligand is quite similar to that of the enzyme-methotrexate complex.

The effects of folate binding on the ^{19}F spectrum are somewhat different from those of methotrexate. Resonances Y_N^F and Y_O^F are affected quite differently in the two cases, clearly indicating a difference in the mode of binding of folate and methotrexate. The large downfield shift observed for Y_N^F in the methotrexate.enzyme complex could arise from a direct interaction between the ligand and the tyrosine corresponding to Y_N^F: large downfield shifts would result from second-order electric field effects from London dispersion forces present when an interacting group is close to the fluorine nucleus.

Addition of NADPH to the binary complex of the enzyme with methotrexate causes a large upfield shift of Y_K^F: it seems likely that this arises from a ligand induced conformational change which removes an interacting group away from the fluorine nucleus. These results, together with those obtained with inhibitor fragments are given in Table 1 (11).

Similar experiments have been carried out using 6-fluorotryptophan labelled dihydrofolate reductase (11). In the ^{19}F spectra of the enzyme and its complexes with methotrexate and NADPH (shown in Figure 3) we again observe the line narrowing in the presence of strongly binding ligands. Four signals are resolved in the spectrum of the complex with methotrexate. One of the resonances (W_M^F) is

The ^{19}F chemical shifts† of the 3-fluorotyrosine resonances of 3-fluorotyrosine-labelled dihydrofolate reductase and its complexes with ligands

Ligand	Y^F_K	Y^F_L	Y^F_M	Y^F_N	Y^F_O
None	2.07	0.68	-0.17*	-0.38*	-0.59
Folate	1.91	0.68	0.16	-0.68	-0.55
Methotrexate	1.94	0.68	0.20	2.30	-0.35
2,4-diaminopyrimidine	1.99	0.64	0.20	1.59	-0.28
2,4-diaminopyrimidine + p-amino-benzoyl-L-glutamate	1.97	0.68	0.18	1.83	-0.36
NADPH	0.64 ($\pm$0.5)	0.64 ($\pm$0.5)	0.64 ($\pm$0.5)	0.64 ($\pm$0.5)	-0.89
NADPH + p-aminobenzoyl-L-glutamate	0.64 ($\pm$0.5)	0.64 ($\pm$0.5)	0.64 ($\pm$0.5)	0.64 ($\pm$0.5)	-0.89
NADPH + 2,4-diaminopyrimidine + p-aminobenzoyl-L-glutamate	0.72	0.72	0.72	2.56	-0.60
NADPH + methotrexate	0.52	0.68	0.12	2.94	-0.61

* $\pm$ 0.10 ppm

† Measured in ppm ($\pm$ 0.05 ppm) from the ^{19}F signal of 3-fluorotyrosine in denatured enzyme: positive shifts to low field. See text for comments on assignments.

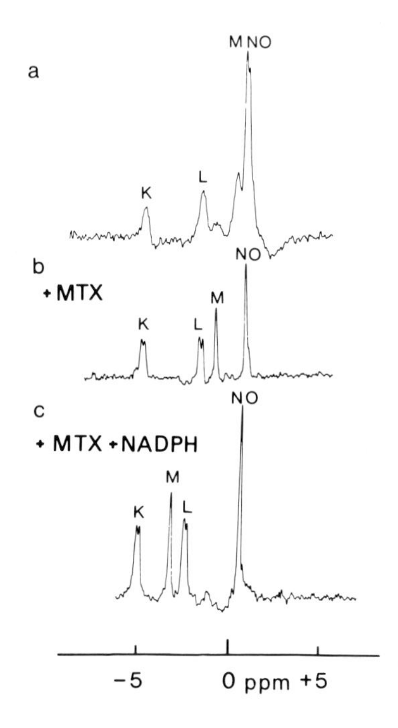

Figure 3. The 94.1 MHz ^{19}F spectra (proton noise decoupled) of 6-fluorotryptophan containing dihydrofolate reductase (11), a) alone, b) with 1 equivalent of methotrexate, c) with 1 equivalent of methotrexate and 1 equivalent of NADPH.

shifted downfield by 1.1 ppm and two others (W_K^F and W_L^F) show smaller downfield shifts (0.28 and 0.13 ppm respectively). Addition of NADPH causes a further large downfield shift for W_M^F (2.34 ppm) and a smaller shift for W_L^F (0.61 ppm). It is tempting to suggest that the tryptophan corresponding to W_M^F is interacting directly with the binding ligands although one cannot rule out the possibility that ligand induced conformational changes are producing the large chemical shifts. The addition of folate causes the signals W_M^F and W_K^F to be shifted to high field (0.3 and 0.1 ppm respectively) which contrasts with the downfield shifts observed with methotrexate. The signals W_K^F and W_L^F show doublet splittings (17 $\pm$ 1 Hz) which we have shown to be due to spin-spin interactions by examining the sample at different field strengths. From a comparison of the proton-decoupled ^{19}F spectra at 94.1 MHz and 254 MHz of the enzyme-methotrexate-NADPH complex (see Figure 4) the splittings are seen to be independent of field strength, thus characterising them as spin-spin splittings rather than chemical shift differences (12). This spin-spin coupling can only arise from ^{19}F-^{19}F through space spin-spin interaction between nuclei in two different tryptophan residues in the molecule. The concept of through

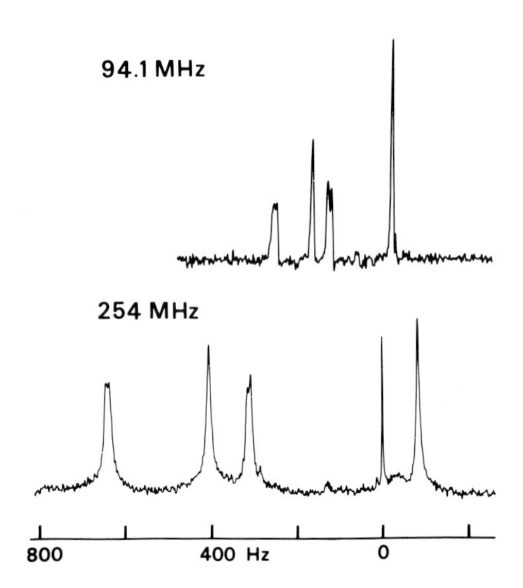

Figure 4. The ^{19}F spectra (proton noise decoupled) of 6-fluorotryptophan containing dihydrofolate reductase in the presence of methotrexate (1 equivalent) and NADPH (1 equivalent) at 94.1 and 254 MHz recorded on the same frequency scale (12).

space spin-spin coupling is well known in small molecules and from results on model compounds it would appear that an internuclear separation of approximately ~3 A between the interacting nuclei would be required to give a 17 Hz spin coupling constant. Thus the study of the ^{19}F labelled enzyme has provided detailed conformational information about the proximity of two tryptophan residues in the structure (12). Clearly this method could be extended to studies of interactions of fluorine-containing ligands with fluorine-containing proteins.

Selectively Deuteriated Enzymes

This method of spectral simplification relies on replacing most of the protons on the enzyme by deuterium and then examining the simplified ^{1}H spectrum given by the few remaining protons. Jardetzky and coworkers (13) outlined the general procedure of this method in their NMR studies of the nuclease isolated from Staphylococcus aureus grown on a mixture of deuteriated amino acids (isolated from algae grown on 99% D_2O) and selected ^{1}H labelled amino acids. A more economical modification of this experiment is to limit the deuteriation to only the aromatic amino acids which can be prepared from the normal amino acids by simple and inexpensive hydrolysis procedures (20).

Using a growth medium for L.casei containing fully deuteriated Trp, Phe and His amino acids supplemented with the remaining normal amino acids and the 3,5-deuteriated Tyr we have prepared and purified a selectively deuteriated dihydrofolate reductase in which all the aromatic protons except the 2,6-protons of tyrosine have been replaced (5). The ^{1}H spectra of these enzymes have simplified aromatic regions which can be studied in the presence of different ligands. This can be seen in the ^{1}H spectra from the 2,6-[^{1}H]-Tyr containing enzyme and some of its complexes shown in Figure 5. Again there are distinct differences between the spectra of the substrate and inhibitor complexes indicating different modes of binding in the two cases. Methotrexate and trimethoprim affect three of the five tyrosine resonances while folate affects only one of them: however, the chemical shift changes are fairly small. In some of the complexes such as those with methotrexate and trimethoprim all five tyrosine signals are resolved. In no case is there evidence for more than five signals which indicates that each of the tyrosine rings is rapidly flipping about the C_γ-C_β bond such that protons at the 2- and 6-positions become equivalent from a magnetic shielding point of view.

We have also prepared 2-[^{1}H]-Trp containing selectively deuteriated enzyme and examined its spectrum in the presence of various ligands (J. Feeney, G.C.K. Roberts, R.W. King, D.V. Griffiths, B. Birdsall and A.S.V. Burgen, unpublished results). Four signals are observed in the aromatic region (see Figure 6) of the enzyme alone: on the addition of methotrexate two signals (W_C and W_D) are shifted upfield and the signal W_D sharpens considerably. In this case, folate causes

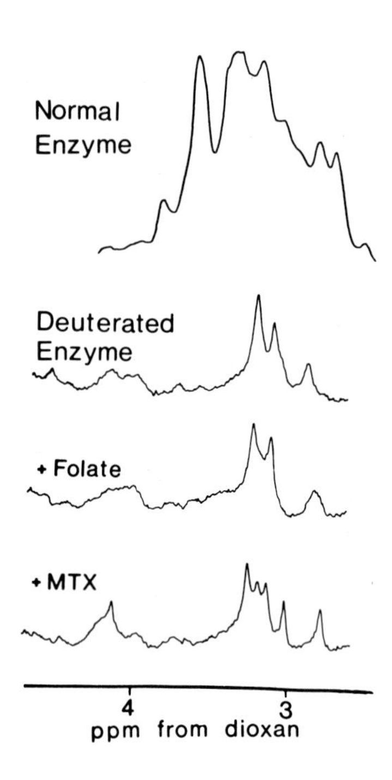

Figure 5. The aromatic region of the 270 MHz nmr spectra of 2,6-[^{1}H]-Tyr labelled selectively deuteriated dihydrofolate reductase and its binary complex with folate and methotrexate.

exactly the same chemical shift changes as does methotrexate. Addition of NADP^{+} results in an upfield shift (0.45 ppm) of one tryptophan (W_C) and similar shifts are seen in the complex with NADPH. It seems likely that this resonance arises from the same tryptophan residue which was assigned to W_M^F in the 6-fluorotryptophan containing enzyme which also showed large chemical shift changes in the presence of methotrexate and coenzyme. In the spectrum of the NADPH.2-[^{1}H]-Trp enzyme complex (Figure 7) we also see additional resonances from the bound coenzyme. This provides an additional advantage of studying deuteriated enzymes.

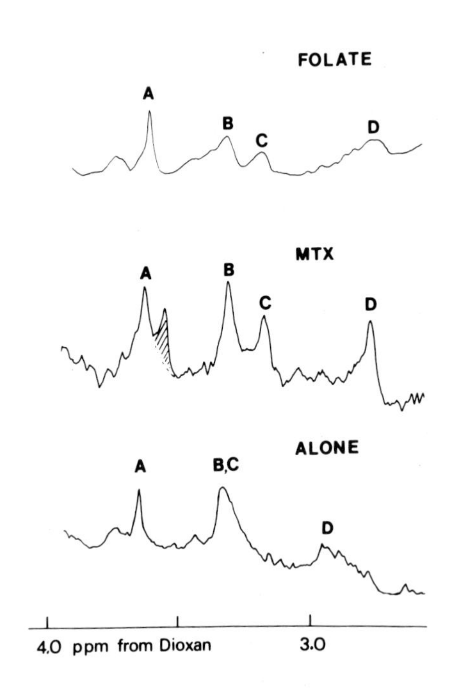

Figure 6. The aromatic regions of the 270 MHz ^{1}H spectra of 2-[^{1}H]-Trp containing selectively deuteriated dihydrofolate reductase and its binary complexes with folate and methotrexate. The shaded peak corresponds to a small amount of denatured enzyme.

EASE OF OBSERVATION OF SIGNALS FROM BOUND LIGANDS

The simplicity of the aromatic region of the spectra of the deuteriated enzymes makes it relatively easy to detect resonances from bound ligands which absorb in this region. For example, additional resonances from bound coenzyme can be clearly seen in spectra of the complexes of the coenzyme with both the 2-[^{1}H]-Trp enzyme (see Figure 7) and the 2,6-[^{1}H]-Tyr enzyme (Figure 8). The signal at 3.64 ppm in all the spectra in Figure 8 has been assigned to the adenine-2 proton in either the oxidised or reduced forms of the bound coenzyme (5,9). In the case of the $NADP^{+}$.enzyme complex the assignment can be made unequivocally by transfer of saturation experiments between the free and bound adenine-2 signals of $NADP^{+}$. It is interesting that for both $NADP^{+}$ and NADPH the signal from the adenine-2 proton in the bound state is shifted to high field by the same amount (0.9 ppm) which clearly indicates that the adenine ring is binding in the same ionisation

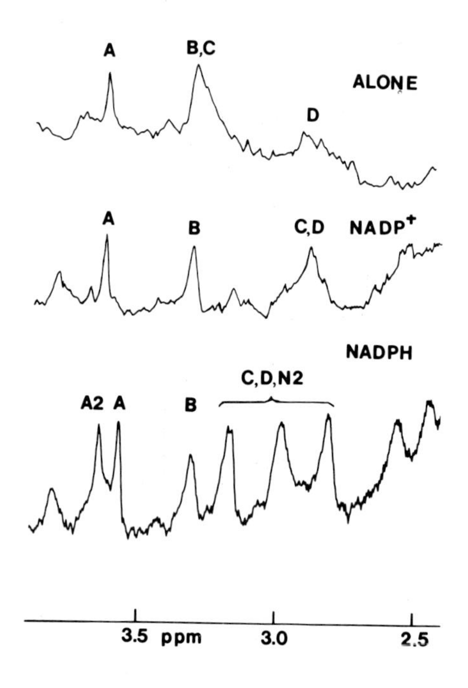

Figure 7. The aromatic regions of the 270 MHz ^{1}H spectra of 2-[^{1}H]-Trp containing selectively deuteriated dihydrofolate reductase and its binary complexes with NADPH and $NADP^{+}$.

state and at the same binding site for both forms of the coenzyme (9). The signal of the adenine-2 proton of bound $NADP^{+}$ is considerably broader than that from the same proton in the NADPH complex. The additional line broadening arises from an exchange contribution to the signal which is not fully in the slow exchange condition (dissociation rate constant estimated to be ~ 45 sec^{-1} from the line broadening). Addition of methotrexate to the enzyme.$NADP^{+}$ complex is seen to sharpen the signal of the adenine-2 proton which indicates that methotrexate has increased the binding of $NADP^{+}$ by reducing the dissociation rate constant (9).

Other signals from protons in bound ligands (N2 in NADPH, H6 in trimethoprim) have been detected in the aromatic regions of the spectra of complexes with the deuteriated enzymes. The detection and assignment of such signals should enable more detailed kinetic and conformational information to be obtained from studies of transfer of

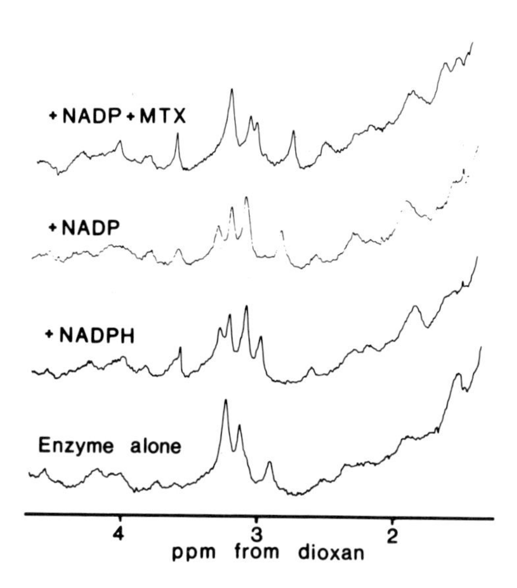

Figure 8. The aromatic region of the 270 MHz ^{1}H spectra of 2,6-[^{1}H]-tyrosine containing selectively deuteriated dihydrofolate reductase in the presence of (a) NADPH, (b) $NADP^{+}$ (c) $NADP^{+}$ + methotrexate.

saturation and nuclear Overhauser effects. The measurement of intensity changes observed in such experiments is simplified if the observed signals are not overlapped by other signals. The use of deuteriated enzymes can often provide the required spectral simplification.

EASE OF TRANSFER OF SATURATION EXPERIMENTS

Transfer of saturation experiments involving ligands bound to deuteriated enzymes can be readily carried out to give signal assignments and lifetime measurements for the complexes. This can be illustrated by considering the spectrum of the complex of trimethoprim with the enzyme where it is difficult to detect directly signals from bound trimethoprim (J.P. Albrand, J. Cayley, J. Feeney, G.C.K. Roberts and A.S.V. Burgen, unpublished results). When we examined the ^{1}H spectrum of the 1:1 complex of trimethoprim with the 2,6-[^{1}H]-Tyr deuteriated enzyme we were unable to see any signals corresponding to the aromatic protons of bound trimethoprim. These signals are either very broad or are overlapped by the 2,6-tyrosine signals in the spectrum. Addition of a second molar equivalent of trimethoprim gives a spectrum (Figure 9a) in which, in addition to

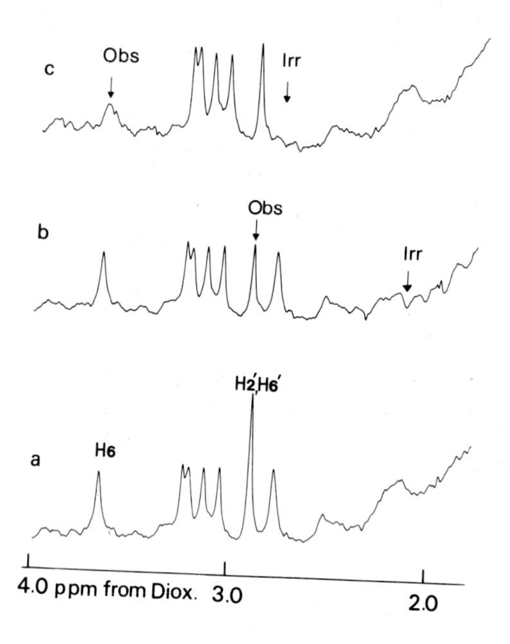

Figure 9. The aromatic region of the 270 MHz ^{1}H spectra of 2,6-[^{1}H]-Tyr containing selectively deuteriated dihydrofolate reductase in the presence of two molar equivalents of trimethoprim under different irradiation conditions.

the five tyrosine resonances, we observe sharp signals for the H2', H6', and H6 protons of the additional equivalent of free trimethoprim. These signals are not overlapped by any enzyme signals and thus it is easy to measure the intensities of the free trimethoprim signals in the presence of a second irradiating field. This irradiation was located at different frequencies at intervals of 10 Hz over a frequency range of ± 800 Hz around the free trimethoprim signals. Transfer of saturation for both signals was observed at the frequency positions indicated in Figures 9b and c; these positions provided us with the chemical shifts of the bound trimethoprim signals. On binding, large upfield shifts are observed for both the H2', H6' (0.78 ppm) and H6 (0.92 ppm) nuclei.

The H6 signal of bound trimethoprim is seen to overlap the high field tyrosine resonance which explains why it was not detected directly in the spectrum. The H2', H6' signal of bound trimethoprim is obviously too broad to be detected directly; this line broadening probably arises from exchange broadening caused by flipping of the aromatic ring of bound trimethoprim about the C_γ-C_β bond. The exchange broadening cannot be related to the dissociation rate constant of the enzyme-trimethoprim complex because this would have also caused line broadening effects on the free trimethoprim signals.

Using Hoffman and Forsen (21) type transfer of saturation experiments we have determined the dissociation rate constants for the complex. A long selective saturation pulse is applied at the frequency of the bound proton for a variable time, t. At the end of the pulse the remaining magnetisation of the proton in free trimethoprim is measured using a non-selective 90° pulse. The rate of decay of magnetisation of the free proton signal can be analysed to give the residence time and relaxation time in the free trimethoprim state.

In this case where we have essentially an equal amount in the free and bound states the residence time in each state will be the same. At 45^oC the residence time for the enzyme-trimethoprim complex was found to be 0.3 sec.

CONCLUSIONS

The NMR spectral simplification resulting from isotopic labelling of proteins can be of assistance (i) in assigning signals to individual amino acid residues, (ii) in monitoring the effects of ligand binding on these signals, (iii) in obtaining kinetic information (e.g. ring flipping in Tyr residues) and conformational information (proximity of Trp residues), (iv) in the detection of signals from strongly bound ligands and (v) in the study of transfer of saturation leading to signal assignments and the lifetimes of complex species.

Other potential uses of isotopic labelling techniques which we plan to explore in the future include measuring signal assignments from heteronuclear decoupling and NOE studies and the use of selective deuteriation to isolate dipolar relaxation contributions.

Acknowledgements

The experiments described here were carried out in collaboration with A.S.V. Burgen, G.C.K. Roberts, B. Birdsall, D.V. Griffiths, R.W. King, J.P. Albrand, B.J. Kimber, J. Cayley and G. Ostler.

References

1. Blakley, R.L.: The Biochemistry of Folic Acid and other pteridines, New York, N.Y. Elsevier, 1969.

2. Hitchings, G.H., and Burchall, J.J.: Adv. Enzymology 27, 417, 1965.
3. Roberts, G.C.K., Feeney, J., Burgen, A.S.V., Yuferov, V., Dann, J.G. and Bjur, R.: Biochemistry 13, 5351, 1974.
4. Birdsall, B., Griffiths, D.V., Roberts, G.C.K., Feeney, J., and Burgen, A.S.V., Proc. Roy. Soc. London.B 196, 251, 1977.
5. Feeney, J., Roberts, G.C.K., Birdsall, B., Griffiths, D.V., King, R.W., Scudder, P. and Burgen, A.S.V., Proc. Roy. Soc. London B. 196, 267, 1977.
6. Way, J.L., Birdsall, B., Feeney, J., Roberts, G.C.K. and Burgen, A.S.V., Biochemistry 14, 3470, 1975.
7. Feeney, J., Birdsall, B., Roberts, G.C.K. and Burgen, A.S.V., Nature 257, 564, 1975.
8. Roberts, G.C.K., Feeney, J., Birdsall, B., Kimber, B.J., Griffiths, D.V., King, R.W., and Burgen, A.S.V. in "NMR in Biology" (Editors Dwek, R.A., Campbell, I.D., Richards, R.E. and Williams, R.J.P.) Academic Press 95, 1977.
9. Feeney, J., Birdsall, B., Roberts, G.C.K. and Burgen, A.S.V. in "NMR in Biology" (Editors Dwek, R.A., Campbell, I.D., Richards, R.E. and Williams, R.J.P.) Academic Press 111, 1977.
10. Birdsall, B., Roberts, G.C.K., Feeney, J. and Burgen, A.S.V. FEBS Letters 80, 313, 1977.
11. Kimber, B.J., Griffiths, D.V., Birdsall, B., King, R.W., Scudder, P., Feeney, J., Roberts, G.C.K. and Burgen, A.S.V. Biochemistry 16, 3492, 1977.
12. Kimber, B.J., Feeney, J., Roberts, G.C.K., Birdsall, B., Griffiths, D.V., Burgen, A.S.V. and Sykes, B.D., Nature 271, 184, 1978.
13. Putter, I., Markley, J.L. and Jardetzky, O. Proc. Natl. Acad. Sci. U.S. 65, 395, 1970: Science 161, 1249, 1968.
14. Crespi, H.L., Rosenberg, R.M. and Katz, J.J. Science 161, 795, 1968.
15. Hull, W.E. and Sykes, B.D. Biochemistry 13, 3431, 1974; 15, 1535, 1976.
16. Browne, D.T. and Otvos, J.D. Biochem. Biophys. Res. Commun., 68, 907, 1976.
17. Anderson, R.A., Nakashima, Y., and Coleman, J.E. Biochemistry 14, 907, 1975.
18. Pratt, E.A., and Ho, C. Biochemistry 14, 3035, 1975.
19. Dann, J.G., Ostler, G., Bjur, R.A., King, R.W., Scudder, P., Turner, P.C., Roberts, G.C.K., Burgen, A.S.V. and Harding, N.G.L. Biochem. J. 157, 559, 1976.
20. Griffiths, D.V., Feeney, J., Roberts, G.C.K. and Burgen, A.S.V. Biochimica et Biophys. Acta 446, 479, 1976.
21. Forsen, S. and Hoffman, R.A. J. Chem. Phys. 39, 2892, 1963.

STUDIES OF ^{19}F LABELED MET-192 CHYMOTRYPSIN: AN NMR STUDY OF AN ACTIVATING MOIETY NEAR THE CATALYTIC SERINE

Lawrence J. Berliner and Bryan H. Landis
Department of Chemistry, The Ohio State University,
Columbus, Ohio, U.S.A. 43210

1. INTRODUCTION

The crystallographic structure of α-chymotrypsin suggests that the aliphatic side chain of Met-192 might be involved in substrate binding and orientation by, in part, acting as a lid over the specificity pocket (Steitz *et al.* 1969). In fact model building studies suggest that a covalently linked aromatic or apolar moiety on the Met-192 side chain can reach into the aromatic specificity pocket ('tosyl hole').

Since several modifications of Met-192 had been shown to activate certain synthetic substrate hydrolytic rates (Schramm and Lawson 1963), we decided to extensively examine the following ^{19}F labeled bromoacetanilides by NMR and kinetic approaches:

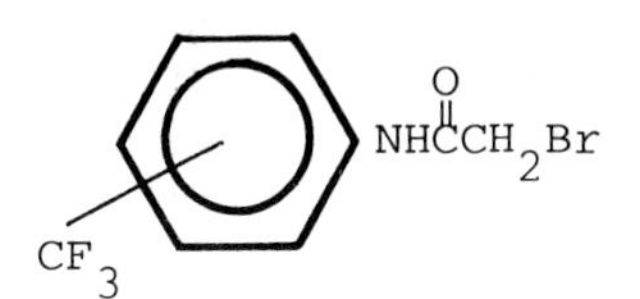

where the CF_3 moiety is either *ortho* (o-CF_3), *meta* (m-CF_3) or *para* (p-CF_3) substituted on the acetanilide ring. The resulting enzyme derivative from the specific reaction is:

$$\text{Met-192-}(CH_2)_2\text{-S-}CH_3 + BrCH_2CONHR \rightarrow \text{Met-192-}(CH_2)_2\text{-}\overset{+}{\underset{\underset{CH_3}{|}}{S}}\text{-}CH_2CONHR$$

where R is the trifluoromethylphenyl moiety.

While these acetanilides had been studied previously by NMR relaxation time measurements of the unpurified labeled enzyme derivatives (Bittner and Gerig 1970) the latter authors were unable to observe the spectra directly nor characterize these derivatives biochemically or kinetically.

B. Pullman (ed.), Nuclear Magnetic Resonance Spectroscopy in Molecular Biology, 311-322.

2. NMR STUDIES

Enzyme samples (150 - 200 mg) were dialyzed extensively after labeling and chromatographed on Whatman CM-52 (2.5 x 25 cm) with 0.1 M phosphate, 0.100 M K^+, pH 5.7. Without the chromatographic purification step these samples also contained significant amounts of autolyzed forms which, while catalytically inactive, gave distinct spectral peaks in the NMR. Furthermore, the subsequent kinetic experiments required pure Met-192 alkylated enzyme, free of unalkylated native enzyme which was always residual in the alkylation reaction mixture.

Figure 1 shows a composite of the ^{19}F NMR spectra (measured on a Bruker HFX-90 at 84.669 MHz) of o-CF_3 labeled α-chymotrypsin a) at pH 4.2, b) in the presence of 8.5 mM indole, and c) in 6 M urea. Note the minor narrower peak slightly upfield from the 'native' peak in a) and b), respectively, which was found to represent ^{19}F labeled enzyme which was subsequently autolyzed. While this 'impurity' peak was removable by CM-52 chromatography, a sample left in solution for several days developed this additional 'autolysis' peak which was always distinctly different from the urea denatured form of the enzyme, spectrum c). Similar spectra were observed for the m-CF_3 and p-CF_3 labeled derivatives as well. Only the 'native' peak was sensitive to indole binding as exemplified in Figure 1a-b.

2.1 Indole effects

In all cases the addition of the chymotrypsin reversible inhibitor, indole, caused an upfield shift of the 'native' line, i.e. in the direction of the denatured enzyme towards the bulk solution environment. Since we found that all of these spectra met fast chemical exchange conditions an indole titration was possible, as shown in Figure 2, where we have shown the upfield shift *vs* total indole concentration and the theoretical hyperbolic fit to these data for the o-CF_3 labeled enzyme at pH 4.2. The dissociation constants for the three derivatives were 5.2, 3.2, and 3.7 mM for the o-CF_3, m-CF_3 and p-CF_3 labeled chymotrypsins, respectively. When these same experiments were performed at pH 5.7 a completely different set of shifts were obtained, reflecting the aggregation (dimerization) behavior of α-chymotrypsin at this pH: all chemical shifts were downfield from those at pH 4.2 (native and indole saturated forms) while the derivative in 6 M urea was identical at either pH as expected. However the almost order of magnitude increase in the K_d for indole reflected the difficulty in binding indole to the dimerized form. The pertinent chemical shift data are found in Table 1.

Since indole had been shown crystallographically to bind in the tosyl pocket, the concomitant shift of the trifluoromethyl aromatic moieties was consistent with a competitive displacement by indole, assuming that the modified enzyme, like the native structure, did not undergo indole induced conformational changes. It should also be noted that the apolar ligand dioxane, another molecule which binds in the tosyl hole, gave similar NMR changes (Landis 1976).

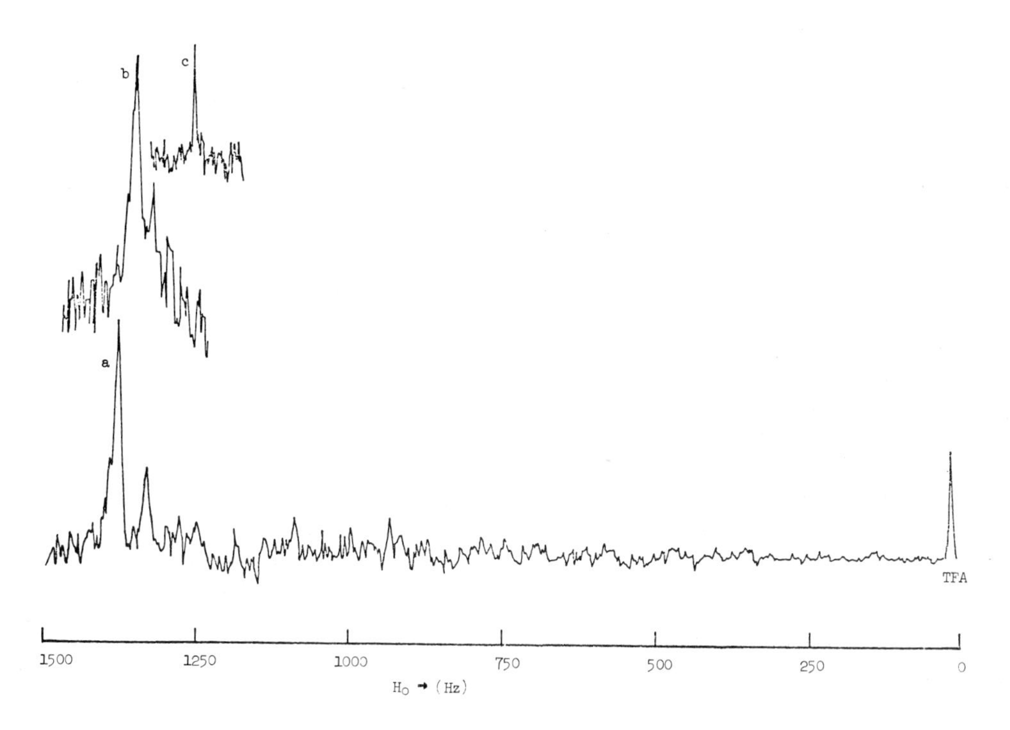

Figure 1. ^{19}F Fourier transform NMR spectra for o-CF_3 labeled chymotrypsin, pH 4.2, 0.008 M phosphate, 0.08 M KCl, 20%(v/v) D_2O, 25 °C. All resonances are expressed in Hz downfield from external trifluoroacetate (TFA) in the same buffer: a) 'native' enzyme, b) in the presence of 8.5 mM indole, c) in 6 M urea.

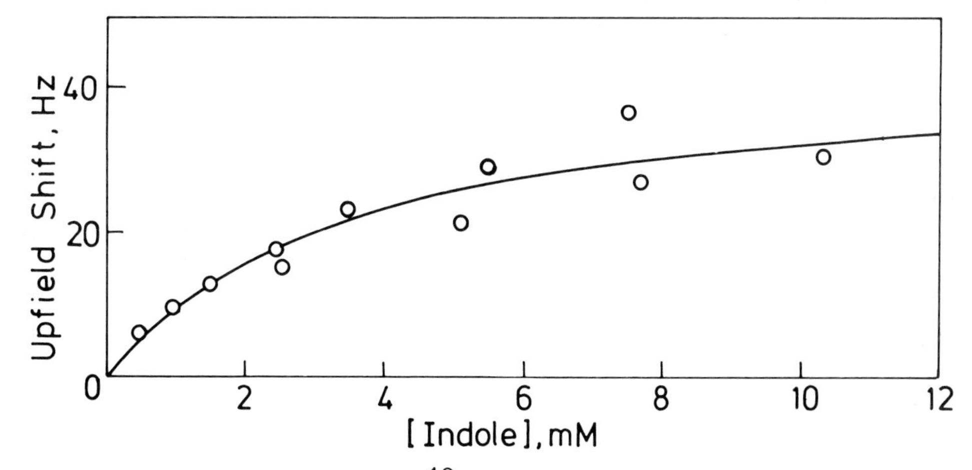

Figure 2. Upfield shift in the ^{19}F resonance of o-CF_3 labeled chymotrypsin as a function of indole concentration. Solid line is the theoretical fit to a single dissociation constant, K_d, of 5.3 ± 2.3 mM.

TABLE 1. ^{19}F NMR DATA[a]

Enzyme Derivative	pH 4.2	pH 5.7
o-CF_3		
Native	-1361 ± 3	-1344 ± 3
6 M urea	-1243 ± 2	-1241 ± 2
ν_B, indole[b]	-1322 ± 11	-1270 ± 13
	(5.3 ± 2.3 mM)	(15 ± 5 mM)
m-CF_3		
Native	-1196 ± 3	-1191 ± 3
6 M urea	-1117 ± 2	-1117 ± 2
ν_B, indole[b]	-1147 ± 6	-1096 ± 13
	(3.2 ± 0.7 mM)	(19 ± 3 mM)
p-CF_3		
Native	-1326 ± 3	-1305 ± 3
6 M urea	-1160 ± 2	-1158 ± 2
ν_B, indole[b]	-1172 ± 9	-1105 ± 25
	(3.7 ± 0.5 mM)	(16 ± 3 mM)

[a]All shifts are expressed in Hz downfield from external TFA, at 84.669 MHz, 25 $^{\circ}$C, and include solvent bulk susceptibility. Protein concentrations were approximately 1 mM.

[b]Maximum shift for a derivative at infinite indole concentration, i.e. saturated with indole. The indole dissociation constants, K_d, are shown in parentheses.

2.2 The active site environment

In order to assess the symmetry of the solvation environment around the CF_3 moiety we examined each alkylating agent alone in varying solvent types *vs* that in water and then included the labeled derivatives in the native, indole saturated (theoretical), and urea denatured states (see Figure 3). While a family of parallel lines was obtained for the various substituents (o-, m-, p-) with the agents alone, the native and indole saturated forms (○, ■), indicated that the three trifluoromethyl groups sensed different, asymmetric environments. On the other hand the urea denatured derivatives (□) more closely approached a 'normalized' environment. While we did not attempt to assess the types of contributions to the chemical shift from such a factor analysis of solvent-induced shifts (Weiner and Malinowski 1971), it was clear that the trifluoromethyl moieties on the enzyme were not exposed to the external solvent, while the slight deviation from linearity in the urea denatured case was confirmed by the deuterium solvent isotope shift studies of Lauterbur *et al.* (1978).

While the aromatic moiety on Met-192 was in each case sensitive to the binding of indole at the 'tosyl hole' (Steitz *et al.* 1969), we could not distinguish from the NMR experiments alone whether the trifluoromethyl moiety resided completely *in* or was in some rapid equilibrium between a site *in* and external to this pocket. The ^{19}F results in Figure 3 indicated, however, that upon completely saturating the enzyme with indole the (magnetic) environment of this moiety was not like the external solvent, but approached that of the urea denatured enzyme. Secondly the substantially downfield environment of the 'native' labeled forms suggested contributions from this hydrophobic binding region of a deshielding nature (ring current, hydrogen bonding, van der Waals interactions, charge transfer, etc.) not experienced in the bulk water solvent environment. Such downfield shifts for enzyme bound ^{19}F ligands were not uncommon in other examples from the literature (Gerig 1978).

3. KINETIC INVESTIGATIONS

Since the kinetics of hydrolysis of the specific substrate acetyl-tyrosine ethylester (ATEE) had been shown to vary with modifications at Met-192 (Schramm and Lawson 1963, Kezdy *et al.* 1967) we examined our derivatives kinetically with both specific and nonspecific chymotrypsin substrates in order to more exactly describe a model for the orientation of these moieties in the active site which correlates the NMR results as well.

3.1 *para*-Nitrophenyl acetate (PNPA)

This 'nonspecific' substrate for α-chymotrypsin was chosen since its kinetic properties were uniquely sensitive to ligands which bound in the aromatic specificity pocket. In particular Foster (1961) had shown that the normally competitive inhibitor indole accelerated the observed deacylation rate constant, k_{cat}, by a factor of 1½ to 2.

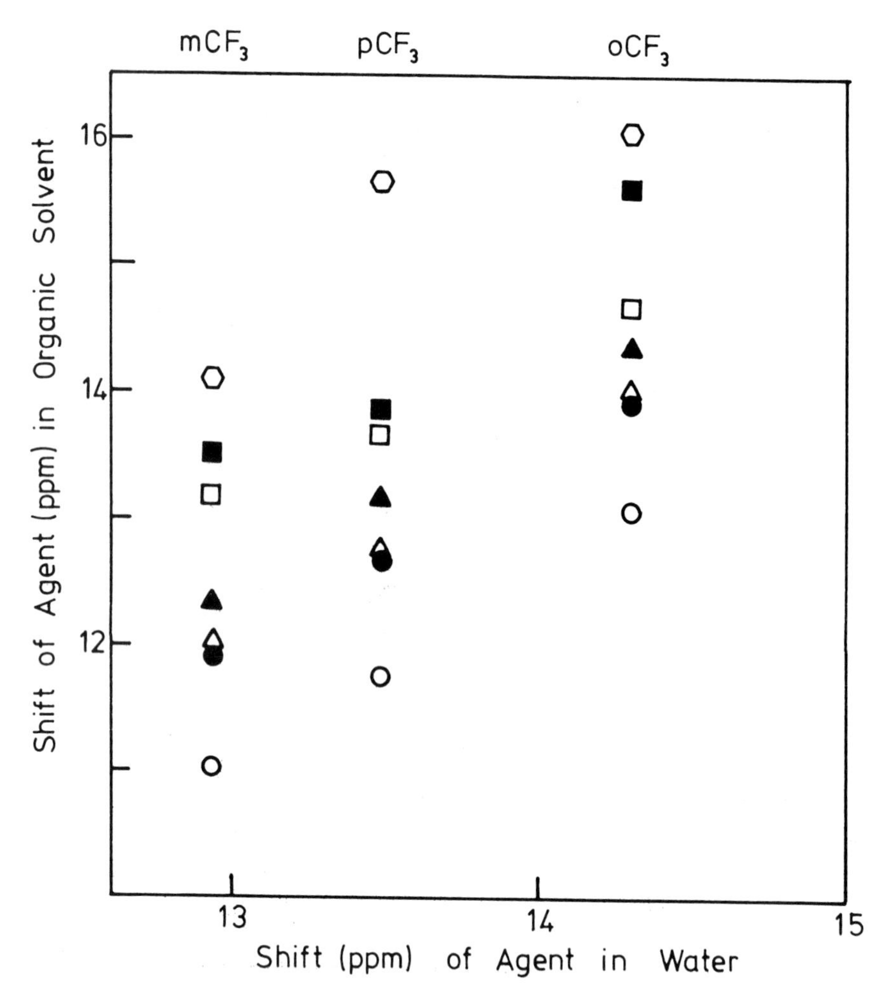

Figure 3. Chemical shifs (downfield from TFA) of the trifluoromethyl-, acetanilide agents in organic solvents vs. their chemical shift in water. The free bromoacetanilides were at 10 mM while labeled proteins were at pH 4.2, ~ 1 mM. All solutions contained 20% (v/v) D_2O. No corrections were made for bulk susceptibility. Symbols are: O, methanol; ●, acetone; Δ, acetonitrile; ▲, dioxane; □, proteins, 6 M urea; ■, proteins, maximum shift at saturating indole; ⬡, proteins alone.

PNPA Kinetics[a]

Derivative	$\frac{k_{cat}(10\text{ mM indole})}{k_{cat}}$	$\frac{k_{cat}}{k_{cat(acet)}}$	$\frac{k_{cat}(10\text{ mM indole})}{k_{cat(acet)}(10\text{ mM indole})}$
Native	1.93	0.80	0.96
Acet.	1.60	1.00	1.00
o-CF_3	1.14	2.62	1.86
m-CF_3	1.23	2.72	2.08
p-CF_3(b)	1.96	1.42	1.74
H-	1.09	14.3	9.77

ATEE Kinetics[c]

Derivative	K_m(app), mM	$\frac{k_{cat}}{k_{cat(acet)}}$	K_I, mM	K_d^{NMR}, mM
Native	1.27 ± 0.05	0.89	0.66 ± 0.37	
Acet.	3.93 ± 0.29	1.00	1.97 ± 0.29	
o-CF_3	3.99 ± 0.22	0.33	2.57 ± 0.08	5.3 ± 2.3
m-CF_3	6.80 ± 0.34	0.66	4.01 ± 0.51	3.2 ± 0.7
p-CF_3(b)	3.50 ± 0.09	0.74	2.57 ± 0.31	3.7 ± 0.5
H-	3.16 ± 0.17	0.11	2.93 ± 0.70	

N-AcAlaGlyGlyOMe Rates[d]

Derivative	$\frac{k_{obs}}{k_{obs(acet)}}$	$\frac{k_{obs}(6.1\text{ mM indole})}{k_{obs}}$
Native	0.62	0.64
Acet.	1.00	0.59
o-CF_3	0.12	2.52
m-CF_3	7.04	0.61
p-CF_3(b)	1.18	0.95
H-	50.5	0.99

(a) pH 7.16, 0.10 M potassium phosphate, 6.9% (v/v) CH_3CN, 25 °C.
(b) the complete chromatographic separation of native enzyme from this derivative was virtually impossible; therefore k_{cat} and K_m values are lower limits.
(c) pH 8.0, 0.100 M KCl, 25 °C, determined by pH stat.
(d) solubility limitations precluded a full concentration study. Data presented are observed rates at a maximum possible substrate concentration of 30 mM, pH 8.0, 0.100 M KCl, 25 °C, pH stat **method.**

Consequently, it was not surprising to discover a rate acceleration for the alkylated enzyme derivatives here if the suspected mechanism was similar to that for indole. As controls we examined acetamido (Met-192-CH_2CONH_2)-chymotrypsin, which lacks the aromatic moiety (called acet.-chymotrypsin), acetanilido-chymotrypsin, which lacks the $-CF_3$ group on the phenyl ring (called H-), and native chymotrypsin. In order to assess the indole (rate accelerating) effects of these derivatives we measured kinetics for all samples in the presence and absence of 10 mM indole. The data, which are found at the top of the preceding page, display several features;

a) The aromatic alkylated derivatives (o-CF_3, m-CF_3, p-CF_3, H-) exhibited enhanced catalytic rate constants (k_{cat}) relative to either native or acet.-chymotrypsin. (Even though pH rate profiles showed a general lowering of the catalytic pK_a in the 7.0 region, the derived pH independent k_{cat}'s revealed that there must have been other factors besides pK shifts to account for these rate increases.)

b) The accelerating effect of indole binding in the tosyl pocket was apparently mimicked by the aromatic moieties since 10 mM indole exerted little additional enhancement of these rate constants.

c) If the accelerating effect of these aromatic moieties were due purely to tosyl pocket binding, then in 10 mM indole these moieties would be displaced and the resultant k_{cat} would be close to that for acet.-chymotrypsin in 10 mM indole. Such was not the case as the aromatic derivatives were approximately two- to tenfold the k_{cat} of the acet. enzyme. Therefore interactions at sites extra to the tosyl pocket were likely.

3.2 N-Acetyltyrosine ethylester (ATEE)

The data are shown at pH 8.0 in the center table and include the competitive inhibition constants for indole and the respective indole dissociation constants obtained from the NMR data. Again several features of these data are notable:

a) If the aromatic moieties were binding in the tosyl pocket, an intramolecular competitive inhibition would result. However, since k_{cat} was changed (even relative to the acet. derivative) other factors were responsible for these decelerated k_{cat} values. Lowering of the pK_a's similar to that observed with PNPA would only account for 5% of the observed decrease in k_{cat} at this pH.

b) Intramolecular inhibition can be used to describe binding phenomena. Since the K_d values were in such good agreement with the K_I values for indole, an intramolecular competitive inhibition, or 'partition' constant, K_A was derived for the aromatic moiety between the tosyl pocket, E(in), and some external environment, E(out). It was easily shown that, assuming the *intrinsic* affinity of the tosyl pocket was identical for an acetamido or acetanilido derivative:

$$K_I(app) = K_I(1 + 1/K_A),$$

where $K_I(app)$ is that (apparent value) measured for the aromatic derivatives, K_I is that for the acetamido derivative, and

$$K_A = E(out)/E(in).$$

From the K_I ratios of the trifluoromethyl acetanilides *vs* the acet. derivative we obtained K_A values of 3.3 ± 0.6 for o-CF_3, 0.96 ± 0.19 for m-CF_3, 3.3 ± 0.6 for p-CF_3, and 2.1 ± 0.6 for H-, respectively, or a range of 23% to 51% partitioning *in* the tosyl pocket.

c) Applying the K_A term to the $K_m(app)$ expressions, and considering that the deacylation rate constant, k_3, dominates k_{cat} for this substrate, calculations indicated that both k_3 and k_2 (acylation rate) were lowered to about the same extent by these modifications using the equation:

$$K_m(app)/k_{cat} = K_s(1 + 1/K_A)/k_2,$$

assuming that the *instrinsic* K_s was the same for E(in) or E(out).

3.3 N-Acetyl-L-alanylglycylglycine methylester (N-AcAlaGlyGlyOMe)

This tripeptide substrate, prepared by standard methods, was examined kinetically since it should bind analogously to a natural polypeptide substrate (Segal 1972) while lacking the side chain to bind in the tosyl pocket. These data (bottom of table, preceding page), while difficult to interpret in detail due to the limited substrate concentration range possible, again reflected participation of the aromatic moiety at a site other than the tosyl hole, since this pocket is not used in substrate binding.

4. STRUCTURAL MODELS

Since indole altered the ^{19}F chemical shifts of the trifluoromethyl acetanilido chymotrypsins it was affecting a change in their chemical and magnetic environments. While xray studies on these modified chymotrypsins have not been published to date, the report that the p-CF_3 moiety at Met-192 does not alter the T_1 or chemical shift of a trifluorotosyl moiety at Ser-195 (J. T. Gerig, personal communication) and the xray results with the native form (Steitz *et al.* 1969) both strongly suggest that indole binding produces no enzyme conformational changes. Thus it seems reasonable, based both on the NMR and ATEE results, to conclude that the aromatic moieties were binding, in part, in the tosyl pocket. On the other hand, the kinetic results for all the substrates examined indicated that the alterations in catalytic rate constants were not solely accounted for by steric or other interactions mediated from this specificity pocket alone. For example, for the ATEE kinetics of the H-derivative, the 68% which partitioned *outside* the tosyl hole must have been binding at another site(s) which overlapped the substrate binding region in order to account for the 90% reduction in k_{cat}. That is, the species E(out) for the aromatic moieties was different from acet.-chymotrypsin, the latter moiety most probably situated in the bulk solvent environment.

A potential possibility for a second binding locus, which was suggested earlier by Kezdy *et al.* (1967) on the basis of a kinetic analysis of a Met-192-(acetyl-2-aminoisobutyrate)-chymotrypsin, is near the acylamido binding site which binds the peptide backbone of mono and oligo peptide substrates (Segal 1972). Model building studies showed that the aromatic moiety could reach to this region without much difficulty. Binding in the tosyl pocket was probably not heavily favored due to the slight degree of increased strain placed on the Met-192 side chain in order to accommodate this orientation. While we cannot predict from these data whether there were additional binding interactions possible, the kinetic evidence suggested that the second binding site(s) must be close to the substrate binding loci and active center residues. A depiction of the simplest models discussed here is shown below.

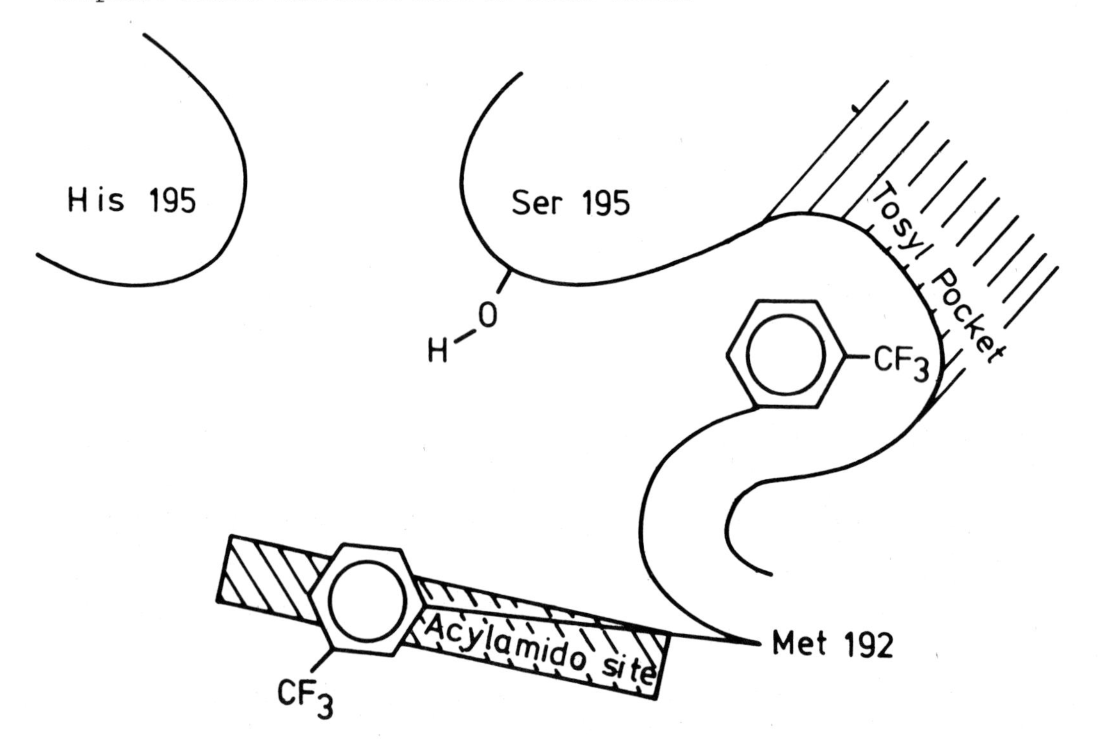

5. CONCLUSIONS

The extreme sensitivity of ^{19}F NMR in probing interactions at the molecular level in α-chymotrypsin was exemplified in the results presented above. Significantly, minor autolyzed forms of the enzyme were easily resolved in the NMR spectra, yet were likely overlooked by other approaches. Chemical shift differences were apparent at the quaternary structure level where enzyme dimers were found at pH 5.7. Most outstanding, however, was the ability to monitor and quantitatively measure the binding of specific ligands to a well characterized binding pocket in the active site.

In order to characterize in more detail the role played by these aromatic moieties on the Met-192 side chain, the complementation of NMR and kinetic investigations were exploited. While an ambiguity remained as to the component states which comprised the observed chemical shift for each labeled enzyme (o-CF_3, m-CF_3, and p-CF_3) the combined results allow a calculation of the partitioning constant between binding *in* and out of the tosyl pocket as well as the respective chemical shifts for the two extreme states. Most significant to note, however, was the absolute necessity of these kinetic data in deducing the importance of a binding mode other than the tosyl hole. in the observed rate alterations with each substrate. We await the xray crystallographic results of these derivatives as a confirmatory approach to the binding models discussed above.

Support for this research was derived, in part, from grants from NSF, NIH and Research Corporation.

6. REFERENCES

Bittner, E. W., and Gerig, J. T.: 1970, *J. Amer. Chem. Soc.*, 92, pp. 2114 - 2118, 5001.-5003.

Foster, R. J.,: 1961, *J. Biol. Chem.*, 236, pp. 2461 - 2466.

Gerig, J. T.,: 1978, in L. J. Berliner and J. Reuben (eds.), *Biological Magnetic Resonance*, Plenum, New York, chap. 4.

Kezdy, F. J., Feder, J. and Bender, M. L.: 1967, *J. Amer. Chem. Soc.*, 89, pp. 1009 - 1016.

Landis, B. H.: 1976, *PhD. Dissertation*, The Ohio State University, Univ. Microfilms Order No. 77-10,556.

Lauterbur, P. C., Kaufman, B. V., and Crawford, M. K.: 1978, in P. F. Agris, B. Sykes, and R. Loeppky (eds.), *Cellular Function and Molecular Structure*, Academic Press, New York.

Schramm, H. J., and Lawson, W. B.: 1963, *Z. Physiol. Chem.*, 332, pp. 97-100.

Segal, D. M.: 1972, *Biochem.*, 11, pp. 349 - 356.

Steitz, T. A., Henderson, R., and Blow, D. M.: 1969, *J. Mol. Biol.*, 46, pp. 337 - 348.

Weiner, P. H., and Malinowski, E. R.: 1971, *J. Phys. Chem.*,75, pp. 1207-1210.

PRINCIPLES IN SUBSTRATE AND CONFORMATION SPECIFICITY OF NUCLEOTIDE-CONVERTING ENZYMES

Hartmut Follmann
Fachbereich Chemie der Philipps-Universität
D-3550 Marburg (Germany)

INTRODUCTION

More than 100 enzymes are involved in the biosynthesis,modification,polymerisation, or degradation of nucleosides and nucleotides, many more act upon the polynucleotides, DNA and RNA, or they require nucleotide coenzymes and nucleotides like ATP as cosubstrates or allosteric effectors. Besides their specific catalytic functions all these proteins should exhibit certain common modes of protein - nucleotide interactions because the amino, imino, carbonyl, hydroxyl, and phosphate groups of nucleotides represent but a limited array of substituents capable to interact with protein residues (Figure 1). There are, however, surprisingly few systematic studies on the structural origins of nucleotide specificity.

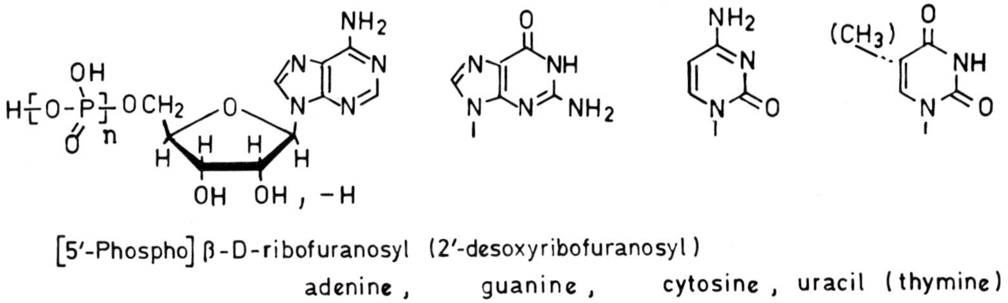

[5'-Phospho] β-D-ribofuranosyl (2'-desoxyribofuranosyl)
adenine, guanine, cytosine, uracil (thymine)

Fig.1. Chemical structures of the common ribo- and deoxyribonucleosides (n = 0) and 5'-nucleotides (n = 1-3).

An especially interesting case of nucleotide - protein recognition is presented by a group of enzymes which use all four pyrimidine and purine nucleotides as substrates or effectors of the same process; key enzymes of cell proliferation such nucleotide kinases, ribonucleotide reduc-

B. Pullman (ed.), Nuclear Magnetic Resonance Spectroscopy in Molecular Biology, 323-337.

tases, DNA and RNA polymerases are among this group. They differ from the enzymes which act specifically on one or two nucleotides in that their binding sites must be delicately balanced to accomodate the four rather different base structures (Fig.1) while retaining some basic nucleotide specificity which excludes sugar phosphates lacking the aglycone. The number of amino acid side chains suited for such variable binding will be small. It is therefore worth asking whether the purine and pyrimidine nucleotides themselves share more subtle structure similarities in addition to the obvious ones,like the charged phosphates and the capacity to form intermolecular hydrogen bonds.

NMR spectroscopy has provided a great number of detailed descriptions of the threedimensional structures of individual nucleosides and nucleotides in solution. In view of the important physiologic functions mentioned above we are, however, particularly interested in the general nature of the forces which determine, and limit, the individual properties. It is attempted to correlate some aspects of nucleotide structures and conformations, obtained from NMR work, with nucleotide activities in enzyme systems.

INTRAMOLECULAR STABILISATION OF NUCLEOSIDE CONFORMATIONS

Let us first consider why all the common nucleosides and nucleotides, as well as most modified ("minor") derivatives are locked in rigid conformations at room temperature and even up to 80°C [1,12]. This "rigidity" concept of nucleotides, later extended to nucleosides [2,3,20] is well documented. Numerous NMR studies, especially by proton NMR [4-16] have shown that the substructures of a typical nucleoside are predominantly arranged in the anti conformation of the pyrimidine or purine base, and in the gauche, gauche (gg) position of the exocyclic C5' group as depicted in Figure 2.

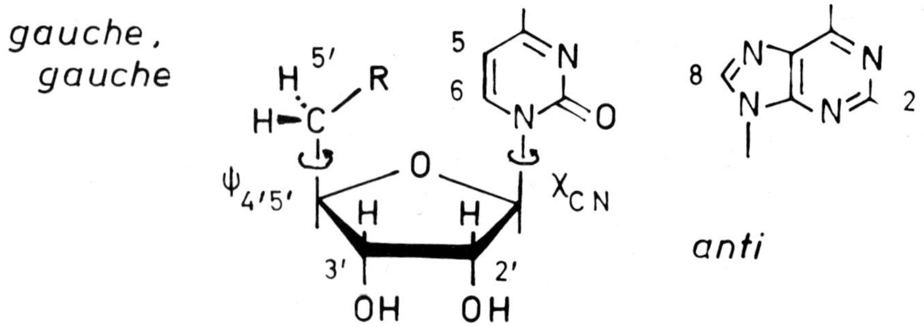

Fig.2. Preferred orientations of base and C5'-substituent in nucleosides (R = OH) and nucleotides (R = OPO_3H_2).

The conformational equilibrium of the central pentofuranose ring, N ⇄ S, or 3' ⇄ 2'-*endo* [17] is frequently coupled with the position of the aglycone [18] but at least in the case of unsubstituted ribose this is thought a consequence, not a cause of the observed *anti*/*gg*-preference and will not be discussed here.

There is no obvious reason for that conformational stability. In quantum mechanical calculations of molecular potential curves [16,19] more than one energy minimum is frequently obtained by variation of the torsion angles χ and ψ about the glycosic and exocyclic bonds (Fig.2). The steric hindrance of rotation about the glycosidic bond ascribed to H2' appears not very convincing in molecular models and because of the sugar's flexibility, but it is difficult to assess experimentally due to lack of 2'-modified reference compounds. In an alternative approach to the problem we have evaluated the various chemical, or electronic interactions which may exist between a purine or pyrimidine and the C5'-moiety of nucleoside molecules over the ribose ring [20-22]. This is possible if the base proton chemical shifts δ are measured in dilute aqueous solutions under strictly comparable conditions, and if large series of related nucleoside derivatives are available in which structure variations at C5' sensibly affect the environment of the base protons. Note that I shall not discuss individual torsion angles (which are per se of little interest for unphysiologic compounds), but the forces which determine these angles. The excellent previous PMR and conformation analyses of common nucleosides and nucleotides [8,10,12,15] will serve as a basis.

In Table 1 are summarized the δ values of the H6 and H5 resonance signals of uracil and thymine, and the H8 and H2 signals of adenine, respectively, in C5'-modified uridine, thymidine, and adenosine derivatives. Cytosine nucleosides (not listed) behave in a completely analogous way. It is evident that the chemical shifts of H6 and H8 vary much more strongly than those of H5 or H2 upon C5' substitution. This effect is not limited to the phosphates where it was originally observed [4] but extends from 5'-deoxynucleosides to 5'-uronic acids. The data directly reflect the universal preference of *anti*/*gg* conformer populations because H6 of pyrimidine and H8 of purine are juxtaposed to and under the influence of C5'-substituents only when these conformations are realized.

Some sort of normalisation of the chemical shift values is required for more generally applicable analysis. The difference δH8-δH2 was found a good indicator of C5'/base interactions in the adenosine series [20] where C-H8 and

C-H2 are chemically equivalent, but this is not the case for C-H6 and C-H5 of the enamine structure in pyrimidines. A better expression is the chemical shift difference of a given proton in a nucleoside and the same proton in the corresponding base,

$$\delta H_x(\text{nucleoside}) - \delta H_x(\text{base}) = \Delta\delta H_x \ .$$

This treatment of data is justified because ribose or deoxyribose substitution of the heterocyclic bases results in deshielding of their protons by a constant amount, due to the electronegativity of the sugar residue and a specific contribution from the bridge oxygen O4', plus a C5'-dependent amount. The first effect is seen by a comparison of adenine with aristeromycin and 5'-deoxyadenosine (Fig. 3;a,b): The carbocyclic analog lacking O4' is not in a rigid conformation [21] and its small, equal Δδ values represent the inductive effect of deoxy sugar as a whole, while in 5'-deoxyadenosine (which has the anti conformation) the inductive effect of O4' next to the heterocycle becomes evident. The same tendency is seen in uracil, uridine trialcohol (Fig.3;f), and 5'-deoxyuridine. In either series, Δδ values for both protons of the normal nucleosides which contain both O4' and the 5'-OH group in a defined alignment are larger than those of the reference compounds.

From an inspection of the Δδ values and the chemical nature of X (Table 1) the mutual interactions between C4'-X and the bases can now be recognized. The double bonds

$$\text{H-}\overset{+\longrightarrow}{\text{C6=C5}}\text{-H} \quad \text{and} \quad \text{H-}\overset{+\longrightarrow}{\text{C8=N7}}\text{-}$$

are polarized as indicated. Taking the 5'-deoxynucleosides (no C5'/base interaction) as basis, ion-dipole interactions with cations (protonated 5'-aminonucleosides, or S-adenosylmethionine) lead to increased shielding of H6 or H8. Vice versa, phosphate or carboxylate anions deshield these protons; in case of pyrimidines the neighboring H5 atom and even the methyl group of thymine are also deshielded but in adenine some shielding at H2 is observed, apparently by induced polarisation across the heterocycles. Dipole-dipole and polarisation (induced dipole) forces are seen for the neutral, yet polar or polarisable nitrate, hydroxyl (and O-methyl or O-acetyl [20]), halogen, thio, or ester groups. These substituents do not significantly affect δH5 or δH2, while H6 or H8 are increasingly deshielded in the above order. An exception is the hydroxyl group which deshields H6 of uracil but causes a small upfield shift of δH5; this is attributed to more efficient polarisation of the double bond by the "hard" OH group, compared to "soft" sulfur or iodine atoms. Moreover the OH function differs from other substituents in its capacity to interact with the solvent. It is likely from marked Δδ changes observed in water and

X base O OH OH,H

Table 1. Chemical shifts δ (in ppm from internal (trimethylsilyl) propionate) and Δδ values of the base protons in uridine, thymidine, and adenosine derivatives. Sample concentrations in D_2O, ⩽ 0.01 M; pD, 7-8. PMR spectra were recorded at room temperature on a Varian XL-100 spectrometer operating in the FT mode.

Compound, or X	uridines		thymidines		adenosines		uridines, thymidines			adenosines	
	δH6	δH5	δH6	δCH_3	δH8	δH2	ΔδH6	ΔδH5	$\Delta\delta CH_3$	ΔδH8	ΔδH2
uracil, thymine	7.54	5.81	7.37	1.87							
adenine					8.20	8.16					
uridine trialcohol	7.72	5.93					0.18	0.12			
aristeromycin					8.26	8.22				0.06	0.06
CH_3	7.74	5.94			8.31	8.27	0.20	0.13		0.11	0.11
CH_2OH	7.88	5.89	7.65	1.90	8.32	8.23	0.28-0.37	0.08	0.03	0.12	0.07
CH_2Cl	7.85	5.90			8.37	8.27	0.28	0.09		0.17	0.11
CH_2J	7.81	5.95			8.38	8.26	0.27	0.14		0.18	0.10
CH_2ONO_2	7.70	5.92			8.32	8.26	0.16	0.11		0.12	0.10
$CH_2NH_3^+$			7.44	1.89	8.26	8.26	0.07		0.02	0.06	0.10
CH_2S^+methionine					8.28	8.28				0.08	0.12
CH_2SAlkyl					8.35	8.24				0.15	0.08
CH_2COO^-					8.32	8.20				0.12	0.04
$CH_2OPO_3H^-$	7.89	5.95			8.47	8.20	0.35-0.44	0.14		0.27	0.04
$CH_2OPO_3^{--}$	8.13	5.99	7.88	1.94	8.59	8.20	0.51-0.62	0.17	0.08	0.39	0.04
COO^-	8.44	5.96	8.35	1.91	8.63	8.25	0.90-1.0	0.15	0.04	0.43	0.09
COOAlkyl	8.05	5.95			8.43	8.23	0.51	0.14		0.23	0.07
range (ppm) from	7.70	5.89	7.44	1.89	8.26	8.20	0.07	0.08	0.02	0.06	0.04
to	8.44	5.99	8.35	1.94	8.63	8.28	1.0	0.17	0.08	0.43	0.12

in dimethyl sulfoxide solution that the CH_2OH group of nucleosides holds a water molecule next to the base [22]. Finally, intramolecular hydrogen bonding is seen between H6 or H8 and the carboxylate anion in nucleoside 5'-uronic acids: These compounds differ from the 5'-phosphates in that downfield shifts of δH6 or δH8 are unusually large (up to 1 ppm) and are strictly limited to one hydrogen, not affecting H5, H2, or the methyl group.

These conclusions are illustrated in Figure 3 for some representative compounds. They are compatible with the PMR data of more than 80 nucleosides and derivatives. We deduce from this an obligatory intramolecular structure stabilization in all nucleosides which contain in juxtaposition polar or ionic 5'-substituents and heterocycles with an asymmetric electron density distribution; the vast majority of natural nucleotides, nucleoside antibiotics, etc., are built in that way.

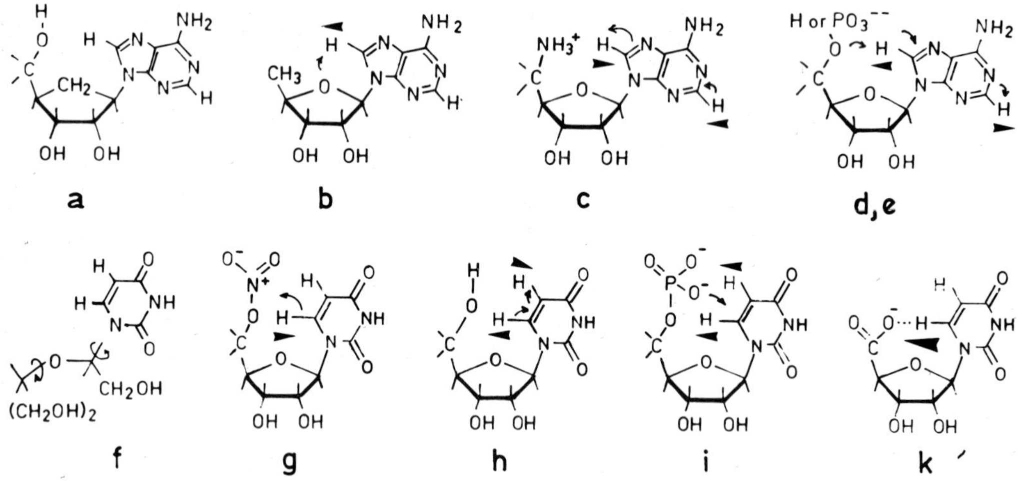

a b c d,e

f g h i k

Fig.3. Polarisation effects (curved arrows) and shielding or deshielding of base protons (►,◄) observed in the PMR spectra of: a, aristeromycin; b, 5'-deoxyadenosine; c, 5'-deoxy-5'-aminoadenosine (pH 6); d,e, adenosine and 5'-AMP; f, uridine trialcohol; g, uridine 5'-nitrate; h, uridine; i, 5'-UMP; k, uridine 5'-uronic acid.

It should be emphasized that these simple observations provide a key for understanding many other, individual nucleotide properties. For example, a surprising number of correlations with circular dichroism spectra was noted [21]. The increased accessibility of syn conformations observed for 2':3'-isopropylidene nucleosides [5,13] or 3':5'-cyclic phosphates [23] is reflected in smaller base proton Δδ values (not listed here), indicating reduced C5'-to-base interactions in presence of the bicyclic, inflexible sugar structures. Our concept predicts that purine or pyrimidine substituents have little influence on conformations; in fact, great similarities are observed for 5-substituted pyrimidines [22] and adenosine,guanosine,inosine,xanthosine, nebularine (=purine riboside),2-aminopurine riboside, and 2-azaadenosine [13,24]. In contrast, conformation changes occur when the possibility of "normal" C5'/base attraction is grossly disturbed by chemical structure changes,e.g. in guanosine protonated at N7, in formycin, or in the 6-aza-pyrimidine and 8-azapurine ribosides (C6- or C8-substituted compounds will be discussed in the following section). Such nucleosides possess flexible anti⇌syn, or syn conformations, the syn form most likely being stabilized through $CH_2OH..N3$ hydrogen bonds [24,26]. In 5'-nucleotides loss of intra-molecular attraction, or the repulsion between phosphate and an electron pair of the extra N atom in aza-nucleotides lead to destabilisation of the gg C5'-orientation in favor of gauche,trans rotamers [25,26], except in 5'-GMP at low pH where the anti/gg preference is retained [11] due to an alternative, zwitterionic interaction.

STRUCTURE AND CONFORMATION SPECIFICITY OF ENZYMES

Turning to biochemical consequences of intramolecular structure stabilisation and the far-reaching similarities among nucleotides it may be expected that enzymes which have not developed specific interactions with unique substituents (e.g., the 2-carbonyl group in pyrimidine-specific ribonuclease) will act on any nucleoside or nucleotide of the same phosphorylation level. This is an important aspect for the biological action of synthetic nucleotide antagonists, or of natural modified nucleotides [28]. Judging from the few examples which have been studied that situation appears indeed to be true; they include a hydrolytic enzyme, ribonuclease T_2 from *Aspergillus*, the ribonucleotide reductases of *Lactobacillus* and *Escherichia coli* which catalyse a redox reaction, and the polymerising enzyme, polynucleotide phosphorylase from *Micrococcus*. These enzymes accept large series of natural and synthetic nucleotide substrates with entirely unsubstituted or highly modified heterocycles. By this property, polynucleotide phosphorylase has become a well-known tool for preparation of modified polynucleotides.

As another excellent example we have investigated ribonucleoside diphosphate reductase from *E.coli* (EC 1.17.4.1) [27]. Despite the very complex catalytic mechanism required for reduction of the 2'-OH group we find that in presence of an allosteric effector, dTTP, the apparent K_m values of numerous pyrimidine, purine, or benzimidazole ribonucleotides vary only by a factor of 12, and the rate constants by a factor of 3, between the best and the slowest substrates (Table 2). The *anti* conformation is known for the nucleosides or 5'-monophosphates of this series and is assumed for the 5'-diphosphates, too.* If the enzyme had to accept the various nucleotide structures without that common conformation much larger K and V differences should exist. The observed, small variations are attributed to steric limitations of proper hydrogen bonding between base substituents and amino acid side chains at the active site.

Table 2. Reduction of ribonucleoside diphosphates by ribonucleotide reductase (*E.coli*)/thioredoxin/NADPH. Conditions: 25°C, pH 7.6, 0.04 mM dTTP, 0.4 mM NADPH, 11 mM Mg^{++}.

Substrate	$K_{m(app.)}$	V_{max}	Conformation
n^2n^6PuDP	$0.36 \cdot 10^{-4}$ M	2.7 $\frac{nmole}{min}$	anti
n^2PuDP	0.36	3.2	anti
GDP	0.54	4.8	anti
BzDP	0.70	2.3	predicted: anti
ADP	0.86	1.6	anti
CDP	1.0	4.0	anti
IDP	1.1	3.7	anti
PuDP	4.3	2.2	anti
br^8ADP	-	0	syn
formycin-DP	-	0	intermediate (anti⇄syn)

n = amino, Pu = purine riboside, Bz = benzimidazole riboside

Are then nucleotides which do *not* fulfil the requirements of intramolecular *anti/gg* structure stabilisation excluded as enzyme substrates? The inactive compounds listed at the bottom of Table 2 and other data available in the literature (Table 3) again support this expectation. Thus, nucleotides of the antibiotic formycin show ambiguous behavior towards enzymes ranging from inactivity to good substrate activity, in keeping with its flexible or intermediate conformation [36]. In contrast, 8-bromopurine and 6-methylpyrimidine nucleotides are strikingly inactive as enzyme substrates. This cannot be due to electronic effects of the halogen or methyl residues because the same enzymes tolerate unphysiologic substituents in other base positions.

* This is especially justified in presence of Mg^{++} ions but was recently questioned for metal-free solutions (except ATP) on the basis of relaxation time studies [29].

Rather the inactivity of these nucleotides must be caused by their *syn* base conformation which has been described in detailed NMR studies [9,13]. We interpret the *anti*→*syn* change not as a consequence of the substituents' bulkiness (as frequently mentioned) but by the loss of C5'-to-H6, or -H8, interactions, and by mutual repulsion of the electronegative C5', C8, and O4' substituents or atoms in the bromopurine derivatives. The recent finding of 8-aminoadenosine and its nucleotide in the *anti* conformation [38, 47], its substrate property with cAMP phosphodiesterase and competition with NADH in lactate dehydrogenase [35,37] is a good example that C8 substitution with one or few extra atoms is not sterically critical but that *anti* stabilisation is retained if chemically feasible. It is obvious that unnatural substituents introduced at the bases or at C5' of nucleotides fit all our chemical and energetic considerations as well.

Table 3. Conformation specificity of nucleotide-converting enzymes. NDP etc. indicates the type of phosphorylation required for enzymatic activity.

Nucleotides of	A,G	formycin	br^8A, br^8G	m^6U, m^6C	ref.
as substrates of					
ribonucleotide reductase					
NDP	+	-	-		[27]
NTP	+	-	-		[28]
polynucleotide phosphorylase					
NDP	+	slow	-	-	[30,32]
RNA polymerase					
NTP	+	+	-	-	[31,32]
ribonuclease T_2					
N2':3'p	+	+	-	-	[33]
as activators of					
ribonucleotide reductases					
dNTP	+		+		[27,28]
glycogen phosphorylase b (rabbit muscle)					
AMP	+		+		[34]
protein kinase (bovine brain)					
3':5'-cAMP	+		+		[35]

In view of these consistent correlations of primary structure, conformation, and enzyme activity in nucleotides it was surprising to find that 8-bromo-dATP acts as a good activator of ribonucleotide reductase like other deoxyribonucleotides [28]. One explanation would be that the deoxyribotide has greater structural freedom than the ribotide, br^8ATP. The available NMR studies of deoxyribose derivatives indicate no basic differences but slightly higher flexibility of CH_2-R and the sugar than in ribosides [10,48]. We have therefore compared the PMR spectra of brdA and brdAMP with those of the corresponding ribosides and unsubstituted compounds (Table 4).

Table 4. Chemical shifts δ (ppm) of adenine and 8-bromoadenine nucleosides and nucleotides. See Table 1 for experimental conditions (D_2O, pD = 8).

Compound	δH8	δH2	ΔδH2	δH1'	δH2'	δH2"	δH4'	δH5'
adenine	8.20	8.16						
bromoadenine		8.12						
A	8.32	8.23	0.07	6.03	4.80		4.29	3.89
br^8A		8.21	0.09	6.15	5.08		4.28	3.90
dA	8.32	8.23	0.07	6.48	2.86	2.57	4.18	3.80
br^8dA		8.18	0.06	6.55	3.08	2.41	4.20	3.89
AMP^{--}	8.62	8.25	0.09	6.14	4.81		4.36	4.01
$brAMP^{--}$		8.23	0.11	6.11	5.32		4.24	4.02
$dAMP^{--}$	8.56	8.24	0.08	6.51	2.87	2.56	4.26	3.94
$brdAMP^{--}$		8.18	0.06	6.52	3.39	2.51	4.14	3.91

No conspicuous variations are seen in the chemical shifts of base and sugar protons when comparing the AMP → brAMP and dAMP → brdAMP structure changes. In particular, the characteristic downfield shift (deshielding) of H2' (Δ=0.5 ppm, or 0.25 ppm in nucleosides) as well as the upfield shift of H4' (Δ=0.12 ppm, not observed in nucleosides) are identical in both nucleotide couples, indicating comparable rotation of the heterocycle and loss of the *gg* preference of the 5'-phosphate. Differences exist in the trends of δH1' and ΔδH2; both change into opposite directions when going from the parent ribo- or deoxyribonucleotide to their 8-bromo derivatives. This is compatible with different base conformer populations in brAMP and brdAMP, H2 of the latter being less influenced by the phosphate. However, because the same trend is present in the nucleosides and the influence of *gt*-oriented phosphate upon H2 cannot be strong in either case these small variations are insufficient to demonstrate major χ_{CN} changes.

Table 3 contains two other cases in which 8-bromoadenine nucleotides possess enzyme affinity (although to varying degree), namely br^8AMP as activator of glycogen phosphorylase, and bromo-cAMP with protein kinase; in contrast, the latter compound is not a substrate of cAMP phosphodiesterase. It thus appears that lack of conformation specificity is a characteristic of nucleotide effector sites. They differ from catalytic sites in their special function to trigger conformation changes. It is not unreasonable to assume that this can be done by interaction with nucleotides in various conformational states as long as other structures (the 3':5'-cyclic phosphate, or 2'-deoxyribonucleoside triphosphate moieties) are intact. On the other hand it is an attractive speculation whether conformation changes occur upon binding to enzyme sites. This possibility has been considered to account for inhibitory effects of *syn* nucleotides [30-32,37] but experimental proof is difficult; inhibition of enzyme activity may have more complex reasons. In a recent ^{1}H and ^{31}P relaxation time study it has been shown that the glycosyl torsion angle χ of dTTP undergoes a change of about 50° when the nucleotide is bound to DNA polymerase I, apparently to adjust the thymine base for proper base pairing; no change of χ is seen in dATP [39]. Such "induced fit" phenomena may well play a role in other enzymes, and in case of ambiguous structures like formycin nucleotides. I would argue, however, that it contradicts the normal functions and capacities of enzyme proteins to induce the major conformation changes which would be required for stable *syn* nucleotides to mimic the natural compounds; obviously this does not happen at substrate sites.

NUCLEOSIDE CARBOXYLATES AS NUCLEOTIDE MODELS

In a final example of NMR-aided biochemical interpretation I wish to comment on the enzymatic activity of nucleoside carboxylates of the type I-III:

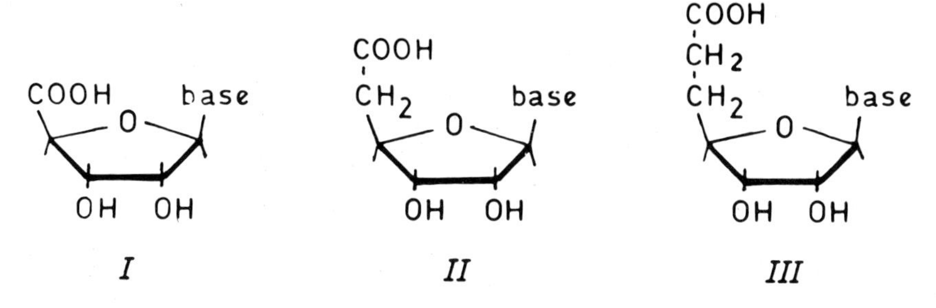

Nucleotide analogs in which the anionic phosphate is replaced by chemically equivalent groups are scarce, but they are of interest for probing intramolecular distance requirements in enzyme-bound nucleotides. For this purpose we have synthesized and investigated adenosine uronic acid (I), 5'-deoxyadenosine 5'-carboxylic acid (II), and 5'-deoxyadenosine 5'-acetic acid (III) [40,41].

Two enzymes were selected in which proper alignment of nucleotide C5' and base moieties is necessary for their catalytic action: AMP aminohydrolase (AMP deaminase, EC 3.5.4.6) from rabbit muscle requires substrates with an anion at C5' and catalyses deamination at C6-NH_2 of adenine [42]. Vice versa, in snake venom 5'-nucleotidase (EC 3.1. 3.5) hydrolysis occurs at the 5'-phosphate which has to be at a defined distance from the base, as judged by the inactivity of 3'-phosphates [43].

The activity of the three adenosine carboxylates, compared with AMP, can be explained with the simple $\Delta\delta$H8 parameter as a measure of C5'/base interaction (Table 5). At pH 6.4 both AMP and the analogs carry a monoanion. The two "elongated" acids are substrates of AMP deaminase and inhibitors of 5'-nucleotidase (they cannot be substrates here because of the C-C-linkages in place of an ester group), but the "short" uronic acid of adenosine is inactive or the poorest inhibitor. In the deaminase system, relative reaction rates and affinities ($1/K_m$) of acids II, III, and AMP clearly parallel the degree of anion/adenine interaction ($\Delta\delta$H8) suggesting that the molecules' intrinsic solution conformation is retained as such at the active site. As nucleotidase inhibitors the order of II and III is reversed but the differences in $\Delta\delta$ and K of these two are small, compared with AMP. The data indicate slightly dissimilar active site geometries of the two enzymes, the spatial requirements for deaminase substrates being more stringent. These conclusions are in agreement with previous studies using various cycloadenosine 5'-phosphates [44].

In obvious contrast, uronic acid I lacks good enzyme affinity. The same is true for thymidine uronic acid which is a very poor inhibitor of thymidylate kinase [45]. This cannot solely be due to the shortened distance between the anion and the base moieties, making coordinated interaction with two or more protein groups less likely, because electrostatic interaction of a positively charged enzyme group with an anion in a flexible molecule should be sterically not as critical as, *e.g.*, hydrogen bonding. We therefore ascribe the relative inactivity of nucleoside 5'-uronic acids to their very rigid molecular conformation which is stabilised by the intramolecular $-COO^-$..H(8or6)- hydrogen bond observed in aqueous solutions of both purine and pyrimidine derivatives, preventing any "induced fit" for attachment to enzyme sites. Again this aspect is of relevance to the design of nucleoside derivatives as potential drugs or nucleotide antagonists. Studies of the effect of elongated thymidine carboxylates as effectors of thymidylate metabolism are in progress.

Table 5. Enzymatic activity of adenosine carboxylic acids as AMP analogs (pH 6.4, 25°C).

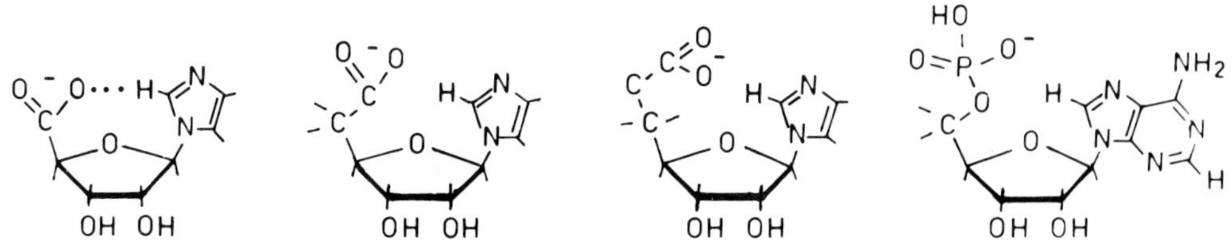

	I	II	III	5'-AMP	
C5'/adenine interaction					
ΔδH8	0.43	0.12	0.16	0.27	ppm
AMP aminohydrolase					
K_m	inactive	3.4	0.76	0.29	mM
rel.V	0	1	21	100	(%)
5'-nucleotidase					
K_i	0.13	0.05	0.10	K_m : 0.014	mM

CONCLUSION

Chemical interpretations of the PMR data of nucleosides and nucleosides and their preferred conformations have been stressed in this review. Recent computer-aided relaxation time studies yield preferential *syn* base orientations [29, 46] but do not provide such chemical interpretations;these discrepancies must await experimental clarification. It may appear oversimplified to draw general conclusions from our data to an extremely complex matter like enzyme specificity. However nucleotides, nucleic acids, and their enzymes have a special innate relationship: They were most likely the first two groups of compounds on a primordial earth which engaged in catalytic and self-replicating processes. They should, therefore, have conserved simple, yet efficient types of chemical interactions. The correlations found above explain, for example, the equivalent affinities of purine and pyrimidine nucleotides in many enzyme systems which in turn is a prerequisite for all their important complementary functions in nucleic acids. Identification of the patterns of amino acids commonly engaged in interaction with nucleotides at enzyme sites [49] has to be a main target of future research.

I am indebted to Drs W.Ludwig and W.Meyer for their cooperation in the enzyme studies, and to Mr. A.Mbonimana for expert technical advice. This work has been supported by Deutsche Forschungsgemeinschaft.

REFERENCES

1 O.Röder, H.D.Lüdemann & E.von Goldammer (1975) Eur.J. Biochem.53, 517-524.
2 N.Yathindra & M.Sundaralingam (1973) Biopolymers 12, 297-314.
3 F.E.Evans & R.H.Sarma (1976) Nature 263, 567-572.
4 M.P.Schweizer, A.D.Broom, P.O.P.Ts'o & D.P.Hollis (1968) J.Amer.Chem.Soc.90, 1042-1055.
5 R.E.Schirmer, J.P.Davis, J.H.Noggle & P.A.Hart (1972) J.Amer.Chem.Soc.94, 2561-2572.
6 T.D.Son & C.Chachaty (1973) Biochim.Biophys.Acta 335,1-13.
7 D.J.Wood, F.E.Hruska & K.K.Ogilvie (1974) Can.J.Chem. 52, 3353-3366.
8 F.E.Evans & R.H.Sarma (1974) J.Biol.Chem.249, 4754-4759.
9 R.H.Sarma, C.H.Lee, F.E.Evans, N.Yathindra, M.Sundaralingam (1974) J.Amer.Chem.Soc.96, 7337-7348.
10 D.B.Davies & S.Danyluk (1974) Biochemistry 13,4417-4434.
11 T.D.Son & W.Guschlbauer (1975) Nucl.Acid Res.2, 873-886.
12 T.Schleich, T.R.Lusebrink, B.P.Cross & N.P.Johnson (1975) Nucl.Acid Res.2, 459-467.
13 E.Westhof, O.Röder, I.Croneiss & H.D.Lüdemann (1975) Z.Naturf.30c, 131-140.
14 E.Westhof, H.Plach, I.Cuno & H.D.Lüdemann (1977) Nucl. Acid Res.4, 939-953.
15 R.G.S.Ritchie & A.S.Perlin (1977) Carboh.Res.55, 121-128.
16 Further experimental and theoretical analyses of nucleotide conformations are found in: E.D.Bergmann & B.Pullman (eds.), Conformation of Biological Molecules and Polymers. Jerusalem Symposia on Quantum Chemistry and Biochemistry, vol.5 (1973), pp.209-467.
17 C.Altona & M.Sundaralingam (1973) J.Amer.Chem.Soc.95, 2333-2344.
18 H.D.Lüdemann, O.Röder, E.Westhof, E.von Goldammer & A. Müller (1975) Biophys.Struct.Mechanism 1, 121-137.
19 B.Pullman & A.Sran (1976) Progr.Nucl.Acid Res.Mol.Biol. 18, 215-322.
20 H.Follmann & G.Gremels (1974) Eur.J.Biochem.47, 187-197.
21 H.Follmann, I.Kuntz & W.Zacharias (1975) Eur.J.Biochem. 58, 31-41.
22 H.Follmann, R.Pfeil & H.Witzel (1977) Eur.J.Biochem.77, 451-461.
23 C.H.Lee & R.H.Sarma (1976) J.Amer.Chem.Soc.98,3541-3548.
24 H.D.Lüdemann & E.Westhof (1977) Z.Naturf.32c, 528-538.

25 F.E.Hruska, D.J.Wood, R.J.Mynott & R.H.Sarma (1973) FEBS Lett.31, 153-155.
26 C.H.Lee, F.E.Evans & R.H.Sarma (1975) J.Biol.Chem.250, 1290-1296.
27 H.Follmann & W.Ludwig (1978) Hoppe-S.Z.Physiol.Chem. 359, 264-265.
28 W.Ludwig & H.Follmann (1978) Eur.J.Biochem.82, 393-403.
29 T.D.Son & C.Chachaty (1977) Biochim.Biophys.Acta 500, 405-418.
30 A.M.Kapuler, C.Monny & A.M.Michelson (1970) Biochim. Biophys.Acta 217, 18-29.
31 A.M.Kapuler & E.Reich (1971) Biochemistry 10, 4050-4061.
32 D.C.Ward, A.Cerami, E.Reich, G.Acs & L.Altwerger (1969) J.Biol.Chem.244, 3243-3250.
33 P.M.Kaiser, L.Bonacker, H.Witzel & A.Holy (1975) Hoppe-S. Z.Physiol.Chem.356, 143-155.
34 M.Morange, F.Garcia Blanco, B.Vandenbunder & H.Buc (1976) Eur.J.Biochem.65, 553-563.
35 K.Muneyama, R.J.Bauer, D.A.Shuman, R.K.Robins & L.N. Simon (1971) Biochemistry 10, 2390-2395.
36 P.Prusiner, T.Brennan & M.Sundaralingam (1973) Biochemistry 12, 1196-1202.
37 F.E.Evans & N.O.Kaplan (1976) J.Biol.Chem.251, 6791-6797.
38 F.Jordan & H.Niv (1977) Biochim.Biophys.Acta 476,265-271.
39 D.L.Sloan, L.A.Loeb, A.S.Mildvan & R.J.Feldmann (1975) J.Biol.Chem.250, 8913-8920.
40 W.Meyer, E.Böhnke & H.Follmann (1976) Angew.Chemie Int. Ed.Engl.15, 499-500.
41 W.Meyer & H.Follmann (1978) Hoppe-S.Z.Physiol.Chem. 359, 299.
42 C.L.Zielke & C.H.Suelter (1971) J.Biol.Chem.246,1313-1317.
43 E.Sulkowski, W.Björk & M.Laskowski (1963) J.Biol.Chem. 238, 2477-2486.
44 A.Hampton & T.Sasaki (1972, 1973) Biochemistry 11, 4965-4970; 12, 2188-2191.
45 J.J.Baker, P.Mellish, C.Riddle, A.R.Somerville & J.R. Tittensor (1974) J.Med.Chem.17, 764-766.
46 C.Chachaty, T.Zemb, G.Langlet, T.D.Son, H.Buc & M.Morange (1976) Eur.J.Biochem. 62, 45-53.
47 A.Pohorille, D.Perahia & B.Pullman (1978) Biochim.Biophys.Acta 517, 511-516.
48 D.M.Cheng & R.H.Sarma (1977) J.Amer.Chem.Soc.99, 7333-7348.
49 C.Helene (1977) FEBS Lett.74, 10-13.

COOPERATIVITY IN THE BINDING OF LIGANDS TO DIHYDROFOLATE REDUCTASE

Berry Birdsall
Division of Molecular Pharmacology,
National Institute for Medical Research, London NW7 1AA

Dihydrofolate reductase is the target enzyme for the important 'anti-folate' drugs such as methotrexate (MTX) and trimethoprim [1,2]. The chemotherapeutic potency of methotrexate arises from its extremely tight binding to dihydrofolate reductase; MTX binds 10,000 times more tightly than the structurally very similar substrate, folate. We are studying dihydrofolate reductase from Lactobacillus casei by a variety of techniques including nmr spectroscopy in an attempt to understand the interactions responsible for this tight binding and to establish a structural basis for the striking difference in affinity between the substrate and inhibitors [3-7].

One approach to this problem is to study the binding of 'fragments' of the ligands. The fragments of methotrexate, 2,4-diaminopyrimidine (DAP) and p-aminobenzoyl-L-glutamate (PABG) (Figure 1), both bind weakly to the enzyme and the binding can be followed easily by nmr spectroscopy. Figure 2 shows the effects of the binding of DAP and p-nitrobenzoyl-L-glutamate (PNBG) (closely related to PABG) on the 2,6 protons of the five tyrosine residues of a selectively deuterated dihydrofolate reductase [5,7]. Both ligands bind in fast exchange and the gradual shift of the peaks with increasing concentration of ligand enables one to connect the signals in the spectrum of the enzyme alone with those in the enzyme.ligand complexes. PNBG and DAP bind simultaneously as can be seen by the fact that the spectrum of the ternary complex, E.PNBG.DAP, is not the same as either of the binary complexes.

The effects of binding PNBG and DAP on the 2,6 tyrosine protons are non-additive. For example, the tyrosine resonance to highest field (2.97 ppm) is shifted upfield by DAP binding but no further change in chemical shift occurs on the subsequent binding of PNBG. However, in the absence of DAP, the binding of PNBG does affect this signal (Figure 2). It is therefore unlikely that the change in chemical shift of this signal is due to a direct interaction of the corresponding tyrosine residue with DAP or PNBG but more likely due to a conformational change in the enzyme induced by the binding of either DAP or PNBG.

B. Pullman (ed.), Nuclear Magnetic Resonance Spectroscopy in Molecular Biology, 339-349.

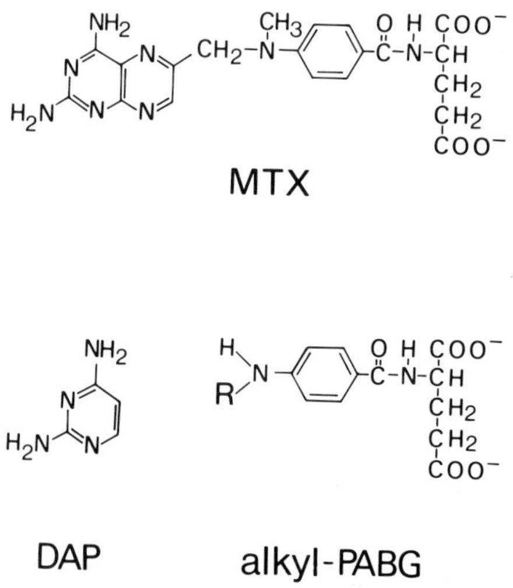

MTX

DAP alkyl-PABG

Figure 1. Structure of methotrexate (MTX) and its 'fragments' 2,4-diaminopyrimidine (DAP) and p-aminobenzoyl-L-glutamate (PABG), R=H. For the PABG analogues R is an n-alkyl chain, 1 to 6 carbon atoms long.

Further non-additive effects are observed in the tyrosine, histidine and high field methyl regions of the spectrum.

For the binary complexes, the shifts of the tyrosine signals as a function of ligand concentration can also be analysed to give the affinity constants for binding of the fragments. However, PNBG binds in a stoichiometric manner to the E.DAP complex (Figure 2). This indicates that PNBG binds more tightly to the E.DAP complex than to the enzyme alone and demonstrates positive heterotropic cooperativity in this system.

The inherent low sensitivity of the nmr technique limits the binding constants that can be measured to the range 10^4 to 10^1 M^{-1}. Even within this range estimates of binding constants are subject to large errors if the signals are not truly in the fast exchange region [8].

A more sensitive and flexible technique for measuring binding constants is fluorescence spectroscopy. Using this technique it has been possible to measure the binding constants of PABG and DAP in the binary and ternary complexes (Figure 3). PABG binds 60 ($\pm$ 9) times more tightly to the E.DAP complex than to the enzyme alone and in agreement with the reciprocal nature of the binding scheme DAP binds 47 ($\pm$ 7) fold tighter to the binary complex than to the enzyme alone. The mean K_{coop} of these two independent measurements is very

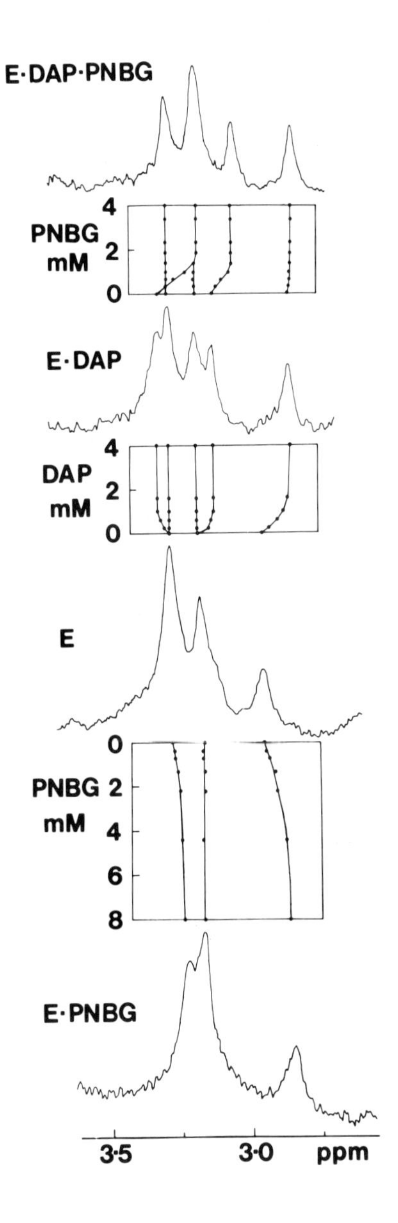

Figure 2. The effects of adding increasing concentrations of 2,4-diaminopyrimidine (DAP) and p-nitrobenzoyl-L-glutamate (PNBG) on the resonances of the tyrosine 2,6 protons in a selectively deuterated dihydrofolate reductase. The 270 MHz ^{1}H nmr spectra of the tyrosine region of the enzyme alone, the binary complexes enzyme.DAP and enzyme.PNBG, and the ternary complex, enzyme.DAP. PNBG. are shown. The enzyme concentration is 1.3 mM.

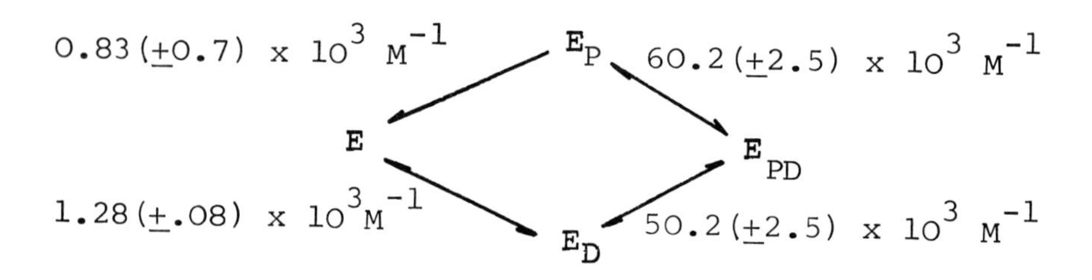

Figure 3. Scheme and equilibrium constants for p-amino-benzoyl-L-glutamate (P) and 2,4-diaminopyrimidine (D) binding to dihydrofolate reductase (E). The equilibrium constants were determined by measuring the changes in the fluorescence of the ligand or the enzyme which accompany complex formation. All measurements were made at 25°C in 15 mM Bis-Tris buffer in the presence of 0.5 M KCl.

substantial (54) and equivalent to a ΔG^{o}_{coop} of -2.35 kcal/mole

As previous evidence has shown that these fragments do bind to the enzyme in a manner similar to that in which MTX itself binds [4, 5,7,10] it appears that one is observing the cooperativity between sub-sites of the inhibitor binding site. In order to investigate the structural basis for these cooperative interactions a series of n-alkyl analogues of PABG (Figure 1) were synthesised and the binding constants in the binary and ternary complexes measured (Figure 4) [9].

The effect of increasing the length of the alkyl chain is to increase the affinity of the analogue for the enzyme. There is a notably large increase in binding constant (27 fold) between the methyl and ethyl analogues. In contrast, all the alkyl PABG analogues bind to the E.DAP complex with the same binding constant (within a factor of 2). The presence of DAP therefore alters both the affinity and the structural specificity of the binding site for the PABG analogues. The observed cooperativity is therefore dependent on the length of the alkyl chain of the PABG analogue. Thus for PABG and the short chain analogues there is a large cooperativity; for butylPABG there is no cooperativity; and for hexylPABG there is negative cooperativity.

From these results and from binding studies of other PABG and DAP analogues we have suggested a model which can be used to visualise this cooperativity [9]. The model (Figure 5) assumes that (i) there are several conformationally different forms of the enzyme, E, of which only two, E_A and E_B bind DAP and the PABG analogues and (ii) both PABG and DAP bind more strongly to E_B than to E_A. The binding of the first ligand, say PABG, increases the proportion of E_B and hence increases the binding constant of the second ligand (DAP). Thus one observes positive cooperativity. In contrast, hexylPABG binds

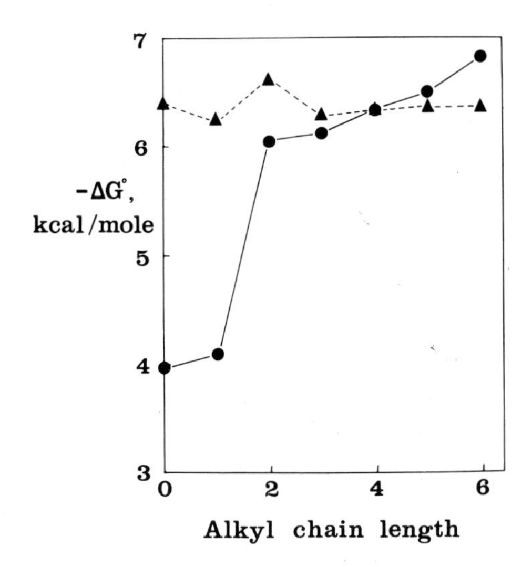

Figure 4. Change in Gibbs free energy on binding of N-alkyl-p-aminobenzoyl-L-glutamates to dihydrofolate reductase, alone (● —— ●) and in the presence of 2,4-diaminopyrimidine (DAP) (▲ ---- ▲) as a function of the number of carbon atoms in the n-alkyl chain.

somewhat more tightly to the A form of the enzyme and therefore the subsequent binding of DAP is weaker and one observes negative cooperativity. Any ligand-induced change in the proportions of the two conformational states will result in cooperative binding either positive or negative. ButylPABG binds equally well to E_A or E_B and therefore exhibits zero cooperativity. This simple model rationalises the results of the fluorescence binding studies and would gain credence if it were possible to demonstrate directly the existence of the A and B forms of the enzyme.

Since the non additive effects in the nmr spectra (e.g. Figure 2) suggest that a conformational change takes place when the fragments bind to the enzyme, the nmr spectra of the binary and ternary complexes of the alkyl PABG analogues have been examined. The complexity of the changes which take place when an alkyl PABG analogue binds to the enzyme is illustrated in Figure 6. The nmr spectra of dihydrofolate reductase and its complex with hexylPABG are shown in B) and A) respectively. The difference spectrum C) shows that there are many changes throughout the spectrum. It is not possible from the examination of the spectrum of a single complex to determine which changes are due to direct interactions with a ligand and which result from conformational changes in the enzyme.

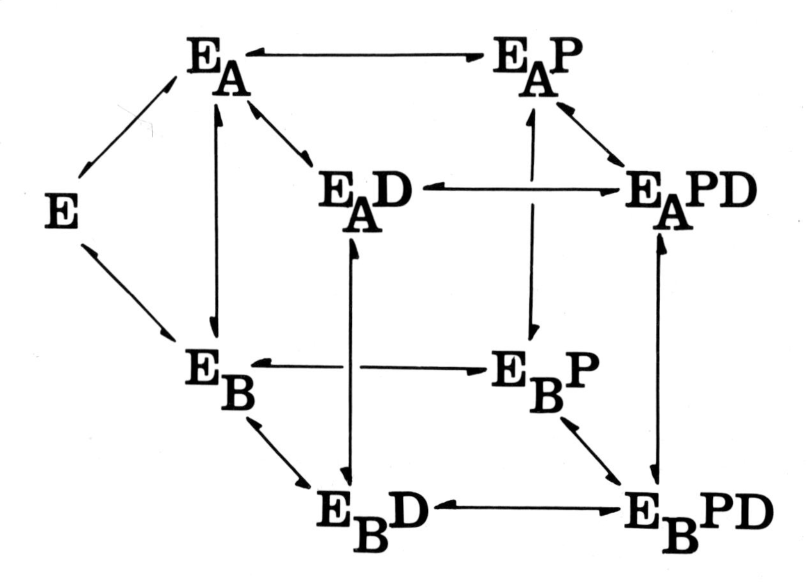

Figure 5. A scheme for describing the binding of 2,4-diamino-pyrimidine (D) and p-aminobenzoyl-L-glutamate (P) to two conformational states E_A and E_B of dihydrofolate reductase.

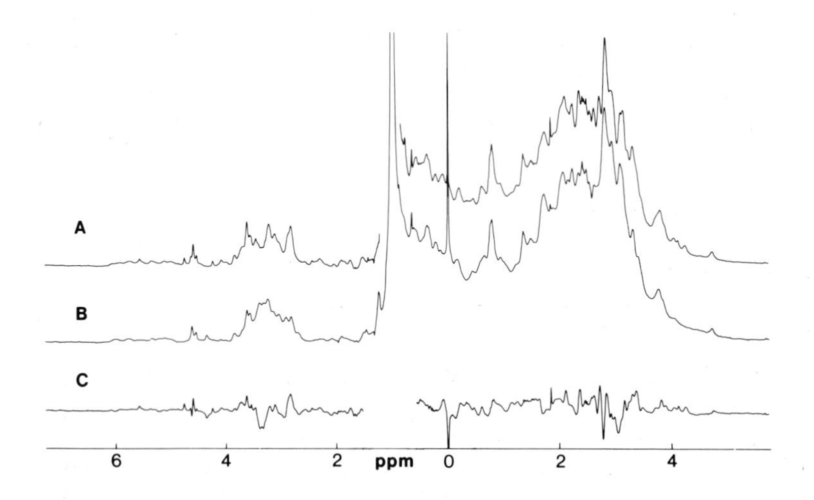

Figure 6. Effects of binding hexylPABG on the 270 MHz ^{1}H nmr spectrum of dihydrofolate reductase. A) the enzyme.hexylPABG complex, B) the enzyme alone and C) the difference spectrum of A-B.

Accordingly the nmr spectra of a series of complexes E.alkyl-PABG.DAP, have been determined and the difference spectra, (E.alkyl-PABG.DAP) - (E.methylPABG.DAP), calculated. The aromatic regions of the difference spectra are shown in Figure 7. In view of the complexity of the aromatic region caused by the large number of overlapping peaks, it has not yet been possible to connect the positive and negative peaks in the difference spectra and assign them to specific amino acid residues in the enzyme. Nevertheless, a qualitative description is possible.

The top five difference spectra in Figure 7 are extremely similar. The high field region of the spectra are essentially identical: a positive doublet (2.45 ppm), a positive and negative peak close together (2.29 ppm), and another positive and negative peak near 2.09 ppm. Other characteristic features are a positive peak at 3.10 ppm and a negative doublet at 3.52 ppm. These peaks are present in the (E.ethylPABG.DAP)-(E.methylPABG.DAP) difference spectrum but they are smaller than in the difference spectra of the longer chain analogues. Although the general silhouettes are similar in this group there are small progressive changes through the series; from the ethyl to hexyl ternary complexes the peak at 3.10 ppm gets larger with increasing chain length, the pattern at 3.61 ppm changes and there is movement of peaks in the region 2.93 to 2.79 ppm. The negative peaks at 3.79 and 2.72 ppm are due to the aromatic protons of excess unbound methylPABG in the E.methylPABG.DAP sample.

In contrast to the similarities exhibited in the top five difference spectra of Figure 7 the difference spectrum (E.PABG.DAP) - (E.methyl-PABG.DAP) is quite distinct from the others. The two large shaded peaks are from the protons of excess free PABG and not relevant to this analysis of the difference spectrum. Excluding these two peaks the remaining difference spectrum only shows very minor deviations from a horizontal line. This demonstrates that the aromatic regions of the E.PABG.DAP and E.methylPABG.DAP complexes are almost identical. An indication of the magnitude of the deviations is that the area of the shaded peak (3.81 ppm) is equivalent to 2 protons. Within this homologous series the other regions of the difference spectra show the same qualitative similarities and differences as described here for the aromatic region.

The difference spectrum (E.DAP) - (E.methylPABG.DAP) is included at the bottom of Figure 7 to illustrate that the characteristic features shown in the top five spectra are common to this homologous series of ternary complexes but not to other complexes. This difference spectrum exhibits large positive and negative deviations and is quite different in many regions from the other difference spectra. This would be expected as the spectrum of the ternary complex incorporates both the direct and indirect effects of binding a PABG analogue.

The differences between the spectra of the E.alkylPABG complexes have been quantitated by measuring the sum of the areas of the positive

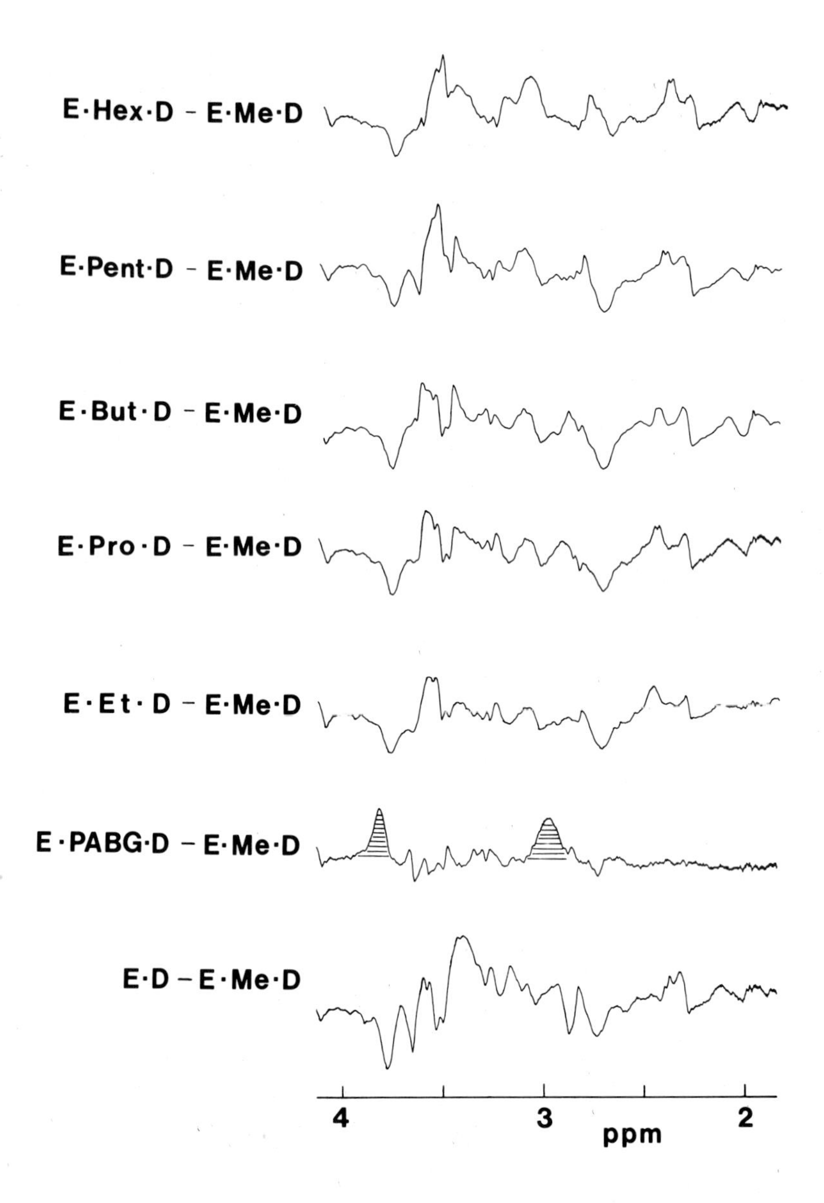

Figure 7. The aromatic region of a series of 270 MHz ^{1}H nmr difference spectra. The spectrum of the ternary complex, enzyme.methylPABG.DAP (E.Me.D) has been subtracted from the spectra of the other ternary complexes.

and negative deviations in the difference spectrum. The areas for the difference spectra in Figure 7 are listed in Table 1. They have been normalised to the total area of the aromatic regions in the original nmr spectra of the two complexes used to generate the difference spectrum and set equal to 100. This number is very close to the actual number of aromatic protons in the enzyme complex. Therefore the numbers in the table represent the minimum number of proton resonances which differ in the spectra of the two complexes. The results in Table 1 confirm the previous qualitative description of the two groups of similar ternary complexes. It should be emphasised that if there were no qualitative similarities between the difference spectra then the area measurements would have no validity in quantitating these similarities.

Table 1. The area of each of the difference spectra (E.alkylPABG.DAP) - (E.methylPABG.DAP)

Difference spectrum	Area difference
E.Hex.D - E.Me.D	7.1
E.Pent.D - E.Me.D	6.5
E.But.D - E.Me.D	6.1
E.Pro.D - E.Me.D	6.6
E.Et.D - E.Me.D	4.9
E.Me.D - E.Me.D	0.0
E.PABG.D - E.Me.D	2.0
E.D - E.Me.D	10.1
Total aromatic region	100

The information that is evident in the difference spectra may be summarised as follows: The ternary complexes of PABG and methylPABG appear to be very similar to each other and different from the ternary complexes of the series ethylPABG to hexylPABG. The longer chain analogues form a separate group, the members of which are similar to each other. However, within this group there is a progressive change through the series with the ethylPABG complex being more similar to the methylPABG complex than the longer chain analogues are to the methylPABG complex.

The existence of two groups of ternary complexes is also predicted by the model derived from the fluorescence binding studies. The short chain analogues of PABG are postulated to bind preferentially to the B form of the enzyme and exhibit a large positive cooperativity on binding DAP whereas the longer analogues have a higher affinity for the A form and exhibit zero or negative cooperativity on binding DAP. In addition the major change in the proportions of the B and A forms of the enzyme takes place between the methylPABG and ethylPABG analogues.

The correlation between the cooperativity of binding the alkyl-PABG analogues and DAP and the area difference measurements calculated from the nmr difference spectra is shown in Figure 8. There is excellent agreement suggesting that both sets of data are monitoring the same phenomenon, namely the change in the proportions of the A and B forms of the enzyme.

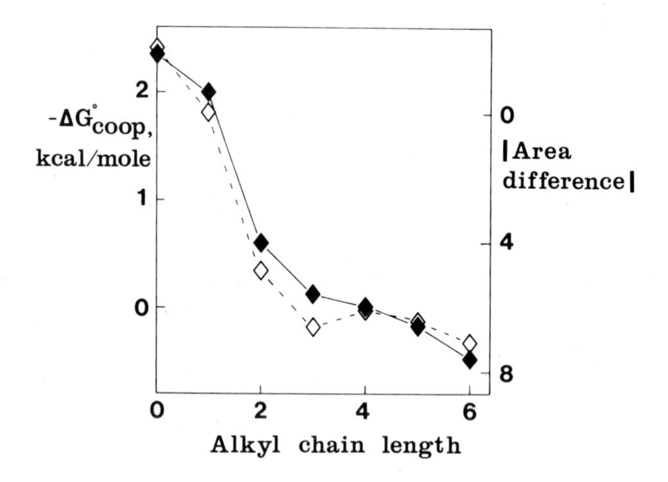

Figure 8. Change in the Gibbs free energy of cooperativity from the fluorescence binding studies (◆——◆) and the area difference in the nmr difference spectra (◇ ---- ◇) as a function of the number of carbon atoms in the n-alkyl chain.

In conclusion, the nmr spectra have provided evidence for two kinds of enzyme.alkylPABG.DAP complexes. The observation of a different enzyme conformation for the complex with longer alkyl chain analogues (ethylPABG to hexylPABG) as compared with the conformation of the ternary complexes of methylPABG and PABG is consistent with the model derived from the fluorescence experiments. In addition the nmr data has been related to the energetics of binding. As only three equilibrium binding constants can be measured using fluorescence spectroscopic methods, the suggested model for cooperativity, with eight independent binding constants, is underdetermined. From the fluorescence data alone it is not possible to predict the proportions of the two conformational forms of the enzyme in the ternary complexes of the alkylPABG analogues. However, the nmr difference spectra show that the long chain analogues produce a considerable change in the proportion of the A and B forms of the enzyme; a change of only 10%, for example, would scarcely be detectable in the nmr spectra. From the extent and amplitude of the changes in the difference spectra it is also clear that the two conformations differ in many respects and that the difference between conformation A and conformation B is not just a subtle change involving one residue. When the signals in the difference spectra are assigned it will be possible to get more specific information about the conformation of the different complexes.

Acknowledgements

I would like to thank my colleagues involved with this study, A.S.V. Burgen, J. Rodrigues de Miranda, J. Feeney, G.C.K. Roberts and N.J.M. Birdsall for their help and advice.

References

1. Hitchings, G.H. and Burchall, J.J.: Adv. Enzymol., 27, 417, 1965.
2. Blakley, R.L.: The Biochemistry of Folic Acid and Related pteridines, North Holland, Amsterdam, 1969.
3. Roberts, G.C.K.: in (Roberts, G.C.K. Ed.) Drug Action at the Molecular Level, Macmillan, London 1977, p.127.
4. Birdsall, B., Griffiths, D.V., Roberts, G.C.K., Feeney, J. and Burgen, A.S.V., Proc. Roy. Soc. Lond.ser.B. 196, 251, 1977.
5. Feeney, J., Roberts, G.C.K., Birdsall, B., Griffiths, D.V., King, R.W., Scudder, P. and Burgen, A.S.V. Proc. Roy. Soc. Lond. ser.B., 196, 267, 1977.
6. Hood, K. and Roberts, G.C.K. Biochem. J., in press.
7. Roberts, G.C.K., Feeney, J., Birdsall, B., Kimber, B., Griffiths, D.V., King, R.W. and Burgen, A.S.V. in Dwek, R.A., Campbell, I.D., Richards, R.E. and Williams, R.J.P. (eds), NMR in Biology, Academic Press, London, 1977, p.95.
8. Feeney, J., Batchelor, J.G., Albrand, J.P. and Roberts, G.C.K., submitted to J. Mag. Res.
9. Birdsall, B., Burgen, A.S.V., Rodrigues de Miranda, J. and Roberts, G.C.K. Biochemistry, in press.
10. Kimber, B.J., Griffiths, D.V., Birdsall, B., King, R.W., Scudder, P., Feeney, J., Roberts, G.C.K. and Burgen, A.S.V. Biochemistry 16, 3492, 1977.

STUDIES OF GENE-5 PROTEIN NUCLEIC ACID INTERACTION.

C.W. Hilbers[x], G.J. Garssen[x], R. Kaptein[¶], J.G.G. Schoenmakers[+] and J.H. van Boom[†].
[x]Department of Biophysical Chemistry and [+]Department of Molecular Biology, University of Nijmegen, Nijmegen; Department of Physical Chemistry, University of Groningen, Groningen; [†]Vakgroep Organische Chemie, University of Leiden, Leiden, The Netherlands.

INTRODUCTION

In the living cell recognition reactions between proteins and nucleic acids very often are key events in the cellular metabolism. A well known example is given by the interactions between transfer RNA's and a number of proteins during protein synthesis. The interaction of DNA replicase and DNA in the course of the replication process is equally well known. Although there is a general knowledge of the events taking place during these processes, very little is known about the interactions determining specificity and mechanism. This is equally true for other nucleic acid protein recognition processes. A fundamental understanding of these interactions requires knowledge of the structure and dynamics of the complexes formed at a submolecular and atomic level. Only in a few instances has NMR been brought to bear upon these type of problems (1, 2, 3).
We have started to investigate the interaction of gene-5 protein with small well defined oligonucleotides (4). This protein, which may serve as a model system for studying protein nucleic acid interactions, is encoded by the filamentous coliphage M13; it plays a role during the replication process of the single stranded viral DNA (for a review see ref. 5). The gene-5 protein has been the subject of a number of physico-chemical studies. It binds strongly and cooperatively to single stranded DNA and consequently it is able to destabilize double stranded DNA (6, 7). From circular dichroism and fluorescence studies it was inferred that tyrosyl residues participate in the protein nucleic acid complex formation (8-10). Moreover, it was found that the protein covers four nucleotides when bound to DNA (10). The primary sequence of the protein has been determined (11, 12); for the discussion in this paper it is relevant

B. Pullman (ed.), Nuclear Magnetic Resonance Spectroscopy in Molecular Biology, 351-364.

to know that it contains one histidyl, three phenylalanyl and five tyrosyl residues. Its molecular weight is 9700.

We are particularly interested in the mechanism of double helix destabilization and the interactions leading to DNA binding. In this contribution the emphasis will be on the latter aspect.

ASSIGNMENTS

Fig. 1 shows the 360 MHz proton NMR of a 0.6 mM solution of native gene-5 protein in D_2O. The spectrum is characteristic of that of a globular protein. Between 8.5 and 6 ppm the resonances of the non-exchangeable aromatic ring protons are found. The resonance at 7.9 ppm can be assigned to the C-2 proton of the single (His-64) residue present in the protein (11, 12). The assignment of other resonances in this spectral region will be discussed below.

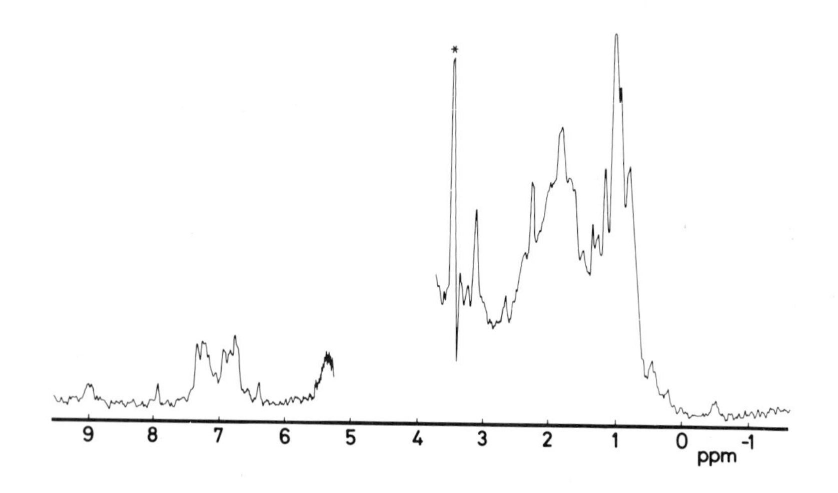

Figure 1. The 360 MHz proton NMR spectrum of gene-5 protein in D_2O recorded at 31°C. The peak indicated with an asterisk is from EDTA.

The C-α protons of amino acid residues resonate between 6 and 3.5 ppm (not shown because of interference with the HDO resonance). The resonances of the aliphatic side chains fall between 3.5 and -0.6 ppm. The well resolved resonances between 0.5 to -0.6 ppm, shifted outside the bulk resonance, most likely come from methyl groups of leucine, isoleucine and/or valine, shifted by the ringcurrents of proximate aromatic residues. Since the influence of the binding of DNA fragments to the protein is most easily studied through the aromatic region of the spectrum, assignments will be restricted to this part of the spectrum. We have approached this problem in two ways a) by incorporating deuterated phenylalanyl residues into the protein b) by employing a recently developed technique of the observation of photo CIDNP effect of tyrosine residues (13).

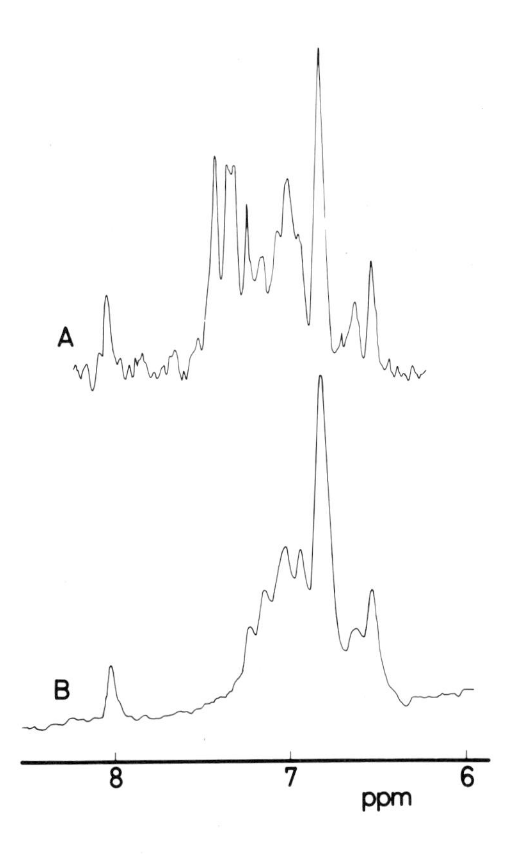

Figure 2. The aromatic part of the 360 MHz proton NMR spectrum of gene-5 protein, recorded at 37°C; (A) with normal phenylalanines; (B) with deuterated phenylalanines.

In Fig. 2 the aromatic part of the gene-5 protein spectra of the normal protein and the protein,containing deuterated phenylalanyl residues,are compared. The incorporation of deuterated phenylalanyl residues was achieved by suppressing the aromatic amino acid synthetic pathway in E.Coli by adding large amounts of tryptophan, tyrosine and deuterated phenylalanine to the growth medium. Although the resonances in the spectrum of the gene-5 protein with deuterated phenylalanines are somewhat broader, it is clear that resonances around 7.3 ppm have disappeared. Integration of spectrum 2B, taking the resonance of the C-2 proton of histidine at 7.9 ppm as unity, yields a total of 22 $\pm$ 1 protons. On the basis of the amino acid composition one expects 22 protons so that the introduction of the deuterated phenylalanyl residues has been performed quantitatively. Thus the resonances around 7.3 ppm in Fig. 2A can be assigned to the phenylalanine ring protons.

Using the photo CIDNP method, resonances from particular amino acid side chains (tyrosine, histidine and tryptophan) can be selectively enhanced when these residues are situated at the surface of the protein. To this end a flavin dye, added to the sample, is photo excited in the NMR probe by an argon laser. In this way triplet state flavin is generated, which in the case of tyrosine residues is able to abstract the phenolic hydrogen atom. Consequently a radical pair is formed (reaction 2 below) which reversibly yields flavin and tyrosine (reaction 3 below).

$$F \xrightarrow{h\nu} {}^{1}F \longrightarrow {}^{3}F \qquad (1)$$

$$^{3}F + \text{Tyr OH} \longrightarrow \overline{FH^{\bullet} + \text{Tyr O}^{\bullet}} \qquad (2)$$

$$\overline{FH^{\bullet} + \text{Tyr O}^{\bullet}} \longrightarrow F + \text{Tyr OH}^{*} \qquad (3)$$

^{1}F and ^{3}F represent the flavin in the excited singlet and triplet state respectively. Nuclear spin polarization arises from the spin selective recombination of the radical pair (reaction 3) (14). In this way a strong emission effect has been observed for the 3,5 ring protons (ortho with respect to the hydroxyl group) of N-acetyl tyrosine and a weak enhanced absorption for the 2,6 protons and the β-CH_2 protons (for a more extensive description of the method see the contribution of Dr. R. Kaptein to this volume).

In Fig. 3 the effects of a photo CIDNP experiment in the 360 MHz spectrum of the deuterated phenylalanine substituted gene-5 protein are shown. Fig. 3A represents the spectrum recorded immediately after the laser pulse (designated light spectrum), while Fig. 3B shows the spectrum without laser irradiation (dark spectrum). The CIDNP effect is represented in Fig. 3C as the difference between the spectra 3A and 3B. The same CIDNP results were obtained with gene-5 protein, containing normally protonated phenylalanines. In Fig. 3C the emission signals at 6.5 and 6.8 ppm are the most outspoken features of the spectrum; the signal at 4.9 ppm arises from a slight shift of the HDO resonance due to heating of the sample by the laser pulse. The broad signal at about 3 ppm comes from the β-CH_2 pro-

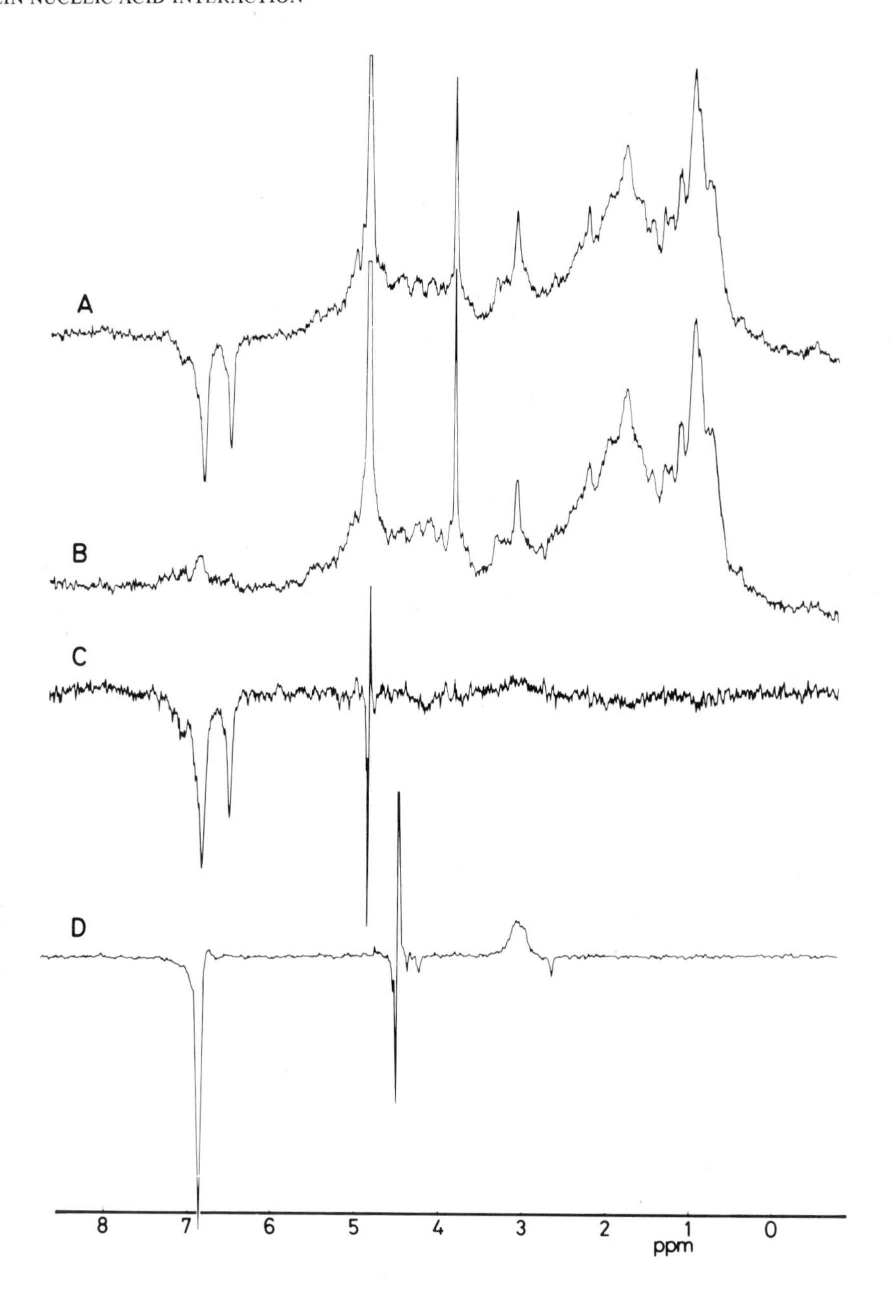

Figure 3. Photo CIDNP effects observed in the 360 MHz proton spectrum of gene-5 protein. (A) the light spectrum; (B) the dark spectrum; (C) photo CIDNP difference spectrum i.e. spectrum A minus spectrum B; (D) photo CIDNP difference spectrum of gene-5 protein dissolved in 5.6 M guanidine HCl.

tons. This is more clearly visible in Fig. 3D, where the CIDNP difference spectrum of the protein, denatured in a solution containing 5.6 M guanidine HCl, is shown. Under these conditions only one strong emission signal is found at a position of 6.8 ppm, where the 3,5 ring protons (protons ortho to the hydroxyl group) of "isolated" tyrosine resonate (15).

The CIDNP effects in the gene-5 protein are markedly dependent on the protein concentration. This is demonstrated in Fig. 4 where the aromatic part of the CIDNP difference spectra is presented for different protein concentrations. At a concentration of 1.54 mM gene-5 protein three resonances at 6.5, 6.75 and 6.9 ppm of about the same intensity are observed in emission; moreover, a shoulder is present at 7.03 ppm (Fig. 4A). Upon decreasing the protein concentration the peak at 6.9 ppm shifts upfield and coincides with the resonance which was originally at 6.75 ppm.

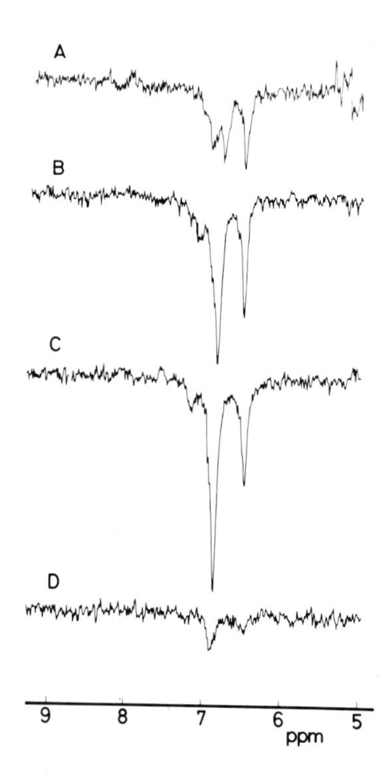

Figure 4. Aromatic part of the photo CIDNP difference spectrum of gene-5 protein at different concentrations of the protein (A) 1.54 mM; (B) 0.78 mM; (C) 0.38 mM; (D) 0.09 mM.

Interestingly the intensity of the emission signals increases at decreasing protein concentrations. In Fig. 4C the intensity ratio of the two major emission signals becomes 2:1. At the lowest concentration (Fig. 4D) the amount of protein present in the sample becomes the limiting factor to the signal intensity.

Since the aromatic spectrum of the gene-5 protein in Figs. 3 and 4 contains only resonances from a single histidine and five tyrosines, we can on the basis of the results in Fig. 3 and experiments on N-acetyl tyrosine (13) conclude that the 3,5 ring protons of tyrosines give rise to the observed emission signals. In addition the results in Fig. 4 show that only three out of the total of five tyrosines give rise to emission signals. These residues are situated at the surface of the protein and are about equally accessible to the photo excited dye molecule. The present results are in excellent agreement with the experiments of Anderson et al. (8), who found that in the native gene-5 protein three out of the five tyrosine residues can be modified chemically by tetranitromethane and by N-acetyl imidazole. In addition these investigators showed that these tyrosines correspond with tyr 26, 41 and 56 in the primary sequence of the protein and on this basis we assign the resonances at 6.52 and part of the resonances at 6.8 ppm to the 3,5 ring protons of these tyrosines. The resonance at 7.05 ppm most likely comes from the 2,6 ring protons of these tyrosines. Although for small molecules like N-acetyl tyrosine these protons are in enhanced absorption, they may be observed in emission for molecules with longer rotational correlation times due to polarization transfer from the 3,5 protons by dipolar cross relaxation.

Under most solution conditions the protein is present as a dimer with the tendency (at low ionic strength and at higher concentrations) to form higher aggregates (10, 16). The concentration dependence of the CIDNP effect and of the shifts observed in Fig. 4 we consider indicative of the association of gene-5 dimers. When the protein concentration is lowered one would expect a decrease of the CIDNP effect, since the effect arises from a second order reaction with the triplet flavin. Therefore the observed increase of the emission signal most probably reflects a deblocking of the tyrosines by a shift from tetramers or polymers to dimers. Consequently this indicates that in the formation of higher aggregates that side of the protein is involved where the tyrosines are situated. On the other hand these tyrosine residues are not involved in dimer formation.

INTERACTION OF THE GENE-5 PROTEIN WITH DNA

In vivo the gene-5 protein is involved in the regulation of viral DNA replication by binding to single stranded DNA (5). In addition circular dichroism and fluorescence studies indicated that tyrosine residues participate in the protein nucleic acid binding process (8-10). It is therefore of interest how binding of DNA affects the protein NMR spectrum and the observed CIDNP effects. Complexation with long stretches of DNA will be impractical because this will completely broaden the protein spectrum and therefore we studied the binding of short well defined DNA oligomers. The changes brought about in the aromatic region of the protein spectrum by complexation of the hexanucleotide d(pC-G-C-G-C-G) are shown in Fig. 5.

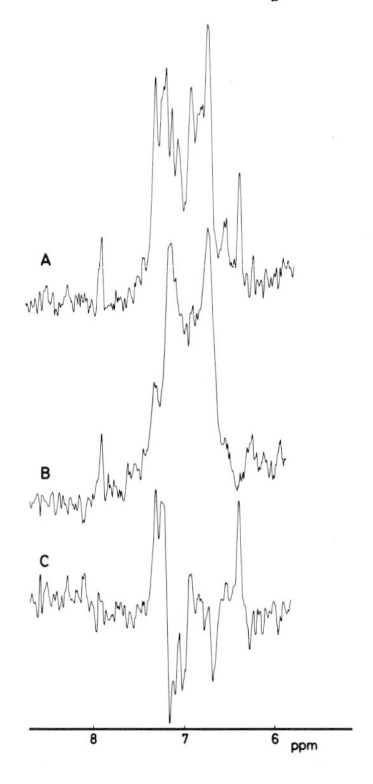

Figure 5. Effect of the binding of the hexanucleotide d(pC-G-C-G-C-G) on the aromatic part of the 360 MHz proton NMR spectrum of gene-5 protein. (A) spectrum of the native protein; (B) spectrum of the hexanucleotide protein complex; (C) difference spectrum A-B.

Fig. 5A represents the extended form of the aromatic part of the spectrum of the native protein; Fig. 5B the spectrum of the protein hexanucleotide complex. For comparison in Fig. 5C the difference spectrum 5A-5B is given. To this end the spectra 5A and 5B were scaled by equating the histidine C-2 proton resonance at 7.9 ppm. From the difference spectrum it is clear that the main changes, occurring upon binding, are the shifts of resonances around 7.3 ppm in the native protein upfield to about 7.0 ppm. Furthermore, in the spectrum of the protein-hexanucleotide complex an additional resonance is found at 6.7 ppm while the resonance at 6.5 ppm has disappeared. Similar difference spectra have been observed by Coleman et al. (3). Somewhat remarkebly the resonances of the ring protons of the DNA bases do not show up in Fig. 5B. Similar effects have been obtained for the tetranucleotide d(pC-G-C-G). Fast exchange between the oligonucleotides free in solution and complexed to the protein can explain the absence of DNA ring proton resonances in the spectrum 5B (4). Comparison of the results presented in Fig. 5 with those of Fig. 2 shows that the resonances, that are shifted from 7.3 ppm, belong to the phenylalanyl residues. Two interpretations of these shifts are possible. First, the phenylalanine residues directly participate in the interaction with oligonucleotide; for instance the phenylalanine rings are involved in stacking interactions with the bases of the DNA and thereby the ring proton resonances are shifted upfield. An alternative explanation is that upon complex formation the protein changes its conformation, which may give rise to the observed shifts.

The resonance at 6.5 ppm,which disappears after the protein is complexed to the hexanucleotide,has been assigned above to the 3,5 ring protons of a tyrosine at the surface of the protein. Apparently also this tyrosine residue is influenced by the protein DNA interaction. This has been investigated in more detail by looking at the influence upon the observed CIDNP effects of oligonucleotide binding to the protein. For two reasons we choose the selfcomplementary tetranucleotide d(pC-G-C-G) as a substrate. First, it is known that the protein covers four nucleotides, when bound to DNA (10). Secondly, many of the binding and unwinding properties of the protein with respect to this tetranucleotide are already known (4).

In Fig. 6A and 6B the aromatic region of the normal and the photo-CIDNP difference spectrum of gene-5 protein containing deuterated phenylalanyl residues is shown. The photo-CIDNP difference spectrum obtained from the gene-5 protein tetranucleotide complex is given in Fig. 6C. The tyrosine emission signals are completely quenched. To ensure that the total amount of protein was complexed to the

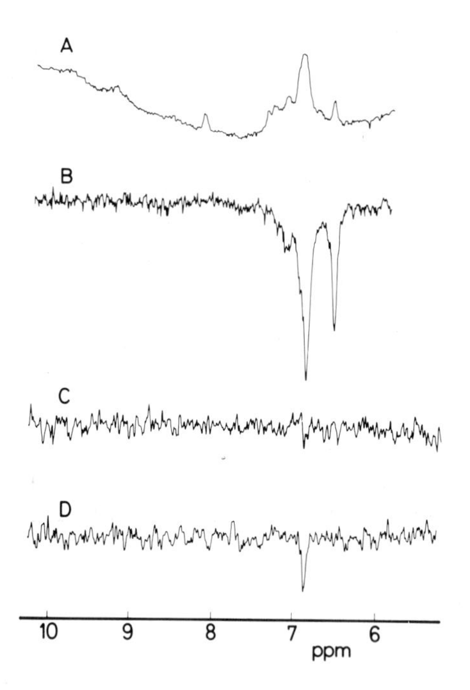

Figure 6. Effect of the binding of the tetranucleotide d(pC-G-C-G) to gene-5 protein. Aromatic part of the 360 MHz spectrum of gene-5 protein; (A) dark spectrum; (B) photo-CIDNP difference spectrum; (C) CIDNP difference spectrum of protein tetranucleotide complex; (D) CIDNP difference spectrum recorded after adding N-acetyl tyrosine to sample of spectrum C.

DNA fragment an excess of tetranucleotide was present in solution (0.30 mM protein versus 0.46 mM d(pC-G-C-G). To exclude the possibility that the flavin dye is rendered inactive by the presence of the tetranucleotide extra N-acetyl tyrosine (0.28 mM) was added to the solution. Under these conditions the emission signal of a free tyrosine appears in the CIDNP difference spectrum (Fig. 6D). This observation makes an inactivation of the flavin dye by the presence of the tetranucleotide very unlikely. In a separate experiment it was shown that the tetranucleotide does not show any CIDNP effects. Thus the experiments presented in Fig. 6 demonstrate that upon binding of the tetranucleotide the hydroxyl groups of the tyrosines 26, 41 and 56 are no longer available for reaction with the flavin dye. This we interprete to mean that these tyrosines are involved in the protein-DNA interaction. The present results again are in excellent agreement with those of Anderson et al. (8), who showed that in the in vitro complex of gene-5 protein with single stranded DNA the reaction of these tyrosyl residues with tetranitromethane is completely abolished.

Since the protein covers approximately four nucleotides in the in vitro protein-DNA complex (10), we expected that its DNA binding site is not wholly covered when a dinucleotide is bound to the protein. This is indeed demonstrated by the experiments presented in Fig. 7. At increasing ApA concentrations the tyrosine emission, originally at 6.5 ppm, shifts upfield (0.2 ppm) and its intensity decreases, while the emission signal at 6.8 ppm is virtually unaffected. Because of its low association constant a considerable excess of ApA had to be added in order to observe its shielding effect. From these experiments we conclude that the tyrosyl residue giving rise to the emission signal at 6.5 ppm is involved in binding while the other CIDNP active

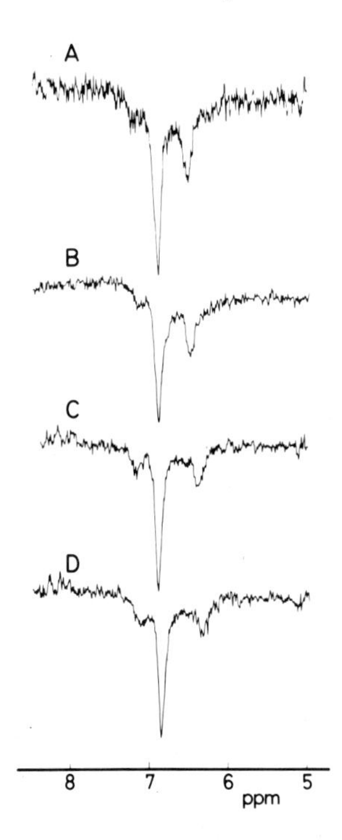

Figure 7. The effect of ApA binding upon the photo-CIDNP difference spectrum of the aromatic part of the 360 MHz proton NMR spectrum of gene-5 protein. Ratio ApA/gene-5 protein: (A) 0.0; (B) 4.6; (C) 11.4; (D) 17.3.

tyrosines do not bind to ApA. In a way it is rather surprising that this fragment binds preferentally to a specific part and not randomly over the total DNA binding site. This suggests to us, that it is just this part of the protein molecule, which first hooks on to the DNA in the course of the association process. Further experiments to establish this mechanism are presently under way.

CONCLUSIONS

The application of the photo-CIDNP method in conjunction with the introduction of deuterated amino acids holds great promise for the study of protein-protein and protein-nucleic acid interactions with NMR. In this contribution it has been shown that this allows partial assignment of the aromatic part of the protein spectrum in a convenient way. Binding of DNA fragments affects both the resonances of the phenylalanyl residues as well as the resonances of the tyrosines at the surface of the protein. The disappearance of the emission signals of the latter provides direct evidence for their involvement in the interaction with DNA. The changes in resonance positions of the phenylalanine ring protons may come from a change of the conformation of the protein upon DNA binding in which case these resonances may provide convenient markers for studying cooperative effects of DNA binding. Using DNA fragments of different chain lengths the protein DNA binding site and the binding mechanism can be investigated in detail. Also it is suggested by the present experiments that the protein-protein interactions in the dimer are of a different nature than those in the higher aggregates. It is evident that the techniques presented here are also applicable to other systems e.g. the gene-32 protein (melting protein from the T4 bacteriophage), the T7 RNA polymerase and the lac repressor where tyrosyl residues have been implicated in the interaction with nucleic acids (17, 18).

ACKNOWLEDGEMENTS

We are indebted to Dr. R.N.H. Konings for the gift of the bacteriophage. This research was supported by the Netherlands Foundation of Chemical Research (S.O.N.) with financial aid from the Netherlands Organization for the Advancement of Pure Research (Z.W.O.). We wish to acknowledge Z.W.O. for support of the 360 MHz NMR facility at Groningen.

REFERENCES

1. Shulman, R.G., Hilbers, C.W. and Miller, D.L. (1974) J. Mol. Biol. 90, 601-607.
2. Shulman, R.G., Hilbers, C.W., Söll, D. and Yang, S.K. (1974) J. Mol. Biol. 90, 609-611.
3. Coleman, J.E., Anderson, R.A., Ratcliffe, R.G. and Armitage, I.M. (1976) Biochemistry 15, 5410-5430.
4. Garssen, G.J., Hilbers, C.W., Schoenmakers, J.G.G. and Van Boom, J.H. (1977) Eur. J. Biochem. 81, 453-463.
5. Denhardt, D.T. (1975) CRC Crit. Rev. Microbiol. 4, 161-223.
6. Oey, J.L. and Knippers. R. (1972) J. Mol. Biol. 68, 125-138.
7. Alberts, B., Frey, L. and Delius, H. (1972) J. Mol. Biol. 68, 139-152.
8. Anderson, R.A., Nakashima, Y. and Coleman, J.E. (1975) Biochemistry 14, 907-917.
9. Day. L.A. (1973) Biochemistry 12, 5329-5339.
10. Pretorius, H.T., Klein, M. and Day. L.A. (1975) J. Biol. Chem. 250, 9262-9269.
11. Cuypers, T., Van der Ouderaa, F.J. and De Jong, W.W. (1974) Biochem. Biophys. Res. Commun. 59, 557-564.
12. Nakashima, Y., Dunker, A.K., Marvin, D.A. and Konings-berg, W. (1974) FEBS Lett. 40, 290-292 and 43, 125.
13. Kaptein, R., Dijkstra, K., Müller, F., Van Schagen, C.G. and Visser A.J.W.G. (to be published).
14. a. Closs, G.L. (1969) J. Am. Chem. Soc. 91, 4552-4554.
    b. Kaptein, R. and Oosterhoff, L.J. (1969) Chem. Phys. Letters 4, 195-197.
    c. Kaptein, R. and Oosterhoff, L.J. (1969) Chem. Phys. Letters 4, 214-216.
    d. Closs, G.L. and Trifunac, A.D. (1970) J. Am. Chem. Soc. 92, 2183-2184.
15. McDonald, C.C. and Phillips, W.D. (1969) J. Am. Chem. Soc. 91, 1513-1521.
16. Cavalieri,S.J., Neet, K.E. and Goldthwait, D.A. (1976) J. Mol. Biol. 102, 697-711.
17. Anderson, R.A. and Coleman, J.E. (1975) Biochemistry 14, 5485-5491.
18. Geisler, N. and Weber, K. (1977) Biochemistry 16, 938-943.

DISCUSSION

Malherbe-Davanloo: Using UUCA, Geerdes, Van Boom and myself obtained results essentially identical to the one you presented. By monitoring the hydrogen bonded ring N protons we did not find changes in the hydrogen bonded structure of yeast $tRNA^{Phe}$ upon codon binding. This rules out the disrupture of tertiary structure (disrupture of DMU loop TΨC loop interactions) at least for the codons studied by NMR.

STRUCTURAL ELUCIDATION STUDIES OF AURINTRICARBOXYLIC ACID, A POTENT INHIBITOR OF PROTEIN NUCLEIC ACID INTERACTIONS

Thomas Schleich, R. Gilberto González and Barry J. Blackburn
Division of Natural Sciences
University of California
Santa Cruz, CA 95064 USA

INTRODUCTION

The triphenylmethane dye, aurintricarboxylic acid (ATA), whose commonly accepted structure is shown in Figure 1, is a potent inhibitor of protein nucleic acid interactions. First employed by the metals industry for the quantitative determination of the aluminion ion (1), this dye in recent years has been utilized extensively by molecular

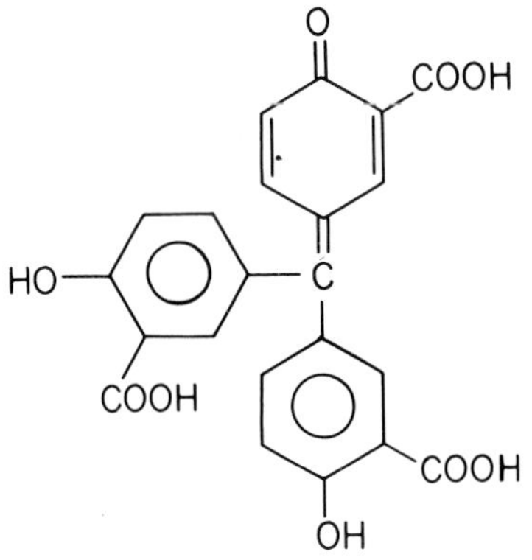

Figure 1. The commonly accepted structure of ATA.

biologists for the inhibition of a wide variety of important cellular processes that ultimately depend on the formation of a protein nucleic acid complex. Numerous studies employing both procaryotic (2-8) and eucaryotic systems (9-24) have clearly demonstrated that the initiation phase of protein synthesis is extraordinarily sensitive to the inhibitory action of ATA; by contrast, the subsequent step of chain elongation is much less sensitive to the dye. Two review articles (25-26) summarizing the inhibitory effects of ATA have been published. Commercial grade ATA also inhibits the synthesis of ppGpp (27), the poly rA, poly rC, and viral RNA directed DNA synthesis by detergent disrupted virions of Moloney murine leukemia virus (28), and the elaboration of progeny influenza virus (29). Studies employing isolated macromolecular systems

B. Pullman (ed.), Nuclear Magnetic Resonance Spectroscopy in Molecular Biology, 365-379.

demonstrate that ATA inhibits E. coli RNase V (30), Qβ replicase, E. coli RNA polymerase, and T7 RNA polymerase at the initiation step (31), the Qβ-replicase, E. coli polymerase, and lac-repressor binding of templates (31), Rauscher leukemia virus reverse transcriptase (32), the EF-Ts stimulatory effect of GDP exchange with EF-Tu bound GDP (33), and the binding of poly rU to ribosomal protein S1 (34). Other investigations have shown that ATA binds strongly to bovine serum albumin (35), thus presumably accounting for its ability to reverse or mask ATA inhibitory effects (15,28,31), and to many enzymes, regardless of their specific catalytic function, thereby causing inhibition (36). Curiously, seryl tRNA synthetase is not sensitive to ATA inhibition (36) while phenylalanyl tRNA synthetase from rabbit reticulocytes is inhibited by the dye (22). Because of its strong interaction with ribosomal proteins, ATA has been proposed as a marker stain for this type of protein in gel electrophoresis experiments (37).

The inhibitory potency of commercial preparations of ATA has been found to vary considerably (14), and, indeed, samples subjected to various chromatographic separation procedures have revealed the presence of numerous components, not all of which are active in eliciting the inhibition of protein nucleic acid interactions (28,29,31,38). Supposedly chemically pure ATA has only one tenth the inhibitory activity of commercially available material and undergoes gradual decomposition (14).

Given the widespread use and interest in ATA for the inhibition of particular cellular processes, and the well documented fact that commercial lots of ATA are impure, we felt it was necessary to structurally characterize the active components present in "ATA" preparations. We achieved fractionation by the combined use of dialysis and ultrafiltration, and structural elucidation by carbon-13 magnetic resonance spectroscopy. Evidence will be presented that ATA is a mixture of polymers of the phenol-formaldehyde type whose inhibitory potency increases with the average molecular weight of the polymer fraction.

MATERIALS AND METHODS

Materials

"Aluminon" grade ATA was purchased from the following companies: Fisher Scientific Co. (lot no. 752556); Sigma Chemical Co. (lot no. 8-2-74); Mallinkrodt Chemical Co. (lot no. 33); Aldrich Chemical Co. (lot no. 061757). Salicylic acid was USP grade. ^{13}C-enriched formaldehyde was obtained from Merck Sharpe and Dohme Canada Ltd. Methylenedisalicylic acid (MDSA) and "ATA" were synthesized using published procedures (1,14,39). The reduction of ATA was achieved using a catalyst of 10% palladium on charcoal under a 900 psi atmosphere of hydrogen. Poly L-lysine hydrochloride ($\overline{M}_n$ = 2300) and calf thymus DNA were obtained from Miles Laboratories and Worthington Biochemical Co., respectively. [^{3}H] labelled poly rU was supplied by P-L Biochemicals. Unwashed 70S ribosomes were obtained from E. coli MRE 600 following customary procedures. Ribosomal protein S1 was isolated from a 1 M NH_4Cl wash of 70S ribosomes employing poly rC-cellulose affinity chromatography (40).

Fractionation of ATA

A quantity of "Aluminon" grade ATA was dissolved in water and subjected to dialysis using Spectrapor membrane tubing with a molecular weight cutoff of ca. 3500. Dialysis was usually allowed to continue for two days. The retentate was designated LATA. The dialysate was subjected to ultrafiltration using an Amicon pressure cell fitted with a Diaflo UM-05 membrane (molecular weight cutoff = 500). The retentate and ultrafiltrate were designated MATA and SATA, respectively.

Physical measurements

^{13}NMR spectra were obtained using Varian XL-100, Varian CFT-20, and JEOL FX-60 spectrometers operating in the Fourier transform mode. Most spectra were obtained under conditions of proton noise decoupling. Chemical shift values are expressed relative to external tetramethylsilane (TMS). Deuterated dimethylsulfoxide (DMSO) held in a concentrically mounted capillary tube served both as an intermediate external reference standard and the "locking" compound. In those instances where no capillary was employed, deuterated water served as the lock. Circular dichroism (CD) spectra were recorded using a Durrum-Jasco J-20 spectropolarimeter. Number average molecular weights ($\bar{M}_n$) were determined using a Hewlett-Packard model 301 vapor phase osmometer. Spectral grade acetone was employed as the solvent for the free acid form of the dye; calibration was achieved using benzil.

ATA inhibitory assays

(i) DNA-poly L-lysine complex formation. This assay determines the ability of ATA to inhibit the formation of the calf thymus DNA-poly L-lysine complex. Complex formation was assessed by monitoring CD from 310 to 240 nm. Approximately 1.0×10^{-4} M DNA (base residues) was allowed to interact with an equivalent amount of poly L-lysine in the presence of a known amount of ATA. The buffer contained 0.05 M NaCl and 1 mM sodium citrate, adjusted to pH 7.0. The degree of inhibitory activity was determined in terms of the amount of ATA required for the re-establishment of the CD spectrum equivalent to that for DNA alone. (ii) poly rU binding to ribosomal protein S1. This assay measures the inhibitory potency of an ATA sample in terms of preventing the binding of labelled poly rU to ribosomal protein S1. Polynucleotide binding was assessed by the Millipore filter binding technique previously described (41). The buffer contained 0.1 M NaCl, 10 mM $MgCl_2$ and 10 mM sodium cacodylate, adjusted to pH 7.0. (iii) poly rU binding to 70S ribosomes. This assay determined the efficacy of ATA inhibition of labelled polynucleotide binding to ribosomes monitored by use of the Millipore filter binding technique (41). The buffer was the same as employed in the previous assay.

RESULTS AND DISCUSSION

Fractionation, physical characteristics, and inhibitory activity

A representative sample of commercially available "Aluminon" grade ATA (Fisher Sci. Co., lot no. 752556) yielded upon fractionation by dialysis and ultrafiltration 13% SATA ($\bar{M}_n$ = 396), 65% MATA ($\bar{M}_n$ = 1800, and 22% LATA ($\bar{M}_n$ = 6000). These percentage values reflect the amount of material present on a weight basis. Table I lists the ultraviolet and visible spectral characteristics of each of the fractions. Commercial

TABLE I. Ultraviolet and visible spectral characteristics of ATA fractions in water.

FRACTION	λ_{max} (nm)	$E_{1\%}^{1cm}$
SATA	305	223
MATA	308	181
	450 (shoulder)	12
	535	15
LATA	312	192
	460 (shoulder)	46
	520	64

samples obtained from other suppliers also contained significant amounts of polymeric material. For example, ATA furnished by Mallinkrodt (lot no. 33) was found to be composed of 7% SATA, 39% MATA, and 54% LATA.

Each of the fractions derived from the ATA sample supplied by Fisher was assayed for protein nucleic acid complex formation inhibitory activity. We first examined the ability of the three ATA fractions to inhibit DNA-polylysine complex formation as assessed by CD. The spectra shown in Figure 2 indicate that SATA, MATA, and LATA have no, intermediate, and high inhibitory activities, respectively, at the listed concentrations. Even at significantly higher SATA concentrations no inhibitory activity was detected, and 20 μgm/ml of MATA was sufficient to completely prevent complex formation. We also utilized two biologically relevant assays, based on the binding of poly rU to ribosomal protein S1 and to *E. coli* 70S ribosomes. Ribosomal protein S1 has been found to be indispensible for the binding of mRNA to ribosomes (42). In both assays (Figures 3 and 4) we found that SATA has very little or no inhibitory activity, MATA possesses significant activity, and LATA exhibits substantially greater inhibitory activity. By comparison, unfractionated ATA has somewhat less inhibitory activity than MATA. We also investigated whether SATA could mask the inhibitory activity of LATA. As shown in Figure 3 (bottom panel) the presence of increasing amounts of SATA does not influence the activity of LATA. On the basis of these experiments we conclude that ATA in the form of the commonly accepted structure is ineffective as an inhibitor of protein nucleic

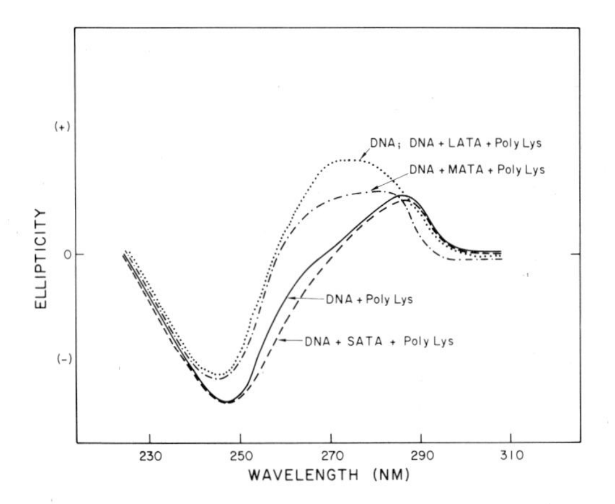

Figure 2. ATA inhibition of DNA-polylysine complex formation. The concentrations of the ATA fractions were as follows: SATA, 10 μgm/ml; MATA, 10 μgm/ml; LATA 6.7 μgm/ml.

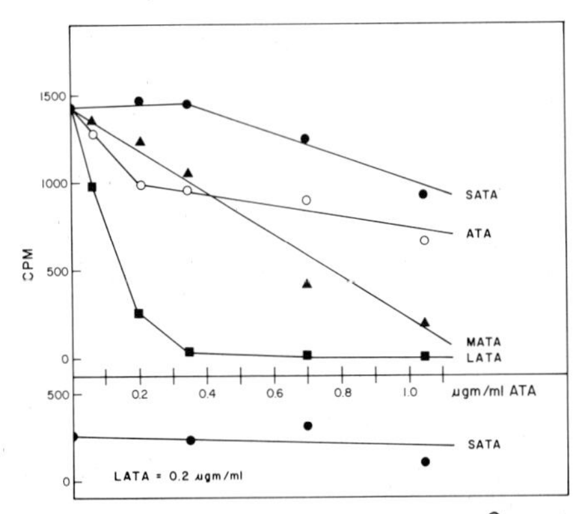

Figure 3. (Top Panel) ATA inhibition of [^{3}H] labelled poly rU binding to ribosomal protein S1. (Bottom Panel) The inhibitory activity of LATA in the presence of increasing amounts of SATA. In this experiment SATA was added prior to the LATA.

acid interactions, and that the inhibitory activity of an "ATA" preparation is proportional to the degree of polymerization. A significant number of studies have alluded to the variable activity of commercial ATA preparations (1,14,28,29,31) and that the commonly accepted structure might not be the true inhibitor (31,38). Furthermore, a number of different workers have observed heterogeneity by paper chromatography, thin layer chromatography, and silica gel column chromatography (28,29, 38), and that not all fractions are active in the inhibition of protein nucleic acid complex formation. Our results are fully in accord with these observations and provide new insight into the structural differences between active and inert ATA fractions. The remainder of this paper is devoted to the structural elucidation of ATA preparations by carbon-13 NMR spectroscopy.

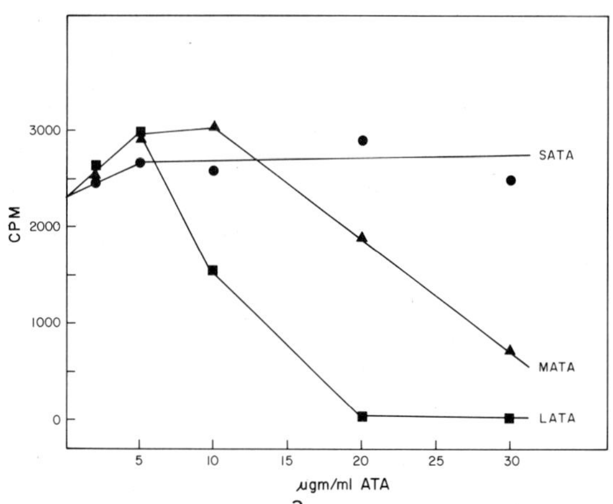

Figure 4. ATA inhibition of [^{3}H] labelled poly rU to E. coli 70S ribosomes.

NMR assignments and interpretation

Using the published ^{13}C assignments for salicylaldehyde (43) as a starting point in conjunction with a spectrum of salicylic acid we made assignments for the aromatic ^{13}C resonances of two of the three possible MDSA isomers (see Table II). Three isomeric linkage possibilities for the two aromatic rings of MDSA are possible: *para-para*, *ortho-para*, and *ortho-ortho*. That at least two of these isomers are present in our sample of MDSA is shown by the two methylene carbon resonances located at 40.3 and 34.9 ppm. Employing the methylene carbon assignments for the isomeric dianisylmethanes (44) we assign these resonances to the methylene carbons of the *para-para* and *ortho-para* isomers, respectively. We have not observed methylene carbon resonances due to the *ortho-ortho* isomer in our MDSA preparations. Typical ratios for the relative amounts of the *para-para* isomer to the *ortho-para* form are *ca.* 4:1. The aromatic assignments for the *para-para* isomer are consistent with the pattern and magnitude of both one- and three-bond carbon-proton spin-spin coupling effects (see Table II). (For benzene the CH coupling constants have been reported (45): $^{1}J_{CH}$ = 157.5 Hz; $^{2}J_{CH}$ = 1.0 Hz; $^{3}J_{CH}$ = 7.4 Hz; $^{4}J_{CH}$ = -1.1 Hz.) Because of the substantially lower amount of the *ortho-para* isomer in our preparation it was not possible to evaluate coupling constants for this isomeric form. Our MDSA preparations are characterized by a $\bar{M}_n$ of 285 ± 12 (calculated molecular weight is 288) and therefore contain insignificant amounts of polymer.

The natural abundance proton decoupled ^{13}C NMR spectrum of ATA synthesized according to Smith (1) using an MDSA preparation whose characteristics were described above as the starting material is shown in Figure 5. Eight primary resonances are observed plus *ca.* 6 resonances of significantly lower intensity. The 8 primary resonances are consistent with the commonly accepted structure of ATA (see Figure 1) if tautomerism is assumed. This sample has a $\bar{M}_n$ of 580, and such a

TABLE II. ^{13}C NMR parameters for MDSA.

para-para isomer

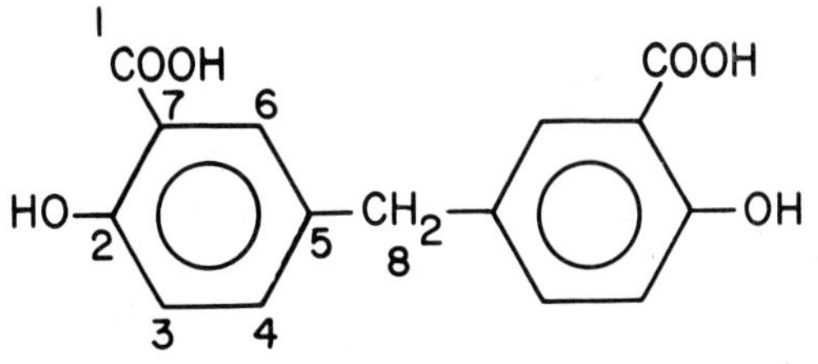

ortho-para isomer

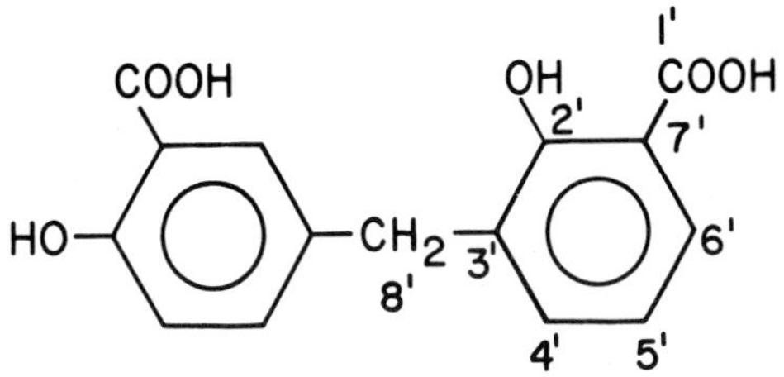

Position	δ (ppm)[a]	$^{1}J_{CH}$ (Hz)	$^{3}J_{CH}$ (Hz)	Position	δ (ppm)[a]
1	176.0	-	4.5	1'	176.3
2	159.1	-	8.3	2'	157.6
3	117.3	161.3	-	3'	132.5
4	135.0	161.1	7.1	4'	135.2
5	133.3	-	7.5	5'	118+
6	130.9	158.1	4.8	6'	129.4
7	118.5	-	5.0	7'	118.8
8	40.3	126.5	8.8	8'	34.9

[a] Relative to TMS; solvent is water

value is compatible with the presence of equal amounts of the structures depicted in Figure 5. The "extra" resonances observed in the spectrum are consistent with the presence of the higher molecular weight species in the sample and the existence of two MDSA isomers in the starting material used for the synthesis. Assignments were made by analogy to MDSA and triphenylcarbinol, and are listed in Table III. These assignments are further supported by the pattern and magnitude of one- and three-band carbon-proton spin-spin coupling effects, and by the qualitative estimation of relative relaxation times for the various aromatic carbon resonances. We found that the relaxation times for resonances 3, 4, and 6 were much shorter than the others and therefore must correspond to carbon atoms bearing directly attached hydrogens. Furthermore, a proton coupled ^{13}C NMR spectrum reveals that these resonances are split into doublets characterized by a coupling constant of *ca.* 160 Hz (see Table III), indicating one-bond carbon-proton coupling. Smaller splittings arising from three-bond coupling effects are evident for all the resonances with the exception of the one located at

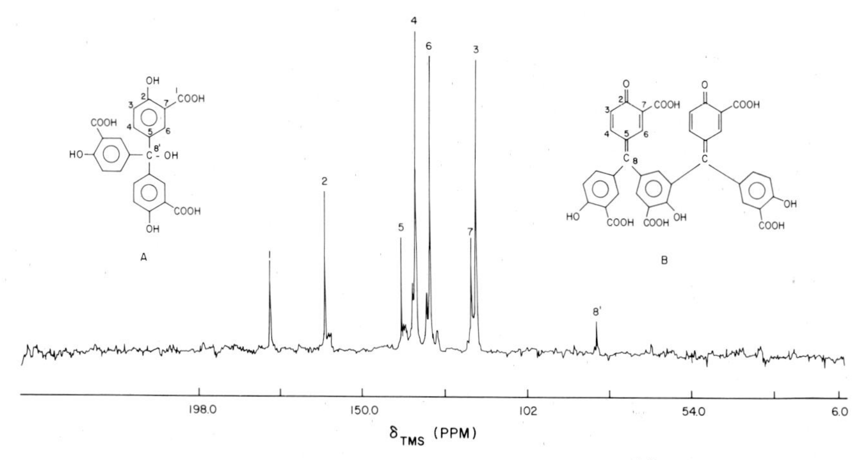

Figure 5. The natural abundance proton decoupled ^{13}C NMR spectrum for ATA synthesized according to Smith (1). This synthesis utilized MDSA that was not polymerized. The number average molecular weight is compatible with the presence of equal amounts of structures A and B.

TABLE III. ^{13}C NMR parameters for the 8 primary resonances of ATA prepared according to Smith.

Position	δ (ppm)[a]	$^{1}J_{CH}$ (Hz)	$^{3}J_{CH}$ (Hz)
1	176.1	-	3.7
2	160.2	-	7.9
3	116.8	161.5	-
4	131.4	159.9	7.6
5	138.2	-	7.6
6	130.3	159.2	6.7
7	118.3	-	4.8
8 $\rightleftarrows$ 8'	82.0	-	2.3

[a]Relative to TMS; solvent is H_2O.

116.8 ppm (position 3). Examination of the structures shown in Figure 5 reveals that carbon 3 is the only carbon with no possibility for engaging in three-bond carbon-proton coupling, thereby corroborating this assignment. Resonance 8', assigned to the central carbon of ATA, is somewhat broader than the other resonances and a coupling constant of 2.3 Hz is observed. On the basis of the chemical shift value for the carbinol carbon of triphenylcarbinol dissolved in acetone-DMSO (2:1) (δ_{TMS} = 80.5 ppm) we propose that an equilibrium occurs in aqueous solution between the carbinol and unsaturated central carbon forms, i.e., the carbinol is merely the hydrated form of the unsaturated central carbon. In Figure 5 we have drawn structures A and B in the carbinol and unsaturated central carbon forms, respectively, and it should be understood that both of these structures participate in the equilibrium process described above. It is interesting to note that no methylene carbon resonances are observed in this preparation shortly after synthesis thereby ruling out polymerization by methylene bridges, and thus supporting the oligomeric structure of the type represented by structure B. The lower relative intensities of the "extra" resonances are undoubtedly due to some of the carbons present in structure B, although unequivocal assignment is not possible. Smith (1) has associated the "faded form" of ATA with the carbinol, and as stated above we have assigned the resonance located at 82.0 ppm to represent the carbon atom in equilibrium between carbinol and unsaturated carbon forms. Upon catalytic reduction this resonance disappears, and a new resonance located at 54.8 ppm appears representing a methine carbon ($^{1}J_{CH}$ = 126.0 Hz) further supporting the central carbon assignments made above. On storage (dry and in darkness) for two months new resonances appear at 160.3, 139.3, 65.7, and 35.0 ppm corroborating previous observations that ATA preparations are prone to gradual deterioration (14). The last resonance corresponds to an _ortho-para_ methylene bridge carbon; however, within the accuracy of vapor phase osmometry molecular weight determinations no significant additional polymerization could be detected.

In an attempt to gain further insight into the structural complexity of ATA we synthesized material following the procedure of Heisig and Lauer (39) employing ^{13}C-enriched formaldehyde. Assuming the structure for ATA shown in Figure 1 we expect "tagging" of only the central carbon. Figure 6 depicts the ^{13}C NMR spectrum of the enriched unfractionated reaction mixture. As shown, a diverse number of resonances are observed; it is not known at present to what extent the natural abundance "background" ^{13}C contributes to this spectrum. A minor contribution from this source would imply the occurrence of much complex chemistry. The following assignments were made: 194.0 ppm - benzaldehyde carbon; 117 to 135 ppm - aromatic carbon atoms of salicylic acid bearing directly attached hydrogens; 82.2 ppm - the central carbon atom in equilibrium between carbinol and unsaturated forms; 68.9 ppm - the central saturated carbon atom flanked by _four_ salicylic acid moieties (a value of 69.3 ppm is predicted for the central carbon atom of this structure by employing the chemical shift values for the methylene carbon in _para-para_ MDSA (40.3 ppm) and the methine carbon of the reduced form of ATA (54.8 ppm) and assuming additivity); 54.8 ppm - the methine carbon; 40.4, 35.1 and 29.8 ppm - _para-para_, _ortho-para_ and _ortho-ortho_ linked methylene carbon atoms, respectively. The non-decoupled spectrum is shown in the bottom panel of Figure 6; the observed spin-spin coupling effects are in full accord with the assignments made above. The data are tabulated in Table IV. It is quite clear that the synthetic scheme used to prepare what is thought to be ATA results in a complicated mixture of products. Our experience with these syntheses reveals that the type of product(s) obtained is highly dependent upon the exact reaction conditions. Quite unexpected was the observation of aromatic salicylic acid resonances. If it can be shown that these resonances arise from the enriched formaldehyde used in the synthesis it would imply the occurrence of hitherto

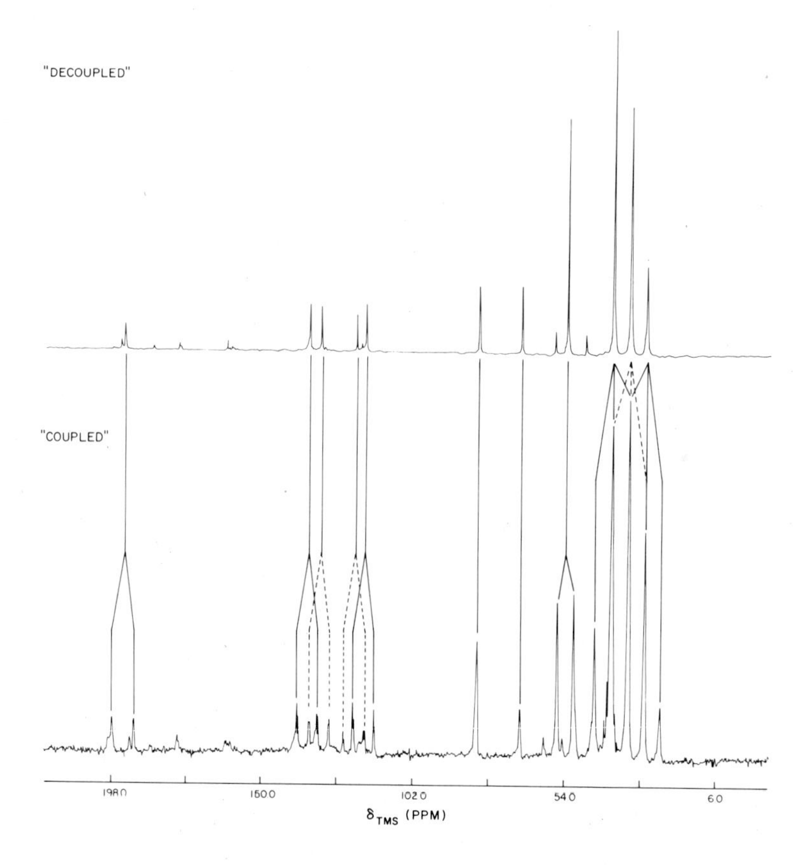

Figure 6. ^{13}C NMR spectra for ATA prepared by the procedure of Heisig and Lauer (39) employing ^{13}C-enriched formaldehyde. (Top Panel) - proton decoupled spectrum. (Bottom Panel) - proton non-decoupled spectrum. NMR spectral parameters are listed in Table IV.

unsuspected novel insertion reactions. The presence of polymerization products in the reaction mixture was investigated by ultrafiltration using an Amicon Diaflow UM-2 membrane (molecular weight cutoff = 1000). As shown in Table IV the only remaining resonances in the retentate are those arising from the central unsaturated carbon atom in equilibrium with the hydrated form, a methine carbon, and the methylene carbons of the three positional isomers. The relative intensity of the _para-para_ methylene carbon resonance decreases significantly after ultrafiltration, suggesting removal of _para-para_ linked species, possibly MDSA or low molecular weight oligomers of ATA interconnected by methylene bridges. Upon catalytic reduction the resonance at 82.2 ppm completely disappears with a concomitant increase in intensity of the methine carbon located at 54.8 ppm. A new resonance appears at 48.4 ppm and probably represents a methine carbon of a methylene positional isomer of ATA.

TABLE IV. ^{13}C NMR spectral parameters for the reaction products of ATA prepared according to ref. (39)

δ (ppm)[a]	$^{1}J_{CH}$ (Hz)	$^{3}J_{CH}$ (Hz)
194.0	179.5	4.6
135.1	161.6	9.0
131.5	160.3	8.5
120.3	164.5	8.9
117.4	161.5	7.4
82.2*	-	n.o.
68.9	-	8.4
58.4	139.5	n.o.
54.8*	127.9	2.7
48.8	masked	masked
40.4*	132.4	n.o.
35.1*	131.1	n.o.
29.8*	129.3	n.o.

[a] Relative to TMS; solvent is H_2O.
n.o. - not observed
* Remains after ultrafiltration; see text.

With this background and insight into the structural complexity of ATA we are able to interpret the natural abundance ^{13}C NMR spectra of SATA ($\bar{M}_n$ = 396), MATA ($\bar{M}_n$ = 1800), and LATA ($\bar{M}_n$ = 6000). These spectra appear in Figure 7. The effect of increasing molecular weight of the three fractions is clearly manifested by the increasing broadening of the resonances. The spectrum for SATA is compatible with a mixture of low molecular weight reactants, by-products, and the virtual absence of ATA in the structural form depicted in Figure 1 (i.e., the absence of a resonance at 82.2 ppm). Reactants and by-products include salicylic acid, isomeric MDSA species, benzaldehyde functional groups, and possibly formaurindicarboxylic acid. Comparing the spectrum of SATA with those obtained for MATA and LATA we note a consistency in the aromatic region (with the exception of fine structure that is smeared out due to line broadening), the clear emergence of resonances at 82, 69, 55, 49, 35, and 30 ppm which were also observed in our ^{13}C enriched preparation and assigned to specific carbon atoms (see above). More recent spectra of LATA with improved signal to noise clearly reveal the presence of all the central carbon resonances observed in MATA. The lack of significant resonances in the central carbon spectral region implies that "monomeric" ATA does not exist, at least in this preparation.

The results obtained in this study clearly argue for a revision in the commonly accepted structure for ATA (see Figure 1). We propose that ATA is best represented in terms of a heterogeneous population of polymeric structures, i.e., in terms of size, type of central carbon

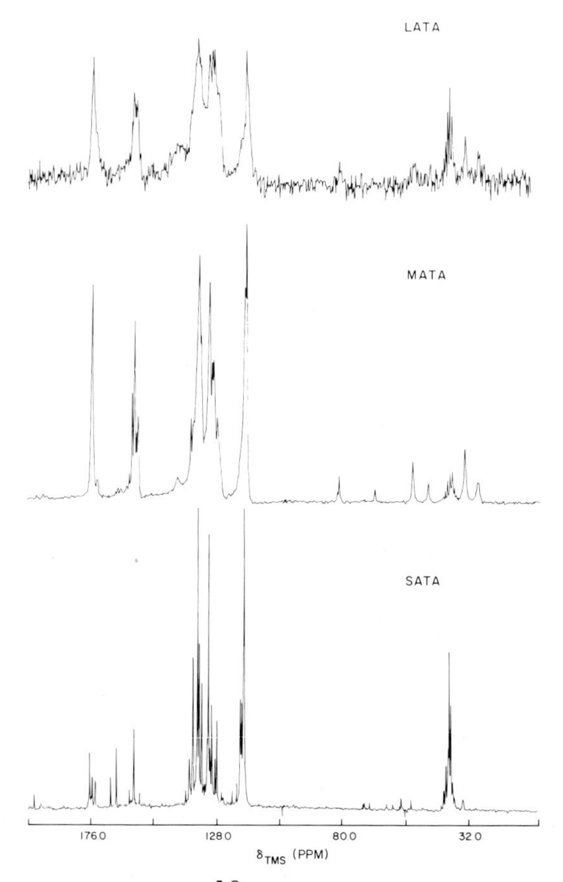

Figure 7. Natural abundance ^{13}C NMR spectra for SATA ($\bar{M}_n$ = 396), MATA ($\bar{M}_n$ = 1800), and LATA ($\bar{M}_n$ = 6000). These fractions were obtained by dialysis and ultrafiltration as described in the text from a sample of "Aluminon" grade ATA supplied by Fisher Sci. Co. (lot no. 752556). Other samples of commercially available ATA gave similar results. The sharp multiplet at 40 ppm is from an external reference of DMSO.

atom present in a particular residue, and placement of methylene linkages on the aromatic salicylic acid moiety. A schematic structure is illustrated in Figure 8. It is possible that forms such as structure B (see Figure 5) also contribute to the polymeric structure, although we do not have direct evidence for this. Furthermore, the presence of "monomeric" ATA structures in commercial preparations of this dye is doubtful.

Our findings also help to explain the variable activity often observed in different ATA preparations. We have demonstrated that the inhibitory activity of an ATA preparation depends on the degree of polymerization, thus a different blend of polymeric sizes would result in altered activity. Furthermore, the inhibitory activity of ATA depends on the integrity of the central unsaturated carbon atom (14),

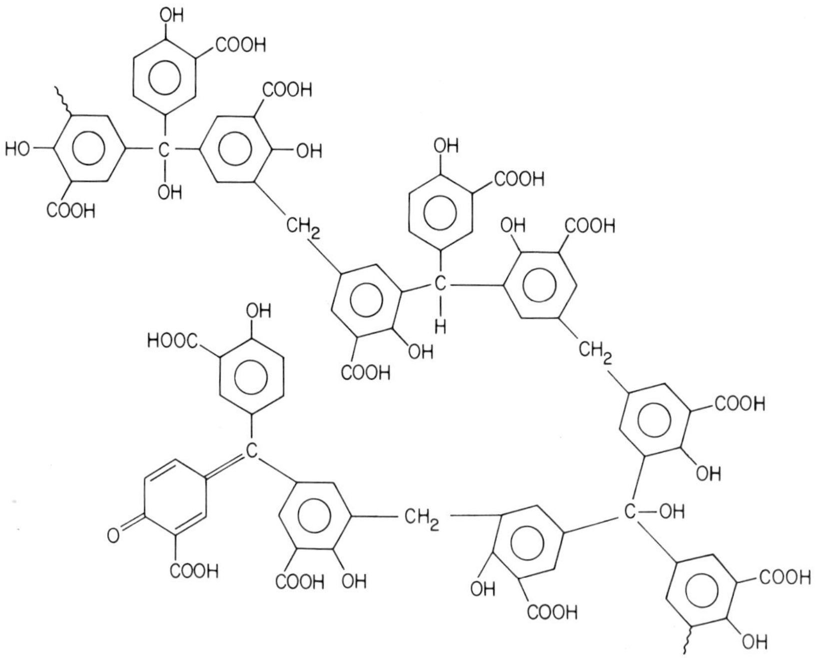

Figure 8. Proposed schematic structure for ATA.

and thus the percentage of central carbon atoms present in the methine form in a given fraction would also serve to define the inhibitory activity of a preparation. The results of these studies also illustrate the need for cautious and judicious interpretation of ATA inhibitory effects.

SUMMARY

Commercially available as well as synthetically prepared samples of ATA consist mostly of a heterogeneous collection of polymers as revealed by fractionation schemes utilizing both dialysis and ultrafiltration, and by molecular weight measurements. Carbon-13 NMR studies suggest that the polymeric material is of the phenol-formaldehyde type, and inhibitory assays that depend on the formation of a protein-nucleic acid complex reveal that potency varies directly with the molecular weight of the polymer. ATA fractions of molecular weight 400 are essentially inactive.

ACKNOWLEDGEMENT

This research was supported by grants from the National Science Foundation (BMS 75-17114), the National Institutes of Health (1 R01 GM 23951-01), and the Faculty Research Committee of the University of

California, Santa Cruz. R.G.G. was supported by a Minority Biomedical Support Grant awarded to the University of California, Santa Cruz, by the National Institutes of Health. We also wish to thank Dr. G. Matson of the University of California, San Francisco, Magnetic Resonance Laboratory for providing help and spectrometer time. This facility is supported by grant No. RR 00892-01A1 from the Division of Research Resources, National Institutes of Health. The FX-60 NMR spectrometer was purchased with funds awarded to the Chemistry Board of Studies at UCSC by the National Science Foundation (CHE 77-08911). B.J.B. was on sabbatical leave from the University of Winnipeg. We also wish to thank J. Maciulis and D. Bear for their help with the inhibitory activity assays.

REFERENCES

1 W. H. Smith, E. E. Sager and I. J. Siewers (1949) Anal. Chem., 21:1334.
2 A. P. Grollman (1968) in: G. L. Hobby (ed.), Antimicrobial Agents and Chemotherapy, Amer. Soc. Microbiol., Bethesda, p. 36.
3 A. P. Grollman and M. L. Stewart (1968) Proc. Nat. Acad. Sci. (U.S.A.), 61:719.
4 S. Pestka (1969) ibid., 64:709.
5 R. E. Webster and N. D. Zinder (1969) J. Mol. Biol., 42:425.
6 J. M. Wilhelm and R. Haselkorn (1970) Proc. Natl. Acad. Sci. (U.S.A.), 65:388.
7 F. Siegelman and D. Apirion (1971) J. Bact., 105:902.
8 M. L. Stewart, A. P. Grollman and M-T. Huang (1971) Proc. Natl. Acad. Sci. (U.S.A), 68:97.
9 B. Leblew, G. Marbaix, J. Werenne, A. Burny, and G. Huez (1970) Biochem. Biophys. Res. Comm., 40:731.
10 A. Marcus, J. D. Bewley and D. P. Weeks (1970) Science, 167:1735.
11 E. Baṭtaner and D. Vazquez (1971) Biochim. Biophys. Acta, 254:316.
12 W. Hoerz and K. S. McCarty (1971) ibid., 228:526.
13 M. B. Mathews (1971) FEBS Lett., 15:201.
14 M-T. Huang and A. P. Grollman (1972) Mol. Pharm., 8:111.
15 D. P. Leader (1972) FEBS Lett., 22:245.
16 C. S. Wang, R. B. Naso and R. B. Arlinghouse (1972) Biochem. Biophys. Res. Comm., 47:1290.
17 M. Ayuso-Parilla, C. A. Hirsch and E. C. Henshaw (1973) J. Biol. Chem., 248:4394.
18 G. I. Dettman and W. M. Stanley, Jr. (1973) Biochim. Biophys. Acta, 299:142.
19 C. Baglioni, M. Jacobs-Lorena and H. Meade (1974) ibid., 277:188.
20 N. Borgese, W. Mok, G. Kreibich and D. Sabatini (1974) J. Mol. Biol., 88:559.
21 J. M. Whitelam and H. Naora (1974) Biochim. Biophys. Acta, 349:178.
22 S. J. Igarashi and J. A. Zmean (1975) Can. J. Biochem., 53:120.
23 S. J. Igarashi and J. A. Zmean (1975) ibid., 53:124.
24 M. H. Sarma, E. R. Feman and C. Baglioni (1976) Biochim. Biophys. Acta, 418:29.

25 D. Apirion and D. Dohner (1975) in: J. W. Cocovan and F. E. Hahn (eds.), Antibiotics, III, Springer, New York, p. 327.
26 A. P. Grollman and M-T. Huang (1976) in: E. H. McConkey (ed.), Protein Synthesis, II, Marcel Dekker, New York, p. 125.
27 L. Beres and J. Lucas-Lenard (1975) Biochim. Biophys. Acta, 395:80.
28 J. F. Givens and K. F. Manley (1976) Nucleic Acids Res., 3:405.
29 D. L. Steward, J. Martin and A. P. Grollman (1977) in: 3rd Conf. Antiviral Substances, 284:638.
30 M. Kuwano, D. Schlessinger and D. Apirion (1970) Nature, 226:514.
31 T. Blumenthal and T. A. Landers (1973) Biochem. Biophys. Res. Comm., 55:680.
32 L-L. Liao, S. B. Horwitz, M-T. Huang, A. P. Grollman, D. Steward and J. Martin (1975) J. Med. Chem., 18:117.
33 H. Weissbach and N. Brot (1970) Biochem. Biophys. Res. Comm., 39:1194.
34 M. Tal, M. Aviram, A. Kanarek and A. Weiss (1972) Biochim. Biophys. Acta, 281:381.
35 A. Lindenbaum and J. Schubert (1956) J. Phys. Chem., 60:1663.
36 M. Bina-Stein and T. R. Tritton (1976) Mol. Pharm., 12:191.
37 T. Hultin and H. Nika (1975) FEBS Lett., 51:184.
38 K. Tsutsui, S. Seki, K. Tsutsui and T. Oda (1978) Biochim. Biophys. Acta, 517:14.
39 G. B. Heisig and W. M. Lauer (1928) Org. Syn., 9:8.
40 G. C. Carmichael (1975) J. Biol. Chem., 250:6160.
41 M. Smolarsky and M. Tal (1970) Biochim. Biophys. Acta, 199:447.
42 W. Szer, J. M. Hermoso and S. Leffler (1975) Proc. Natl. Acad. Sci. (U.S.A.), 72:2325.
43 L. F. Johnson and W. C. Jankowski (1972) spectrum no. 231 in: Carbon-13 NMR Spectra, John Wiley and Sons, New York.
44 W. E. Smith (1972) unpublished results quoted in: G. C. Levy and G. L. Nelson, Carbon-13 Nuclear Magnetic Resonance Spectroscopy for Organic Chemists, Wiley-Interscience, New York, p. 84.
45 F. J. Weigert and J. D. Roberts (1967) J. Amer. Chem. Soc., 89: 2967.

MAGIC ANGLE SAMPLE SPINNING IN INHOMOGENEOUSLY BROADENED SPIN SYSTEMS-- APPLICATION TO UNSONICATED PHOSPHOLIPID BILAYERS.

R.A. Haberkorn*, J. Herzfeld[P], and R.G. Griffin*
Francis Bitter National Magnet Laboratory*[†], Massachusetts Institute of Technology, Cambridge, MA 02139 and Biophysical Laboratory[P], Harvard Medical School, Boston, MA 02115, USA

I. Introduction

Nuclear magnetic resonance has been used extensively in the last decade to study the structure and dynamics of model and biological membranes.[1] However, the complexity of these systems, which should manifest itself in a corresponding richness of their NMR spectra, has in most cases not been observed because of the substantial breadth of the NMR lines. It is now understood that this breadth is due primarily to residual chemical shift anisotropy and dipole-dipole interactions. For dilute spins, such as ^{13}C and ^{31}P, the dipolar broadening can be removed by sufficiently intense rf irradiation at the proton resonance frequency.[2-4] Nevertheless, a substantial broadening due to the anisotropy of the chemical shift remains. In order to obtain "high resolution" NMR spectra, it has become customary to subject multilamellar dispersions to prolonged ultrasonic irradiation.[5] This process, which results in particles of reduced size with reduced reorientational correlation times, does indeed improve the resolution of the NMR spectra; however, its exact physical and chemical consequences are a subject of much debate.[6] We describe below a method whereby high resolution spectra can be obtained without resorting to sonication.

II. The Technique

Almost two decades ago Lowe[7] and Andrew et al.[8] suggested that rapid spinning of a sample about an axis inclined at the "magic angle", 54° 44', with respect to H_0 would suffice to narrow dipolar broadened NMR lines. The criterion to achieve complete narrowing is that the rotation rate, ν_R, must be greater than the dipolar line width.[9] For this reason ^1H spinning experiments on phospholipids have not been very successful.[10] With the development of pulse methods for obtaining high resolution NMR spectra in solids,[11-14] there has been renewed interest in this technique because chemical shift tensors, like dipolar tensors, are of rank two, and can be averaged to their trace by magic angle spinning. Recently Schaefer and Stejskal[15] demonstrated that high resolution ^{13}C spectra of several polymeric samples could be obtained by combining dilute spin double resonance and sample spinning.[16] However, it was thought that ν_R must be greater than the breadth of the shift

B. Pullman (ed.), Nuclear Magnetic Resonance Spectroscopy in Molecular Biology, 381-391.

powder spectrum, $\Delta\sigma$, in order to achieve substantial narrowing. Thus, it appeared that this approach would be of utility only at relatively low magnetic fields--i.e., < 2.0 T.

Recently we, and others,[17,18,19] have noticed that for inhomogeneous interactions such as the chemical shift, the condition $\nu_R > \Delta\sigma$ need not be fulfilled in order to achieve a significant degree of spectral narrowing. Specifically, magic angle sample spinning (MASS) will narrow chemical shift powder patterns even when $\nu_R < \Delta\sigma$, as can be seen by an examination of Fig. 1. Shown in this figure are proton decoupled ^{31}P spectra of barium diethyl phosphate, $Ba[C_2H_5PO_4]_2$ (BDEP), as a function ν_R. At $\nu_R = 0$ we obtain an axially asymmetric powder spectrum of ~190 ppm breadth (Fig. 1a) which amounts to ~23kHz at our 6.8T field.[4] Upon spinning the sample at 0.94 kHz (Fig. 1b), we observe that the spectrum breaks up into a series of sidebands spaced at the spinning frequency. At our maximum ν_R (2.92 kHz) the linewidths are 0.5 ppm and we believe they are limited by imperfect alignment of the spinning axis with the magic angle. Thus, with this technique we can effect a dramatic narrowing of a powder spectrum. Maricq and Waugh[18] have explained this phenomenon in terms of coherent averaging theory.[11,12] Briefly, it happens that for the case of an anisotropic chemical shift the lowest order average Hamiltonian and all higher order corrections to it vanish identically. Thus, the linewidths in the spectra, such as Fig. 1, should be limited by inhomogeneous broadening not removed by the sample rotation and by homogeneous broadening, arising from incomplete decoupling, molecular motion, etc.

Figure 2, which is a series of ^{13}C spectra of glycine 90% ^{13}C enriched at the carboxyl carbon, illustrates another interesting effect. Figure 2a was obtained at $\nu_R = 0$ and shows what we believe are two overlapping shift tensor powder patterns, but, at high ν_R we find doublets instead of single lines. One might attribute these doublets to intermolecular dipolar coupling which is not removed by sample rotation; however, the distances in the glycine lattice appear to be too large to make this explanation plausible. Furthermore, the $-CO_2$ centerband in natural abundance glycine spectra is coincident with one of, rather than between, the lines of the doublet. Thus, we believe the structure in Fig. 2d. arises from glycine molecules with different isotropic shifts. The final step in the synthesis of the ^{13}C enriched samples is recrystallization from ethanol and this is also the method for preparation of β glycine. Thus, we believe the structure in Fig. 2d. arises from the presence of, for example, the α,β and/or γ crystallographic forms of this molecule.

As might be expected, the sideband intensities in this "slow spinning" regime are related to the ratio of the shift anisotropy to the spinning rate. And, as illustrated above for the case of glycine, if two ^{13}C, ^{31}P, etc. atoms exhibit over-

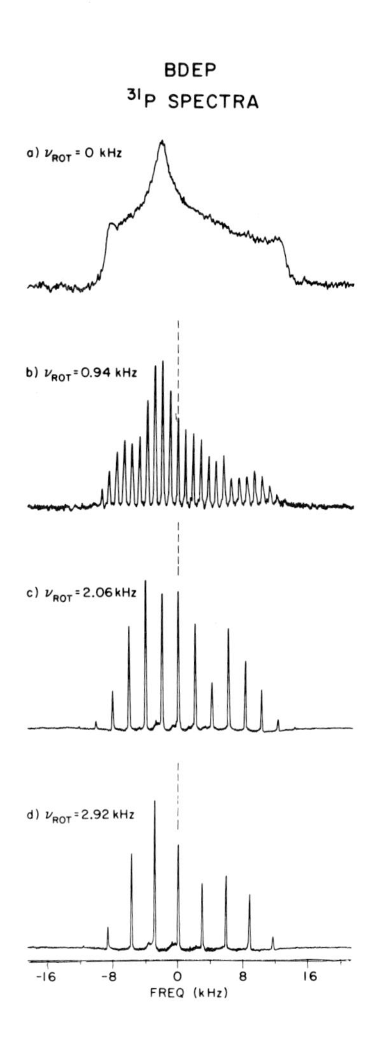

FIGURE 1: Effect of sample rotation on the proton decoupled ^{31}P spectrum of BDEP. The powder pattern in (a) is 23 kHz wide and narrows to a spectrum exhibiting sidebands each of 60 Hz full width at ν_r=2.92 kHz. The dashed line marks the isotopic shift. ν_{31P}=119.05 MHz.

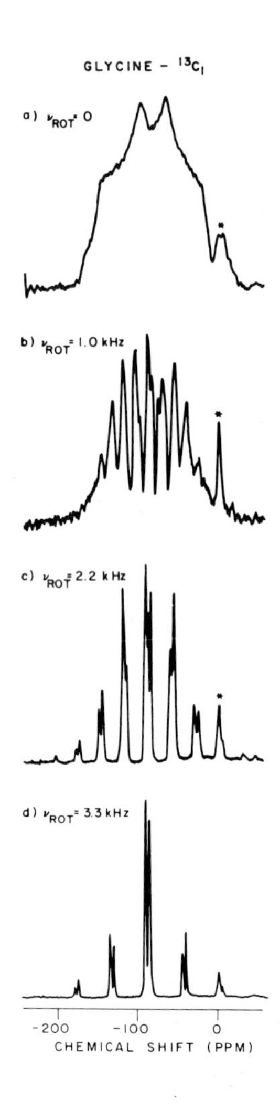

FIGURE 2: Effect of sample rotation on the ^{13}C NMR spectrum of glycine-$^{13}C_1$ (90% enriched). The line marked with an asterisk arises from the ^{13}C nuclei in the plastic rotor. ν_{13_C} = 73.966 MHz.

lapping powder spectra, but possess different isotropic chemical shifts, then upon spinning the center and side band spectra will be resolved. This would allow measurement of the principal values of the shift anisotropies without resorting to a single crystal study. Lippma and co-workers[7] have calculated the sideband intensities for an axially symmetric shift tensor in the limit that $\beta = \nu_o\delta/2\nu_r<1$, where ν_o is the resonance frequency, $\delta = \sigma_{zz}-\frac{1}{3}\,Tr\tilde{\sigma}$, and ν_r is the rotation frequency. We are currently working to extend these results to account for axially asymmetric shift tensors and higher order sidebands such as those we observe. In summary, our work indicates MASS together with dilute spin double resonance is an effective means for obtaining high resolution NMR spectra of a number of different nuclear spins in solids. Contrary to popular beliefs, it is effective at high as well as low magnetic fields. We now discuss an application of the technique to biophysics; namely, to the study of model membranes.

III. Application to Model and Biological Membranes

As mentioned above, early NMR studies of unsonicated model and biological membranes established that these systems do not yield high resolution spectra. It is now understood that the source of this broadening is the slow tumbling of these macromolecular aggregates which only partially averages the chemical shift, dipolar, and quadrupolar interactions. Although it has been shown that with proper decoupling, proton dipolar broadening can be eliminated[2], one still must, in the case of ^{31}P and ^{13}C spectra, contend with the residual shift anisotropy, [3,4]. Thus, despite its drawbacks, sonication has been necessary when high resolution spectra are desired. The MASS technique described above allows us to circumvent the need for studying sonicated species.

Figure 3 shows ^{31}P spectra of dipalmitoylphosphatidylcholine (DPPC) dispersed in excess water ($\geq$50 wt % H_2O). Here motional averaging narrows the rigid lattice, axially asymmetic ^{31}P spectrum of $\sim$190 ppm breadth[11] to an axially symmetric one of $\sim$60 ppm, which amounts to 7.2 kHz at our ^{31}P frequency of 119.05 MHz.[3] Upon spinning this sample in an Andrew[20] type rotor at 2.6 kHz, we observe a single line of 95 Hz full width at half height flanked by two sets of sidebands. This linewidth is essentially identical to that observed in sonicated DPPC vesicles at roughly an equivalent magnetic field and it may be determined by shift anisotropy relaxation effects.[21]

The effect of sample spinning is more dramatically illustrated by the proton decoupled ^{13}C spectra of dimyristoylphosphatidylcholine (DMPC) and DPPC, dispersed in excess water, displayed in Fig. 4. Figure 4a shows a spectrum obtained with a stationary sample of DMPC and again the rigid lattice ^{13}C shift tensors are partially averaged to the point that assignment of bands to the carbonyl carbons, the $-N(CH_3)_3$, glycerol backbone, and head group $-CH_2-$carbons, and the acyl chain carbons can be accomplished; but, there remains a considerable amount of shift anisotropy broadening. Upon spinning, the spectral bands narrow dramatically as is shown in Fig. 4b and at a spinning rate of 2.6 kHz, we resolve all of the lines that are observed when the lipid is dissolved

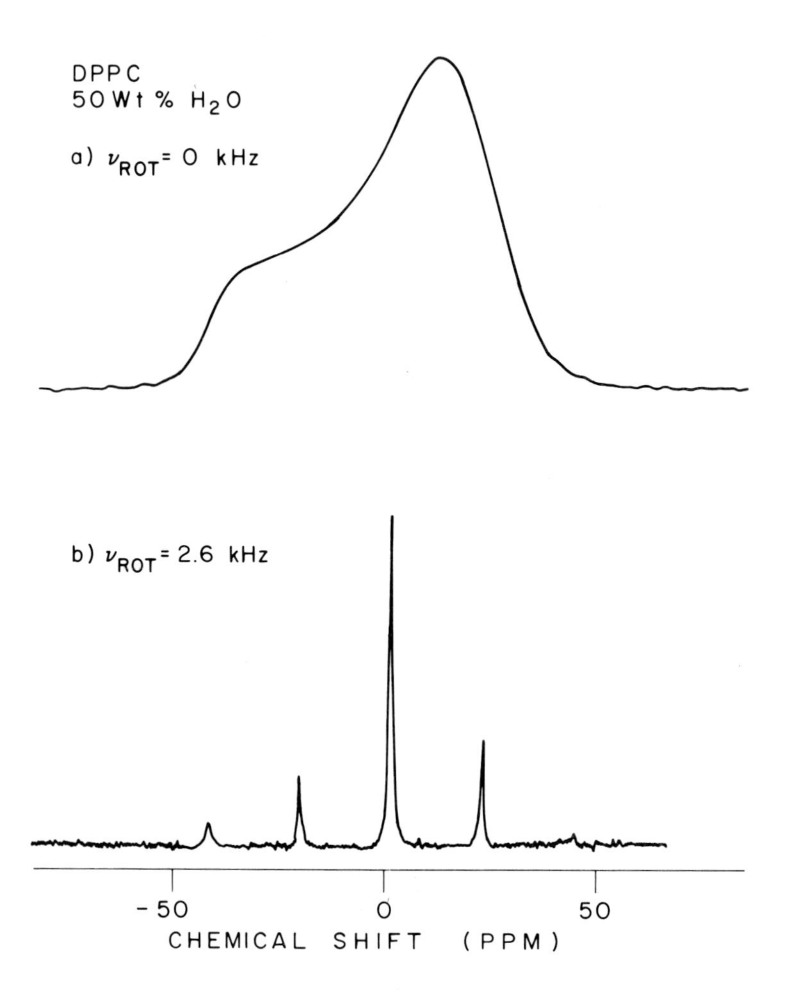

FIGURE 3: Proton decoupled ^{31}P spectra of DPPC in 50 wt % H_2O (T=21°C) (a) axially symmetric powder spectrum, $\Delta\sigma$=-60 ppm (7.2 kHz), ν_{rot}=0; (b) ν_{rot}=2.6 kHz. In (b) the side bands are spaced at the spinning frequency and the full width of the centerband is 95 Hz. Shifts are referenced to external 85% H_3PO_4.

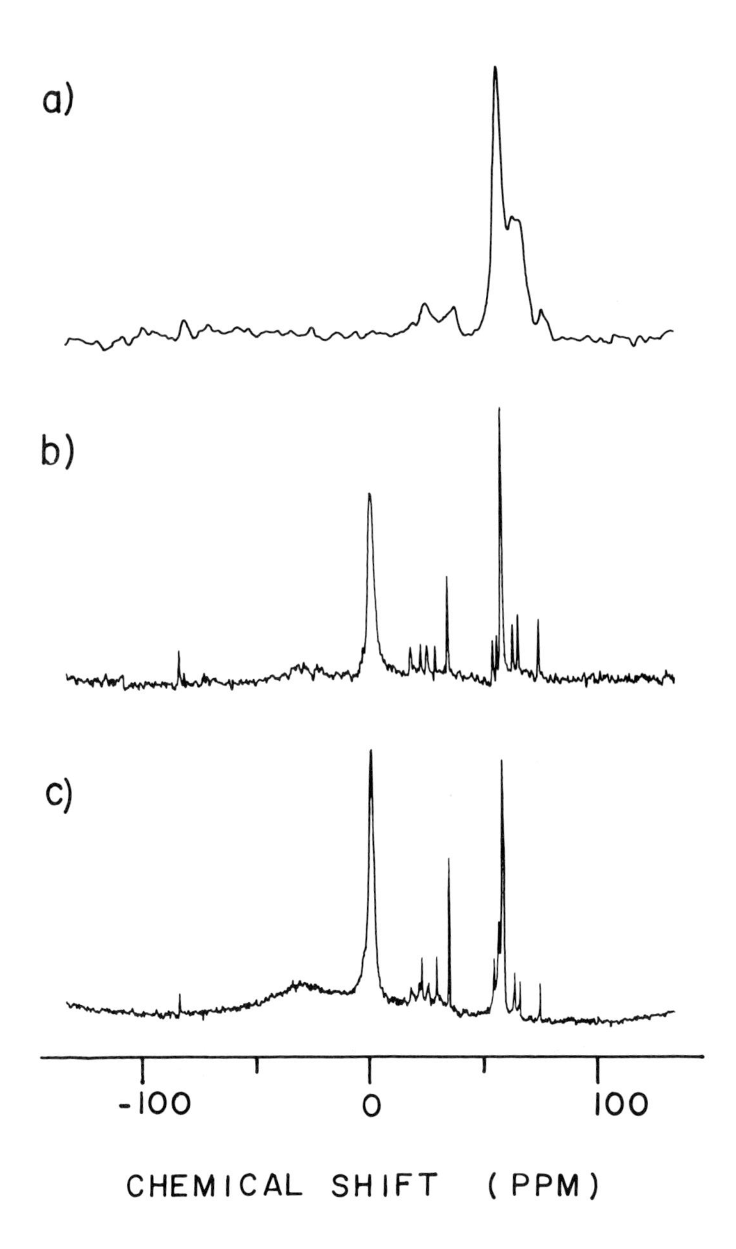

FIGURE 4: Proton decoupled ^{13}C spectra of DMPC and DPPC in 50 wt % H_2O (T=21°C); (a) DMPC, $\nu_{rot}=0$; (b) DMPC, $\nu_{rot}=2.6$ kHz; (c) DPPC, $\nu_{rot}=2.6$ kHz. Line assignments are based on those given in ref. 23 and 26 and shifts are referenced to external delrin. Full widths of the single carbon resonances are 20 Hz. The broad line at -30 ppm also arises from the delrin rotor.

in $CHCl_3$.[22,23] The assignment of the lines is given in Table 1 referenced to the strong line in the center of our spectrum which arises from the ^{13}C nuclei in our plastic (Delrin) rotor. Since we have observed the broad line spectra of Fig. 4a both prior to and following spinning, it appears that sample rotation at 2.5 kHz does not degrade the structure of the liposomes.

Based on a measurement of the temperature of the gas employed to drive the spinner we believe the sample temperature is close to 21°C. For DMPC the pretransition temperature, T_p, is 14°C and the gel-liquid crystalline transition temperature, T_c, is 24°C, so we believe the lipid is in its monoclinic phase.[24,25] However, if there were a slight amount of heating due to ^1H decoupling, then the spectrum of Fig. 4 b could be due to the liquid crystalline phase. In order to obtain a gel phase spectrum, we have studied DPPC which has T_p = 35°C and T_c=41°C and its spectrum is shown in Fig. 4c; the chemical shifts are also given in Table 1. A comparison of Figures 4b and 4c shows that the acyl chain and carbonyl regions of the spectra are rather similar, but the glycerol and choline regions are somewhat different. Specifically, in the DPPC spectrum the α and β choline carbon lines sharpen slightly and all three glycerol carbon lines are now resolved. On the basis of DPPC spectra obtained in methanol,[26] we assign the glycerol lines, in order of decreasing shielding, to C1, C3 and C2, respectively. The fact that the three glycerol lines are resolved in DPPC, but not in DMPC, suggests the possibility of slightly different conformations in the glycerol segment below and above T_c.

The ^{13}C spectra in Figures 4b and 4c do not exhibit rotational sidebands, in spite of the fact that they were obtained at a rather high magnetic field. This can be understood by noting that the apparently large breadths of the powder spectra of Fig. 4a are due in part to overlap of residual shift anisotropy patterns. In fact the powder spectra from individual carbons are probably $<$1 kHz wide so that we are in the limit where the narrowing is complete. This also suggests that ^{13}C lipid spectra which are free of rotational sidebands could be obtained at 2-3 times higher magnetic fields with currently available rotor designs. On the other hand, information on the residual shift anisotropy should be available by decreasing the spinning rate to the point where sidebands appear. This is an appealing approach to studying these parameters because the lines will be sharp and consequently, the problem of signal overlap can be partially circumvented.

IV. Acknowledgements

This work was supported in part by the National Institutes of Health through Grants GM 23316 and GM 23289 and by the National Science Foundation through contract C-670 to the Francis Bitter National Magnet Laboratory.

TABLE 1. ^{13}C Chemical shifts for unsonicated DMPC and DPPC in excess water at T = 21°C with respect to external Delrin reference[a].

Carbon position	Chemical shifts, ppm DMPC	DPPC
Acyl chain		
C-1 (C=O)	-84.9	-84.6
C-2	54.4	54.7
C-3	64.0	63.7
C-4-C-11 (13)	58.0	58.3
C-12 (14)	56.1	56.6
C-13 (15)	65.6	66.0
C-14 (16)	74.5	75.0
Glycerol		
C-1 (CH_2OP)	25.2	25.4
C-3 (CH_2OCOR)	25.2	21.5
C-2 (CHOCOR)	17.7	17.9
Choline		
CH_2OP	28.8	29.3
CH_2N	22.3	22.7
$N(CH_3)_3$	34.3	34.6

[a]In the acyl chains C=O and $-CH_3$ are labeled C-1 and C-16, respectively, and in the glycerol C-1 and C-3 are esterified with the acyl chain and PO_4, respectively. The numbers in parentheses refer to the carbon in the DPPC acyl chains.

V. References

1. For reviews see for instance A.G. Lee, N.J.M. Birdsall and J.C. Metcalfe, "Methods of Membrane Biology", Vol. II, E.D. Korn, Ed., Plenum Press, New York, NY, 1974, pp. 1-156; J. Seelig, Quart. Rev. Biophysics, 10, 353 (1977).

2. J. Urbina and J.S. Waugh, Proc. Natl. Acad. Sci. USA, 71, 5062 (1974).

3. R.G. Griffin, J. Amer. Chem. Soc., 98, 851 (1976).

4. S.J. Kohler and M.P. Klein, Biochemistry, 15, 967 (1976); J. Herzfeld, R.G. Griffin, and R.A. Haberkorn, ibid, (in press).

5. C.H. Huang, Biochemistry, 8, 344 (1969).

6. For recent references see for instance, N.O. Petersen and S.I. Chan, Biochemistry, 16, 2657 (1977); G.W. Stockton, C.F. Polnaszek, A.P. Tulloch, F. Horsan, and I.C.P. Smith, ibid, 15, 954 (1976).

7. I.J. Lowe, Phys. Rev. Letters 2, 285 (1959).

8. E.R. Andrew, A. Bradbury, and R.G. Eades, Nature (Lond) 182, 1659, (1958).

9. U. Haeberlen and J.S. Waugh, Phys. Rev. 185, 420 (1969).

10. D. Chapman, E. Oldfield, D. Doskocilova, and B. Schneider, FEBS Letters, 25, 261 (1972).

11. U. Haeberlen, "High Resolution NMR in Solids - Selective Averaging", Adv. in Magnetic Resonance, Supplement 1, J.S. Waugh, Ed., Academic Press, New York, NY (1976).

12. M. Mehring, "High Resolution NMR Spectroscopy in Solids", Springer-Verlag, New York, NY (1976).

13. J.S. Waugh, L.M. Huber, and U. Haeberlen, Phys. Rev. Letters, 20, 180 (1968).

14. A. Pines, M.G. Gibby, and J.S. Waugh, J. Chem. Phys. 59, 569 (1973).

15. J. Schaefer and E.O. Stejskal, J. Amer. Chem. Soc., 98, 1031 (1976).

16. Combined multiple pulse and sample spinning experiments have also been reported--B.C. Gerstein, R.B. Pembleton, R.C. Wilson, and L.M. Ryan, J. Chem. Phys. 66, 361 (1977).

17. E. Lippmaa, M. Alla, and T. Tuherm, Proc. Congr. Ampere, 19th, 1976, 113-118 (1976).

18. M. Maricq and J.S. Waugh, Chem. Phys. Lett., 47, 327-329 (1977).

19. E.O. Stejskal, J. Schaefer, and R.A. McKay, J. Magn. Reson., 25, 569-573 (1977).

20. E.R. Andrew, L.F. Farnell, M. Firth, T.D. Gledhill, and I. Roberts, J. Magn. Reson., 1, 27-34 (1964).

21. J.A. Berden, P.R. Cullis, D.I. Hoult, A.C. McLaughlin, G.K. Radda, and R.E. Richards, FEBS Letters, 46, 55-58 (1974).

22. The ^{13}C spectrum of DMPC in $CHCl_3$ is essentially identical with Figure 2b so it is not included here.

23. Y.K. Levine, N.J.M. Birdsall, A.G. Lee, and J.C. Metcalfe, Biochemistry, 11, 1416-1421 (1972); E.Oldfield and D. Chapman, Biophys. Biochem. Res. Comm. 43, 949-953 (1971).

24. M. Janiak, D.M. Small, and G.G. Shipley, Biochemistry, 15, 4575-4580 (1976).

25. E.J. Luna and H.M. McConnell, Biochem. Biophys. Acta 466, 381-392 (1977).

26. N.J.M. Birdsall, J. Feeney, A.G. Lee, Y.K. Levine and J.C. Metcalfe, J. Chem. Soc., Perkin Trans. 2, 1441 (1972).

‡ Supported by the National Science Foundation

IONIC PERMEABILITIES OF MEMBRANES; NMR KINETIC STUDIES

Hadassa Degani
Isotope Department, Weizmann Institute of Science, Rehovot, Israel

INTRODUCTION

Ionic concentration gradients across membranes serve a variety of functions in cellular activity such as maintainance of resting potential, transmission of nerve impulse, driving cotransport, and secretion or activation of enzymes, hormones and other substances. Thus, studies of the kinetics and mechanism of ion transport across membranes are essential for the understanding of many physiological processes.

In this report, the application of NMR spectroscopy for studying ionic flows across membranes is presented. NMR has two inherent advantages over the other methods used to study transport processes (mainly the isotope tracing techniques). 1) It is applicable for both fast and slow kinetics with half lives times ranging between milliseconds to hours. 2) The transport process can be followed while occuring without destructing or perturbing the system. We shall concentrate on describing the methodology by which the transport measurements are carried out. Also, several examples will be described concerning ionophore mediated ion translocation through lipid bilayers.

A. NMR MEASUREMENTS OF IONIC FLUXES ACROSS VESICULAR MEMBRANES

This method is based on following, in real time, the flux of ions into (or out of) a compartment by monitoring the increase (or decrease) in the signal intensity of the transported species or nuclei affected by it. Such measurements are applicable for rates of half lives ranging between seconds and hours.

In the following example we have monitored changes with time in the intensity of the NMR signal of the membrane (^{1}H and ^{31}P) affected by the transport of paramagnetic ions.

Polar head groups of phospholipids arranged in a vesicular bilayer are pointing either toward the outer bulk or the intravesicular medium.

B. Pullman (ed.), Nuclear Magnetic Resonance Spectroscopy in Molecular Biology, 393-403.

The difference between the outer and inner polar moeities is revealed in the ^1H and ^{31}P NMR spectra of phosphatidylcholine (PC) vesicles by a shift of the resonances belonging to the inner groups, to a higher field. This is shown in Fig. 1A for vesicles composed of egg PC and dipalmitoyl phosphatidic acid (9:1). For both lipids the inner phosphates are separated from the outer ones. When paramagnetic ions are added to a vesicle suspension they interact with the polar head groups on the outer surface and can cause a broadening and/or a shift of the NMR signals due to the outer nuclei. For instance, in Fig. 1B the addition of 1mM $CoCl_2$ to the vesicle suspension caused a marked broadening of the outer phosphate signals belonging to the choline and

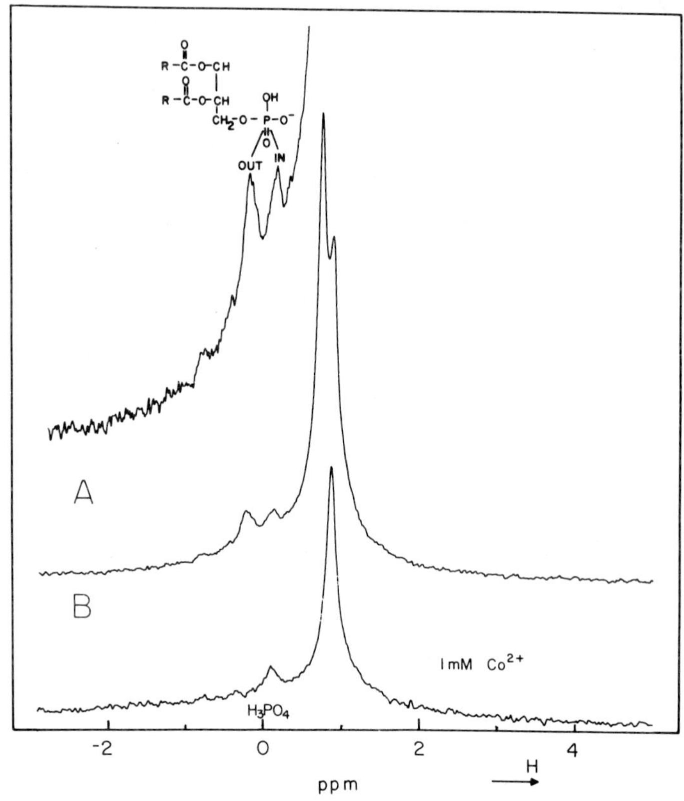

Fig. 1. 109 MH_z ^{31}P-^1H decoupled (5W) spectra of lipid vesicles (5%) in ^2H_2O, p^2H=6.9 at 32°C. Vesicles were prepared by sonicating egg phosphatidylcholine (Makor) and L-α-phosphatidic acid dipalmitoyl (Sigma) at 9:1 weight ratio. Shifts were referenced to external 8% phosphoric acid in ^2H_2O.

phosphatidic acid moeities. It should be pointed out that the presence of the negative phosphatidic acid enhances the binding of Co^{2+} to the bilayer surface thus causing a much larger broadening of the choline phosphate than that observed in pure phosphatidylcholine. The inner phosphate signals shown in this figure remain sharp without change for several hours indicating that the membrane is impermeable to cobaltous ions through the time range of our studies. However, upon addition of the antibiotic X-537A which acts as an ionophore for divalent ions (1) the membrane became permeable to the Co^{2+} ions. The ionophore mediated flow of the cobaltous ions into the inner vesicular medium induced a broadening of the signal due to the interacting inner phosphates and caused a decrease in the intensity of this signal (Fig.2).The

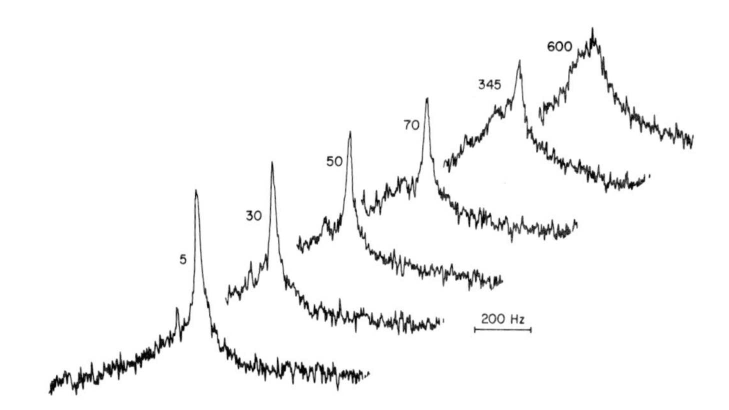

Fig. 2. ^{31}P spectra of lipid vesicles (The same as in Fig. 1) containing 50 μM X-537A and 1.25mM $CoCl_2$ added after sonication. The inserted numbers refer to the time in minutes after the addition of the $CoCl_2$ to the outer medium. 60 transients were accumulated with a recycle time of 1.0 sec.

rates of Co^{2+} and Mn^{2+} transport across the vesicles membrane has been determined by following the changes with time of the intensity of the inner phosphate signal. Similar X-537A mediated transport of manganous ions has been performed by monitoring with time changes in the intensity of the proton signal of the inner methyl cholines as can be seen in Fig. 3.

Using this method ionophore mediated transport rates have been studied as a function of ionophore concentration (2,3) pH (2) and temperature (2,3) (Table 1) At an ionophore to vesicles concentration ratio higher than two (for a constant concentration of Mn^{2+} of lmM) the transport rate depends upon the square power of the X-537A concentration thus indicating that the dominant transporting species is a 1:2 Mn:X-537A complex. However, at ionophore to vesicles concentrations ratio below two, the concentration dependence of the transport rate changes. The plot of the logarithmus of the rate versus the logarithmus of the ionophore concentration has a slope of 1.2 ± 0.2 revealing that the 1:1 Mn^{2+}:X-537A complex is the dominant transporting species at this concentration ratio.

The pH dependence of the transport rates indicates

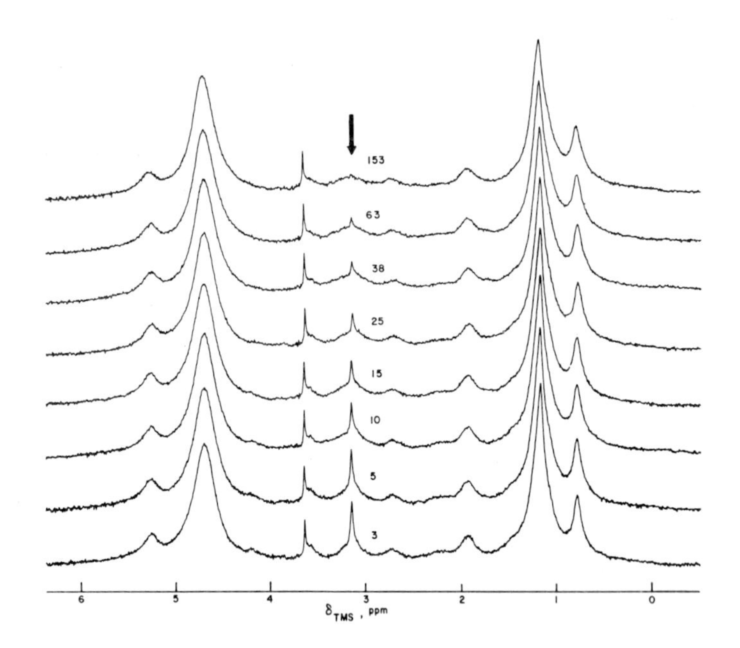

Fig. 3. 270 MH_Z 'H spectra of egg phosphatidylcholine vesicles (5%) in ^2H_2O p^2H=6.7 containing 30μM X-537A and 1mM $McCl_2$ added after sonication. Temp. 28°C. *

Table 1. X-537A concentration dependence of Mn^{2+} transport rates across L-α-phosphatidylcholine dipalmitoyl vesicles p^2H=6.8 at 43.8°C

[X-537A] μM	[X-537A] / [vesicles] **	half life time min.
3.8	0.15	660
9.8	0.4	260
15	0.6	120
36	1.4	38
71	2.8	18.5
142	5.7	5.5
214	8.6	2.0

* These results were obtained using a 360 MH_Z spectrometer in collaboration with A.C. McLaughlin (Brookhaven) and S.R. Simon (SUNY at Stony Brook).

** This ratio was calculated using a vesicular molecular weight of $2x10^6$ (C. Huang (1969), Biochemistry 8, 344).

that only the complexes with the ionized form of the ionophore (X^-) are of importance in inducing the mediated transport. Assuming that at pH 6.7 all the ionophore is actually ionized (4) and taking into account the dependence of the rates on the square power of X^-, the pKa of X-537A in the vesicle suspension was calculated to be 5.0±2 (in ^2H_2O). This value is significantly higher than the pKa in water of 3.7 (5) but very similar to the value obtained by U.V. and fluorescence spectroscopy (4). Thus indicating that the ionophore is mainly partitioned into the membrane.

Table 2. pH dependence of X-537A mediated Mn^{2+} transport across egg phosphatidylcholine vesicles at 28°C [X-537A]/[vesicles]=4

p^2H *	Rate μM/min
6.7	7.2
5.9	4.9
4.9	1.9
4.4	1.3

* The vesicles were prepared in ^2H_2O containing 10mM Succinic acid. The pH was adjusted with NaO^2H.

The temperature dependence of the transport rates (Fig. 4) studied with synthetic PC vesicles exhibits a discontinuity at ∿38°C, which corresponds to the phase transition temperature of those vesicles (6). The relative high activation energy (∿15 kcal/mol) above the phase transition and the enormous slow down of the transport rate below the phase transition reflect the intrinsic dependence of mediated transport processes upon the physical state of the lipids in the membrane.

In summary, it has been shown that the flux of paramagnetic ions moving through vesicular membranes can be followed by NMR spectroscopy thus providing a method to analyse the mechanism of the flow.

Although flux measurements can also be performed by monitoring with time changes in the intensity of the signal of the transported ions (i.e. ^{23}Na and ^{7}Li), we shall demonstrate the use of direct measurements in describing the second method below.

B. NMR RELAXATION STUDIES OF ION EXCHANGE ACROSS VESICULAR MEMBRANES

This method is unique for magnetic resonance studies and applies when the transport rate is of the order of the nuclear relaxation rate (m-sec^{-1} - sec^{-1}). It is based on following nuclear relaxation rates of ions exchanging between two compartments which are differentiated magnetically (i.e. by probing one compartment with a

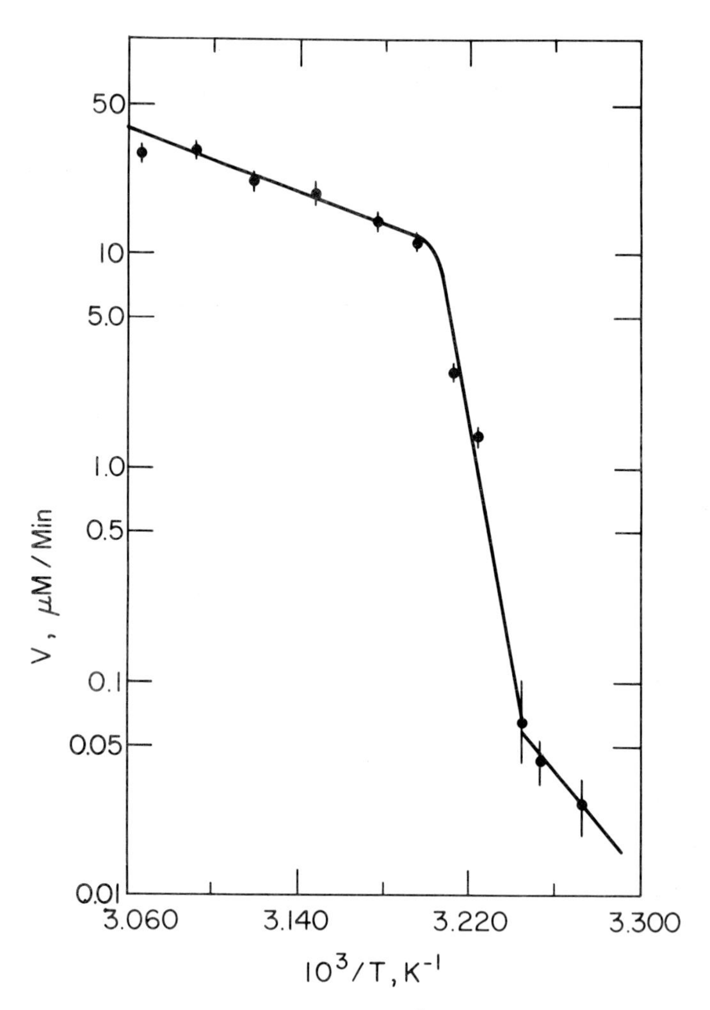

Fig. 4. Temperature dependence of Mn^{2+} transport rates. Vesicles were prepared by sonicating of L-α-phosphatidylcholine dipalmitoyl (Sigma) at 50°C. 71μM X-537A and 1mM $MnCl_2$ were added after sonication. These measurements were performed with a 360 MH_z spectrometer in collaboration with A.C. McLaughlin (Brookhaven Lab.) and S.R. Simon (SUNY at Stony Brook).

paramagnetic relaxation reagent). A similar approach has been used previously to study water diffusion across cells and vesicles (8,9). Here we extend the method by using an anionic paramagnetic relaxation reagent for positively charged transporting species for which the usual relaxation reagents are not suitable. Specifically we consider the transport of alkali metal ions (sodium and lithium) through phospholipid vesicles using ^{23}Na and ^{7}Li NMR and employ $Gd(EDTA)^-$ as a relaxation reagent for cations (7).

In the upper trace of Fig. 5 the NMR spectrum of ^{23}Na in a suspension of PC vesicles is shown. In this preparation 10% of the sodium ions are intravesicular while the rest are dissolved in the bulk solvent. Thus, the single line observed in the spectrum consists of a super position of both the signals of the inner and outer sodium ions. A discrimination between the two signals can be achived by using paramagnetic shift or relaxation reagents. However, the common reagents Mn^{2+}, Gd^{3+} and Pr^{3+} had only a very small effect on the sodium ions' resonance. This is not surprising in view of the fact that these reagents are positively charged and will not bind to the alkali cations. On the other hand, the ethylenediamine tetracetate (EDTA) complex of gadolinium (III), which is negatively charged, and can bind reversibly sodium ions, was found to be quite effective in broadening the alkali ion resonances. The trace in Fig. 5B corresponds to the same solution used to record the upper trace but to which 11.6mM of $Gd(EDTA)^-$ was added. Although the relaxation effect on the sodium resonance is quite remarkable, no discrimination between the signals due to inner and outer sodium is apparent. There may be several reasons for that: (i) The $Gd(EDTA)^-$ complex diffuses through the membrane and thus affects both compartments. (ii) There is a very fast transport

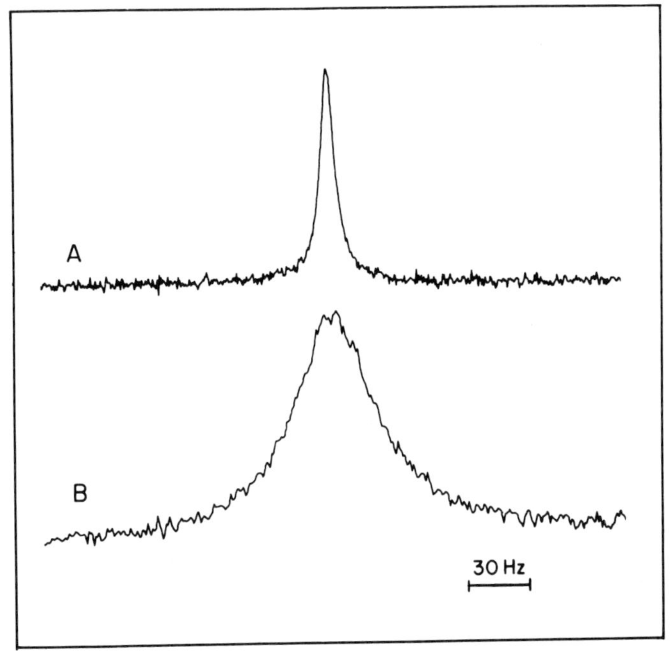

Fig. 5. ^{23}Na NMR spectra of: A. 150 mM NaCl dissolved in egg phosphatidylcholine vesicle suspension, p^2H 9.4 (10 mM Tris) at 28°C. B. The same as A after addition of 11.6 mM $Gd(EDTA)^-$. The vesicles were prepared by sonicating 10% pure egg phosphatidylcholine (Makor Chemicals, Ltd.) at 4°C under nitrogen in H_2O. The spectra were obtained by the pulse-FT method at 21.14 kG (Bruker WH-90) using 10 mm diameter tubes. 40 transients were accumulated with a recycle time of 1.1 sec.

of sodium across the membrane causing the relaxation of the inner and outer sodium to average. (iii) the inner signal is too weak to observe. We shall show shortly that (i) and (ii) may be ruled out and we are left with the problem of separating the weak signal from the strong superimposed signal. For that purpose rather than Fourier transforming the free induction decay (FID) signal following single pulses we used a pulse sequence of 180°-τ-90° and Fourier transformed the FID signal following the 90 pulses. The interval was chosen so that $\tau \simeq T_1^{ex} \ln 2$ where T_1^{ex} is the longitudinal relaxation time of the sodium ions in the bulk solution containing the Gd(EDTA)$^-$. This procedure essentially abolishes the outer signal but since T_1^{in} of the inner sodium, which is not affected by the Gd(EDTA)$^-$, is longer than T_1^{ex}, the inner signal intensity is only slightly reduced (although its phase is inverted).

An example of spectra obtained using this procedure is shown in Fig. 6. The sharp, inverted signals correspond to the inner sodium and lithium ions. The extracellular ions are responsible for the weak, broad, uninverted signal. This spectrum clearly demonstrate that there is no diffusion of Gd(EDTA)$^-$ into the vesicles and that

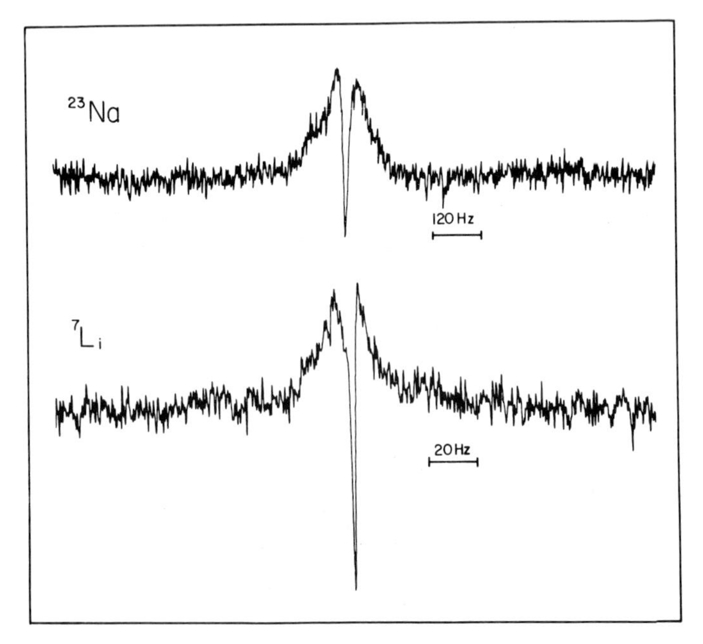

Fig. 6. The NMR signals of sodium and lithium ions entrapped in the inner volume of phosphatidylcholine vesicles. The signal due to the extravesicular ions were eliminated by addition of 12 mM and 8 mM Gd(EDTA)$^-$ (for ^{23}Na and ^{7}Li respectively), and the use of a 180° - τ - 90° pulse sequence as explained in Fig. 5 and contained 150 mM of the metal chloride. 4800 transients with τ =4 msec were accumulated for ^{23}Na, and 100 transients with τ= 49 msec for ^{7}Li.

transport of sodium and lithium through the membrane is slow compared to the nuclear relaxation rates of the ions.

We now demonstrate the application of this latter method to the study of the transport of sodium and lithium ions mediated by the ionophore monensin (10). The effect of adding monensin to the vesicular preparation on th linewidth of the inner signal of sodium is shown in the three traces in Fig. 7. There is a conspicuous increase in the linewidth upon increasing the ionophore concentration which we attribute to enhancement of the transport rate across the membrane.

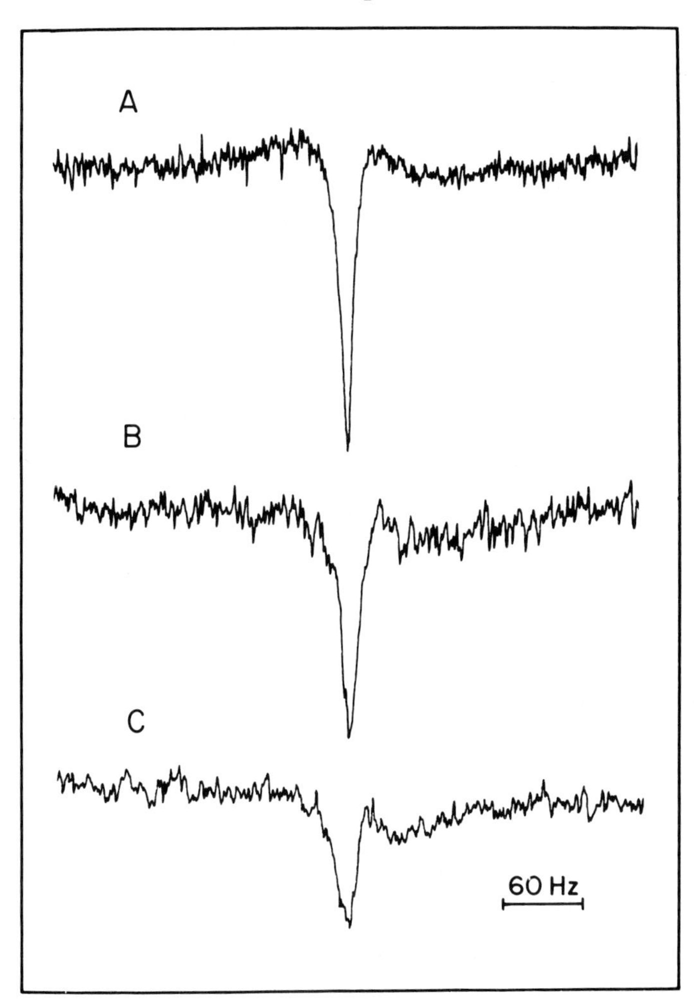

Fig. 7. Sodium signal obtained as in Fig. 6 of ions entrapped inside the vesicle at various monensin concentrations at 28°C. The vesicle suspension was the same as in Fig. 5B. Monensin concentrations: A. 0 μM B. 15 μM C. 37.5 μM. A. 7200 transients B,C 4500 transients.

The exchange rate ($1/\tau$) and the permeability (P) of the membrane is calculated in this case directly from the change in the observed relaxation rate using the following equations (11):

$$\frac{1}{\tau} = \frac{1}{T_{1,2}^{in}} - \frac{1}{T_{1,2}^{in}(o)} \qquad [1]$$

$$P = \frac{V}{A} \cdot \frac{1}{\tau} = \frac{R}{3} \cdot \frac{1}{\tau} \qquad [2]$$

where the index (0) refers to the relaxation rates in the absence of transport, and V, A and R are respectively the internal volume, surface area, and radius, of the vesicles.

Sodium permeabilities were found to be 62, 82, 126 and 158 nm/sec for 15, 22.5, 30 and 37.5 μM monensin respectively and lithium permeabilities were 12 and 33 nm/sec for 400 and 800 μM monensin respectively. Thus, the permeabilities extrapolated to 1 μM of monensin for the same ion and lipid concentration are: for Na^+ 4.0 ± 0.4 nm/sec, for Li^+ 0.035 ±0.005 nm/sec. These results show that within the concentration range studied the sodium transport rate increases fairly linearly with the ionophore concentration, indicating that the dominant transporting species is a 1:1 complex of the sodium ionophore. The much higher value obtained for sodium either indicates that the complex association-dissociation processes determine the overall rate of transport or reflects the difference in the binding constants for these two ions. Both possibilities are in accordance with the significantly higher affinity of monensin towards sodium relative to that of lithium (12).

Two comments concerning the use of $Gd(EDTA)^-$ in such studies are in order. First, we must rule out the possibility that $Gd(EDTA)^-$ is itself transported by monensin into the vesicles thus causing the broadening of the inner signal. This was easily confirmed using the proton NMR signals of the vesicular phosphatidylcholine. Addition of $Gd(EDTA)^-$ caused line broadening of the methyl choline protons of the outer membranal surface, but did not affect the linewidth of the corresponding protons on the inner surface in the presence of 80 μM monensin even after allowing the preparation to stand for several hours. The second comment concerns the effect of $Gd(EDTA)^-$ upon the nuclear relaxation rates of sodium and lithium ions in aqueous solutions (13). At constant pH, temperature and ion concentration, the relaxation enhancement was found to be linear with the $Gd(EDTA)^-$ concentration. However, the $Gd(EDTA)^-$ enhanced relaxation rates of ^{23}Na and ^{7}Li was found to be pH dependent. The pH profiles reveal the existence of two titrations around pH=8 and pH=12. Therefore, the pH of the studied systems should be kept under control.

CONCLUSION

It has been shown that NMR spectroscopy can serve as a tool for studying the kinetics and mechanism of ion transport via membranes. The first method applies for studying slow transport processes. The change in the ion concentration in the intravesicular bulk solution is followed in real time by monitoring changes in the intensity of the signal of the ions or nuclei at the membrane inner surface affected by the transported ions. The second method applies when the transport rate is of the order of the nuclear relaxation rates thus providing away for determining fast kinetics. It is based on following changes in the ions' nuclear relaxation rates due to the transport and makes use of paramagnetic relaxation reagents. Here the anionic chelate $Gd(EDTA)^-$ is employed as a relaxation reagent for positively charged ions.

The methods has been used to study the kinetics of ionophore mediated transport across phopholipid vesicles. In this system all the parameters affecting the transport process (i.e. ion and ionophore concentration, lipid concentration and composition, pH and temperature) were quantitatively controlled and studied, thus enabling to analyze the mechanism of the transport.

ACKNOWLEDGEMENT

The encouragement and help of Prof. Z. Luz during the course of this work are greatfully acknowledged.

REFERENCES

1. B.C. Pressman (1976), Annual Rev. of Biochem. 45, 501.
2. H. Degani (1978), Biochem. Biophys. Acta 508, 364.
3. H. Degani, A.C. McLaughlin and S.R. Simon, unpublished results, to be submitted.
4. H. Degani, R.M.D. Hamilton, J.H. Mangold and S.R. Simon, to be submitted.
5. H. Degani and H.L. Friedman (1974) Biochemistry 13, 5022-5032.
6. A.C. McLaughlin
7. H. Degani and G.A. Elgavish, Febs. Letters, in press.
8. T. Conlon and R. Outhred (1972) Biochim. Biophys. Acta 288, 351.
9. M. Shporer and M.M. Civan (1975) Biochim. Biophys. Acta 385, 81.
10. A. Agtarap, J.W. Chamberlin, M. Pinkerton and L. Steinrauf (1967) J. Am. Chem. Soc. 89, 5737.
11. N. Haran and M. Shporer (1976) Biochim. Biophys. Acta 426, 638.
12. G.P. Gertenbach and A.I. Popov (1975) J. Am. Chem. Soc. 97, 4738.
13. G.A. Elgavish and J. Reuben, to be submitted.

INTERACTION OF CANNABINOIDS WITH MODEL MEMBRANES - NMR STUDIES

I. Tamir[a], D. Lichtenberg[b], R. Mechoulam[a]
Department of Natural Products, Pharmacy School[a] and
Department of Pharmacology[b] Hadassah Medical School
Hebrew University, Jerusalem, Israel

SUMMARY

In a attempt to clarify the mode of psychotropic action of Δ^1-tetrahydrocunnabinol (Δ^1-THC), a comparison was undertaken between its effects on the physical properties of model membranes and those of the psychotropically inactive cannabidiol (CBD). The two cannabinoids were incorporated into sonicated dispersions of mixtures of various relative compositions of egg yolk lecithine (EYL) and cholesterol.

Several parameters were determined from the ^{1}H and ^{31}P magnetic resonance spectra which enabled the evaluation of drug induced changes in the size of the vesicles, the cation binding capacity of the membrane surface and the motional state of the hydrocarbon chains of the phospholipids. At high concentrations, both drugs induce fusion of the vesicles, tighten the packing of the paraffinic chains of the phospholipid molecules and reduce the binding of P_r^{+3} ion to the surface. Δ^1-THC is more effective than CBD in causing these changes. At low concentrations, neither drug had a detectable effect when incorporated into vesicles of pure EYL. The introductions of cholesterol into the vesicles caused significant changes in the effect of cannabinoids. While there were no qualitative differences between their effects on the surface properties, Δ^1-THC increased the mobility of the hydrocarbon chains of EYL in the vesicles while CBD had the opposite influence. Thus, it is possible that the psychoactive cannabinoids might exert their activity by fluidizing the hydrophobic core of biological membranes and, perhaps, by interfering with their ability to become curved.

INTRODUCTION

Δ^1-Tetrahydrocannabinol (Δ^1-THC,I), the major active component of hashish (1) has a chemical structure basically different from those of most centrally active drugs. It has no nitrogen atom; it has a terpene moiety and a pentyl side chain which make it highly lipophilic (2). On the basis of these differences, as well as on a critical analysis of the pharmacological actions of Δ^1-THC, it has been suggested (3) that

B. Pullman (ed.), Nuclear Magnetic Resonance Spectroscopy in Molecular Biology, 405-422.

I

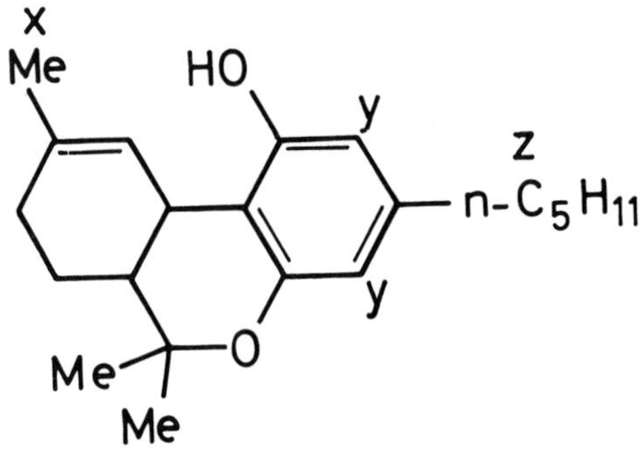

x Me, HO, y, z, $n\text{-}C_5H_{11}$, y, O, Me, Me

II

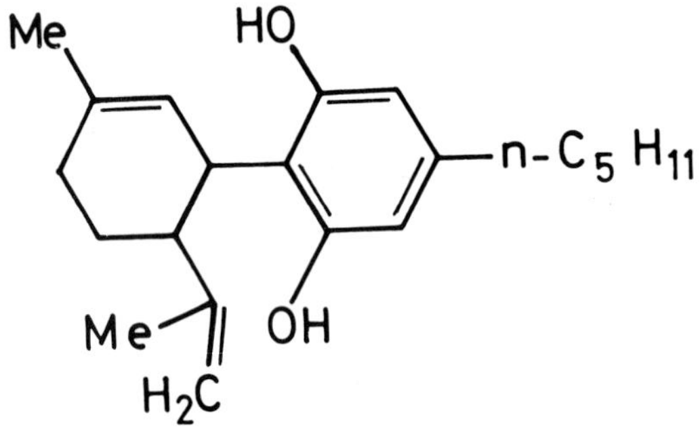

Me, HO, $n\text{-}C_5H_{11}$, Me, OH, H_2C

III

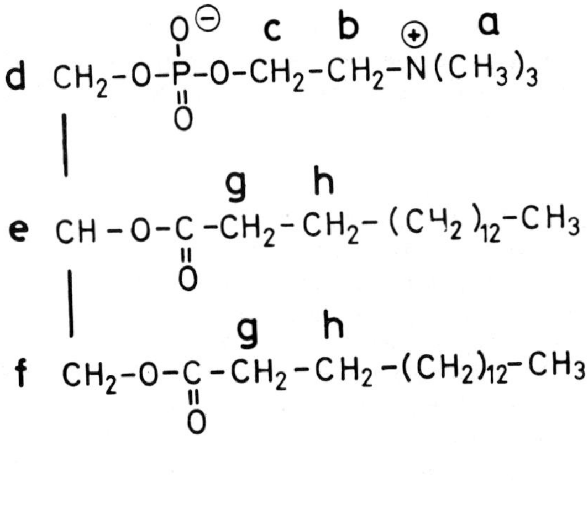

$$
\begin{array}{ll}
 & \quad\quad O^{\ominus}\quad c \quad\;\; b \quad\;\; \oplus \;\; a \\
d & CH_2\text{-}O\text{-}\underset{\|}{\overset{|}{P}}\text{-}O\text{-}CH_2\text{-}CH_2\text{-}N(CH_3)_3 \\
 & \quad\quad\;\; O \\
 & | \qquad\qquad\quad g \qquad h \\
e & CH\text{-}O\text{-}\underset{\|}{C}\text{-}CH_2\text{-}CH_2\text{-}(CH_2)_{12}\text{-}CH_3 \\
 & \qquad\quad\; O \\
 & | \qquad\qquad\quad g \qquad h \\
f & CH_2\text{-}O\text{-}\underset{\|}{C}\text{-}CH_2\text{-}CH_2\text{-}(CH_2)_{12}\text{-}CH_3 \\
 & \qquad\quad\;\; O
\end{array}
$$

unlice other psychotropic drugs, which act on specific receptors, the cannabinoids interact rather nonspecifically with phospholipid constituents of nerve cell membranes.

Due to the complexity of biological membranes, model membranes are often used in investigations in this field of research. Phospholipid liposomes organized in bilayered structures have been used as such models especially for the study of the interactions of various drugs with membranes (4-8). These models are certainly an over simplification of membrans. Nevertheless, since it is possible to vary the composition of the model membranes, it permits deduction of interactions of drugs with the various membrane components. In any case, for model membrane studies to be regarded as relevant to the *in vivo* effects, a correlation should be established between the relative pharmacological activity of these substances and their effect on the model membrane.

A vast number of pharmacological effects have been demonstrated for various cannabinoids (9, 10). Many studies indicate that these compounds have effects on the structural and functional integrity of membranes (10-14). However, no single set of *in vitro* studies can be expected to elucidate the mode of action of cannabinoids for all the *in vivo* effects of these compounds since their relative activity varies for different effects. Only *in vitro* effects of Δ^1-THC which are exhibited to a notably lesser extent by a non psychotropic constituent like cannabidiol (CBD, II) should be considered potentially relevant for the understanding of the mode of the psychotropic action of Δ^1-THC.

Such *in vitro* differential effects have in fact been reported by Lawrence and Gill (4). These authors concluded that the psychoactive cannabinoids may be classified as partial anesthetics producing the same perturbation of the membrane structure as that caused by subanesthetic doses of general anesthetics (15). They employed the technique of spin label electron spin resonance (ESR) developed by Hubbel and McConnell (16) for the determination of the degree of disorder of the hydrocarbon chains in phospholipid liposomes. Introduction of various cannabinoids altered the order parameters (S) within the hydrophobic core, and the change in this factor (ΔS) correlated qualitatively with the psychotropic potency. This method however, involves introduction of a relatively large external probe which can affect considerably the system under investigation, and brings forward the question of relevancy of the observed changes (17). Another characteristic of the above technique is that the probe reflects the structure of its immediate environment and is very sensitive to its location with respect to the surface of the bilayer (18).

Nuclear magnetic resonance (NMR) relaxation times are known to be very sensitive to molecular motions and their measurement does not require the introduction of external probes. The spin-spin relaxation time (T_2) of protons is especially sensitive to local motions with large amplitude (19, 20) which in turn are the most sensitive ones to changes in packing. On the other hand, the major factor which affects the ESR signal of the spin label (SL) is the faster and more restricted motions of the paraffinic chains (21). Unfortunately, the interpretation of proton magnetic resonance (PMR) linewidths ($\nu_{\frac{1}{2}}$), which are inversely proportional to T_2, is far from being straightforward and is still a matter of controversy (ca. ref. 19-21 versus 22,23). The limitation imposed by the possible ambiguities in the interpretation of $\nu_{\frac{1}{2}}$ will be discussed further on. However, if drug induced changes in this parameter are properly considered, PMR studies can provide additional valuable information regarding the interaction and orientation of drugs in the membrane system.

Another valuable information that can be gained from NMR spectra is the average size of the vesicles and its alteration by drugs. This is possible because the ratio between the number of molecules at the outer surface and the number of molecules at the inner surface of the bilayers is a function of the size of the vesicle (24, 25). The numerical values of this ratio change: from greater than 2.0 for small vesicles (125-150 A°) to 1.0 for bilayers with uncurved surfaces and to less than 1.0 for multibilayered aggregates (24). This factor can be obtained experimentally for vesicles composed of EYL (III) by determination of the ratio between the number of inner and outer choline head groups. Addition of a shift reagent like $PrCl_3$, which cannot penetrate the hydrophobic core, causes a separation of the composite choline signal (at $\delta = 3.2$ ppm) into two distinct signals, shifting downfield the signals due to the groups with which it can interact (26, 27). Determination of the ratio of intensities of those two peaks (Ao/Ai) would therefore constitute an estimate for the average vesicle size. The same information can also be obtained from the ratio of intensities of the corresponding phosphate groups in the ^{31}P spectra of the $PrCl_3$-dopped dispersions of vesicles.

The Pr^{+3} induced shift of the signal of the outer head group indeed depends on the relative concentrations of Pr^{+3} to egg yolk lecitihin (EYL) of the outer surface in vesicles of this type. However, it also depends on the interaction between these two components. Drug induced changes in the binding of the cation to the surface, will therefore be indicated by the extent of the Pr^{+3} induced shifts.

This communication presents the results of a study of the influence of Δ^1-THC and CBD on different components of the most simple model systems of a cell membrane consisting of sonicated egg yolk lecitihin (EYL, III) liposomes with various amounts of cholesterol in it. The interaction of both the psychotropic Δ^1-THC and the non-active CBD with such vesicles were studied and their effects on the properties of the model membranes were compared.

MATERIALS AND METHODS

CBD was isolated from hashish. Δ^1-THC was prepared from CBD according to a published procedure (28). Egg yolk lecithin (EYL, grade I, Makor Chemicals, Jerusalem) and cholesterol (Merck) were chromatographically pure and were used without further purification. $PrCl_3$ (99.9%) was obtained from Alfa Products and was dissolved in D_2O (Merck 99.7%) to produce a stock solution of 0.3M. Dimethyl-sulfoxide-d_6 (DMSO) (99% deuterated, Merck) was used.

Liposomes were prepared as follows: chloroformic solutions of EYL and cholesterol at the appropriate molar ratios of these components were mixed, the organic solutions were evaporated to dryness under nitrogen and were then liophylized. The dried mixtures were dispersed in D_2O to the proper concentration and were then sonicated (Heat Systems, 350W sonicator), until clear dispersions were obtained. The final EYL concentration in all aqueous samples were maintained at 50 mg/ml.

Two methods were employed to introduce the drugs into the liposomes. Method A - the cannabinoids, dissolved in small amounts (50μl) of DMSO, were added to aqueous dispersions of the liposomes (0.5 ml) obtained by sonication. Method B - chloroformic solutions of the drugs were mixed with the mixed EYL - cholesterol chloroformic solutions, followed by evaporation to dryness, dispersion in water and co-sonication.

PMR spectra were recorded immediately after sonication on a JEOL JNM-MH-100 spectrometer operating at 100 MHz with a probe temperature of 30 $\pm$ 1°c. All spectra were recorded without accumulation and under the same conditions of RF and modulation.

^{31}P-NMR spectra were recorded on a Bruker WP-60 spectrometer equipped with a FT, operating at 24.2 MHz and probe temperature of 30 $\pm$ 1°C. ^{31}P spectra were measured using spectral width of 15.99 Hz/Cm and were an accumulation of several hundreds scans.

The following spectral parameters were determined.

1) <u>^{1}H Linewidths</u> ($\nu_{\frac{1}{2}}$) were measured using a spectral of 10 Hz/cm and a sweep rate of 1 Hz/sec and are expressed as the difference between their measured values and the width of the HOD signal in the spectra.

Every result is an average of at least four consecutive scannings under the same condition of RF and modulation. The experimental errors which are indicated in the figure are a mathematical average of these measurements. Care was taken to ensure reproducibility in the base line determination. This was achieved by the use of the base line of the corresponding spectrum determined at spectral width four times higher.

2) <u>Intensities</u> were measured by weighing paper cutouts of the signals. The errors in these measurements were at least as large as in ($\nu_{\frac{1}{2}}$) determinations.

3) The intensities ratio Ao/Ai was determined, in both the PMR and ^{31}P-NMR spectra following complete separation of the signals of the molecules of the inner and outer surface by Pr^{+3}. For these measurements a relative molar concentration of Pr^{+3} to EYL of up to 0.9 was used. The ratios given herein are averages of at least 5 measurements each. In the determination of the Ao/Ai ratio by ^{31}P-NMR, care was taken to ensure direct proportion between the intensities and the concentration of the corresponding components. The nuclear Overhauser effect (NOE) was interrupted since this effect can be different for the outer and inner surfaces (32). The delays between consecutive pulses was increased to allow complete relaxation of the phosphate group on the inner surface of the bilayer.

4) The Pr^{+3} induced chemical shifts of groups on the outer surface of vesicles of various composition were determined from PMR and ^{31}P-NMR.

RESULTS

A proton magnetic resonance (PMR) spectrum of a mixed solution of EYL and Δ^1-THC in chloroform is presented in Fig. 1A. The spectrum is a superposition of the spectra of the individual substance (30, 31), showing that there is no PMR evidence for their interaction in organic solution. There was also no evidence for interactions in organic solution between CBD and EYL, nor between either drug and cholesterol.

Replacement of the organic solvent by D_2O yielded multilamellar structures with no high resolution characteristics. Sonication of these suspensions led to the formation of single bilayered vesicles which contained either Δ^1-THC or CBD. In these vesicles no signals from the cannabinoid protons could be detected at concentrations of Δ^1-THC or CBD twice as high as those for which PMR signals are easily determined in $CDCl_3$ solution (Fig. 1B).

Addition of $PrCl_3$ to the aqueous dispersions resulted in a splitting of the choline head group signal into two peaks, i.e. the signal of the head groups on the outer surface shifted downfield, the rest of the spectrum remaining unchanged (Fig. 1C). This enabled us to determine the Ao/Ai ratio, which depended on the presence and relative concentration of the drug to EYL.

In samples prepared by method B, the cosonication for the same length of time of either Δ^1-THC or CBD (9.5 mm) and EYL (67 mm) in D_2O yielded vesicles whose average size was larger than that of pure EYL. After 15 minutes of irradiation, the ratio Ao/Ai for vesicles containing CBD or Δ^1-THC was 1.6 as compared to a value of about 2.0 for the pure EYL ones. Doubling the sonication time reduced the size of the vesicles containing CBD, as indicated by the change in Ao/Ai ratio to 1.9. Such a treatment caused no change in the size of vesicles containing Δ^1-THC and its value remained 1.6.

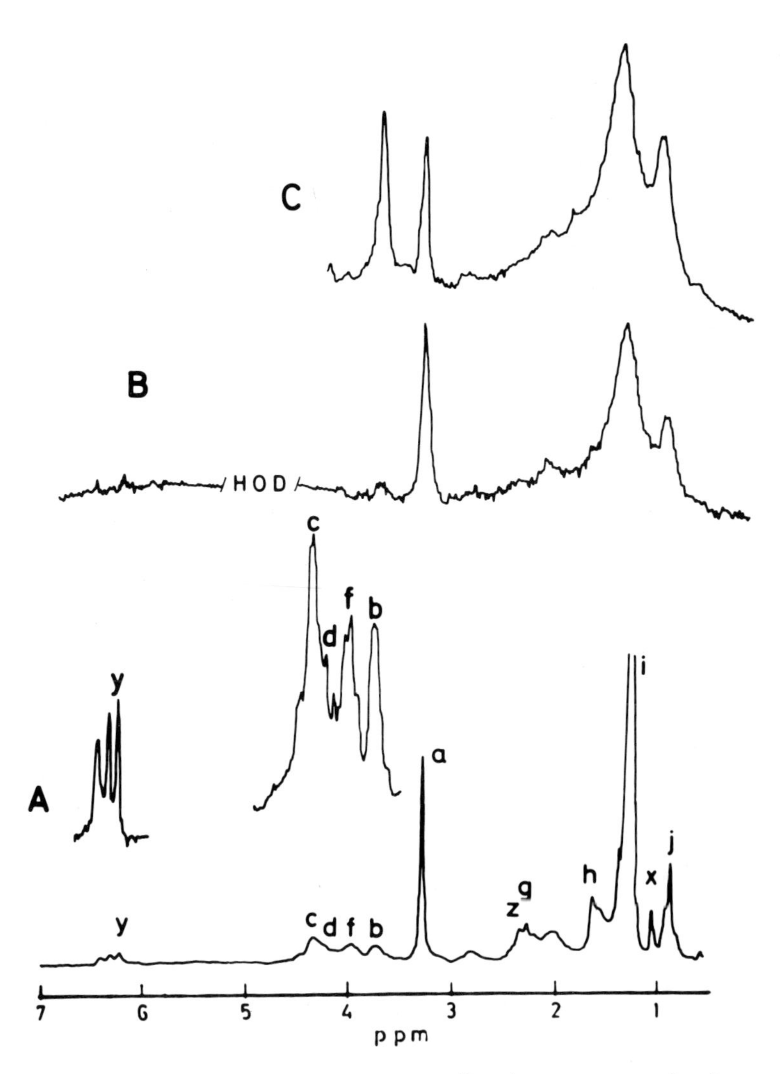

Fig. 1. Proton magnetic resonance (PMR) spectra of mixtures of EYL 67 mM) and Δ^1-THC (9.5 mM)

A. spectrum in $CDCl_3$. The assignment of the various peaks is based on that of the two components β-γ, dipalmitoyl α phospatidyl choline (ref. 30) peaks a-j, as indicated in the formula III. (EYL is a mixture of saturated and unsaturated phospholipids similar to III.) Δ^1-tetrahydrocannabinol (ref. 31) peaks x-z, as shown in formula I. The insert is a magnification of the corresponding groups as indicated by the identical symbols. It was measured with the same amplification as in the spectra of Figs 1B and 1C.

B. spectrum obtained from the same components in ultra sonically irradiated aqueous dispersion.

C. spectrum as in 1B after the addition of $PrCl_3$ (7mM). The amplification is as in 1B.

In samples prepared by method A, drugs were introduced in DMSO solutions to sonicated dispersions. In these experiments, the DMSO used did not exceed 5% (v/v). At this concentration, DMSO by itself did not perturb any of the measured spectral parameters. Initially, addition of DMSO solutions of Δ^1-THC or CBD caused the appearance of cloudiness, probably due to the insolubility of the drugs in the squeous solution. After a few seconds it cleared completely probably indicating absorption of the drugs into the vesicles. The completion of this rapid process was determined by the NMR spectrum, which remained unchanged and was very similar to that of dispersions of the same composition and Ao/Ai values, prepared by method B. Addition of high concentrations of both drugs was accompanied by a reduction of the latter ratio in all vesicle dispersions. Thus a concentration of 3 mole % or more of either cannabinoid (with respect to EYL) reduced this ratio by ca 0.2 - 0.3 (e.g. from 2.0 to 1.7 on preparation of small pure EYL vesicles and from 1.6 to 1.4 when the cannabinoids were added to larger ones). At lower relative concentrations of the cannabinoids, the Ao/Ai ratio remained the same, within experimental error. Some of the other spectral parameters did change. All the above was true for the effects of either drug on vesicles composed of EYL and cholesterol at various molar ratios.

Inclusion of the cannabinoids in the vesicles changed the linewidth of the choline head group signal by less than 2 Hz. On the other hand, the apparent linewidth of the methylene signal changed considerably. The dependence of the latter parameter on the molar ratio of both Δ^1-THC and CBD to EYL is described in Fig. 2. As evident from this figure, the effect of the various additives depends not only on their relative concentrations but also on the composition of the model membrane. In pure EYL vesicles (Fig. 2A), both drugs increased the PMR linewidths of the signals of the phospholipids in the vesicles, the dependence of the broadening of the NMR signals on the drug concentration being biphasic. No appreciable effects on the linewidths can be seen at low concentrations and only at relative drug concentrations higher than about 6 mole % the methylene signals broadened, Δ^1-THC having a larger effect than CBD. It is of interest that the effect of both drugs first became evident at a concentration which is inversely related to the size of the vesicles, i.e. the larger a vesicle is, the more sensitive are the widths of its signals to the drug (Fig. 3). Our results, therefore, indicate size dependence to the effect of the cannabinoids.

When vesicles composed of EYL and cholesterol were utilized, the two drugs affected the linewidth differently. (Figures 2B and 2C). Again at high molar ratios, both cannabinoids increased the apparent linewidth, the broadening being larger than in the absence of cholesterol (Fig. 2A) and directly proportional to the relative concentration of cholesterol in the model membrane. Thus, when 0.05 moles of CBD were added to vesicles of EYL to cholesterol of a ratio 1:0.5, $\Delta(\nu_{1/2})$ app. increased by ca. 8 Hz (Fig. 2B), whereas addition of the same amount of CBD to vesicles of a ratio of 1:0.9 resulted in a 16 Hz broadening.

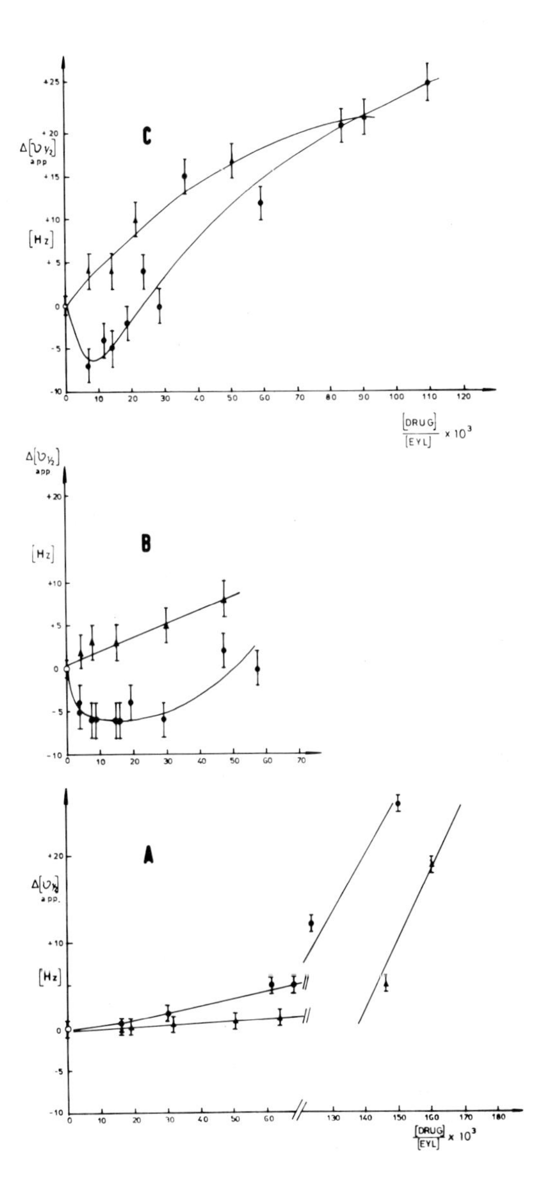

Fig. 2. The dependence of the change in the apparent linewidth of the methylene signal $\Delta(\nu_{\frac{1}{2}})$ app. on the molar ratio of added drug to EYL (method A) for small vesicles composed of cholesterol to EYL at various molar ratio (R). The circles denote Δ^1-THC and the triangle CBD.

A. R = 0
B. R = 0.5
C. R = 0.9

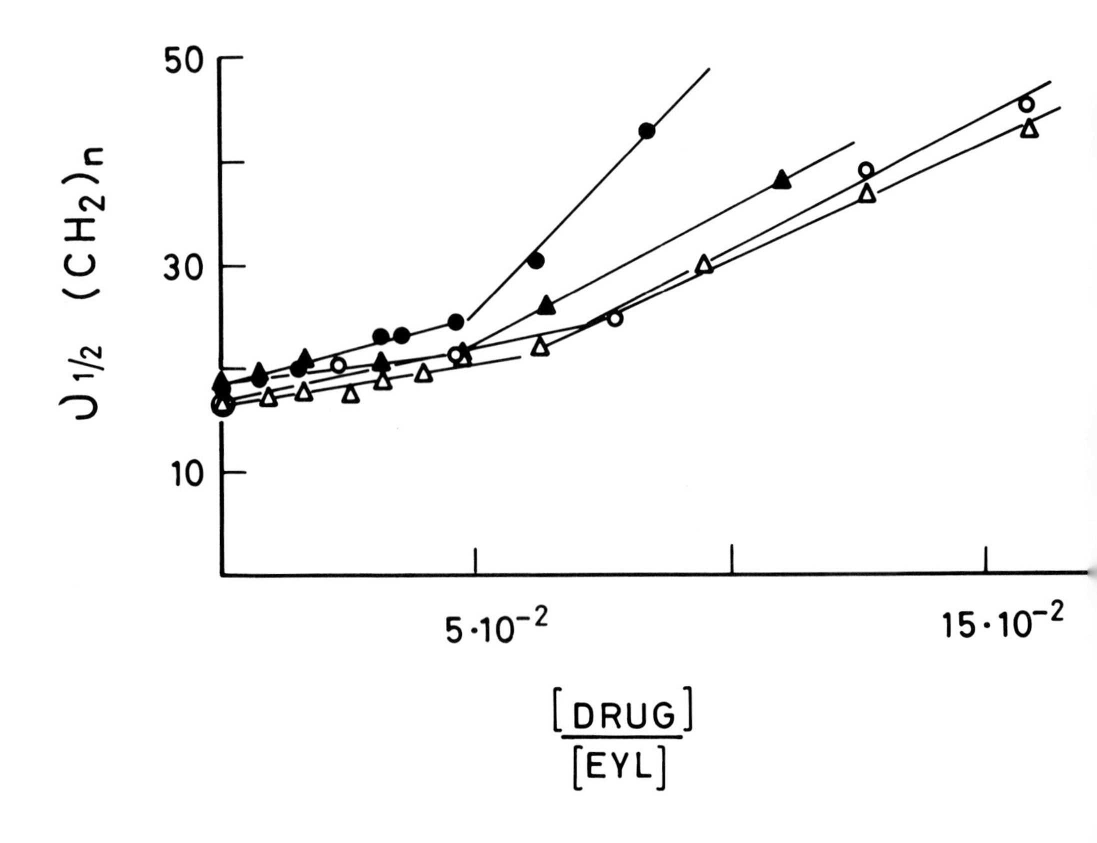

Fig. 3. The dependence of apparent linewidth of the methylene protons signal ($\nu_{\frac{1}{2}}$) on the molar ratio of added drug to EYL vesicles (method A). The circles denote Δ^1-THC and the triangles CBD. Full symbols indicate larger vesicles (Ao/Ai = 1.7); empty ones indicate smaller vesicles (Ao/Ai - 2.0).

A more pronounced difference between the effects of Δ^1-THC and CBD on vesicles of various compositions was recorded at lower molar ratios of these two drugs to EYL. Unlike in pure EYL vesicles, Δ^1-THC caused narrowing of the methylene signal, while a line broadening was obtained by similar concentrations of CBD (Figs. 2B and 2C). In this range of cannabinoid concentrations, the difference between their effects was enhanced with increasing relative cholesterol to EYL molar ratio (compare Figure 2B with 2C).

Another effect of the cannabinoids that depended on the presence of cholesterol in the model membranes was the cannabinoid-induced alteration of the Pr^{+3} paramagnetic shifts of the ^{31}P signal of the molecules on the outer surface. Thus, the latter was not perturbed by the addition of 0.05 moles of either drug to 1

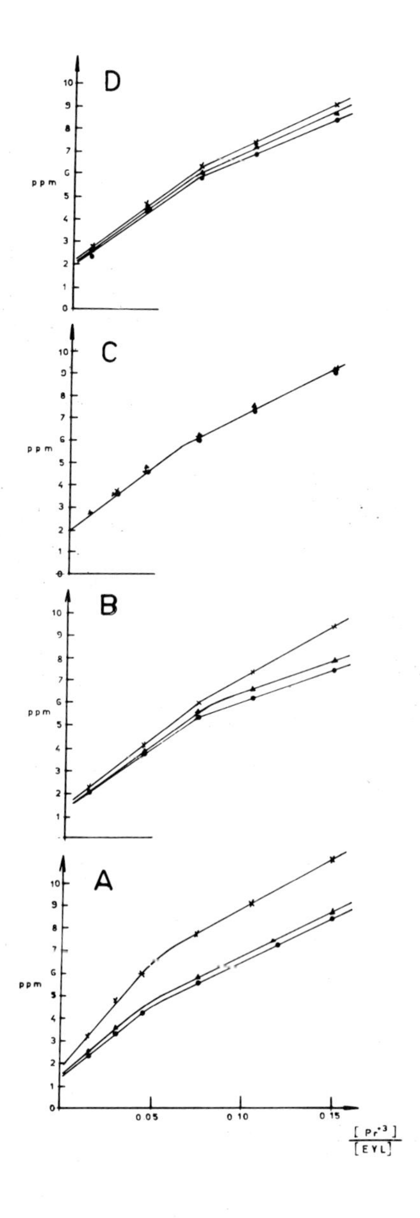

Fig. 4. The relationship between the $PrCl_3$ induced chemical shift of the phosphate group on "the outer surface" (ppm) and the relative molar ratio of Pr^{+3} to EYL. The X denote absence of drugs, the circles Δ^1 THC and the triangles CBD. The composition of the model membrane used were as follows:

A. pure EYL vesicles, the molar ratio of the drug to EYL is 0.24
B. the same as A; the molar ratio of drug to EYL = 0.08
C. the same as A; the molar ratio of drug to EYL = 0.05
D. Vesicles containing cholesterol at a molar ratio of 0.2. The molar ratio of drug to EYL is 0.008.

mole of vesicular lecithin (Fig. 4C) whereas a much smaller dose (molar ratio = 0.008) of either drug reduced the effect of added $PrCl_3$ in vesicles composed of cholesterol and EYL (Fig. 4D). Indeed, when much higher doses of the drugs were introduced to pure EYL vesicles, proportional reduction of the Pr^{+3} effect was obtained (Figures 4B and 4A). It is of interest that with a molar ratio of cholesterol to EYL of 0.9 in the membrane the reducing effect of the drug on the Pr^{+3} induced shifts is similar to that obtained with vesicles containing a ratio of cholesterol to EYL of 0.2.Similar effects of the drugs were observed on the shift of the outer choline head group signal in the PMR spectra. Indeed the magnitude of these effects was much smaller as compared to that obtained in the ^{31}P spectra.

DISCUSSION

This paper presents NMR studies on the interactions of model membranes with the psychotropic Δ^1-tetrahydrocannabinol (Δ^1-THC) as well as with the inactive cannabidiol (CBD). The model membranes used in these studies were composed either of EYL and cholesterol at various ratios of these components or of pure EYL. In the latter case, vesicles of various radii were investigated.

Throughout this work, two methods were used for introduction of the cannabinoids into the model membranes. Co-sonication of large doses of either drug with the model membrane components yielded vesicles whose average size was larger than those obtained in the absence of the drug. The smallest obtainable Δ^1-THC-containing vesicles were larger than the CBD-containing ones. This difference might be a consequence of the rigid structure of Δ^1-THC, which interferes with the formation of very curved surfaces.

In an attempt to reduce possible ambiguities due to vesicle size differences, the effects of drugs added to the same preparation of small vesicles were measured. Cannabinoids dissolved in small amounts of dimethyl sulfoxid (DMSO), were added to sonicated liposomes. This procedure led to the incorporation of the drugs within the hydrophobic core of the bilayers. This is suggested by the broadening (beyond detection) of the signals of the cannabinoids, which most probably is an outcome of the immobilization of the cannabinoids within the membranes. It is also supported by the drug-induced changes of the spectra of the phospholipid.

This method seems to us more relevant to the *in vivo* application than the other method used in this work for mixing the drugs with the model membranes (method B). However, the former method of mixing might result in the formation of systems which do not reach a state of equilibrium in a short period of time. Nevertheless, after five minutes, the close similarity of spectra of samples prepared by the two methods with the same composition and average size (Ao/Ai) indicates that method A leads rather fast to a state of equilibrium.

The decrease of Ao/Ai, observed in the ^{1}H and ^{31}P NMR spectra upon addition of more than 3 mole percent of either drug, probably reflects a drug induced vesicle-vesicle fusion similar to that observed for alamethicin (33). This phenomenon is in accordance with the fact that vesicles of larger size were obtained by co-sonication of the membrane components with either drug, as compared to that obtained by sonication in the absence of drugs. In any case, the fact that low doses of either drug did not alter Ao/Ai, indicates that the size was not changed.

The spectral changes introduced by small amounts of the drugs are therefore most probably an outcome of interactions of the drugs with intact vesicles. The broadening of the choline signal in all vesicle preparations was very small, even with large doses of the drugs, indicating that the drugs are buried inside the hydrophobic core of the membrane. Still with high doses of both drugs some surface properties of the model membrane did change. This is suggested by the fact that even in pure EYL vesicles both drugs reduced the Pr^{+3} induced shifts of the ^{31}P signals of phosphate groups on the outer surface (Fig. 4). Δ^{1}-THC is somewhat more effective in this respect. This effect of the cannabinoids cannot be accounted for by drug induced fusion, which would result in a decrease in the Ao/Ai ratio, with a concomitant increase in the ratio of Pr^{+3} to interacting phosphate groups and consequently result in a larger shift of the outer group. One possible explanation for the reduction of interaction between the phosphate and Pr^{+3} is that hydrogen bonds are formed between the phosphate groups of EYL and the phenolic groups of the drugs (34). The lower sensitivity to drugs of the Pr^{+3} induced shift of the choline head group signals might indeed be a result of its lower sensitivity to praseodimium (35).

The most interesting and only differential effects of Δ^{1}-THC and CBD on the spectra of model membranes are the effects on the apparent linewidth of the bulk methylene signal in the proton NMR. This parameter is a very complex one since it depends on various factors. First, this signal is an envelope of resonances due to at least sixteen magnetically inequivalent methylenes, each split by neighboring groups. This overlapping results in broadening of the bulk methylene signal not only in aqueous medium but in organic solvent as well. In the latter case the linewidth was about 4 Hz. The much larger linewidths in aqueous dispersions is most likely an outcome of the restriction in molecular motions, imposed by the aggregation of the phosphlipid molecules in the bilayer. The dependence, of the obtained super Lorentzian lineshape (and width) on the vesicle size is unknown.

In this work the complete intensity of the methylene signals was observed in all the spectra. This is indicated by the constant ratio of intensities of the methylene to choline signals. It might therefore be concluded that the intensity "in the wings" of the methylene "super Lorentzian" signal (36) is detectable in our spectra.

Typically, the linewidths were from about 18Hz in spectra

of pure EYL vesicles to about 80-90 Hz in the spectra of vesicles containing EYL and cholesterol at a molar ratio of 1 to 0.9. This large difference cannot be a mere consequence of an increase in the vesicle size since in some "uncomplete sonications", of dispersions of pure EYL, the formed vesicles were larger than those obtained with the EYL-cholesterol mixtures as indicated by their lower Ao/Ai, still their PMR signals ($(\nu_{\frac{1}{2}})$ app = 25 Hz) were considerably narrower than in the presence of cholesterol. The excessive broadening is probably a result of the rigidifying effect of cholesterol (37, 38, 39).

It is difficult to tell whether or not the broadening obtained with large doses of both Δ^1-THC and CBD can be solely accounted for by drug induced fusion processes. However at molar ratios of the drugs lower than 3 mole % no such fusion can be seen and it might be concluded rather safely that the changes in the apparent linewidth of the methylene protons is predominantly, if not exclusively, due to an alteration of the packing of the phospholipid molecules within the bilayers. Namely, in cholesterol containing vesicles, low concentrations of Δ^1-THC reduce the restriction of molecular motions of the hydrocarbon chains whereas CBD has the opposite effect. These results are in agreement with those of Lawrence and Gill (4) , who demonstrated a fluidizing effect of Δ^1-THC and a tightening of packing introduced by CBD on the model membranes.

In the absence of cholesterol in the model membrane none of the above effects was observed with low concentrations of either drug (Fig. 2A). There was no indication for specific interactions between cholesterol and either cannabinoid in organic solvents. Thus, the differences in the effects of the drugs with the various model membranes are most likely due to the different packing in these membranes introduced by cholesterol. Possibly, when the phospholipid molecules are loosely packed, there might be enough room for cannabinoid molecules to be incorporated without altering the packing within the bilayer. On the other hand, in more ordered lamellas, introduction of the drugs, even in small amounts, will result in more pronounced effects.

The higher sensitivity of the PMR spectra of large pure EYL vesicles, to introduction of cannabinoids, as compared to that of smaller ones (Fig. 3.) supports the above explanation. In terms of packing, "the large vesicles" are probably an intermediate between small vesicles composed of pure EYL and those which include cholesterol (19, 20, 24). Consequently the sensitivity of these large vesicles to the addition of cannabinoids is an intermediate between the two other types of vesicles. In these large vesicles, the rigidizing effect of both drugs can be noticed at lower doses of the additives; these concentrations being higher than the ones at which the difference between the two drugs can be manifested.

In considering the differential effects of Δ^1-THC and CBD, it should be kept in mind that qualitatively the two drugs might have the same effect on surface properties of the bilayers (Fig. 4). Thus, both drugs are probably anchored at the surface of the bilayer similarly, possibly through hydrogen bonds. Nonetheless, they might have different effects in expanding the membranes (40). Such an expansion could most likely lead to fluidization of the hydrophobic core of the membrane. In contrast, a possible rigidization might occur if the interaction between the drugs and the model membrane components is stronger than that between these components themselves. The biphasic effect of Δ^1-THC on the motions within the hydrophobic core of the bilayer might be the result of a balance between two such contradicting actions. More specifically, up to a concentration of about 3 mole percent Δ^1-THC to EYL (in EYL-cholesterol vesicles), the predominant effect might be an expansion of the membrane induced by this drug. This reduces the restriction on the segmental motion of the hydrophobic chains. At higher doses of Δ^1-THC the membrane expansion is less important than the opposite effect of the drug in tightening of the membrane. In CBD, whose "accommodation area" on the bilayer surface is probably smaller, due to its flexible configuration (41), even in small doses the membrane expansion does not overrule its tightening effect.

This hypothesis is in a way reflective of the rigid structure of Δ^1-THC which would certainly lead to larger membrane expansion. It is also in agreement with the larger size of vesicles, obtained with high doses of Δ^1-THC by method B and with the higher potency of this drug in inducing vesicle-vesicle fusion. The latter might also offer an explanation for the larger rigidizing effect of large doses of Δ^1-THC as compared to CBD.

The above conclusions might be of biological importance for two different reasons. First, Δ^1-THC might exert some of its activity through its membrane fluidizing potency, which might, in turn influence the activity of various membraneous proteins. Secondly, the capability of biological membranes to become very curved, the minimal radii of curvatured structures being of the order of 75° (29), is of vast importance in physiological mechanisms such as microtubles or such microfilament formation (42). Thus, we might speculate that prevention of such processes may be of importance in the mode of action of Δ^1-THC. Further studies aimed at investigating these possibilities are being carried out.

ACKNOWLEDGEMENT

The work was supported, in part, by a generous grant from the U.S. National Institute on Drug Abuse, to whom we are thankful.

REFERENCES

1. Mechoulam R., Shani A., Edery H. & Grunefeld Y. (1970) Science 169, 611-612.

2. Seeman P., Chaw-Wong M. & Moyyen S. (1972) Can. J. Physiol. Pharmacol. 50, 1193-1200.

3. Paton W.D.M. (1975) Ann. Rev. Pharmacol. 15, 191-220.

4. Lawrence D.K. & Gill E.W. (1975) Mol. Pharmacol. 11, 595-602.

5. Hauser H., Finer E.G. & Chapman D. (1970) J. Mol. Biol. 53, 419-433.

6. Cerbon J. (1972) Biochem. Biophys. Acta. 290, 51-57.

7. Hammes G.G. & Tallman D.E. (1971) Biochim. Biophys. Acta, 233, 17-25.

8. Lau A.L.Y. & Chan S.I. (1974) Biochemistry 13, 4942-4948.

9. Paton W.D.M. & Pertwee R.G. (1973) in Marijuana (Mechoulam, R., ed.) pp. 192-362 Academic Press, New York, London.

10. Braude M.C. & Szara, S. (eds) (1976) Pharmacology of Marihuana, Raven Press, New York, N.Y.

11. Alhanaty E. & Livne A. (1974) Biochim. Biophys. Acta 339, 146-155.

12. Schurr, A., Sheffer N., Graziani Y. & Livne A (1974) Biochem. Pharmacol. 23, 2005-2009.

13. Gibermann E., Gothlif S., Sharar A. & Bino T. (1975) J. Reprod. Fert 42, 389-390.

14. Laurent B. & Roy P.E. (1975) Int. J. Clin. Pharmacol. 12, 261-266.

15. Seeman P. (1972) Pharmacol. Rev. 24, 583-655.

16. Hubbel W.L. & McConnell H.M. (1968) Proc. Nat. Acad. Sci. 61, 12-16.

17. Hubbel W.L. & McConnell H.M. (1971) J. Amer. Chem. Soc. 93, 314-326.

18. Devaux P. & McConnel H.M. (1973) Ann. N.Y. Acad. Sci. 222, 489-498.

19. Seiter C.H.A. & Chan S.I. (1973) J. Amer. Chem. Soc. 95, 7541-7553.

20. Lichtenberg D., Petersen N.O., Girardet J.L., Masatsume K., Kroon P.A., Seiter C.H.A., Feigenson, G.W. & Chan S.I. (1975) Biochim. Biphys. Acta 382, 10-21

21. Petersen N.O. & Chan S.I. (1977) Biochemistry, 16, 2657-2667.

22. Bloom, M., Burnell, E.E. Roeder, S.B.W. and Valic, M.I., (1977) J. Chem. Phys., 66, 3012-3020.

23. Stockton, G.W., Polanszek, E.F., Tulloch, A.P., Hassan, R. and Smith, I.C.P. (1976) Biochemistry 15, 954-966.

24. Sheets M.P. & Chan S.I. (1972) Biochemistry 11, 4573-4581.

25. Berden J.A., Barker R.W. & Radda G.K. (1975) Biochim. Biophys. Acta 375, 186-208.

26. Kotelnik R.J. & Castellano S.M. (1972) J. Magn. Resonance 7, 291-223.

27. Lee A.G., Birdsall N.J.M. & Metcalf J.C. (1974) in Methods in Membrane Biology 2, (Korn E., Ed.) pp. 2, Plenum Press, New York, N.Y.

28. Gaoni Y. & Mechoulam R. (1971) J. Amer. Chem. Soc. 93, 217-224.

29. Chrzeszczyk, A. Wishnis A. and Springer, C.S. (1977) Biochim. Biophys. Acta 470, 161-169.

30. Haque R., Tinsley I.J. & Schmedding D. (1972) J. Biol. Chem. 247, 151-161.

31. Gaoni Y. & Mechoulam R. (1964) J. Amer. Chem. Soc. 86, 1646-1647.

32. Sears, B., William, C.H. & Thompson, T.E. (1976) Biochemistry 15, 1635-1639.

33. Lau, A.L.Y. and Chan, S.I. (1975) Proc. Nat. Acad. Sci. USA 72, 2170-2174.

34. Rosenqvist, E. and Ottersen, T. (1975) Acta Chem. Scand. B., 29, 379-384.

35. Michaelson, D.M., Horwitz, A.F. and Klien, M.P. (1973) Biochemistry, 12 2605-2612.

36. Wennerstrom, H. (1973) Chem. Phys. Lett. 18, 41-44.

37. Hemminga, M.A. and Berendsen, H.J.C. (1972) J. Magn. Resonance 8, 1-11.

38. Ladbrooke, B.D., Williams, R.M. and Chapman, D. (1968) Biochim. Biophys. Acta. 150, 333-340.

39. Rothman, J.E. and Engelman, D.M. (1972) Nat. New Biol. 237, 42-44.

40. Raz, A., Schurr, A. and Livne, A. (1972) Biochim. Biophys. Acta 274, 269-271.

41. Tamir, I., Meyer, A.Y., Falvello L., Jones, P.G.,Kennard O., and Mechoulam, R. Manuscript in preparation.

42. Thompson, T.E., Lentz B.R. & Barnholz Y.(1978) Biochemistry, in press.

N.M.R. SPECTRA OF MUCOPOLYSACCHARIDES

G. Gatti
Istituto di Chimica delle Macromolecole
Via A.Corti 12, 20133 Milano, Italy

1. INTRODUCTION

Mucopolysaccharides (MPS, also called glycosaminoglycans) are anionic biopolymers widespread in animal tissues, especially in the intercellular matrix of connective tissues. These polysaccharides are built up from amino sugars, either D-glucosamine or D-galactosamine, together with uronic acids, either D-glucuronic acid or L-iduronic acid and may have N-acetyl, O- or N-sulfate groups.

The highly charged mucopolysaccharide chains are thought to play an important role in controlling water and metal ion exchange into, and from, the cells. Some MPS like chondroitin sulfates and herapan sulfate also seem to hold functional proteins on cell surfaces, and release them in the presence of polysaccharides having a higher charge density, such as heparin. Heparin, a well known natural anticoagulant and antilipemic, actually exerts its biological functions through interactions with plasma proteins (1,2).

In order to understand the molecular basis of the biological activity of MPS a thorough knowledge of their structure is required. Although the major structural features of MPS are now established, heterogeneities at the polymer level hinder the complete elucidation of MPS structure. In spite of such heterogeneities, the general molecular structure of the regular part of most important MPS has been determined in the solid state (3). The conformation in solution has been the subject of numerous investigations by different methods, including NMR spectroscopy (4). The present communication reports recent high field NMR studies of heparin and chondroitin sulfates in water solution, for which proton and carbon-13 spectra have given configurational and conformational information.

2. HEPARIN, ^{1}H-SPECTRUM

Extensive work in recent years (5) has proved that the most of hep

B. Pullman (ed.), Nuclear Magnetic Resonance Spectroscopy in Molecular Biology, 423-438.

arin structure is accounted for by disaccharide repeating units, *i.e.*, α-L-iduronic acid (sulfated at position 2) and α-D-glucosamine (sulfated at position 6 and N-sulfated):

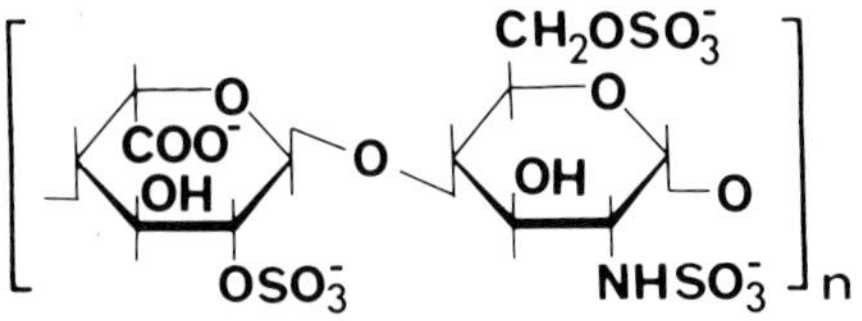

Concerning the conformation, Atkins and Nieduszynski using X-ray diffraction and computer model building have proposed that the glucosamine unit is present in the normal chair $^4C_1(D)$ conformation whereas for the iduronic

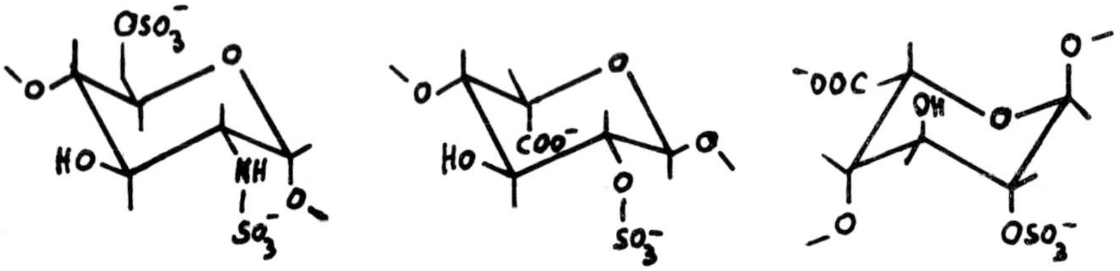

$^4C_1(D)$ $^4C_1(L)$ $^1C_4(L)$

acid unit the $^1C_4(L)$ was preferred over the $^4C_1(L)$ form (6). The $^4C_1(L)$ conformation of the iduronic acid moiety was ruled out by proton NMR studies of Perlin and coworkers (7) although incomplete resolution prevented at that time a conclusive evidence of the actual conformation.

In order to clear this question we have measured the ^1H-NMR spectrum of heparin at 270 MHz in the FT mode. Resolution was increased (fig. 1) by convolution difference technique (8) and by raising the temperature up to 90°C (fig. 2). Systematic proton decoupling was carried out in the pulsed mode leading to the assignments reported in the figures. The first order spectral parameters were refined by computer simulation of the spectra with the iterative program LAOCN 3. Due to the lower resolution obtained at 35°C the parameters for the C_5-C_6 fragment were not determined at this temperature.

The calculated spectra are reproduced in figures 1 and 2 and the parameters are collected in tab. 1. The obtained interproton vicinal

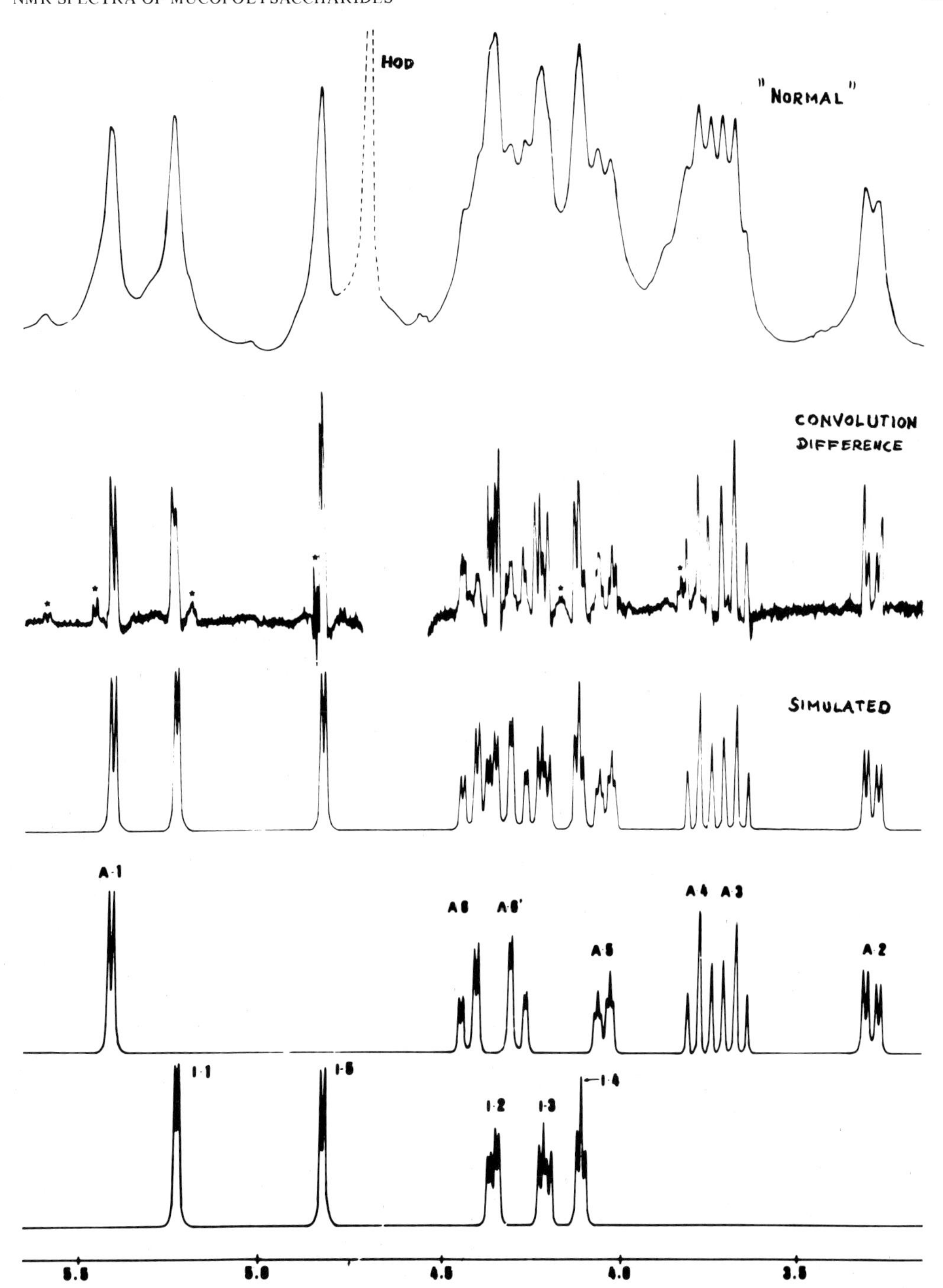

Fig. 1 - 270 MHz FT ^{1}H-spectrum of heparin at 35°C in D_2O solution (10% w/v)

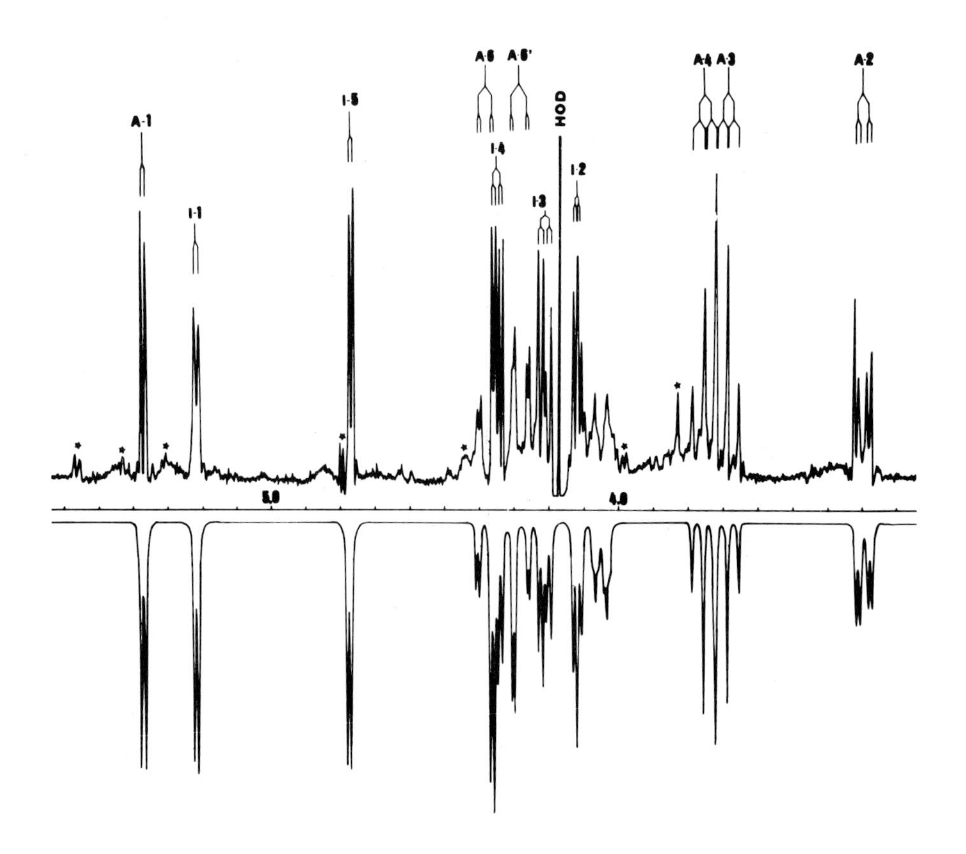

Fig. 2 - Observed (upper trace) and simulated 270 MHz FT ^{1}H-spectrum of heparin at 90°C in D_2O solution (10% w/v)

coupling constants can be interpreted in conformational terms (9). In the glucosamine unit the H_1-H_2 relation is *gauche* (J=3.6) all the others are *trans* diaxial, as shown by the relatively large value of the couplings J_{23}, J_{34} and J_{45}. This observation is in agreement with the $^4C_1$(D) conformation, the same as observed for the monomeric 2-deoxy-2-acetamido-α-D-glucose (10).

In the iduronic acid moiety one notes the absence of high values of J. This fact definitely rules out the $^4C_1$(L) conformation in which all the ring hydrogens but one are axially oriented and would therefore give rise to large couplings (9-10 Hz). The observed couplings are instead in agreement with the alternative conformation $^1C_4$(L) in which all the ring protons are *gauche*.

The observed range of the values of J_g in the iduronic acid unit (2.64-6.10 Hz) can be interpreted by taking into account the stereospecific electronegativity effect rule introduced by Booth (11) and extended to carbohydrates by Anteunis (12). Considering the Newman projections of the iduronic acid unit (fig. 3) one can see that the protons involved in the *gauche* coupling are antiperiplanar to two carbon atoms (J_{23} and J_{34}), to one carbon and one oxygen atom (J_{12}) and finally to two oxygen atoms (J_{45}). The scheme of Anteunis predicts different ranges of J in the three cases, and the observed values are within the ranges

Tab. 1 - Spectral parameters(*) for heparin

Iduronic acid unit			Glucosamine unit		
	35°C	90°C		35°C	90°C
δ_1	5.225	5.218	δ_1	5.401	5.369
δ_2	4.347	4.350	δ_2	3.284	3.298
δ_3	4.205	4.213	δ_3	3.673	3.691
δ_4	4.106	4.120	δ_4	3.771	3.750
δ_5	4.820	4.772	δ_5	4.035	4.049
J_{12}	2.64	3.29	δ_6	4.407	4.378
J_{23}	5.90	6.10	$\delta_{6'}$	4.278	4.284
J_{34}	3.44	3.60	J_{12}	3.66	3.57
J_{45}	3.09	3.14	J_{23}	9.98	9.88
			J_{34}	9.09	8.91
			J_{45}	9.23	9.23
			J_{56}	2.92	2.92
			$J_{56'}$	2.15	2.15
			$J_{66'}$	-11.23	-11.23

(*) Chemical shifts in ppm from internal TSP; coupling constants in Hz

for J_{12} and J_{34} but somewhat larger for J_{23} and J_{45}.

The small deviation of J_{23} and J_{45} might reflect a slight distortion of the chair form $^1C_4$(L) of the iduronic acid unit. Anyway the present data are substantially consistent with the disaccharide repeating residues with glucosamine in $^4C_1$(D) conformation and iduronic acid in

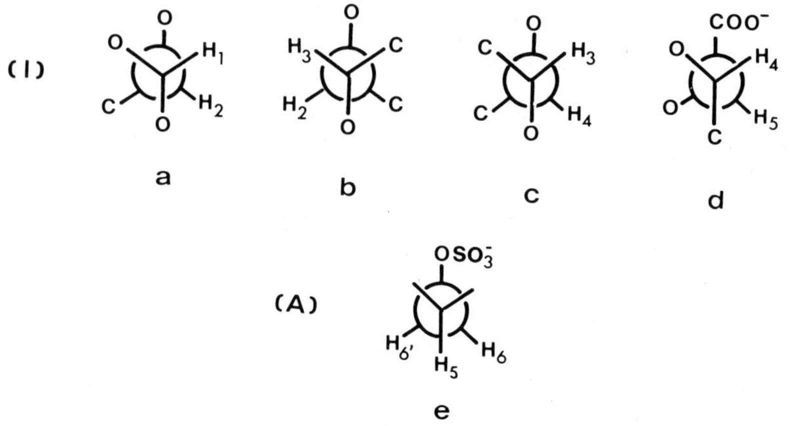

Fig. 3 - Newman projections for ring protons of iduronic acid (I:a,b,c,d) and protons at C_6 of aminosugar (A,e)

$^1C_4$(L) conformation.

The data obtained at 90°C for the fragment C_5-C_6 of the hexosamine unit show that rotamer *e* (tab. 2) is favoured on the basis of the observed small and similar values of J_{56} and $J_{56'}$. These values are substantially different from those observed in D-glucose (13) and in 2-deoxy-2-acetamido-D-glucose (10): apparently in heparin the bulky OSO_3^- group is locked in the *e* form (fig. 3). Inspection of the spectrum at 35°C shows that this situation holds also at room temperature.

3. HEPARIN, ^{13}C SPECTRUM

While the proton spectrum has mainly given conformational information, the ^{13}C spectrum (fig. 4) gives direct evidence both on the primary structure on the configuration of heparin. In the proton broad band decoupled spectrum (upper trace of fig. 4) one can note essentially the eleven signals due to the aliphatic carbons, to which must be added the singlet due to the carboxyl group. This is a further direct evidence of the structure proposed by Perlin (7), *i.e.*, an alternating copolymer with a repeating unit of 12 carbon atoms.

The assignments of signals due to different carbons were obtained by selective single frequency proton decoupling, with the exception of C_6 of glucosamine which is easily recognizable as a triplet in the undecoupled spectrum (fig. 4, lower trace). In order to assure a good selectivity of the decoupling, a irradiation field of about 200 Hz was used. With this procedure it was possible to selectively irradiate pro-

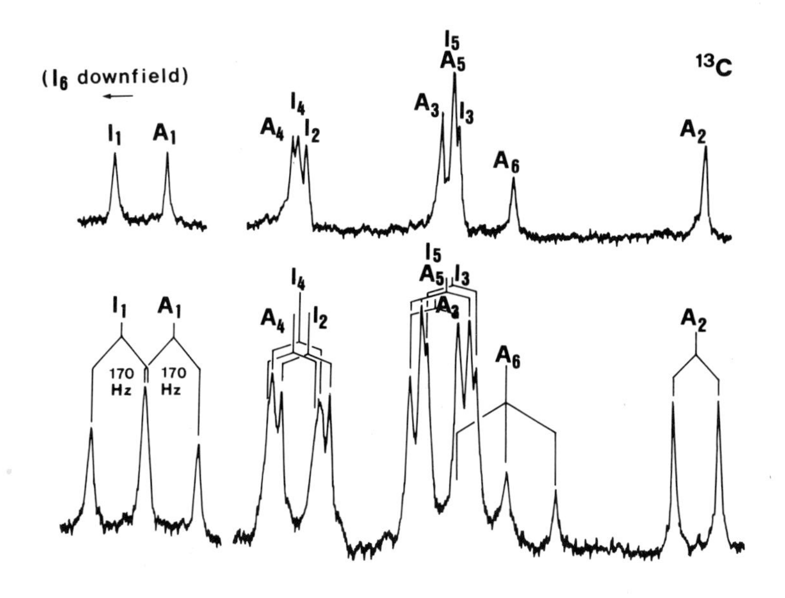

Fig. 4 - ^{13}C-NMR spectrum of heparin measured at 67.9 MHz in D_2O solution (20%) at 40°C. Upper trace: proton broad band decoupled. Lower trace: proton coupled

tons separated by only 0.15 ppm.

The direct proton-carbon-13 coupling constants are reported in tab. 2. The most prominent feature of this table is the size of J_{CH} of both anomeric carbons, which is related to their steric state. In a large series of glycopyranosides it has been observed (14) that α-glycosides with equatorial H_1 give a value of J ∿ 170 Hz whereas β-glycosides with

Tab. 2 - ^{J}C-H Coupling constants of heparin

Glucosamine			Iduronic acid		
C-1	168	Hz	C-1	168	Hz
C-2	138	"	C-2	150	"
C-3	144	"	C-3	150	"
C-4	150	"	C-4	150	"
C-5	144	"	C-5	144	"
C-6	150	"			

axial H_1 give 160 Hz. Thus the value of 168 measured in heparin favours for both the D-glucosamine and L-iduronic acid units the α configuration with equatorial H_1. This finding is in agreement with the conformation obtained from the analysis of the proton spectrum.

Concerning the remaining coupling constants of tab. 2 they are in the rather narrow range of 144-150 Hz with the exception of the J value of C_2 of glucosamine which is reduced to 138 Hz by the presence of the nitrogen atom.

4. SPECTRAL CHANGES INDUCED BY H^+ AND METAL IONS

By adding acids and metal ions to the water solution of heparin it is possible to induce chemical shift changes which reveal specific binding of the polyelectrolyte.

Thus on addition of deuterium chloride the carboxylate group of the iduronic acid moiety is protonated and this fact produces a change in the electronic charge reflected both in the proton and in the carbon-13

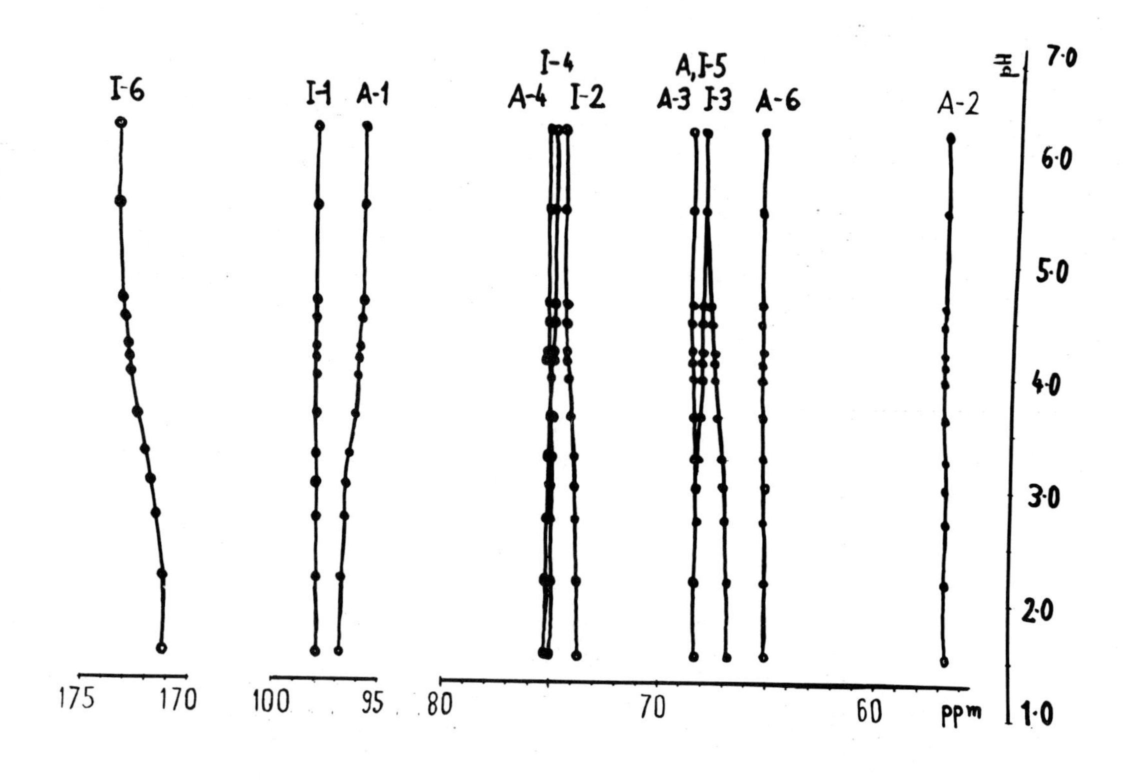

Fig. 5 - ^{13}C-titration shifts of heparin in 20% D_2O solution

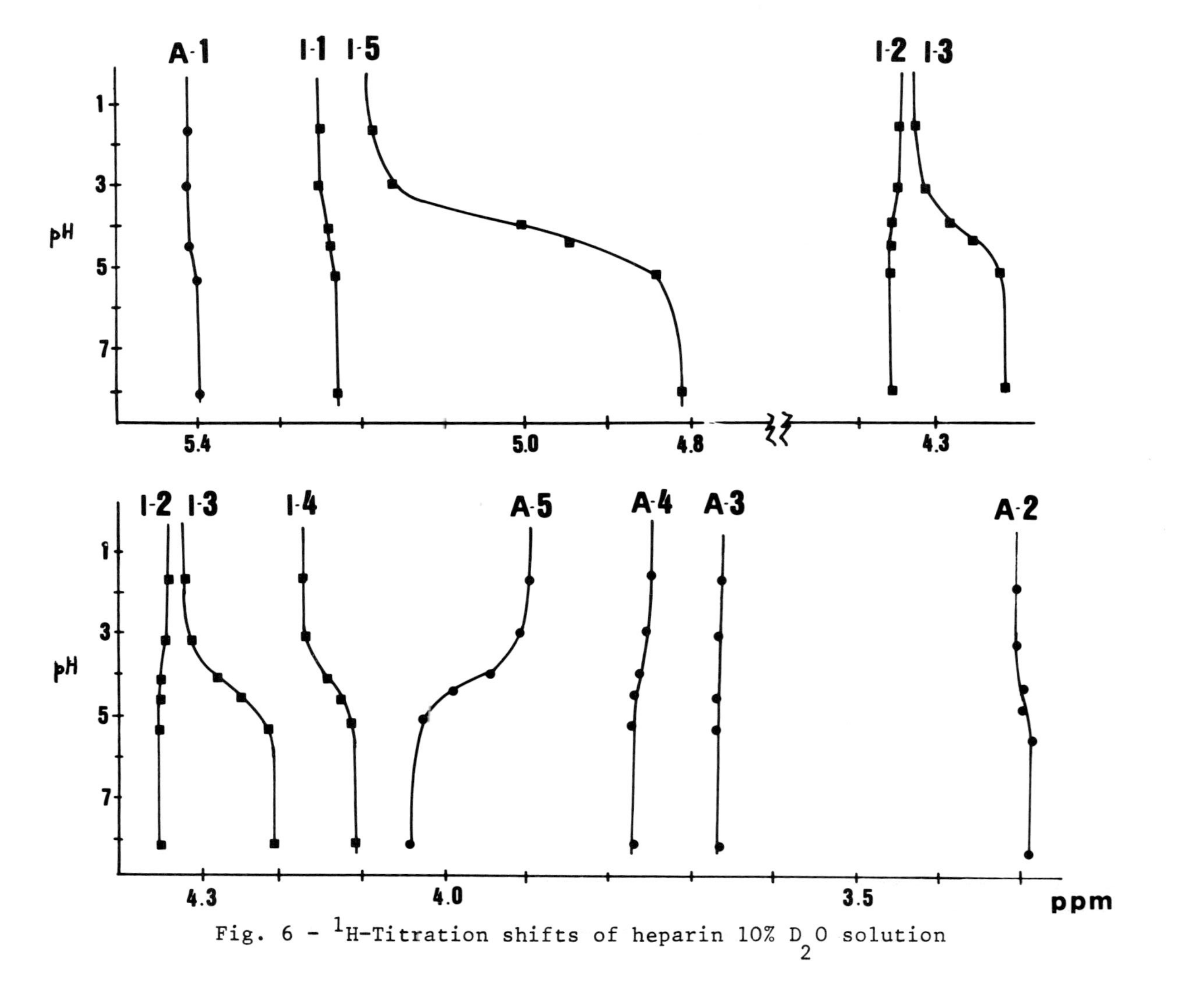

Fig. 6 – ^{1}H-Titration shifts of heparin 10% D_2O solution

spectra. As one can see from the titration curves of figs. 5 and 6 the protonation of the carboxyl group is monitorized by the downfield shift of the H_5 proton and by the upfield shift of the carboxylic carbon and of C_5 of the iduronic acid.

It is known that the effect of protonation on ^{13}C chemical shift of carboxylic acids is attenuated monotonically as a function of the distance from the protonationsite. However, in the present case relatively large shifts are also observed in region of the molecule far from the protonation site, as shown by H_5 proton and C_1 carbon of the glucosamine unit. The latter changes could be attributed to a pH dependent conformational change of the polymeric chain rather then to the carboxylate protonation, also in view of the fact that these shifts are opposite in sign to the protonation shifts.

The titration diagrams of figs. 5 and 6 allow to determine the apparent pK value of the carboxyl group of the iduronic acid. Lowering the concentration of the polymer from 20% to 1% produced an increase of the pK from 3.93 to 5.03. This effect is to be ascribed to the polyelectrolyte character of the heparin, and it must be considered when one makes pK measurements on this kind of polymers.

Another pronounced effect on the spectrum of heparin is that obtained by adding to the solution a complexing agent such as calcium ions. The data reported in tab. 3 are relative to the carbon-13 spectrum, and similar effects

Tab. 3 - Shifts induced in the ^{13}C-NMR spectrum of heparin by addition of calcium chloride

Glucosamine			Iduronic acid		
C-1	+0.06	ppm	C-1	+0.30	ppm
C-2	+0.20	"	C-2	+0.41	"
C-3	+0.12	"	C-3	-0.56	"
C-4	-0.08	"	C-4	+0.30	"
C-5	+0.22	"	C-5	+0.28	"
C-6	+0.09	"	C-6	+0.24	"

are observed also in the proton spectrum (15). Beside the possible application of the calcium ion as a shift reagent (non paramagnetic) it is interesting to note the differential shift induced in the two monomer units: in the average the most prominent effect is detected in the iduronic acid, which suggests a regiospecific complexation in this part of the polymer chain.

This specificity of cation binding was previously (16) noted in the case of gadolinium salt addition to heparin, which resulted to affect more strongly the anomeric proton and carbon resonances of the iduronic acid component. In this case the observed effect was a broadening rather than shift of signals, as the gadolinium ion acts as a paramagnetic relaxation reagent.

In conclusion, one can say that the iduronic acid of heparin is a preferential complexing site for polyvalent cations as shown by a set of different NMR data. This result is in agreement with the recent finding of Laszlo (17) on how heparin binds sodium.

5. CHONDROITIN SULFATES, ^1H SPECTRA

Chondroitin sulfates A and C are mucopolysaccharides having the same components, D-glucuronic acid and N-acetyl-D-galactosamine, but they differ for the position of the sulfate ester group, which is located at C-4 of galactosamine for chondroitin sulfate A and at C-6 for chondroitin sulfate C. Chondroitin sulfate B differs from the others because contains mainly iduronic acid and galactosamine; since its chemical structure and its biological activity are different from A and C the name of dermatan sulfate was proposed (5).

The following formulae

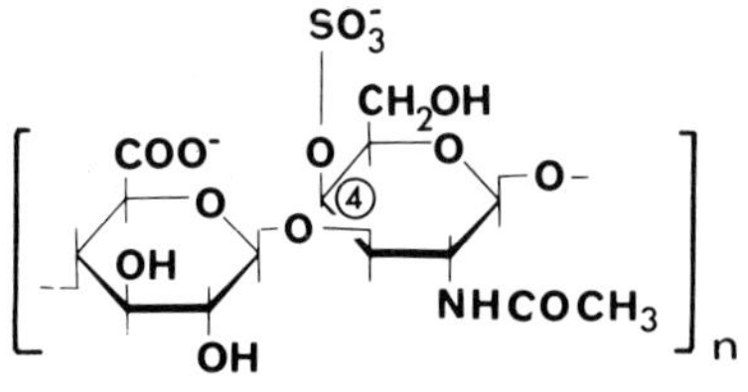

account for the major structural features of these MPS. However heterogeneity is also a common feature. In particular some of the uronic acid in dermatan sulfate is D-glucuronic instead of L-iduronic (1,2).

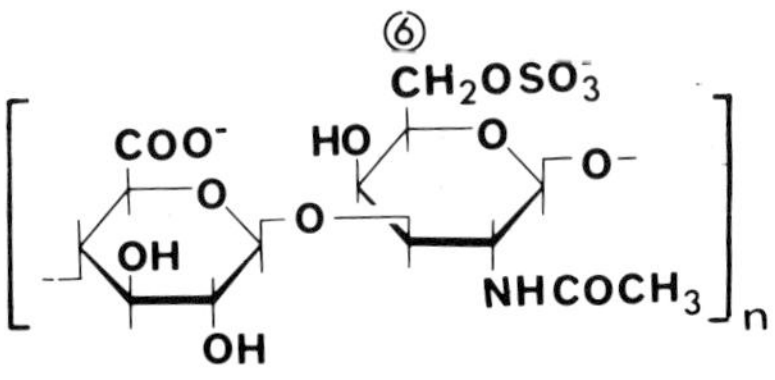

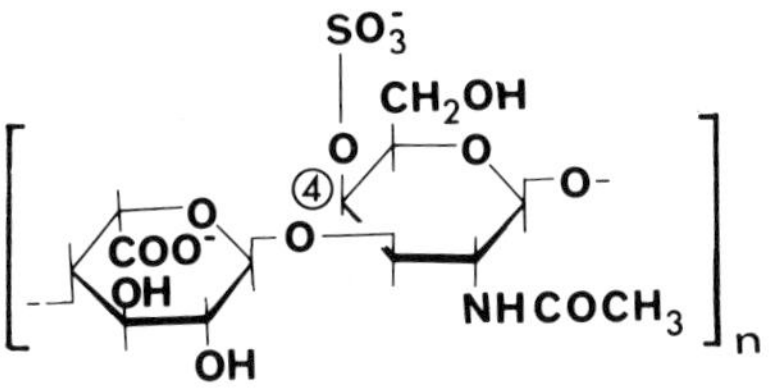

The 270 MHz spectra of the chondroitin sulfates obtained with the same procedure as for heparin are reported in fig. 7.

The resonances of the glucuronic acid in chondroitin-4-sulfate have

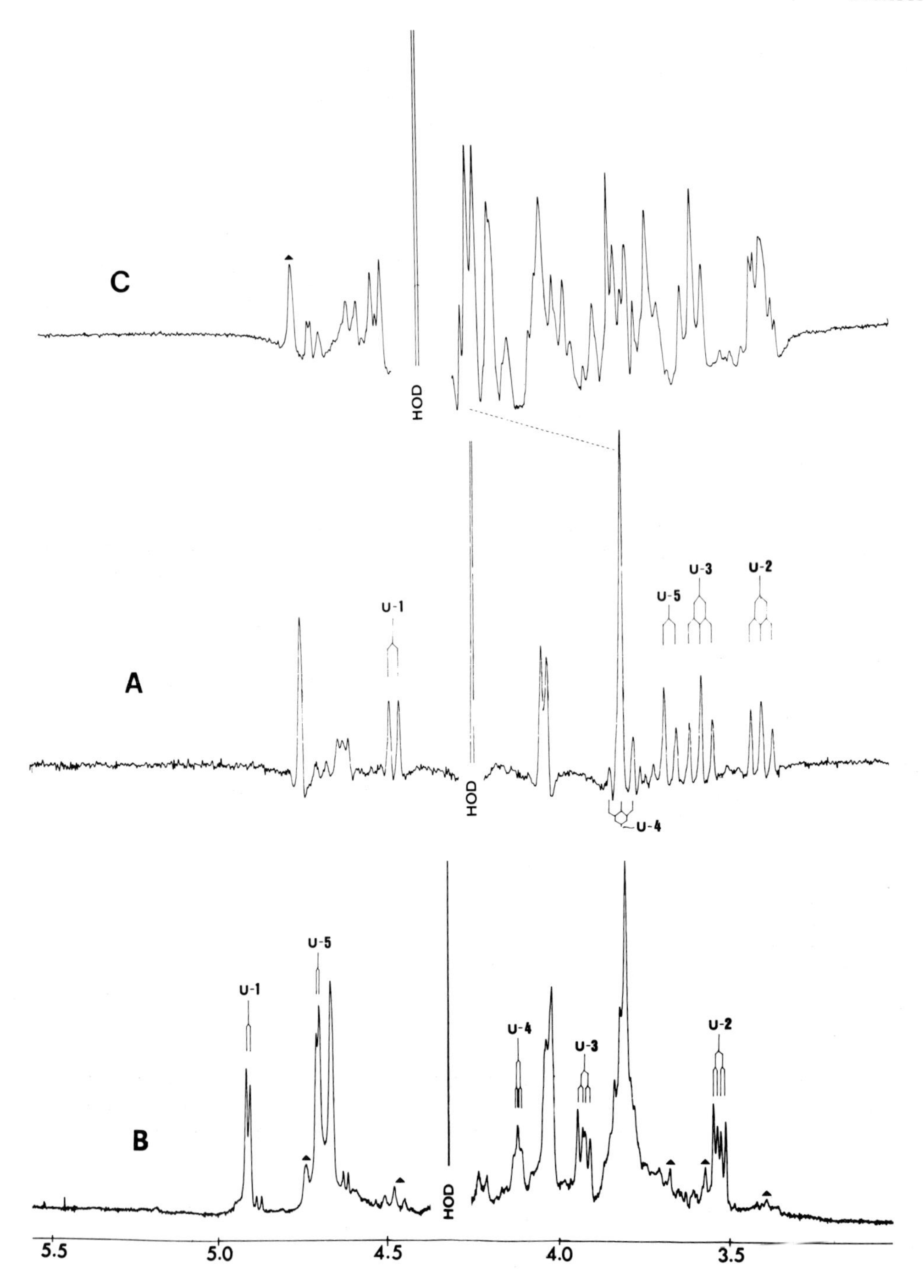

Fig. 7 - ^{1}H-270 MHz spectra of chondroitin sulfates in D_2O solution (10%)

been assigned by homonuclear proton decoupling by starting from the doublet due to the anomeric (H-1) proton of this unit at δ 4.46. The observed large value of the vicinal splittings in the pattern of the uronic acid favours the normal $^4C_1$(D) conformation.

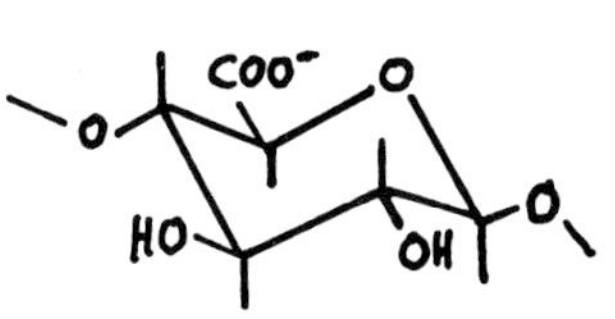

$^4C_1$(D)
glucuronic acid

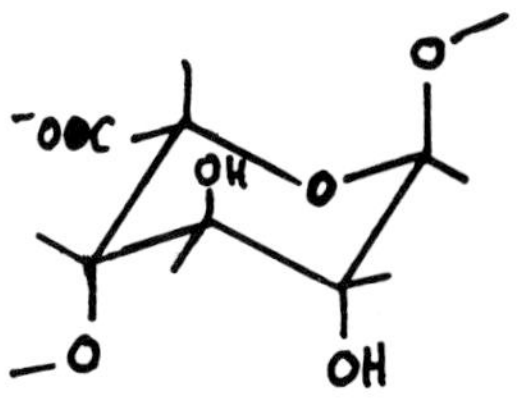

$^1C_4$(L)
iduronic acid

The remaining signals due to the galactosamine on the contrary show small splittings wich suggest that the conformation of this unit is not a perfect $^4C_1$ chair. Clearly, this point needs further investigation.

The conformation of the L-iduronic acid in dermatan sulfate has been the subject of controversy: the proton magnetic resonance spectrum measured at 220 MHz (7) suggested the $^1C_4$(L) conformation, but optical rotatory dispersion gave conflicting results (18). Recent X-ray diffraction (19) studies seem to favour the $^4C_1$(L) form, and further evidence in support of this hypothesis was obtained by periodate oxidation, altough the possibility could not be excluded that a small non sulfated L-iduronic acid unit had the $^1C_4$(L) conformation (20).

The relevant results of the present investigation are reported in the spectrum of fig. 7 where all the signals due to the iduronic acid have been assigned by homonuclear proton decoupling. The pattern of the five protons of the iduronic acid is very similar to that of heparin, with the same couplings but with different chemical shifts. This finding rules out the normal $^4C_1$(L) form and is a direct evidence of the presence of non sulfated L-iduronic acid in the $^1C_4$(L) conformation.

Structural heterogeneity of dermatan sulfate and chondroitin sulfate C is evident from comparison of spectra A-B-C. The peaks labelled with a black triangle are attributable to chondroitin 4-sulfate in B and C.

Work is in progress to establish whether these inhomogeneities are due to cross contamination or are intrinsic (*i.e.* at the polymer level).

6. ENZYMIC DEGRADATION PRODUCTS OF CHONDROITINS

Minor but biologically important structural details of chondroitin

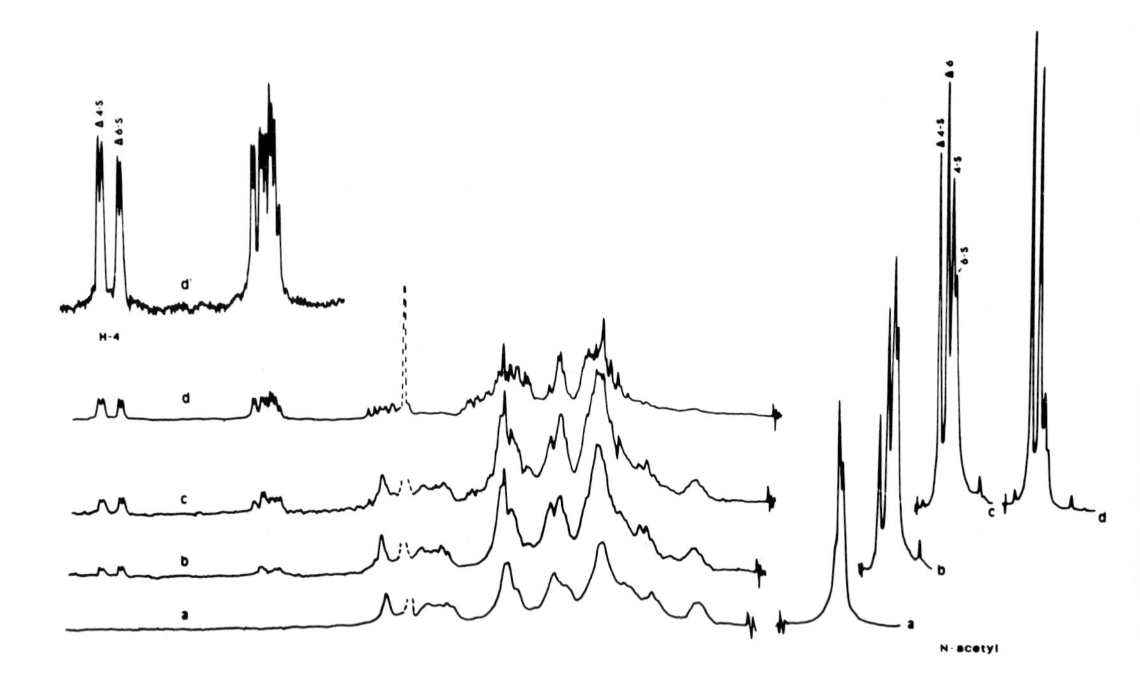

Fig. 8 - Enzymic degradation kinetics of chondroitin from whale cartilage followed by ^{1}H-NMR

sulfates appear to be confined to heterogeneous regions of these polymers (2). In order to unravel such details, a study of the controlled degradation using specific chondroitinases and NMR for monitoring the reaction has recently been started (21).

Fig. 8 illustrates this approach as applied to a sample of chondroitin sulfate from whale cartilage. The enzyme reaction (with chondroitinase AC) was performed directly in the NMR tube starting from a 10% solution in D_2O (a). Immediately after addition of the enzyme the spectrum of the starting material is drastically modified. Signals from the main reaction products, *i.e.*, Δ 4,5-unsaturated-di- and oligosaccharides are especially evident in the lowfield region (olefinic protons) and in the high field region (N-acetyl protons).

ACKNOWLEDGMENTS

This communication contains results of published and unpublished work done in collaboration with Professors A.S. Perlin, B. Casu and J.R. Vercellotti, to whom the author is indebted for stimulating discussions.

REFERENCES

1. Lindahl U., Structure and Biosynthesis of Iduronic Acid-Containing Glycosaminoglycans, in IRS, Organic Chemistry Series Two, 7, 283 (1976) London, Butterworths
2. Muir H., Hardingham T.E., Structure of Proteoglycans in IRS, Biochemistry Series One, 5, 153 (1975) London, Butterworths
3. Atkins E.D.T., Isaac D.H., Nieduszyuski I.A., Phelps F.C. and Sheehan J.K., Polymer, 15, 263 (1974).
4. Rees D.A.,Polysaccharides Conformation in IRS, Organic Chemistry Series One, 7, 251 (1973) London, Butterworths.
5. Jeanloz R.W., Mucopolysaccharides of higher animals in The Carbohydrates, Vol. 2, 589 (1970), New York, Ac. Press.
6. Atkins E.D.T. and Nieduszynski I.A. in Heparin Chemistry and Clinical Usage, 21 (1976) edited by Kakkar V.V. and Thomas D.P., New York, Ac. Press.
7. Perlin A.S., Casu B., Sanderson G.R. and Johnson L.E., Can. J. Chem., 48, 2260 (1970).
8. Campbell I.D., Dobson C.M., Williams R.J.P. and Xavier A.V., J. Magn. Reson., 11, 172 (1973).
9. Gatti G., Casu B. and Perlin A.S., Biochem. Biophys. Res. Comun., submitted
10. Perkins S.J., Johnson L.N., Phillips D.C. and Dweck R.A., Carbohydr. Res., 59, 19 (1977).
11. Booth H., Tetrahedron, 22, 615 (1966).
12. De Bruyn A. and Anteunis M., Org. Magnetic Res., 8, 228 (1976).
13. De Bruyn A. and Anteunis M., Carbohydr. Res., 47, 311 (1976).
14. Bock K. and Pedersen C., J. Chem. Soc. Perkin Trans., 2, 293 (1974).
15. Casu B., Gennaro U., Perlin A.S. and Vincendon M., Communication at Symposium on Glycoconiugates, Brighton 1975
16. Casu B., Gatti G., Natsuko Cyr and Perlin A.S., Carbohydr. Res., 41, C6-C8 (1975).
17. Herwats L., Laszlo P. and Genard P., Nouveau J. de Chimie, 1, 173 (1978).
18. Suzuki S., Saito H., Yamagata T., Anno K., Seno N., Kawai Y. and Furuhashi T., J. Biol. Chem., 243, 1543 (1968);
    Davidson E.A., Biochim. Biophys. Acta, 101, 121 (1965).
19. Atkins E.D.T. and Laurent T.C., Biochem. J., 133, 605 (1973).
20. Frausson L.A., Carbohydr. Res., 36, 339 (1974).
21. Gatti G., Vercellotti J.R., Casu B. and Torri G.G., Communication presented at the Cambridge Symposium of Mucopolysaccharides Club, April 30, 1978.

DISCUSSION

Jones: In the polymer cellulose, which is 1,4-β-linked anhydroglucose, there is intramolecular hydrogen bonding form $(OH)_3$ in one glucose residue to the ring oxygen of the next residue. Is this possible sterically in iduronic acid and could this restriction affect mobility ?

Gatti: We have no experimental information on this but we have found that the ring conformation of the iduronic acid moiety is completely in the $^1C_4$(L) form. Perhaps this is arising because of such bond.

AN *in vivo* ^{15}N NMR STUDY OF BACTERIAL CELL WALLS

Aviva Lapidot and Charles S. Irving
Department of Isotope Research, Weizmann Institute of Science, Rehovot, Israel

INTRODUCTION

Bacterial cells, because of their relatively simple organization, have proved to be useful model systems for studying how the structure and function of cellular organelles are determined by the course of their biosynthesis. Although the sequence of the biochemical events leading to the assembly of cellular structural components determine to a large extent function and supramolecular structure, the physico-chemical properties of the biopolymers themselves might also contribute to shape and function (1). Less is known about the *in vivo* physico-chemical properties of biopolymers, than about their biosynthesis. An understanding of the three-dimensional structures and molecular dynamics of biopolymers is often required to bridge the gap between biochemical information on the one hand, and the structure of a cellular organelle, as revealed by electron microscopy, on the other. An example is the expression of the morphogenetically determined shapes of bacteria. The shape of a bacterial cell is maintained by the rigid peptidoglycan layer in its cell wall (2). It was once believed that correlations could be made between the shapes of bacteria and the chemical composition of the peptidoglycan layer (2). However, it has subsequently been demonstrated that bacteria with different shapes can have identical peptidoglycans and that mutant bacteria can revert from spherical to rod forms without changing the chemical composition of the peptidoglycan layer (3,4). The organization and motional properties of the peptidoglycan layer now appear to be the determinents of cell shape.

Measurements of the *in vivo* physical properties of the biopolymers found in cellular organelles have proved extremely difficult in complex biological systems. Measurements on isolated cellular components cannot always be relied upon, since changes in organization and conformation occur during isolation. Among the biophysical techniques available for characterizing the conformation and dynamic properties of polymeric systems, nuclear magnetic resonance spectroscopy is a promising tool for probing the *in vivo* physico-chemical properties of biopolymers. Nmr resonances originating from specific sites in a polymer can often be

B. Pullman (ed.), Nuclear Magnetic Resonance Spectroscopy in Molecular Biology, 439-459.

resolved and information on the electronic structure, conformation, solvation, and molecular motion of a functional group can be extracted from nmr spectral parameters (5). Recent advances in the correlation of nmr parameters with local and long range segmental motions have shown how nmr measurements can also provide information on the mechanical properties of polymeric materials (5-7). Since the function of many cellular structural components are mechanical in nature, the possibility of correlating dynamic nmr mobility measurements with mechanical properties of polymeric molecules makes nmr an exciting approach to problems in cell structure research.

The ^{1}H and ^{13}C nmr spectra of intact cells are unfortunately complex and consist of envelopes of poorly resolved resonances (8,9). When individual resonances have been observed, they usually originated from small matabolite molecules and not from structural components (8,10). ^{15}N, on the other hand, is a more favorable nucleus for studying complex systems for several reasons. 1) The smaller number of different types of nitrogen functional groups leads to simpler spectra. 2) Spectra are better resolved due to the narrower linewidths, that result from the smaller gyromagnetic ratio of ^{15}N. 3) Additional resolution is provided by the sensitivity of ^{15}N chemical shifts to conformation and solvation effects. 4) Spectra can be further simplified by partial nuclear Overhauser enhancement nulling of broad envelope resonances. In the past, ^{15}N nmr measurements on biological molecules have been hindered by sensitivity problems associated with the low natural abundance of ^{15}N (11). Isotope enrichment techniques, and to a lesser extent the use of large natural abundance samples, have succeeded in overcoming this problem. The amount of physico-chemical information that can be obtained from ^{15}N nmr measurements on polypeptides has increased, as a result of recent progress made in correlating ^{15}N chemical shifts with electronic, conformation, and solvation effects in amino acid and peptide model systems (12-15). The sensitivity of ^{15}N chemical shifts to secondary and tertiary structure has permitted the resolution of several types of glycyl ^{15}N resonances in globin polypeptide chains (16). This enables residual structures to be detected in unfolded proteins (17). The scope of ^{15}N nmr has been extended to purified bacterial cell membranes, envelopes, and walls; where ^{15}N nmr measurements have provided insights into the effect of cell envelope organization on the dynamic properties of the constituent molecules (18,19). ^{15}N nmr resonances have been obtained from systems as complex as intact bacteria, yeast, and mammalian cells (20,21).

In this paper, we show how the ^{15}N nmr resonances of one cellular organelle, the gram positive bacterial cell wall, can be studied *in vivo*. Although these measurements are far from complete, *in vivo* ^{15}N nmr has begun to provide insights into the physico-chemical basis of cell wall structure and function, which cannot be obtained from other spectroscopic measurements, electron microscopy, or biochemical analyses.

A SURVEY OF INTACT CELL AND CELL ENVELOPE SPECTRA

In order to determine what types of cells systems lend themselves to ^{15}N nmr spectroscopy and whether ^{15}N resonances originating from cell envelope components can be unambiguously identified in whole cell spectra, we obtained the proton decoupled ^{15}N nmr spectra of uniformily ^{15}N enriched *Eschericia coli*, *Bacillus licheniformis*, Baker's yeast, and murine Friend leukemic cells (Figure 1, left) and compared them to the corresponding cell envelope spectra (Figure 1, right). Intact cell spectra were found to be relatively simple and consist of only a few resonances in the amide and amino regions. The amide region can be dominated by an envelope of protein backbone peptide resonances (PP1 and PP9 of yeast, PP2 and PP8 of Friend leukemic cells) or the protein backbone resonances can be nulled by proton decoupling, leaving only a few narrow amide resonances, as in the case of *E.coli*. In gram positive bacteria, the protein backbone resonances are broadened beyond detection and only two relatively narrow resonances (P5 and P9 of *B.licheniformis*) are left which we later show originate from cell wall polymers. In all cell spectra, the amino region is dominated by the resonance of the Lys-N_{ω} nitrogen, which is the protein nitrogen group with the highest degree of mobility. The amino resonances of small metabolite molecules are not observed. The amino resonances that are observed, originate from higher molecular weight cellular structural components, which include cell wall teichoic acid (AM4 of *B.licheniformis*) and peptidoglycan (AM4 of *E.coli* and AM5 of *B.licheniformis*). The amino group of phosphatidylethanolamine (PE), originating from cell membranes has been observed in *E.coli*.

A straightforward approach to assigning intact cell resonances to particular cell organelles is to determine whether the intact cell resonances occur in the spectrum of the isolated organelle and whether all the resonances of the isolated organelle occur in the intact cell spectra. When these criteria are applied to the spectra shown in Figure 1, we find that only in the case of the gram positive bacteria, *B. licheniformis*, can intact cell resonances be unambiguously assigned to cell wall components. Such comparisons are based on the assumption that the physical properties of the organelle, that determine which of its resonances are observed do not change during isolation. In the case of *B.licheniformis*, precautions were taken to inactivate cell wall autolysins before harvesting. Even when the covalent structure of a cell organelle can be preserved, the inevitable alteration of supramolecular structure, that occur during isolation, can lead to changes in relative resonance intensities. The separation of the cytoplasmic membrane from the cell wall, that takes place during the mechanical disruption of gram positive bacteria (22), can certainly contribute to quantitative differences between the intact cell and isolated cell wall spectra. In gram positive bacteria, it is possible to study the ^{15}N nmr spectrum of one cellular organelle in its native environment. The remainder of this paper will be devoted to the study of the gram positive cell wall in the intact cell and in purified preparations.

BACTERIAL CELL WALL PROPERTIES

Having shown that it is possible to study the ^{15}N nmr resonances of cell wall components *in vivo*, it is worthwhile to review what is known about gram positive cell walls, with the aim of determining what contributions ^{15}N nmr can make to cell wall research. Gram positive bacteria are encased within an amphoteric polyelectrolyte gel (23-25) that is composed of peptidoglycan and acidic polysaccharides, and accounts for 20-30% of the dry weight of the cell. Cell wall peptidoglycan consists of glycan strands of β(1→4) linked N-acetylmuramic acid and N-acetylglucosamine that have an average length of 15-20 disaccharide units (26). Peptide chains, that contain iso-amino acid residues, are attached to the muramic acid residues and serve as covalent cross-links. The primary sequences of the peptidoglycan peptide chains of five gram positive bacteria (27) are summarized in Figure 2. Since the monomeric peptides are T-shaped, it is useful to divide the peptide into a stem and a cross-bar region, in addition to the cross-linking bridge region (28). The N-acetylmuramic acid residues of the glycan strands can also be substituted with teichoic acid, a polymer of glucose, phosphate, and ribitol substituted with ester-linked D-alanine (29). In *B. licheniformis*, the glycan strands are also substituted with teichuronic acid, a polymer of N-acetylgalactosamine and glucuronic acid (30). Practically nothing is known about 1) the secondary and tertiary structure of peptidoglycans, 2) the arrangement of the acidic polymers, 3) the organization of the polymers into fibers or sheets, or 4) the mode of packing of peptidoglycan within the cell wall (1).

The cell wall is somewhat elastic and contracts when the cell tugor is reduced (23). The cell wall is porous to molecules up to 1.200 M.W. (31) and behaves like a polyelectrolyte gel, which can be made to shrink or swell by altering charge balance within the wall or the ionic strength of the media (24). In the living bacteria, the cell wall maintains the morphogenetically determined shape of the cell (rod or coccus) and prevents the fragile cytoplasmic membrane from bursting, under the high osmotic pressures (15-20 atms) associated with hypotonic environmental conditions (1). The cell wall might also serve to concentrate magnesium ions and ionic nutrients (32). During cell growth, new material is constantly being inserted into the wall and rearranged by autolysis, during the course of cell elongation (26). An imbalance in the insertion and degradation of peptidoglycan can result in abnormal cell shapes (33,34) or the loss of cell wall tensile strength, followed by lysis and cell death (35).

Perhaps the most perplexing problem in bacterial cell wall research is the way in which bacteria, which have been deprived of their cell walls and have lost their native shapes (L-forms), assume the shape of the parent bacteria (coccus or rod), when the synthesis of peptidoglycan and accessory polymers is switched on (3). It is believed that the shape of the cell wall is molded by the topological distribution of cross-links, the lengths of the relatively stiff glycan chains, and the

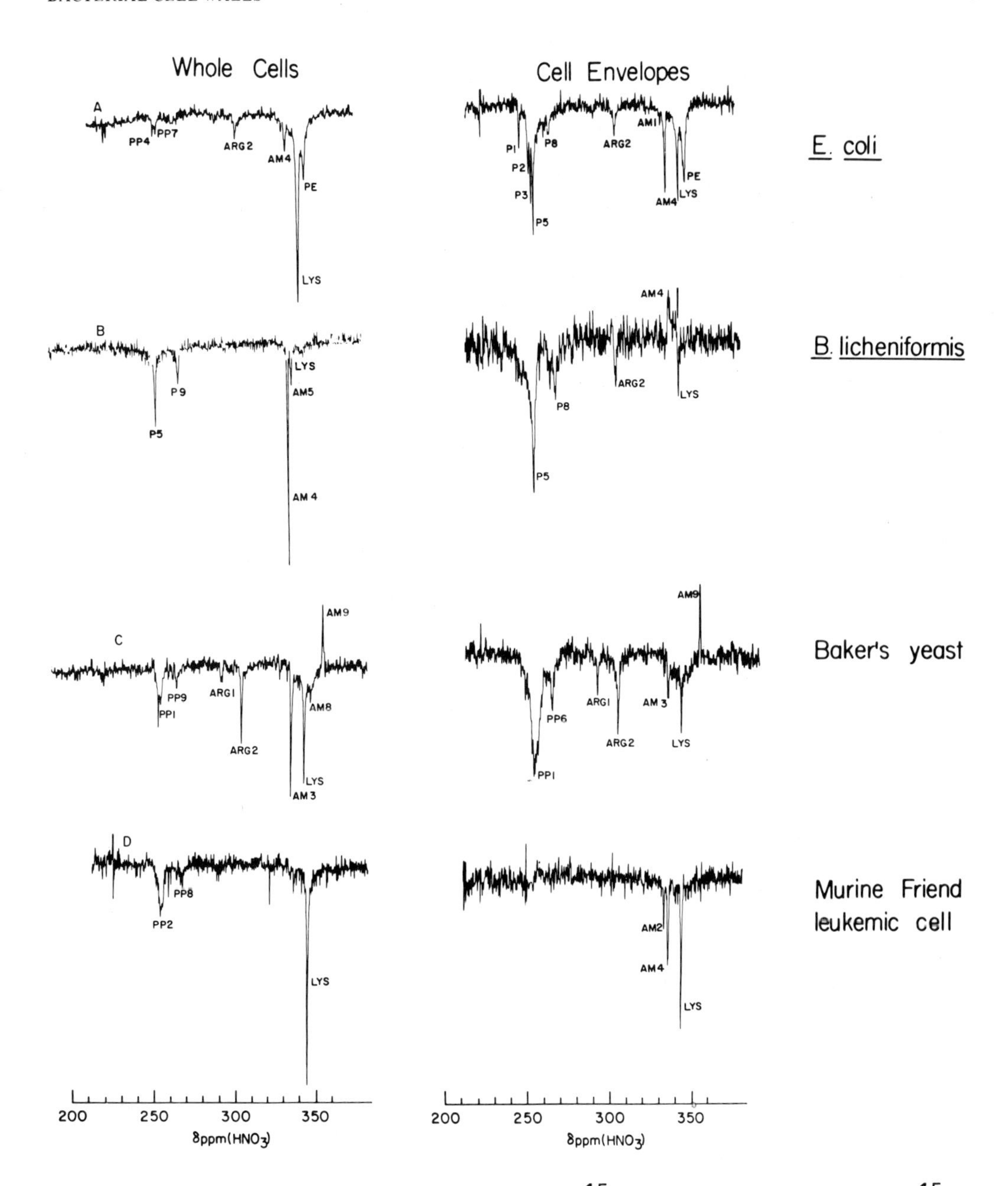

Figure 1.- The proton decoupled 9.12 MHz ^{15}N NMR spectra at 27° of ^{15}N-enriched cells (left) and cell envelopes (right). Spectra of 1.5 cc of packed cells (≃ 300 mg. dry wt.) or cell membranes and walls (≃ 100 mg. dry wt.) were obtained on a Brucker HFX-90 spectrometer using 30,000-60,000 accumulations, 90° pulse angle, 4K data points, 3000 Hz spectral width, 0.94 Hz exponential decay filter.

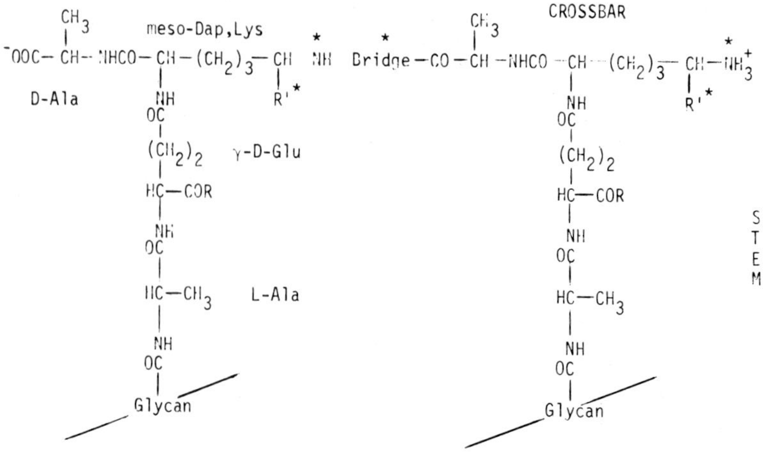

```
       CH3                                          CH3        CROSSBAR
        |         meso-Dap,Lys        *       *      |                               *
 -OOC-  CH--NHCO- CH -(CH2)3- CH   NH  Bridge-CO- CH -NHCO- CH -(CH2)3- CH-NH3+
                   |           | *                          |            | *
 D-Ala            NH           R'                          NH            R'
                  OC                                       OC
                   |                                        |
                 (CH2)2    γ-D-Glu                        (CH2)2
                   |                                        |
                  HC-COR                                   HC-COR
                   |                                        |                    S
                  NH                                       NH                    T
                  OC                                       OC                    E
                   |                                        |                    M
                  HC-CH3    L-Ala                          HC-CH3
                   |                                        |
                  NH                                       NH
                  OC                                       OC
                   |                                        |
                 Glycan                                   Glycan
```

Bacteria	R	R'	Bridge	Accessory Polymer
B.licheniformis	OH, NH_2	$COO^-, CONH_2$	Direct	Teichoic acid Teichuronic acid
B.subtilis	"	"	"	Teichoic acid
S.faecalis	NH_2	H	$-NH-CH(CONH_2)-CH_2-CO-$	Polysaccharides
M.lysodeikticus	$NHCH_2COO^-$	H	$(L\text{-}Ala\text{-}\gamma\text{-}D\text{-}Glu(Gly)\text{-}L\text{-}Lys(D\text{-}Ala))_{1-6}$	Trace amounts of N-acetyl polysacc-harride amines
S.aureus	NH_2	H	$(-NHCH_2CO-)_5$	Teichoic acid

Figure 2.- The primary structure of cell wall peptidoglycan in five gram positive bacteria.

orientations of the polymer in the cell wall (1). All these factors will affect the motions of cell wall polymers and the mechanical properties of the wall. Because there is so little physical information on the configuration and arrangement of macromolecules in the cell walls, no systematic attempt have been made to formulate cell wall functions in terms of the physical chemistry of cell wall polymers.

Before ^{15}N nmr can provide insights into the dynamic basis of cell shape, much basic work remains to be done in assigning cell resonances, determining the mobility gradient along the peptidoglycan peptide chains, characterizing the types of local segmental motions that occur in the peptidoglycan matrix, and outlining the sensitivity of nmr spectral parameters to changes in biological function and integrity of the cell wall. Experiments in this direction are reported below.

BACTERIAL CELL WALL SPECTRA

The proton decoupled ^{15}N nmr spectra of intact cells, isolated cell walls, and cell wall lysozyme digests were obtained for five gram positive bacteria, which have different peptidoglycans and combinations of accessory polymers, as shown in Figure 2. The ^{15}N chemical shifts of each system is tabulated in Table 1. As seen in Figure 3, each bacteria has a unique intact cell ^{15}N nmr spectrum. The uniqueness of the intact cell spectra reflect differences in the chemical composition and degree of mobility of the cell wall.

Partial spectral assignments have been made by specific isotopic labelling and comparison of the lysozyme digest spectra. Specific isotope labelling of S.aureus has provided unequivocal assignments for the peptidoglycan pentaglycine cross-bridge (267.3 ppm), the Lys-N_ω peptide bond (255.1 ppm), one alanyl residue (250.2 ppm), and the C-terminal alanyl residues of cell wall teichoic acid (335.6 ppm). Titration of the M.lysodeikticus lysozyme digest has provided assignments for peptidoglycan C-terminal alanyl (247.0 ppm) and glycyl (259.6 ppm) groups. The free amino group of DAP can be assigned to a resonance at 337.9 ppm. The $-NH_2$ groups of amidated γ-D-Glu residues in B. licheniformis, B.subtilis, and S.faecalis, the amidated DAP in B.subtilis and B.licheniformis, and the amidated γ-Asp in S.faecalis occurs at 268.8, 267.1, and 269.7 ppm, respectively, in the cell wall lysozyme digests. Downfield shifts of the $-NH_2$ groups are observed on going from the lysozyme digests to the intact cell walls. This makes the assignment of the $-NH_2$ groups in the intact cells less certain; however, a tentative assignment is given in Table 1. The 251 and 256 ppm resonances, which occur in the four lysozyme digests measured, must originate from the stem region of peptidoglycan, which is common to all peptidoglycans. The DAP peptide and the glycan and teichuronic acid acetamido resonances are degenerate and occur at 253.4 ppm. Table 2 summarizes current assignments of cell wall resonances.

TABLE 1. OBSERVED ^{15}N RESONANCES IN ^{15}N-LABELLED GRAM POSITIVE BACTERIA

Bacteria	System	Resonances
B. licheniformis	cells	252.8, 266.8, 335.6, 337.9, 343.6
	walls	246.0, 252.7, 264.1 266.6, 335.6, 337.9
	digest	245.5, 250.5, 251.3, 253.4, 256.0, 263.4, 267.06,268.8
B. subtilis	cells	253.4, 266.8, 335.5, 337.8, 343.6
	walls	245.2, 253.0, 267.0, 343.7
	digest	246.1, 249.4, 250.3, 251.1, 252.3(sh), 253.3, 253.9(sh), 256.3, 263.4, 267.06, 268.8, 337.9
S. faecalis	cells	255.7, 260.3, 267.3, 268.4, 335.3, 338.0, 343.3
	walls	247.0, 251.5, 254.6, 255.7, 260.3, 265.3, 267.3, 335.6, 343.3
	digest	249.5, 251.3, 253.7, 254.4, 256.4, 259.6, 268.7, 269.7
M. lysodeikticus	cells	250.2, 251.5, 253.6, 256.2, 260.3, 343.6
	walls	247.0, 250.0, 251.6, 253.5, 257.1, 260.2, 343.5
	digest	247.0, 250.0, 251.6, 253.5, 256.3, 260.2, 343.1
S. aureus	cells	250.2 - 251.7, 255.7, 267.3, 343.9, 349.2
	walls	250.4, 255.6, 267.3
S. aureus-[Lys-$^{15}N_{\omega}$]	walls	255.1, 343.6
	cells	335.6
S. aureus-[Ala-^{15}N]	autolysate	247.0, 250.2, 252.7
S. aureus-[Gly-^{15}N]	cells	267.3, 249.2
	walls	267.3, 249.2

Table 2. ASSIGNMENTS OF CELL WALL ^{15}N NMR RESONANCES

Polysaccharides		
Glycan	(acetamide)	253
Teichuronic acid	(acetamido)	
Teichoic acid	(C-terminal Ala)	335.6
Cross bar		
L-Lys-Nω	(peptide)	255.5
L-Lys-Nω	(amino)	343.5
meso-DAP	(peptide, α-COO^-)	(253)
meso-DAP	(peptide, α-$CONH_2$)	(263.4)
meso-DAP	(amino)	337.8
D-Ala	(C-terminal, peptide)	246
D-Ala	(cross-linked, peptide)	?
Bridge		
pentaglycine	(peptide)	267.3
pentaglycine	(amino)	349.2
iso-aspargine	(peptide)	260.0
iso-aspargine	(amino)	338.0

Stem		
L-Ala		250.5
γ-D-Glu		251.5
γ-D-Gln		256.7
L-Lys-N_α		253
meso-DAP		
meso-DAP	(C-terminal)	?
Carboxylate Substituents		
amide	meso-DAP	267.5
amide	meso-DAP, N-terminal	266.8*
amide	D-Glu	268.8
amide	L-Asp	267.7
amide	L-Asp, N-terminal	267.3*
glycine	D-Glu	260.1

* Position of resonance in the intact cell spectrum.

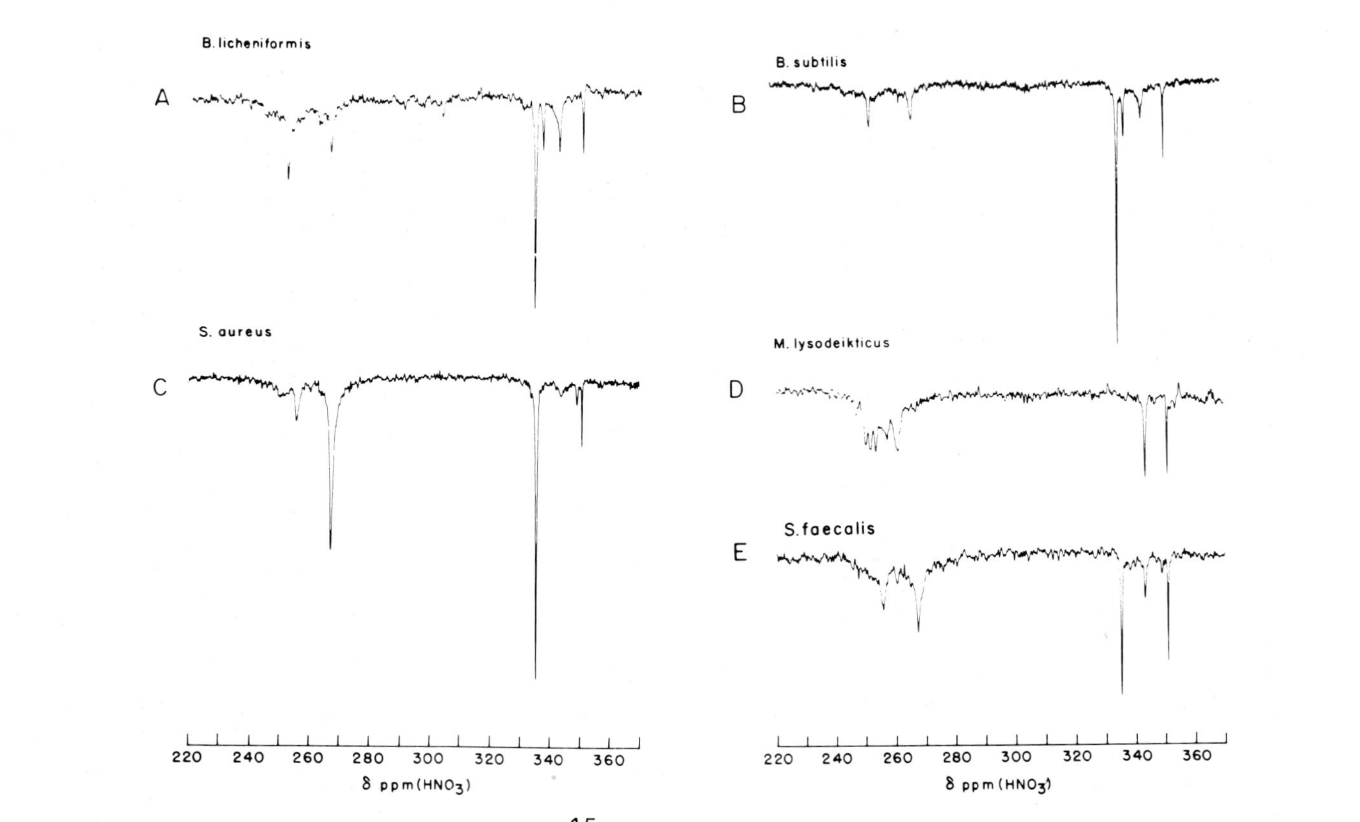

Figure 3.- The proton decoupled 9.12 MHz ^{15}N NMR spectra at 27°C of five ^{15}N-enriched intact gram positive bacterial cells. Spectra of 1.5 cc packed cells (≃ 300 mg. dry wt.) were obtained on a Brucker WH-90 spectrometer using 10,000 accumulations, 90° pulse angle, 4 K data points, 2000 Hz spectral width, 2 Hz exponential filter, with quadrature and alternating phase detection.

Since hydrolytic cleavage of the β(1→4) glycosidic linkage of glycan strands by lysozyme does not affect the covalent structure of the peptidoglycan peptide chains (36), the ^{15}N chemical shifts of the lysozyme digests are representative of the chemical shifts that peptide residues would display in the intact cell, if the peptide residue had water solvated, random conformations. The similarity of many of the intact cell chemical shifts to those of lysozyme digest indicate that intact cell resonances originate from regions of peptidoglycan with random, flexible conformations. Otherwise, the peptide groups would not have the degree of mobility required to produce relatively narrow, inverted resonances. Comparison of spectral amplitudes in the intact cell to those in lysozyme digest (Figures 5,7) relative to an external $^{15}NH_4Cl$ reference signal indicates that a large fraction of the peptidoglycan residues are in a mobile state.

The differences in the five gram positive bacteria intact cell spectra (Figure 3) are readily explained by variations in cell wall composition, as summarized in Figure 2. Substitution of DAP in place of Lys in B.licheniformis and B.subtilis leads to the appearance of a new amino free amino resonance at 337.8 ppm and the replacement of the Lys-N_ω peptide resonance at 255.9 ppm with the resonance of the DAP peptide at 253 ppm. The stronger intensity of the 253 ppm resonance in B.licheniformis, results from the overlap of acetamide resonance of teichuronic acid with the DAP peptide resonance. The substitution of γ-D-Glu carboxylic acid group in M. lysodeikticus with a glycyl residue instead of amide $-NH_2$ leads to the appearance of a new resonance at 260.7 ppm, and the absence of amide resonances in the vicinity of 267 ppm. M.lysodeikticus cell walls do not contain teichoic acid and explains the absence of the 335.6 ppm resonance, which dominates the amino region of the other four intact cell spectra.

The observation of a ^{15}N nmr resonance in the intact cell or cell wall depends on the resonance having a relatively narrow (<70H_z) linewidth and a large negative nuclear Overhauser enhancement. Normally we assume that when a proton decoupled resonance is not observed, it has been broadened beyond detection by the large dipole interaction, that results from a low degree of motional freedom. However, in the case of proton decoupled ^{15}N nmr, normally narrow resonances can be lost, as a result of partial quenching of the nuclear Overhauser enhancement by factors other than mobility (e.g. paramagnetic ions) (37,38). In this case, relatively narrow resonances would be observed in gated decoupled spectra (proton decoupling only during spectral accumulation), but not in broad band noise decoupled spectra. Examination of gated decoupled and broad band noise decoupled spectra of B.licheniformis, B.subtilis, S.faecalis, and M.lysodeikticus cells and cell walls (Figures 4-7) show that the peptidoglycan stem resonances (250, 251 and 256 ppm) have not been nulled by proton decoupling. They probably occur in the low intensity broad envelope, which has nuclear Overhauser enhancement factor of -2.

MOBILITY ALONG THE PEPTIDOGLYCAN PEPTIDE CHAIN

Some insights into the mobility gradient in peptidoglycan can be obtained by examining from which positions along the peptide chain (Figure 2) the resonances of the intact cell orignate. The DAP and Lys-N_{ω} peptide resonances, the pentaglycyl resonance, and the DAP and iso-Asp amide NH_2 resonances of B.licheniformis, B.subtilis, S.aureus, and S.faecalis all originate from either the cross-bar or bridge regions of the peptide. No resonance was observed, which could be shown to originate from the stem region of the peptide. Neither was any resonance observed which could be uniquely assigned to the glycan strands. We can conclude that in a tight peptidoglycan matrix, the glycan strands and the peptide stems are rigid, while considerable mobility exists in the peptide cross-bars and bridge regions.

M.lysodeikticus has a "loose" peptidoglycan matrix, in which not all the N-acetylmuramic acid residues are substituted with peptide. Furthermore, the bridges consist of replicas of the peptide itself $[(L\text{-Ala-}\gamma\text{-D-Glu(Gly)L-Lys(D-Ala)}]_{1-6}$. The low degree of cross-linking is believed to introduce considerable "looseness" into M.lysodeikticus peptidoglycan (27). This is borne out by the intact cell spectrum of M.lysodeikticus, which is the only intact cell spectrum that displays all the resonances observed in the lysozyme digest. In a "loose" peptidoglycan there is an apparently high degree of mobility throughout the entire peptidoglycan peptide chain. It is reasonable that peptidoglycan stem residues, whose resonances were not observed have a low degree of motional freedom, because they engage in interactions that contribute to the function of the cell wall.

MOBILITY AND CELL WALL TENSILE STRENGTH

^{15}N nmr of intact cells can be used to determine whether the principal function of the cell wall, its ability to maintain cell shape and to withstand high osmotic pressures, is directly related to the rigidity of the glycan strands and the peptide stems. The tensile strength of the cell wall is lost upon limited hydrolytic cleavage of glycan $\beta(1\rightarrow4)$ glycosidic bonds by lysozyme or by autolytic cleavage of the muramic acid L-Ala amide linkage (26). When no precautions are taken to inactivate the autolysins, autolysis can occur by anerobic shock of cells during harvesting (39). Autolysis can also occur when antibiotics, such as vancomycin, are used to block cell wall assembly in rapidly growing cells (40).

Mild lysozyme treatment of osmotically protected B.licheniformis cells results in an intact cell spectrum that is practically identical to the spectra of highly mobile lysozyme digests of intact cell wall and purified peptidoglycan (Figure 8). Cleavage of a relatively small number of glycosidic linkages results in an increase of the mobility of peptidoglycan peptide stem residues throughout the peptidoglycan matrix. However, subsequent glutaraldehyde cross-linking of the free amino groups in the lysozyme treated cells leads to the disappearance of the stem

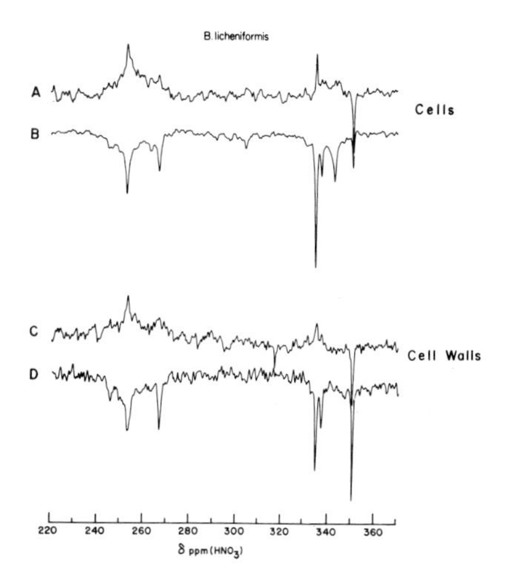

Figure 4.- The gated decoupled (A,C) and broad hand noise decoupled (B,D) ^{15}N nmr spectra of B. lichenformis cells (A,B) and cell walls (C,D). Proton irradiation was off 3 sec between accumulations.

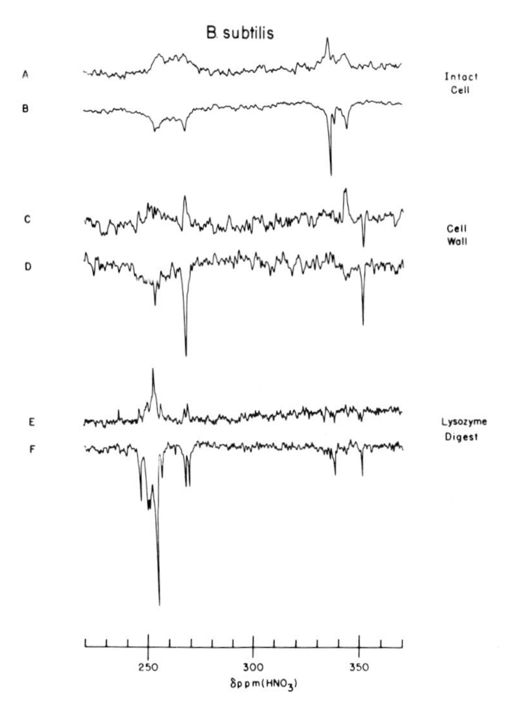

Figure 5.- The gated decoupled (A,C,E) and broad band noise decoupled (B,D,F) ^{15}N nmr spectra of B.subtilis cells (A,B), cell walls (C,D), and cell wall lysozyme digest (E,F).

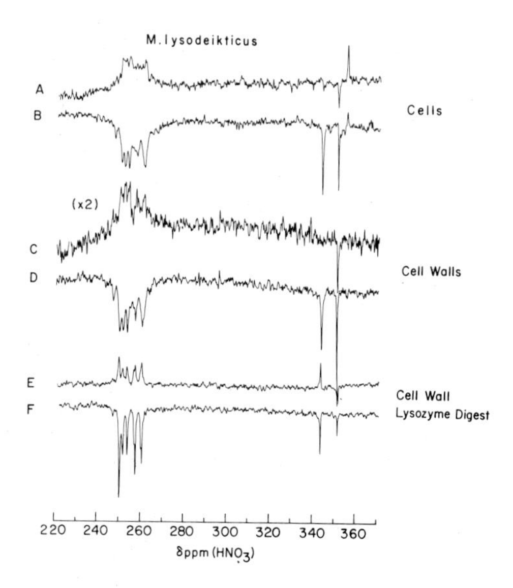

Figure 6.- The gated decoupled (A,C,E) and broad band noise decoupled (B,D,F) ^{15}N nmr spectra of M. lysodeikticus cells (A,B), cell walls (B,C), and cell wall lysozyme digest.

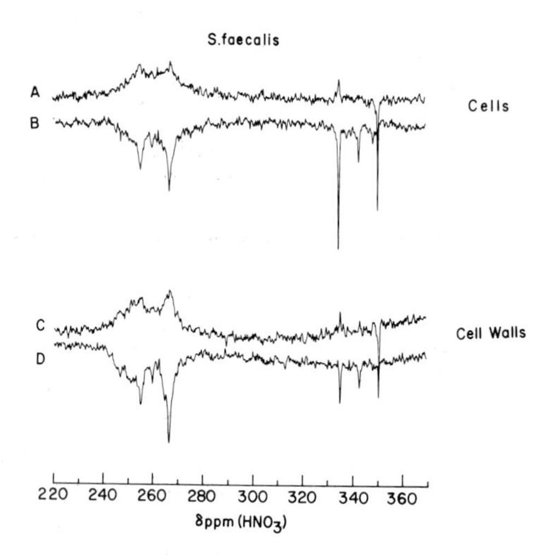

Figure 7.- The gated decoupled (A,C) and broad band noise decoupled (B,D) ^{15}N nmr spectra of S. faecalis cells (A,B) and cell walls (C,D).

resonances (Figure 8). A complete set of peptidoglycan stem resonances can also be observed (Figure 9) in B.licheniformis cells harvested without inactivation of autolysins. Treatment of B.licheniformis cells with bactericidal doses of vancomycin produced a different type of intact cell spectra, which may have resulted from the binding of vancomycin to peptidoglycan that still remained attached to the wall. In summary, the appearance of the peptidoglycan stem resonances paralles the loss of cell wall tensile strength. Limited disruption of the covalent structure of peptidoglycan probably relaxes steric constraints placed on the cell wall polymers at the time of their insertion into the existing wall or during subsequent remodeling of the wall that accompanies cell elongation.

MOBILITY AND NON-COVALENT BONDING

Ionic, hydrophobic and hydrogen bonding interactions have been proposed as stabilizing forces in the cell wall, which might contribute to tensile strength (41). Peptidoglycan residues participating in non-covalent bonding interactions are expected to have a low degree of mobility. It was therefore of interest to determine whether peptidoglycan stem resonances could be observed in cell wall spectra under conditions leading to the unfolding of polypeptides.

Stem resonances were not observed in purified B.licheniformis peptidoglycan, suspended in 6M guanidine hydrochloride either at 27° or 60°C (Figure 10). Stem resonances were not observed in intact B.licheniformis cell walls in aqueous suspension at 80°C (Figure 11). Phenol, another reagent which disrupts intra-molecular non-covalent bonding, had no affect on cell wall spectra (Figure 12). Exhaustive washing of cell walls with EDTA, which probably removes bound metal ions that could cross-link carboxylate groups (42), did not change the cell wall spectra (Figure 12). Although the gram positive bacterial cell wall has many gel properties, it does not resemble polysachharide gels with non-covalent cross-links that undergo reversible thermal denaturation. On the other hand, protonation of B.licheniformis cell wall carboxylate groups at pH 1.5 led to the appearance of additional resonances at 250.3 and 254.3 ppm (Figure 13), which may have originated from stem residues. Among non-covalent bonding interactions, only ionic interactions affect cell wall mobility, as measured by the appearance of additional peptidoglycan resonances.

^{15}N NMR AS A TOOL IN MICROBIAL RESEARCH

Considerable attention was given over the last twenty years to the elucidation of the primary structure of peptidoglycan by biochemical techniques (27). The basis of the changes in cell morphology have been sought in the degree of amidation and cross-linking of peptidoglycan (3,4) as well as in the arrangement of accessory molecules, which are difficult to determine by biochemical degradation techniques. Once the amino acid composition of a peptidoglycan is established by biochemical analysis, ^{15}N nmr spectroscopy of cell wall lysozyme digests provides

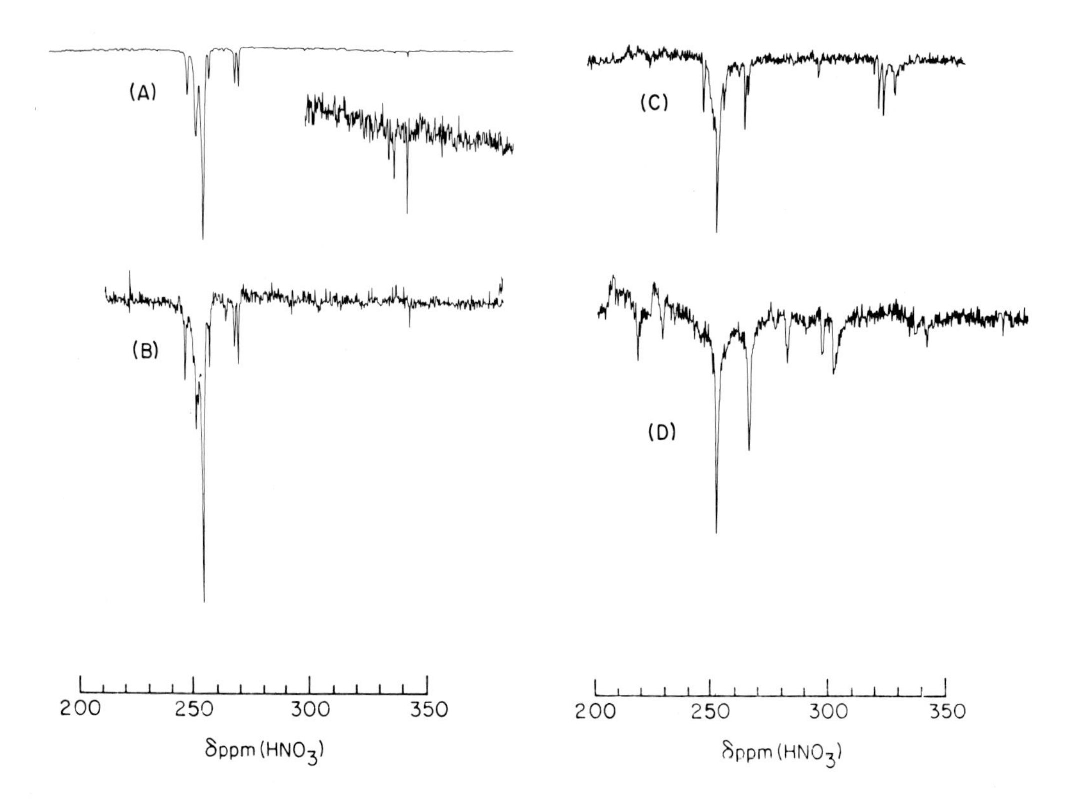

Figure 8.- The ^{15}N NMR spectra of the lysozyme digest of B. licheniformis cell walls stripped of teichoic acid and teichuronic acid (A), the lysozyme digest of an intact cell wall (B), lysozyme treated cells in 2 M sucrose (C), and lysozyme treated cells fixed with glutaraldehyde (D). Spectral conditions as in Figure 1.

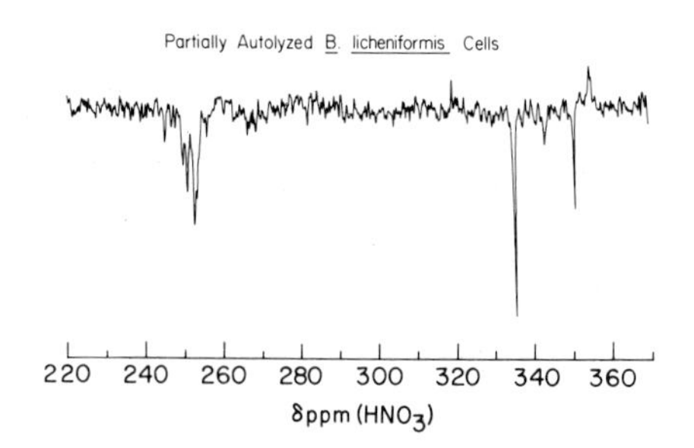

Figure 9.- The ^{15}N NMR spectrum of B. licheniformis cells harvested without addition of 4% SDS to inactivate autolysins. Spectral conditions as in Figure 3.

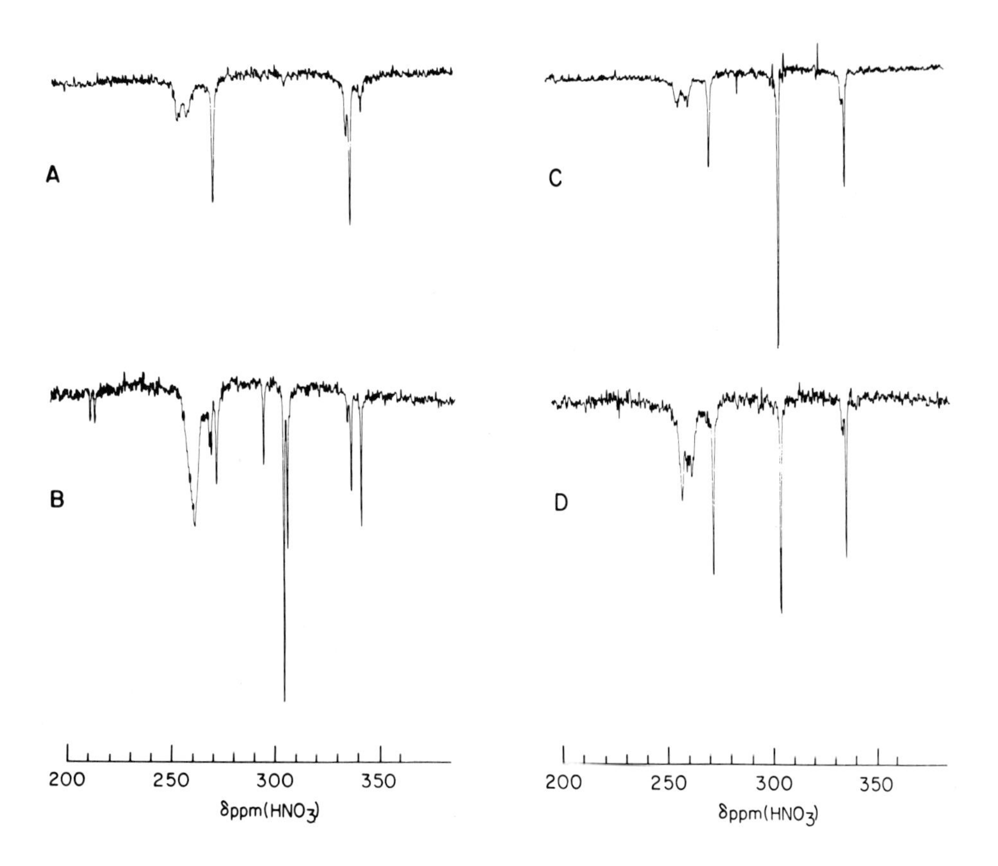

Figure 10.- The ^{15}N NMR spectra of B. licheniformis cell walls stripped of teichoic acid and techuronic acid A in H_2O (A), in 6 M guanidine hydrochloride (B), in 6 M guanidine hydrochloride at 27° (C) and 60°C (D) after boiling in 4% SDS for 5 minutes to remove non-covalently bound protein. Spectral conditions as in Figure 1.

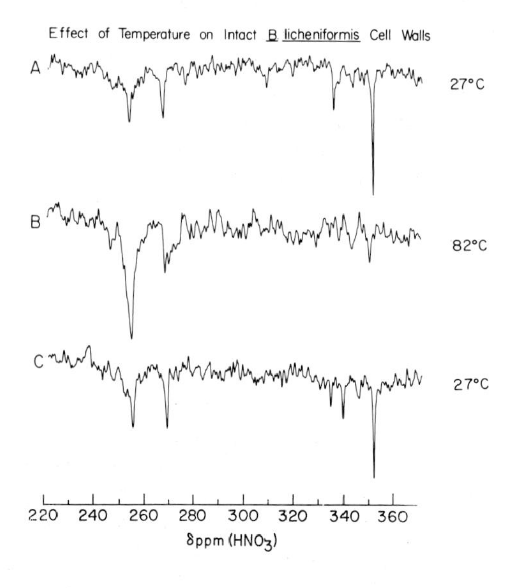

Figure 11.- The ^{15}N nmr spectra of B. licheniformis cell walls at 27°C (A), 80°C (B), and returned to 27°C (C). Spectral conditions as in Figure 3.

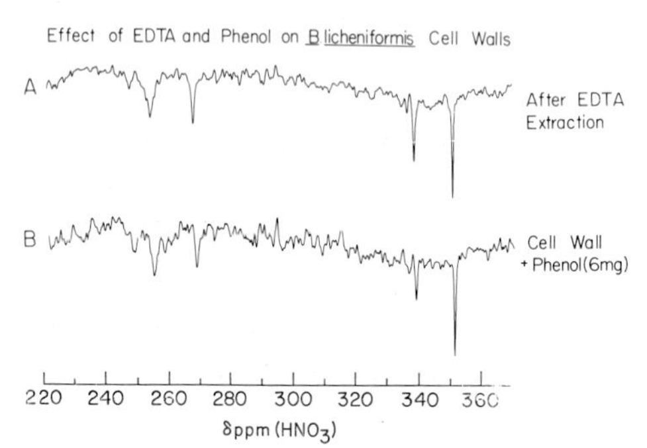

Figure 12.- The ^{15}N nmr spectra of B. licheniformis cell walls (from Figure 11C) extracted with EDTA (A) and then suspended in 0.1 M phenol (B). Spectral conditions as in Figure 3.

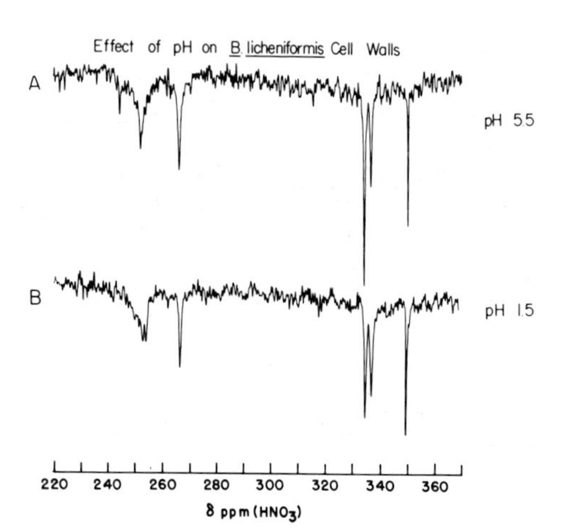

Figure 13.- The ^{15}N nmr spectra of B. licheniformis cell walls at pH 5.5 (A) and pH 1.5 (B). Spectral conditions as in Figure 3.

ready estimates of the degree of cross-linking and amidation, which can be used to screen for changes in cell wall organization. Since there has been so little physical information on bacterial cell walls, several three-dimensional model building studies have sought to determine the mode of packing of peptidoglycan in the cell wall (28,41, 43-45). Aside from measurements of the changes in cell wall volume (23-25) and infra-red spectroscopy (44) and X-ray scattering measurements (44,46) of dried peptidoglycan films, there has been no way to test the validity of the various models proposed. Now models can be judged on the basis of whether they provide for the mobility observed in the intact cell wall ^{15}N spectra. Model building studies have been based on the assumption that stable structures are determined by hydrogen bonding. ^{15}N nmr spectroscopy has shown, however, that ionic and steric effects are the important determinants of packing of the peptidoglycan peptide chains. ^{15}N nmr spectroscopy is a potentially powerful tool for determining whether the arrangement of the peptidoglycan in various cell wall structures (polar caps, elongated regions, cross walls, and septa) differ, and whether changes in organization of polymers occur during cell growth. ^{15}N nmr spectroscopy is probably the only tool available which can provide any insights into the mechanism of the change in cell shape during the reversion of mutant bacteria. Antibiotics are generally considered to inhibit the enzymes engaged in cell wall biosynthesis or to bind to cell wall precursors (26). ^{15}N nmr spectroscopy can be used to test whether the additional possibility that the binding of antibiotics to the bacterial cell wall itself alters the conformation of pre-exisitng cell wall polymers which inhibits the insertion of new wall material. ^{15}N nmr mobility measurements can provide information on how cell wall autolysins affect cell wall mechanical strength. In all these studies the power of ^{15}N nmr, as a tool in microbial research, is considerably enhanced by the fact that the cell wall can be studied in its native environment.

REFERENCES

1. H.J. Rogers (1974) Ann.N.Y.Acad.Sci. 235:29-51.
2. W. Weidel and H. Pelzer (1964) Adv.Enzymol. 26:193-232.
3. H.J. Rogers, M. McConnell and R.C. Hughes (1971) Journal of General Microbiology 66:297-308.
4. U. Schwartz and W. Leutgeb (1971) J.Bacteriol. 106:588-595.
5. J. Schaefer (1974) in: G.C. Levy (ed), Topics in Carbon-13 NMR Spectroscopy, John Wiley & Sons, New York, pp.150-208.
6. F. Heatley and A. Begum (1976) Polymer:399-408.
7. J. Schaefer, E.O. Stejskal and R. Buchdahl (1977) Macromolecules 10:384-405.
8. N.J.M. Birdsall, D.J. Ellar, A.G. Lee, J.C. Metcalfe and G.B. Warren (1975) Biochim.Biophys.Acta 380:344-354.
9. A. Daniels, R.J.P. Williams and P.E. Wright (1976) Nature 261: 321-323.
10. R.T. Eakin, L.O. Morgan, C.T. Gregg and N.A. Matwiyoff (1972) FEBS Letters 28:259-264.

11. D. Gust, R.B. Moon and J.D. Roberts (1975) Proc.Natl.Acad.Sci.USA 72:4696-4700.
12. G.E. Hawkes, E.W. Randall and C.H. Bradley (1976) Nature 257:767-772.
13. V. Markowski, T. Posner, P. Loftus and J.D. Roberts (1977) Proc.Natl.Acad.Sci.USA 94:1308-1309.
14. M. Llinas and K.Würthrich, Biochim.Biophys.Acta, in press.
15. F. Blomberg, W. Maurer and H. Rüterjans (1976) Proc.Natl.Acad. Sci.USA 73:1409-1413.
16. A. Lapidot, C.S. Irving and Z. Malik (1976) J.Am.Chem.Soc. 98:632-634.
17. A. Lapidot and C.S. Irving (1977) J.Am.Chem.Soc. 99:5488-5490.
18. C.S. Irving and A. Lapidot (1977) Biochim.Biophys.Acta 470:251-257.
19. A. Lapidot and C.S. Irving (1977)in: M. Goodman & J. Meienhofer (eds), Peptides, Proc.Fifth Amer.Peptide Symp, John Wiley & Sons, New York, pp.419-422.
20. M. Llinas, K. Wüthrich, W. Schworzer and W. von Philipshorn (1975) Nature 257:817-818.
21. A. Lapidot and C.S. Irving (1977) Proc.Natl.Acad.Sci.USA 74:1988-1992.
22. G.R. Millward and D.A. Raveley (1974) J.Ultrastructure Res. 46:309-326.
23. R.E. Marquis (1968) J.Bacteriol.95:776-781.
24. L.T. Ou and R.E. Marquis (1970) J.Bacteriol. 101:92-101.
25. L.T. Ou and R.E. Marquis (1972) Can.J.Microbiol. 18:623-629.
26. J.M. Ghuysen and G.D. Shockman (1973) in: Loretta Leive(ed), Bacterial Membranes and Walls, Marcel Dekker, Inc., New York, Chp.2.
27. J.M. Ghuysen (1968) Bact.Rev. 32:425.
28. E.M. Oldmixon, S. Glauser and M.L. Higgins (1974) Biopolymers 13:2037-2060.
29. J. Baddiley (1970) Accounts Chem.Res. 3:98-105.
30. R.C. Hughes, J.G. Pavlik, H.J. Rogers and P.J. Tanner (1968) Nature 219:642-644.
31. R. Scherrer and P. Gerhardt (1971) J.Bacteriol. 107:718-735.
32. R.E. Marquis, K. Mayzel and E.L. Carstensen (1976) Can.J.Microbiol. 22:975-982.
33. M.J. Tilby (1977) Nature 266:450-452.
34. N.H. Mendelson (1976) Proc.Natl.Acad.Sci.USA 44:1740-1744.
35. H.J. Rogers (1967) Nature 213:31-33.
36. J.M. Ghuysen (1972) Proc.Lysozyme Conf. pp. 185-193.
37. C.S. Irving and A. Lapidot (1975) J.Am.Chem.Soc. 97:5945-5946.
38. J.F. Farnell, E.W. Randall and A.I. White (1972) J.Chem.Soc.,Chem. Commun. 1159-1160.
39. E. Work (1971) in: J.R. Norris & D.W. Ribbons (eds), Methods in Microbiology, Academic Press, pp.361-418.
40. H.J. Rogers and C.W. Forsberg (1971) J.Bacteriol. 108:1235-1243.
41. M.V. Kelemen and H.J. Rogers (1971) Proc.Natl.Acad.Sci.USA 68:992-996.

42. R.C. Hughes and P.P. Thurman (1970) Biochem.J. 119:925-926.
43. J.T. Tipper (1970) International Journal of Systematic Bacteriology 20:361-377.
44. H. Formanek, S. Formanek and H. Wawra (1974) Eur.J.Biochem. 46, 279-294.
45. V. Braun, H. Gnirke, U. Henning and K. Rehn (1973) J.Bacteriol. 114:1264-1270.
46. H.H.M. Balyuzi, D.A. Reaveley and R.E. Burge (1972) Nature New Biology 235, 252-253.

DISCUSSION

Feeney: Have you tried to extend your studies to investigate, in vivo, the effects of antibiotics known to interfere with vacterial cell wall synthesis ?

Lapidot: B.licheniformis cells treated with vancomycin at bacteriostatic doses show a reduction in the intensity of the 252.8 ppm acetamido resonance. The effect resembles that observed during cell plasmolysis in 2M sucrose, which may be related to the reported plasmolysis of bacilli by vancomycin. Bactericidal doses of vancomycin lead to a general reduction in the intensity of cell wall resonances and the appearance of new peptidoglycan resonances in the whole cell spectra of B.licheniformis cells, harvested when lysis reached 30%. The binding of vancomycin to non-growing cells had no effect on the cell wall resonances; however, the binding of vancomycin to M.lysodeikticus cell walls led to a marked increase in the NOE of cell wall resonances.

THE APPLICATION OF AROMATIC RING CURRENTS IN THE ELUCIDATION OF DRUG-LIGAND AND METALLO-PORPHYRIN COMPLEXATIONS

By R.J. Abraham
The Robert Robinson Laboratories, The University of Liverpool, England.

1. Introduction

The aromatic ring current of the circulating π-electrons in benzenoid systems has been used as a conformational probe to study a wide variety of biologically important molecules in solution or complexed ever since the original suggestion of the ring current shift by Pople.[1]

The equivalent dipole model used by Pople was largely superceded by the current loop model of Johnson and Bovey[2], in which the magnetic field of the two π-electron "current loops" was calculated directly, the spacing between the loops ($\pm$ 0.64Å) being adjusted to give the correct chemical shifts. More recent quantum-mechanical calculations (see references 3 and 4 for recent reviews) have not found such wide acceptance and the Johnson and Bovey tables have been one of the most widely used calculations in the n.m.r. literature for many years. It has the great advantage over the equivalent dipole model that the high field shift of nuclei inside the current loop in the molecular plane (as in the annulenes) is correctly predicted.

In a recent development of a ring current model for the porphyrin ring system (see later) we had occasion to compare the equivalent dipole and current loop models.[5] To our considerable surprise the models were in virtually complete agreement for all inter-molecular positions.

As the equivalent dipole model has considerable advantages in its simplicity and application, we wish to consider here this model of the aromatic ring current and illustrate its use in a variety of systems of biological interest.

B. Pullman (ed.), Nuclear Magnetic Resonance Spectroscopy in Molecular Biology, 461-479.

2. The Equivalent Dipole Model

In a magnetic field (B_o) the six π-electrons of benzene circulating with an angular frequency ω equal $eB_o/2mc$ each produce a current of $e\omega/2\pi$. The total ring current (i) is thus given by

$$i = 3e^2 B_o/2\pi mc \tag{1}$$

The magnetic field of this current loop at any point $P(r,\theta)$ (Figure 1) is given, on the equivalent dipole approximation, by

$$\Delta B = M(3 \cos^2\theta - 1)/r^3 \tag{2}$$

where the equivalent dipole M equals iA/c, A being the area enclosed by the loop. Dividing ΔB by B_o gives the screening constant for this direction, and averaging over all orientations gives the isotropic screening constant ($\Delta\sigma$)

$$\Delta\sigma = \frac{\Delta B}{3B_o} = \frac{e^2 A}{2\pi mc^2} (3 \cos^2\theta - 1)/r^3$$

Inserting the appropriate values gives finally

$$\Delta\sigma = 4.48 \text{ A } (3 \cos^2\theta - 1)/r^3 \text{ p.p.m.} \tag{3}$$

with all the distances in Å.

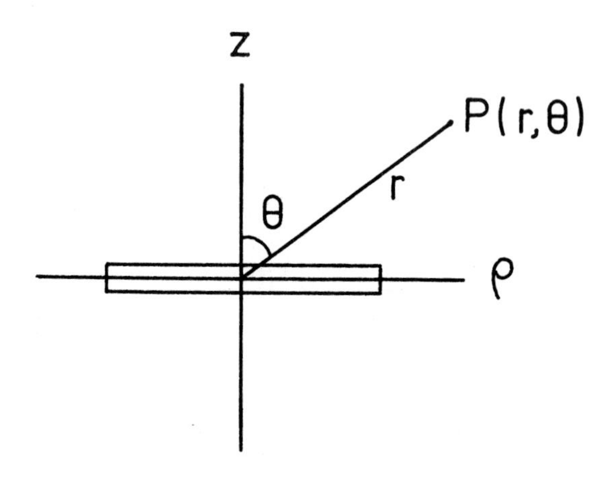

Figure 1. Co-ordinate axes used in the calculations

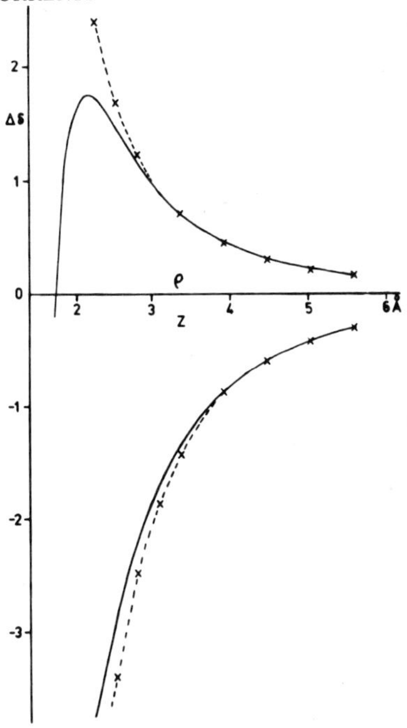

Figure 2. Calculated shielding contributions above (z) and in the plane (p) of the benzene ring. Solid line current loop; broken line, equivalent dipole (from reference 5).

In his original work[1], Pople took the area of the benzene ring A equal πR^2 with R = 1.4Å, i.e. A = 6.16Å^2.

Incorporating this value in equation (3) and adopting the standard convention in which $\Delta\delta$ is the low field shift at any point P, gives finally

$$\Delta\delta = 27.6\ (1 - 3\cos^2\theta)/r^3 \qquad (4)$$

We shall use this equation henceforth. However it is pertinent to note that even this simple calculation is not without its parametrisation. The actual area of the benzene ring is 5.08 Å^2 and if this value of A was used the constant in equation 4 is 22.8, significantly less.

The comparison of the equivalent dipole model (equation 4) and the current loop calculations is shown schematically in Figure 2 and for some selected nuclei in Table 1. On approaching the benzene ring in the molecular plane (the ρ axis) the two curves are identical for all distances $\geqslant$3Å. For the benzene ring protons (p=2.5A) equation 4 gives $\Delta\delta$1.82 p.p.m. versus 1.57 from the current loop model, but for nuclei further removed the shifts are identical. For example for all the nuclei of the toluene methyl group

Table 1: Comparison of Equivalent Dipole and Current Loop Ring Current Shifts (p.p.m.)

System		Equivalent Dipole	Current Loop	Observed Shift
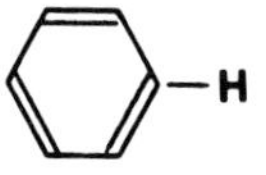		1.82	1.57	1.29, 1.48[a]
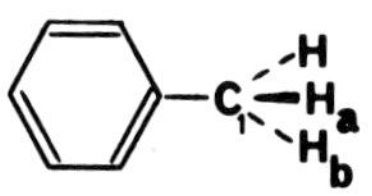	C_1	1.11	1.04	
	H_a	0.50	0.50	
	H_b	0.63	0.64	
	H_{AV}	0.59	0.59	0.57[b]
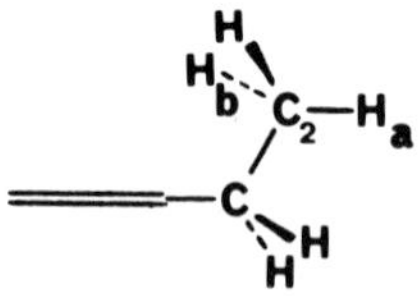	C_2	0.29	0.30	
	H_a	0.19	0.19	
	H_b	0.09	0.09	
	H_{AV}	0.12	0.12	0.28[c]
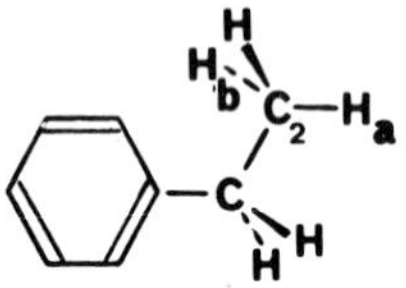	C_2	0.53	0.52	
	H_a	0.26	0.26	
	H_b	0.43	0.44	
	H_{AV}	0.37	0.38	

[a] benzene vs cyclohexa-1,3-diene

[b] toluene vs 2-methyl cyclohexa-1,3-diene

[c] Ethylbenzene vs methyl linolenate (Varian A-60 catalogue, nos. 505 and 337).

(C and H) the predicted shifts are identical (Table 1).

Figure 2 also shows the comparison along the z (normal to the benzene ring) axis. Again the two curves are identical for all distances $\geqslant 3.5$Å. At shorter distances the equivalent dipole model gives slightly larger shifts (e.g. z = 2.5 Å 3.4 vs 2.8 p.p.m.) but even at these close distances the curves are comparable. The important point to note is that for all <u>inter</u>molecular separations and indeed for most <u>intra</u>-molecular positions the dipole and current loop calculations are identical. It may be

Table 2: The Equivalent Dipole Formulation of the Ring Current Effect of Nucleic Acid Bases

I II

$$\Delta\delta \text{ (p.p.m.)} = \sum_{i=5,6} M_i(1-3z_i^2/r_i^2)/r_i^3$$

	Molecule	M_5	M_6
Adenine	I, $R_1 = NH_2$, $R_2 = H$	11.1	19.0
Guanine	I, $R_1 = O$, $R_2 = NH_2$	10.8	6.5
Hypoxanthine	I, $R_1 = O$, $R_2 = H$	10.8	7.3
Cytosine	II, $R_1 = NH_2$, $R_2 = H$	-	6.0
Uracil	II, $R_1 = O$, $R_2 = H$	-	2.4
Thymine	II, $R_1 = O$, $R_2 = CH_3$	-	2.4

argued that the agreement along the p and z axes is fortuitous and that particularly for sites immediately above the current loop there would be significant disagreement. In fact for a position P (z=2.5, p = 1.4 Å) equation 4 gives Δ 1.51 p.p.m. and the current loop tables 1.54 p.p.m.

As equation 4 is simpler, easier and more accurate to use than looking up tables of ring current shifts, and much more convenient than computing them via the current loop model, there is no reason to prefer the current loop model for any situation involving the ring current shift of a simple benzene ring, either alone or in multicyclic molecules (napthalene etc.).

One must of course note that this does not apply to the calculation of ring current shifts of those nuclei within the current loop in the aromatic macrocyclic molecules.

3. Application to Nucleic Acid Bases

One of the more significant recent applications of the current loop model in biological systems has been due to Geissner-Prettre, Pullman and co-workers, who in a series of publications[6], have successively improved these calculations for application to the ring currents of nucleic acid bases. The results of these studies have been widely applied in n.m.r. studies of nucleic acids. It seemed therefore that the equivalent dipole model could be usefully applied in this field also and indeed this is found to be the case.

Table 2 gives the corresponding equivalent dipole model for the nucleic acid bases in which the effective moments M_i of the five and six-membered rings of the various bases have been obtained by direct comparison with the tables of reference 6. Thus the use of the equation and parameters of table 2 will give identical results to the current loop model for all intermolecular positions (reference 6 quotes $z \geqslant 3$ and $p \geqslant 4$ Å as sterically allowed positions). An additional useful characteristic of the equivalent dipole model is that the value of the effective moment M of any ring system is simply proportional to the area and the ring current. Thus comparison of the M_6 values of table 2 with the corresponding value for benzene (27.6 equation 4) gives directly the relative ring currents in the two systems, and when allowance is made for the area of the five membered ring similar comparisons can be made from the M_5 values.

4. Application of the Equivalent Dipole Model to Complex Formation in the Imipramine hydrochloride/Benzyl alcohol system

The form of the equivalent dipole model (equation 4) is exactly the same (for obvious reasons) as the McConnell-Robertson equation[7] for the pseudo-contact shift produced by lanthanide substrates. This equation has now come to be accepted by the great majority of investigations into LIS and all the computer programmes written to deduce the correct position of the lanthanide atom from the observed shifts incorporate the McConnell-Robertson equation[8].

Thus the fact that the form of the equivalent dipole model is identical to that of the McConnell-Robertson equation means that all the iterative LIS programmes may be applied without any modifications to the determination

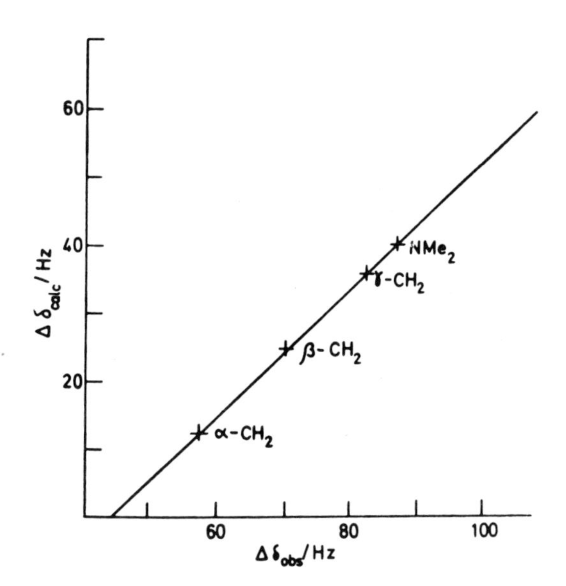

Figure 3. Observed versus calculated complexation shifts ($\Delta\delta$, Hz at 100 MHz) of imipramine hydrochloride-benzyl alcohol (from reference 9).

of the position of an aromatic molecule in a complex from the observed complexation shifts.

Of course there are the important provisos that the complexation shift must be due solely to the aromatic ring current and that for aromatic-substrate complexes in solution a specific well-defined complex is formed. We wish to consider an example of the use of this technique in a drug-ligand interaction.

(1) $CH_2CH_2CH_2\overset{+}{N}HMe_2\ \overset{-}{Cl}$

In the course of an investigation into the conformations of the tricyclic antidepressant imipramine hydrochloride (1) and some related neuroleptics it was observed that benzyl alcohol produced a large high field shift of certain protons in the molecule, notably those on the side chain (see figure 3) either as the solvent or on addition to a solution of (1) in $CDCl_3$.[9]

A complete analysis of the concentration dependence of these high field shifts in terms of a 1:1 complex between (1) and benzyl alcohol in $CDCl_3$ solution gave an association constant of 0.43 l mol^{-1} and complexation shifts for the

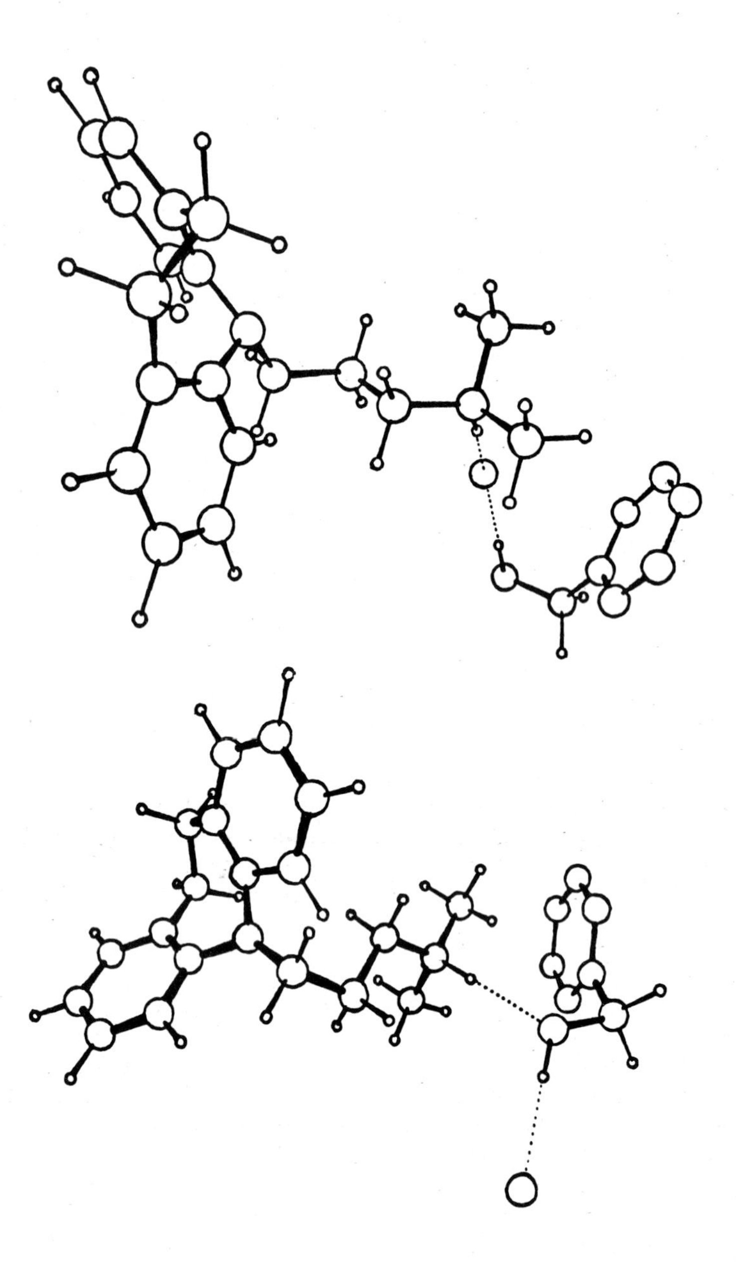

Figure 4. The ligand attached and ligand separated models of the imipramine hydrochloride-benzyl alcohol complex.

side-chain protons shown in figure 3.

These complexation shifts can now be used together with the known geometry of imipramine hydrochloride in solution to determine the position of the aromatic ring of the benzyl alcohol in the imipramine-alcohol complex, using a standard LIS programme.

This however only determines the position of the aromatic ring of the benzyl alcohol. The deduction of the geometry of the complex requires a knowledge of the position of the hydroxyl group as well. This can be inferred from the fact that (1) exists as an ion-pair in solvents of moderate dielectric constant and that further studies showed that a prerequisite for complex formation was that the ligand molecule requires an aromatic ring and an hydroxyl group in proximity.[9]

These results allowed the deduction of the geometry of the imipramine hydrochloride-benzyl alcohol complex. Two models of this complex, in which the aromatic ligand is acting as a ten-plate for the ion-pair are possible. These are the ligand attached and solvent separated ion pair geometries (figure 4). In the former the aromatic ring of the alcohol is close to the NMe_2 of the drug, with the OH proton hydrogen bonded to the Cl counter ion. In the latter the aromatic part still binds to the NMe_2 group, but now the OH group inserts between the NH and the Cl ions.

Further studies would be required to differentiate between these two possibilities. However this example does illustrate the usefulness of the equivalent dipole model in such cases. Finally it should be noted that the results of such iterative LIS programmes to the determination of the structure of aromatic solute complexes must always be considered carefully, particularly in view of the possibility of non-specific ASIS. This is illustrated here by figure 3, which is a linear plot which however does not go through the origin, but has an intercept of ca 40Hz. This is very probably due to non-specific averaging of the benzyl alcohol ring current at all the imipramine protons, which can occur simultaneously with the specific 1:1 complex illustrated.

5. A Ring Current Model of the Porphyrin and Related Rings.

The attempts which have been made to reproduce the ring current shifts of the porphyrin ring are based on either the single loop model or on the network approach.[10]

Table 3. Calculated and observed ring current shifts in metalloporphyrins

Proton		Coordinates	Ring current shift (ppm) Calculated		Observed
meso-H	(4.46, 0.00, 0.00)		5.30		5.2
β-H	(4.51, 2.66, 0.00)		4.40		4.4
β-CH_3 (β-CH_2)	Average position	(5.28, 2.78, 0.00)	2.51		
	H	5.27, 2.79, 1.03	1.97	2.17	2.13
	H H	5.42, 1.91, −0.51	2.48		
		5.13, 3.67, −0.51	2.05	2.27	2.06
β-$CH_2\underline{CH_3}$		6.50, 2.99, 1.45	0.94		
		5.21, 1.87, 1.97	1.06	0.97	0.86
		4.92, 3.63, 1.97	0.91		
Zn—pyridine	α-H	2.07, 0.00, 2.67	−7.93		−6.30
	β-H	2.15, 0.00, 5.12	−2.27		−1.93
	γ-H	0.00, 0.00, 6.40	−1.74		−1.78
Zn—picoline	γ-CH_3	1.03, 0.00, 7.18	−1.29		−1.29

The single loop models so far reported do not give the relative shifts of the porphyrin meso and beta ring protons nor can they be adapted easily to the reduced porphyrin ring systems. The network model breaks the macrocycle down into a number of small current loops, for example into the four pyrrole rings and four hexagons[11] or into the four pyrrole rings and one large ring.[12]

We wished to obtain a porphyrin ring current model for use in metalloporphyrin complexation studies (see later). Insertion of a diamagnetic metal (in our case zinc) into the porphyrin does not affect the proton chemical shifts,[13] thus the ring current is unchanged. However, the application to metalloporphyrins allows the use of the high field shifts of protons on apical ligands (e.g. pyridine) attached to the metal to be used as additional tests of the ring current model.

Good agreement with the observed ring current shifts of protons both in the molecular plane and above it was obtained by the double dipole model.[14] In this model each of the π-electron current loops is replaced by an equivalent dipole. The spacing of the equivalent dipoles was taken from the benzene current loop calculations[2] (±0.64Å), and the porphyrin ring current shifts are the sum of the

Table 4. Calculated and observed ring current shifts ($\Delta\delta = \delta_P - \delta_C$) of chlorins with respect to porphyrins

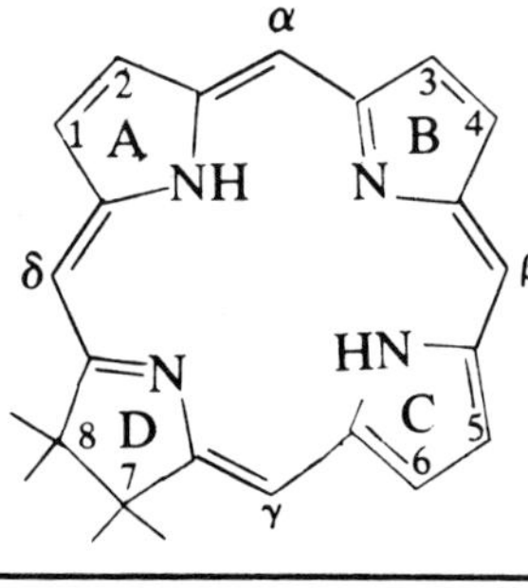

Proton		Ring current shifts ($\Delta\delta$) Calculated	Observed[a]	Observed[b]	Observed[c]
meso-H	α,β	0.35	0.38	0.39 0.36	0.36(α) 0.27(β)
	γ,δ	1.25	1.23	1.20	1.07(δ)
β-H	1,6 / 2,5 / 3,4	0.44		0.46[d]	
β-Me	1,6	0.28		0.20(1)	0.18(1)
	2,5	0.18		0.19(5)	0.15(5)
	3,4	0.15		0.09(3)	0.06(3)
β-C$\underline{H}_2$CH$_3$	1,6	0.29	0.26	0.22(6)	
	2,5	0.19	0.22	0.18(2)	
	3,4	0.16	0.13	0.10(4)	
β-CH$_2$C$\underline{H}_3$	1,6	0.16	0.20	0.10(6)	0.21(4)
	2,5	0.11	0.18	0.10(2)	
	3,4	0.08	0.14	0.07(4)	0.06(4)
γ-CH$_2$					0.68

[a] OEP vs OEC.
[b] Aetio-I vs Aetiochl-I.
[c] 2VC-e_6 vs Chn-e_6.
[d] Deuteroporphyrin-IX dimethyl ester vs deuterochlorin-IX diester.

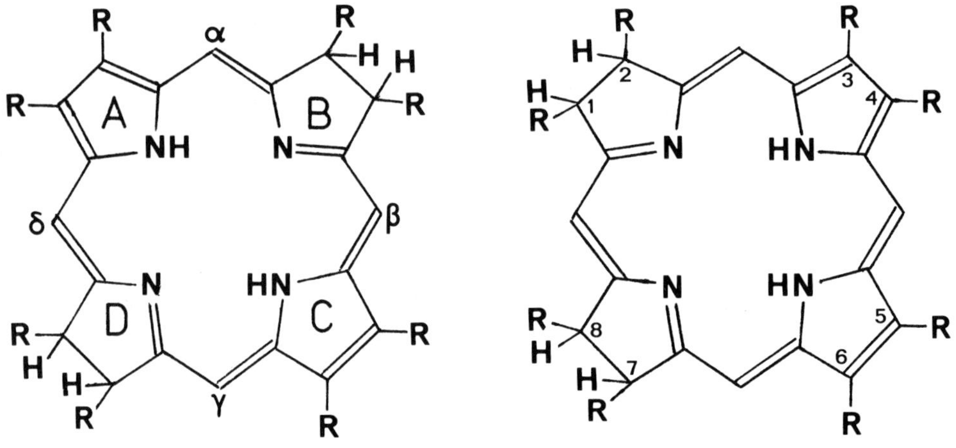

THPA THPB

contributions of these 16 equivalent dipoles. The results (table 3) are encouraging in that the only disagreement is for the pyridine α-protons which are only 2.7Å above the porphyrin plane, well within the Van-der-Waals radius of the porphyrin π-electrons. It is quite likely that other contributions (e.g. steric shifts) could be appreciable of these separations.

The application of the model to reduced porphyrins is straightforward.[14] The important chlorin ring of the chlorophyll system has ring D reduced (table 4), which removes the ring D ring current and reduces the number of π-electrons from 22 to 20. It is convenient to consider both the observed and calculated shifts for this system as the differences from the analogous porphyrin. The general agreement (table 4) is so good that the calculated shifts have been used to assign certain resonances in unsymmetrical chlorins. Note also that the characteristic high field shift of the meso δ proton in the chlorin ring is quantitatively explained.

The application of the model to the tetrahydroporphyrin (THP) ring systems follows similar lines.[15] In octaethyl THPA (R = Et) the observed and calculated (in parenthesis) ring current shifts compared with octaethylporphyrin (OEP) are for the meso and beta CH_2 protons 1.35(1.36) and 0.37 (0.38) p.p.m. respectively. As a further check the ring current shifts of the beta CH_2 protons on the saturated

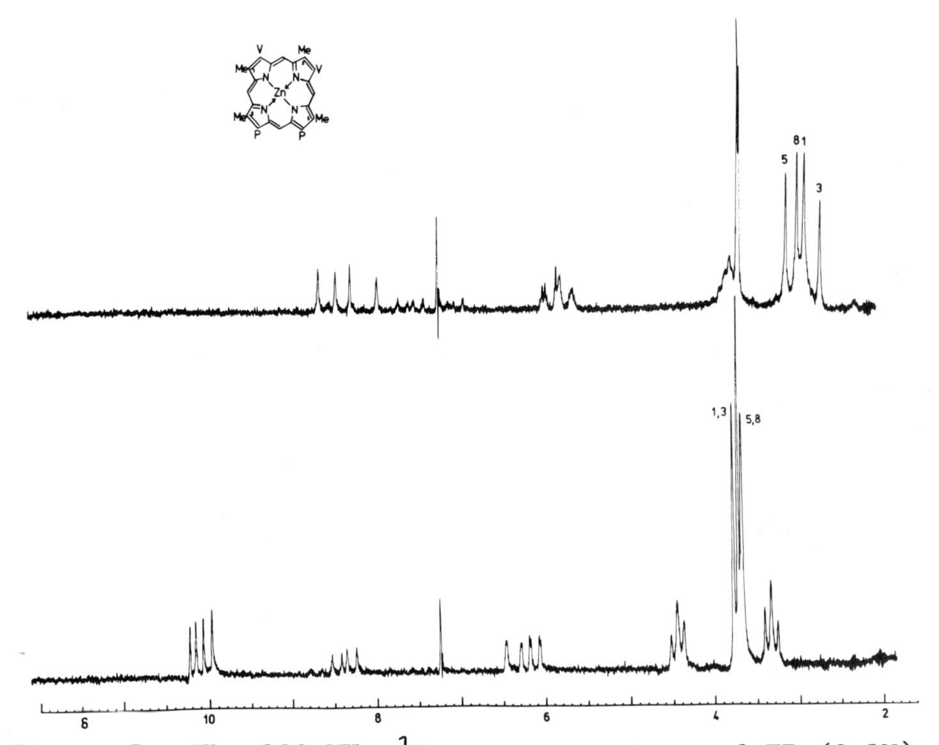

Figure 5. The 100 MHz ^1H n.m.r. spectrum of ZP (0.1M) in $CDCl_3$ (upper) and on the addition of ca 2 equivalents of pyrrolidine (lower).

rings B and D were deduced. These were 2.06 (1.93) p.p.m.

In THPB the lower symmetry enables a more extensive test of the model. The corresponding values, again for the octa-ethyl (R = Et) case are H_β 1.69(1.69); $H_{\alpha,\gamma}$ 2.71(2.80); H_δ 3.32(3.31) and CH_2(3,6) 0.78(0.84) and CH_2(4,5)0.71(0.72). Again the model gives a quantitative interpretation of the observed ring current shifts in these molecules.

Table 5 Equivalent Dipole Moments ($\mathring{A}^3$) for the Macrocycles

	Porphyrin	Chlorin	THPA	THPB
M_H	23.0	22.0	22.0	14.0
M_p	20.0	19.0	19.0	19.0

The values of the equivalent dipole moments for the hexagons (M_H) and pyrrole rings (M_p) used in these calculations are given in table 5. As the areas of the current loops are the same for all these compounds these moments are proportional to the ring current in the loops, and this data would therefore be of use in any quantum-

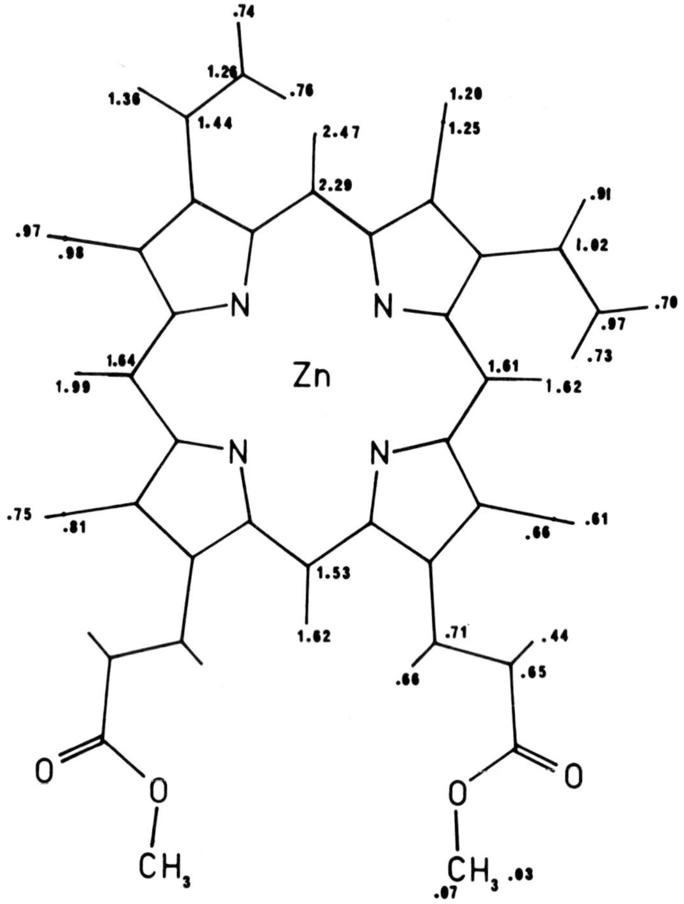

Figure 6. Aggregate shifts (p.p.m.) of zinc protoporphyrin IX.

mechanical calculation of the ring currents in these molecules.

The noteworthy result is for THPB, in which the "main" ring current is considerably reduced whilst that of the pyrrole rings is unaffected. It is probable that steric interactions in the reduced rings and between the adjacent NH protons distort the macrocycle from the planar structure in this case and this would produce the observed effects.

6. Application to Metalloporphyrin Aggregation Studies.

The aggregation properties of chlorophyll, which are responsible for the formation of the "photosynthetic dimer" are due to the interaction of the central magnesium atom

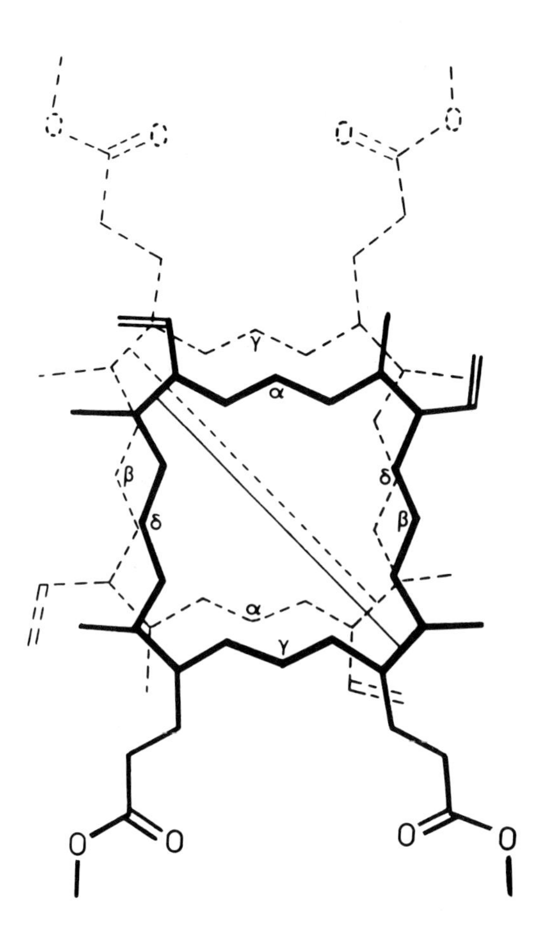

Figure 7. The geometry of the zinc protoporphyrin IX aggregate.

with the carbonyl oxygen of a neighbouring molecule.[10] We have recently identified[16] novel strong aggregation tendencies of zinc and other metalloporphyrins in solution which interestingly are not due to metal-to-ligand side chain interactions, though the aggregates are similarly dissociated by strong Lewis bases such as amines and alcohols. Detailed investigations of the geometry of these complexes in zinc meso trifluoroacetoxy OEP[17] and zinc meso nitro and meso dinitro OEP have been reported.[18] Here we summarise the association shifts of the biologically important zinc protoporphyrin IX (ZP) and show how the equivalent dipole model allows the deduction of the structure of the complex from the observed shifts.[19]

The proton N.M.R. spectrum of ZP in $CDCl_3$ solution shows remarkable concentration shifts. Figure 5 shows the spectrum of a 0.1M solution in $CDCl_3$ and on the addition of a slight excess of pyrrolidine. This immediately dissociates the aggregates and produces the "monomeric" spectrum.[13]

The large high field shifts observed in the aggregate spectrum are so specific that the order of the assignments of the meso-H and beta methyl resonances change with dilution. The assignments of the beta-methyl peaks are shown in figure 5 and the meso-H assignments are from left to right $\alpha, \beta, \delta, \gamma$ (monomer) and $\beta, \gamma, \delta, \alpha$ (aggregate). These assignments have been confirmed by dilution studies, titration with aliquots of pyrrolidine, single frequency decoupling experiments relating the proton and ^{13}C assignments and regiospecific deuteration.[19] Their importance is that labelled protoporhyrin IX is often the isolated material in studies of porphyrin biosynthesis and the biosynthetic pathway is deduced from the position of the labelled atoms.[20]

The ^{13}C n.m.r. spectrum of ZP shows similar dilution shifts, and as the N-H exchange process in the metal-free porphyrins often produces broad, ill-resolved ^{13}C spectra the zinc plus pyrrolidine technique has been recommended for the determination of reliable "monomeric" ^1H and ^{13}C shifts of porphyrins.[13,21] The dilution shifts of the different nuclei in ZP are given immediately from the spectra of figure 5, and the corresponding ^{13}C spectra. These shifts are given in figure 6 and may be used with the ring current model to deduce the structure of the ZP aggregate in solution. It is important to note here that these shifts are not monomer-dimer shifts but are appreciably greater due to the presence of higher aggregates i.e. layered structures in solution. This is shown both by the failure of the monomer-dimer model to reproduce the observed concentration dependence and also by the fact that the values in figure 7 are far too large to be due to the ring current of one porphyrin molecule adjacent to another. For example the average meso-H shift of 1.92 p.p.m. may be compared with the calculated shift of a meso proton in a ZP molecule situated directly above a second molecule. The maximum calculated high-field shift is 0.9 p.p.m. for an inter-ring separation of 4-5 Å. The factor of ca. two between the observed and calculated shifts is indeed that expected for a layered structure in which each ZP molecule experiences the ring current of two adjacent molecules. This simple analysis does not take into account the precise geometry of the aggregate, but the general argument is still valid.

The dilution shifts of figure 6 can now be used with the ring current model to deduce the structure of the ZP aggregate in solution. A possible structure is shown in figure 7 and this structure with an inter-ring separation of 4A and lateral displacements of 0.15 and 0.3A respectively gives calculated relative shifts in good general agreement with the observed shifts. The only exceptions to the general agreement are the meso carbon shifts. These are seen to be generally smaller than the corresponding proton shifts (figure 6).though the meso carbons will normally experience larger ring current shifts. Clearly there is an additional complexation shift at these carbon atoms, which is probably connected with the co-ordination of the pyrrolidine to the zinc atom.

The proposed structure of the complex is of interest in that the electron deficient rings A and B lie over the electron rich rings C and D and this electron imbalance could be an important factor in determining the geometry of these aggregates. Note also that a layered structure of alternate rings is entirely feasible on this basis and this is in accord with the observed data.

In conclusion, this brief summary of the equivalent dipole model of the benzene ring current and its application to both inter and intra-molecular geometries has shown hopefully that this simple concept can be usefully applied to a variety of complex situations to gain detailed conformational information.

Finally I would like to thank the Israel Academy of Sciences and the Fondation Edmund de Rothschild for their invitation and hospitality.

References

1. J.A. Pople, J.Chem.Phys. 24, 1111 (1956).
2. C.E. Johnson and F.A. Bovey, J.Chem.Phys. 29, 1012 (1958).
3. R.B. Mallion in Specialist Periodical Reports on N.M.R. Ed. R.K. Harris Vol.4, Chapter 1, The Chemical Society, London (1975).
4. R. Ditchfield in Specialist Periodical Reports on N.M.R. Ed. R.K. Harris Vol.5, Chapter 1, The Chemical Society, London (1976).
5. R.J. Abraham, S.C.M. Fell and K.M. Smith, Organic Mag.Res. 9, 367 (1977).
6. C. Giessner-Prettre, B. Pullman, P.N. Borer, L.S. Kan and P.O.P. Tso, Biopolymers 15, 2277 (1976) and references therein.
7. H.M. McConnell and R.E. Robertson, J.Chem.Phys. 20, 1361 (1958).
8. O. Hofer, Topics in Stereochemistry, 9, 111 (1976).
9. R.J. Abraham, K. Lewtas and W.A. Thomas, J.Chem.Soc (Perkin II) 1964 (1977).

10. For a review see H. Scheer and J.J. Katz in "Porphyrins and Metalloporphyrins", Ed. K.M. Smith, Chapter 10, Elsevier (1975).
11. R.J. Abraham, Mol.Phys. 4, 145 (1961).
12. C. Giessner-Prettre and B. Pullman, J. Theor.Biol. 31, 287 (1971).
13. R.J. Abraham, F. Eivazi, R. Nayyir-Mazhir, H. Pearson and K.M. Smith, Org.Mag.Res. 11, 52 (1978).
14. R.J. Abraham, S.C.M. Fell and K.M. Smith, Org.Mag.Res. 9, 367 (1977).
15. D.J. Bryan, B.Sc. thesis, University of Liverpool, 1977.
16. R.J. Abraham, F. Eivazi, H. Pearson and K.M. Smith, J.Chem.Soc.Chem.Comm. 698, 699 (1976).
17. R.J. Abraham, G.H. Barnett, G.E. Hawkes and K.M. Smith, Tetrahedron 32, 2949 (1976).
18. R.J. Abraham, B. Evans and K.M. Smith, Tetrahedron, 0000, (1978).
19. R.J. Abraham, S.C.M. Fell, H. Pearson and K.M. Smith, manuscript in preparation.
20. A.R. Battersby and E. McDonald in "Porphyrins and Metalloporphyrins", Ed. K.M. Smith, Elsevier, Amsterdam, (1975), p.87.
21. R.J. Abraham, H. Pearson and K.M. Smith, J.A.C.S. 98, 1604 (1976).

DISCUSSION

Irving: are ring currents sensitive to the interesting structural properties of metalloporphyrins, such as the doming and ruffling of the porphyrin ring ?

Abraham: In our experiments with essentially strain-free porph; ins, the introduction of a diamagnetic metal ion into the porphyrin causes very little change in the proton chemical shifts. The change is less than the experimental error in determining infinite dilution porphyrin chemical shifts.

However there is evidence that more severe steric perturbations (e.g. meso methylation) does affect the proton chemical shifts by mechanisms other than inductive effects etc. This could well be due to ring current changes caused by suffling of the porphyrin ring.

D.W. Jones: You showed a computer drawing, derived from a dipole-type program, for the structure of a dibenzoazepine-based drug complexed with a ring-containing molecule. In some ways, is not the extent of structural detail slightly spurious in that the number of observables is insufficient to determine independently (as distinct from showing consistency with) the main structural framework.

Abraham: The structure of the imipramine hydrochloride was taken from crystal measurements combined with coupling constant data. Only the position of the benzene ring of the benzyl alcohol ligand with respect to the imipramine was obtained from the aromatic shift data considered here.

Giessner-Prettre: Why is the equivalent dipole moment (M_6) of the six membered ring of adenine so much lower than that of benzene ?

Abraham: This is a very interesting question deserving further study. It is possible that the observed shielding effects of the adenine ring, upon which the original ring current models were based, may include effects (anisotropy of lone-pairs, electric field etc.) which are not present in benzene. Also the equivalent dipole model does not take into account any variations in the separation of the π current loops. This would affect the high-field (above the ring) region more than that in the plane of the ring.

SODIUM MAGNETIC RESONANCE IN BIOLOGICAL SYSTEMS. INTERPRETATION OF THE RELAXATION CURVES.

M. Goldberg and H. Gilboa
Department of Chemistry
Technion-Israel Institute of Technology, Haifa, Israel.

The role of sodium in biological systems is quite important, this has been the main reason for the extensive investigations of sodium NMR in the last decade. A comprehensive review on the subject was recently published (1). We would like to emphasize here that the main reason for the determination of sodium NMR in the systems studied was its biological importance and not the vast information one can derive from the measurement. The chemical shifts of sodium are very small so that information can be deduced mainly from relaxation times (T_1 and T_2), line intensity and quadrupole coupling constants or splittings.

The interpretation of sodium NMR spectra may lead to erroneous conclusions if wrong assumptions are used in their interpretation. The first attempt to interpret the spectra has been done by Cope (2). This interpretation was based on the work of Jardetzky and Wertz (3), who attributed the loss of signal intensity in their systems to quadrupolar line broadening, beyond instrumental detection. It has been shown that if one assumes first order quadrupolar interaction the results of Cope and followers may be interpreted differently (4). Other attempts to interprete sodium NMR spectra were done by Berendsen and Edzes (5), Monoi (6) and Lyon _et al._ (7).

Berendsen and Edzes (5) interpreted sodium spectra in terms of the order in the measured system. Assuming medium range order, where diffusion between structurally heterogeneous sites contributes to the spectral density at low frequencies, the experimental results could be explained. In those experiments T_1 measurements show one exponential decay while T_2 relaxation curves show a decay composed of two exponents Monoi (6) assumed a spread of chemical shifts to explain the difference between T_1 and T_2 in biological tissues. Lyon _et al._ (7) used sodium line width measurements to determine its interaction with bacterial surfaces. The interpretation was based on the assumption that the interacting sodium might have a different relaxation rate (or line width) and that there exists fast exchange between free and "bound" sodium.

B. Pullman (ed.), Nuclear Magnetic Resonance Spectroscopy in Molecular Biology, 481-491.

In many biological systems there exist two sites of sodium: one site which may be referred to as the free site whose properties do not differ much from those of sodium ions in aqueous solution. The other site is the "bound" site which, in terms of NMR, means a long correlation time and/or a different quadrupole coupling constant. The bound sodium may be the sodium which interacts with a bacterial surface (7) or the sodium in an inhomogeneous environment (5). In many cases there exists exchange between the "free" and the "bound" sodium. We would like to demonstrate another approach to the interpretation of sodium NMR in biological systems; the method is based on the theory of the relaxation of nuclei with spin I=3/2, and the chemical exchange between two sites. We take into account the rate of the exchange between the two sites, the ratio between free and bound sodium, as well as the frequency dependence of the relaxation curves. We shall demonstrate our method on a system of hallotoleronte bacteria and show the implications to other systems.

1. THEORETICAL BACKGROUND

The sodium nucleus has a spin of I=3/2 and consequently its relaxation is dominated by the quadrupolar interaction of the nuclear quadrupole moment and electric field gradients. The energy levels of a spin 3/2 in a magnetic field with quadrupole interaction (first order perturbation) is shown in figure 1.

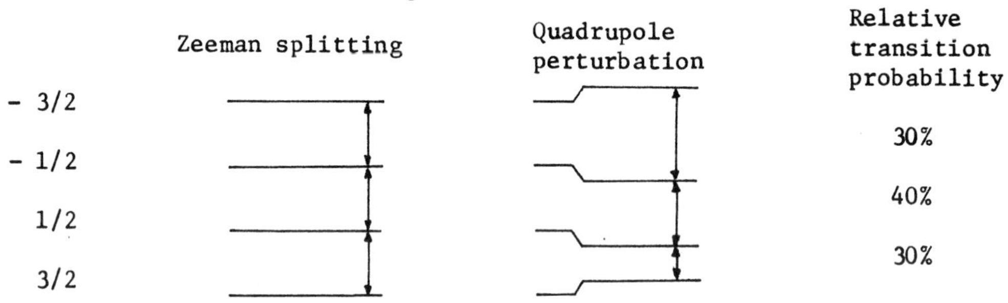

Figure 1: energy levels for spin I=3/2 with quadrupolar perturbation.

A similar description was given by Berendsen and Edzes (5). The quadrupole splitting depends on the angle between the magnetic field and the interaction axis. When the tumbling and diffusion of the molecules are fast or moderate there are two distinct transverse relaxation times (8); one for the 1/2 ↔ -1/2 transition

$$\frac{1}{T_{2s}} = \frac{1}{20}\left(\frac{e^2qQ}{\hbar}\right)^2\left[\frac{\tau_c}{1+w_o^2\tau_c^2} + \frac{\tau_c}{1+4w_o^2\tau_c^2}\right] + \frac{298}{875}\left(\frac{e^2qQ}{4\hbar}\right)^4\frac{\tau_c}{w_o^2} \qquad (1a)$$

and for the transitions ($\frac{3}{2} \leftrightarrow \frac{1}{2}$) and ($-\frac{1}{2} \leftrightarrow -\frac{3}{2}$) the relaxation rate will be

$$\frac{1}{T_{2f}} = \frac{1}{20}\left(\frac{e^2Qq}{\hbar}\right)^2\left(\tau_c + \frac{\tau_c}{1+w_o^2\tau_c^2}\right) \tag{1b}$$

so that 40% of the total magnetization relaxes as T_{2s}^{-1} and 60% as T_{2f}^{-1}, where s and f denote the slow and fast components respectively.

Since one cannot associate a spin lattice relaxation to each of the absorption lines, as in a spectrum of nuclei of I=1/2, it follows that there are two spin spin relaxation components, which can be attributed to the total magnetization only (9)

$$\frac{1}{T_{1s}} = \frac{1}{10}\left(\frac{e^2qQ}{\hbar}\right)^2\left(\frac{\tau_c}{1+w_o^2\tau_c^2}\right)$$

$$\frac{1}{T_{1f}} = \frac{1}{10}\left(\frac{e^2qQ}{\hbar}\right)^2\left(\frac{\tau_c}{1+4w_o^2\tau_c^2}\right) \tag{2}$$

where T_{1s}^{-1} and T_{1f}^{-1} are the relaxation rates of the slow and fast components with 20% and 80% of the total magnetization respectively.

The effect of exchange between two sites on the relaxation times will be derived in the next section.

2. THE DERIVATION OF THE RELAXATION EQUATION WITH EXCHANGE

It was shown that the relaxation curves for T_1 and T_2 may be expressed in terms of the eigenvectors of Redfield's relaxation matrix (10) (11) (12). Bull (13) had shown that for a spin 3/2 exchange between two sites might be written as

$$\frac{d}{dt}\begin{bmatrix} v_{A_1} \\ v_{A_2} \\ \cdot \\ \cdot \\ \cdot \\ \cdot \\ \cdot \\ \cdot \\ v_{B_1} \\ v_{B_2} \end{bmatrix} = \left[\begin{bmatrix} [r_A] & [0] \\ [0] & [r_B] \end{bmatrix} + \begin{bmatrix} [-1/\tau_A] & [1/\tau_B] \\ [1/\tau_A] & [-1/\tau_B] \end{bmatrix}\right]\begin{bmatrix} v_{A_1} \\ v_{A_2} \\ \cdot \\ \cdot \\ \cdot \\ \cdot \\ \cdot \\ \cdot \\ v_{B_1} \\ v_{B_2} \end{bmatrix} \tag{3}$$

where v_{Ai} is an eigenvector of the relaxation matrix in site A, r_A is the relaxation matrix of site A after diagonalization, τ_A is the life time in site A. The same notations are for site B. Solving those equations we obtained for the longitudinal magnetization

$$\frac{\langle I_{ZT}(t)\rangle}{\langle I^\circ_{ZA}\rangle + \langle I^\circ_{ZB}\rangle} = 1 - 2[(K)+(L)+(M)+(N)] \tag{3}$$

where

$$K = \frac{[(1/\tau_B + \lambda_{1+} + 1/T_{1A}) - X(\lambda_{1-} + 1/T_{1A} - 1/\tau_A)]\ \exp(\lambda_{1+}t)}{5(1+X)(\lambda_{1+} - \lambda_{1-})}$$

$$L = \frac{[-(1/\tau_B + \lambda_{1-} + 1/T_{1A}) + X(-1/\tau_A + 1/T_{1A} + \lambda_{1+})]\ \exp(\lambda_{1-}t)}{5(1+X)(\lambda_{1+} - \lambda_{1-})}$$

$$M = \frac{4[(1/\tau_B + 1/T'_{1A} + \lambda'_{1+}) - X(\lambda'_{1-} + 1/T'_{1A} - 1/\tau_A)]\ \exp(\lambda'_{1+}t)}{5(1+X)(\lambda'_{1+} - \lambda'_{1-})}$$

$$N = \frac{4[-(1/\tau_B + \lambda'_{1+} + 1/T'_{1A}) + X(-1/\tau_A + 1/T'_{1A} + \lambda'_{1+})]\ \exp(\lambda'_{1-}t)}{5(1+X)(\lambda'_{1+} - \lambda'_{1-})}$$

The parameters in Eq. (2) are:

$$X = \frac{\langle I^\circ_{ZA}\rangle}{\langle I^\circ_{ZB}\rangle} = \frac{M^\circ_A}{M^\circ_B} = \frac{\tau_A}{\tau_B}$$

$\langle I_{ZA}\rangle$, $\langle I_{ZB}\rangle$ are the magnetizations of sites A and B respectively. τ_A, τ_B the mean lifetime of the spins in sites A and B respectively. T_{1A} is defined as:

$$1/T_{1A} = r_{1A} + 1/\tau_A$$

where

$$r_{1A} = 1/10(e^2qQ/\hbar)^2(\tau_{cA}/1+W_o^2\tau_{cA}^2)$$

T_{1B} is defined by substituting B instead of A. The expressions for $\lambda_{1\pm}$ are

$$\lambda_{1\pm} = 1/2[-1/T_{1B} - 1/T_{1A} \pm \sqrt{(1/T_{1B} - 1/T_{1A})^2 + 4/\tau_A\tau_B}]$$

T'_{1A} and $\lambda'_\pm$ are defined when r_{1A} is substituted by r'_{1A} where:

$$r'_{1A} = 1/10(e^2qQ/\hbar)^2(\tau_{cA}/1+4W_o^2\tau_{cA}^2)$$

The expression for the transverse magnetization $\langle I_+ \rangle$ is:

$$\frac{\langle I_{+T}(t)\rangle}{\langle I_{+A}(o)\rangle + \langle I_{+B}(o)\rangle} = [(O)+(P)+(Q)+(R)] \tag{4}$$

where

$$O = \frac{2[1/\tau_B + \lambda_{2+} + 1/T_{2A}) - X(\lambda_{2-} + 1/T_{2A} - 1/\tau_A)]\exp(\lambda_{2+}t)}{5(1+X)(\lambda_{2+} - \lambda_{2-})}$$

$$P = \frac{2[-(1/\tau_B + \lambda_{2-} + 1/T_{2A}) + X(-1/\tau_A + 1/T_{2A} + \lambda_{2+})] \exp(\lambda_{2-}t)}{5(1+X)(\lambda_{2+} - \lambda_{2-})}$$

$$Q = \frac{3[(1/\tau_B + 1/T'_{2A} + \lambda'_{2+}) - X(\lambda'_{2-} + 1/T'_{2A} - 1/\tau_A)] \exp(\lambda'_{2+}t)}{5(1+X)(\lambda'_{2+} - \lambda'_{2-})}$$

$$R = \frac{3[-(1/\tau_B + \lambda'_{2+} + 1/T'_{2A}) + X(-1/\tau_A + 1/T'_{2A} + \lambda'_{2+})] \exp(\lambda'_{2-}t)}{5(1+X)(\lambda'_{2+} - \lambda'_{2-})}$$

T_{2A} and T'_{2A} are defined in a similar way to T_{1A} and T'_{1A}.
r_{2A} and r'_{2A} are defined as:

$$r_{2A} = 1/20\ (e^2qQ/\hbar)^2\ [(\tau_{cA}/1+W_o^2\tau_{cA}^2)+(\tau_{cA}/1+4W_o^2\tau_{cA}^2)] \ +$$

$$+\ 298/875\ (e^2qQ/4h)^2(\tau_{cA}/W_o^2)$$

$$r'_{2A} = 1/20\ (e^2qQ/\hbar)^2\ [\tau_{cA} + (\tau_{cA}/1+W_o^2\tau_{cA}^2)]$$

T_{2B} and T'_{2B} are obtained in a similar way.
The definitions of $\lambda_{2\pm}$ and $\lambda'_{2\pm}$ are identical in form to those of $\lambda_{1\pm}$ and $\lambda'_{1\pm}$.
Calculated relaxation curves based on these equations and their implication to experimental results are given in the next section.

3. THE RELAXATION CURVES

Longitudinal and transverse relaxation curves for sodium ions undergoing exchange between two sites were calculated, using equations (3) and (4).

The following assumptions were made in the calculations:

a) There exist only two sites, each characterized by a single correlation time (τ_{cA}, τ_{cB}).

b) $(e^2qQ/\hbar)_A = (e^2qQ/\hbar)_B = 5.08 \cdot 10^6$ rad/sec (0.8 MHz).
This assumption has no experimental proof. Though we use this equality in the few examples given in this section, one may adopt different values for the quadrupole coupling constants of the two sites.

c) The electric field gradient q has axial symmetry. This assumption does not affect the general behaviour of T_1 and T_2 as has been pointed out by Shporer and Civan (1).

d) There is no chemical shift difference between the two sites.

Generally sodium chemical shifts are very small.

e) $\tau_B >> \tau_{cB}$, this implies that chemical exchange times are longer than the rotational correlation times. In cases where $\tau_B \simeq \tau_{cB}$ the relaxation mechanism caused by the modulation of the quadrupole coupling constant cannot be ignored.

f) τ_{cB} was arbitrarily chosen in such a way that the spin-spin and the spin-lattice relaxation times would become frequency dependent. (In the examples given here $\tau_{cB} = 1 \cdot 10^{-7}$ sec.).

g) We choose arbitrarily $\tau_{cA}=4 \cdot 10^{-12}$ sec so that the conditions for the free site would suit in the extreme narrowing case i.e. $T_{1A}=T_{2A}$=100 msec.

Figures 2(a) and 2(b) show the relaxation behaviour of the transverse and longitudinal magnetizations for different exchange rates at $W_0=9.79 \cdot 10^7$ rad/sec. (15.8 MHz) and for equal populations of the two sites A and B (X=1), after 90° and 180° pulses respectively. In Fig. 1(a), for the transverse magnetization decay, we note that there is a big difference in the behaviour between the slow and fast exchange cases. For $\tau_B = 2 \cdot 10^{-2}$ sec. the decay curve is composed of three exponentials, one for the site A with half of the total intensity (as obtained by extrapolation of the slow decay line) and two others which are the slow and fast decaying components of site B. It implies that for $\tau_B = 2 \cdot 10^{-2}$ there is slow exchange. For $\tau_B = 5 \cdot 10^{-5}$ sec., the fast exchange limit, we see only two components, one is the average of the slow decays of sites A and B, and the other is the average of the fast decaying components of the two sites. The extrapolation of the slow component yields 0.4 of the total intensity. For $\tau_B = 2 \cdot 10^{-3}$ sec. the averaging between the two sites is incomplete and we obtain a curve which lies between curves for the slow and fast exchange rates. The calculations for a frequency of $5.08 \cdot 10^7$ rad./sec. gives similar results but the relaxation rates at the fast and intermediate exchange ranges are faster.

We now discuss the expected experimental observations for such cases as illustrated in Fig. 2(a). The length of a 90° pulse in our instrument is about 10 μsec. followed by a dead time 50-80 μsec. In the fast exchange case, we would barely observe the average of the slow components with 0.4 of the total intensity. In the slow exchange limit, we would observe a decay composed of more than one component with an extrapolated value of about 0.7 of the total intensity. In the intermediate exchange region, assuming that the signal to noise ratio is not excellent, it would be difficult to distinguish between one or two exponential decays and a total intensity of about 60-70% would be observed (as shown by the dashed line). In the slow exchange limit of the longitudinal relaxation curves (Fig. 2(b)) the curve is composed of more than one exponent, whereas in the intermediate and fast exchange limits, the deviation from linearity is very small. Therefore, the experimental observables for the latter cases would depend predominantly on the signal to noise ratio. Figure 3 gives the transverse relaxation taken at two frequencies where 1% of the sodium is bound in site B. The slow exchange limit shows one exponential decay composed of 99% of the total

Figure 2. Calculated relaxation curves at $W_o = 9.79 \cdot 10^7$ rad./sec. (15.8 MHz) for equal concentrations of "free" and "bound" sites X=1, for various life times $(\tau_B) 2 \cdot 10^{-2}$ s, $2 \cdot 10^{-3}$ s, and $2 \cdot 10^{-5}$ s.

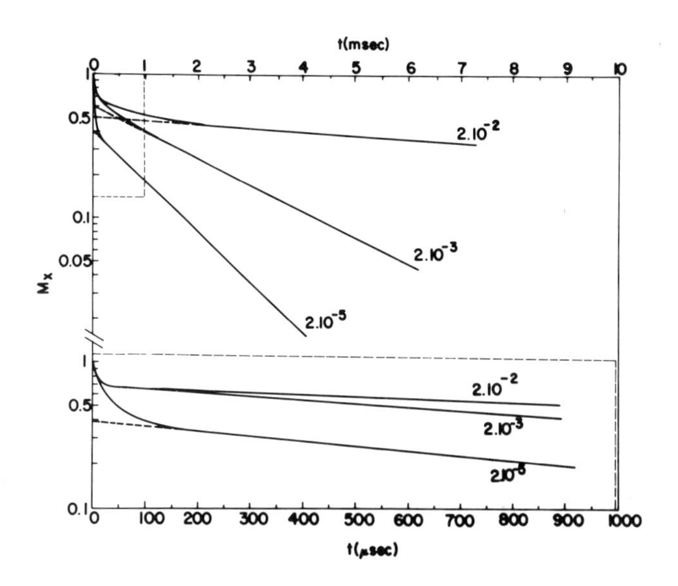

Figure 2(a). The spin-spin relaxation curves.

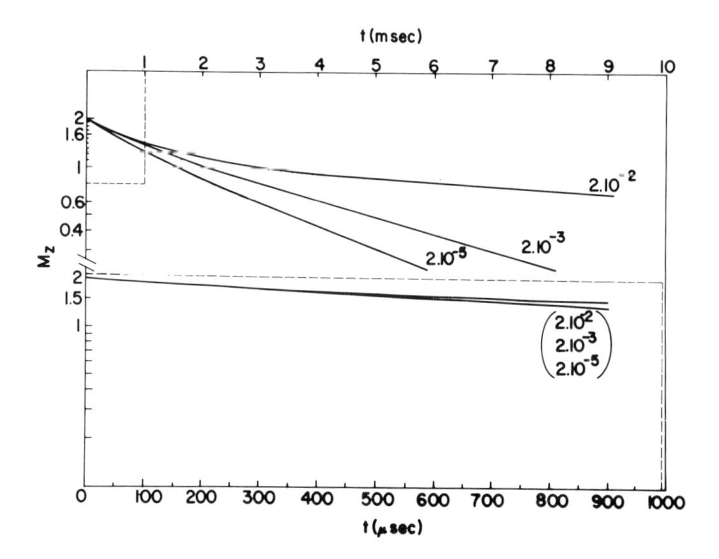

Figure 2(b). The spin-lattice relaxation curves.

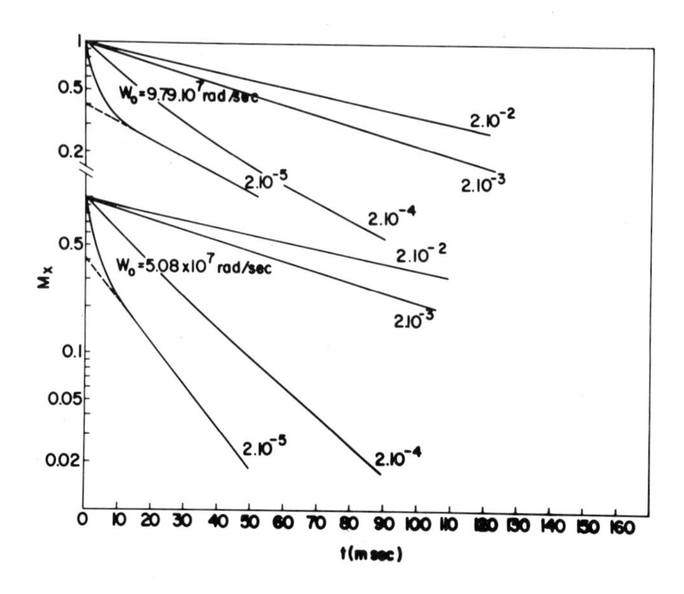

Figure 3. Calculated spin-spin relaxation curves for 1% of "bound" site at two frequencies $9.79 \cdot 10^7$ rad./sec. (15.8 MHz) and at $5.08 \cdot 10^7$ rad./sec. (8.09 MHz), and at various τ_B values.

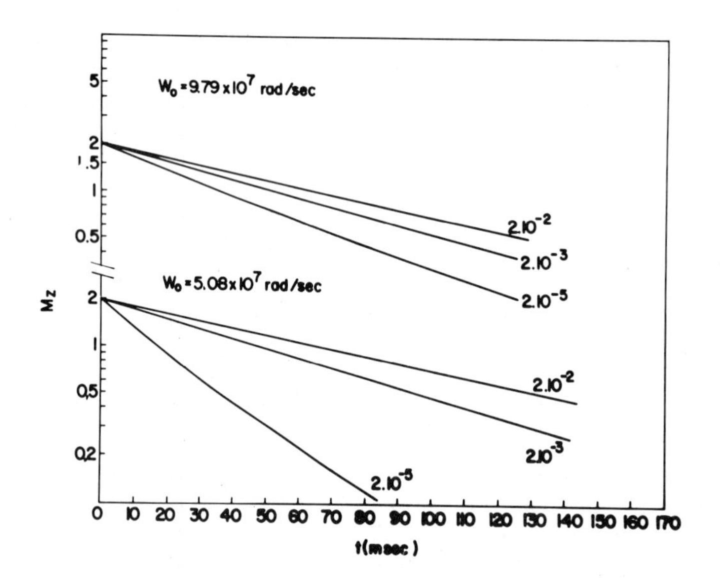

Figure 4. Calculated spin lattice relaxation curves. For details see figure 3.

magnetization. At higher exchange rates there is a deviation from a single exponential decay and the fast and slow components are clearly distinguished. We note that the relaxation curves are frequency dependent only when the life time at site B is sufficiently short (e.g. $\tau_B = 2\cdot10^{-5}$ sec. or $2\cdot10^{-4}$ sec. respectively*). Figure 4 shows the longitudinal relaxation curves for 1% bound site at different frequencies. Note that, it would be very difficult to observe experimentally two exponential decays even in the fast exchange region. Still, the decay is frequency dependent, especially at fast exchange rates.

The few examples presented here illustrate two extreme cases for exchange between two sites that might exist for sodium in biological systems. It also shows that the use of frequency dependent measurements combined with intensity determinations would be useful in the interpretation of the experimental results. Moreover, using these equations one may deal with slow, intermediate, or fast exchange rates, taking into account the experimental limitation such as the dead time of the instrument or its signal to noise ratio. It is shown that even if a small fraction of sodium is bound one may obtain more than one exponential decay curve. However, one must be very carefull in using fast exchange equations because intermediate exchange rates may give similar experimental results while the interpretation of these results assuming fast exchange will result in erroneous conclusions.

4. EXPERIMENTAL VERIFICATION

The application of the method described here was used on a system of halotolerant bacteria B_a isolated from the Dead Sea (14). It was shown (15) that the experimental results could be explained in terms of intermediate exchange between two sites: (A) the intercellular sodium and (B) bound sodium while a fraction (C) of bound sodium inside the cell with very broad lines cannot be detected. The calculated results are given in the following scheme:

site A	site B	site C
$M_A = 0.5$ M	$M_B = 0.23$	$M_C = 0.41$
$\tau_A = 1\cdot10^{-3}$ sec.	$\tau_B = 6\cdot10^{-4}$	τ_C long
$e^2qQ/\hbar = 5\cdot10^6$ rad/sec.	$e^2qQ/\hbar = 9\cdot10^6$	$e^2qQ/\hbar$?
$\tau_{cA} = 8\cdot10^{-12}$ sec.	$\tau_{cB} = 5.5\cdot10^{-7}$	τ_{cC} ?

* The relaxation curve depends on τ_A as well as on τ_B. When the magnetization in the bound site is just 1% of the total, τ_A will be much longer than τ_B and therefore for intermediate and fast exchange rates, τ_B values should be shorter from the case where the bound site is about 50%. For example, for 99% free sodium, a fast exchange average value of the relaxation curve will appear only when τ_B is about $2\cdot10^{-7}$ sec.

An addition of an extracellular solution of sodium to the system in the relaxation curves becomes dependent on the volume of the added solution. The experimental and calculated relaxation times are shown in figure 5.

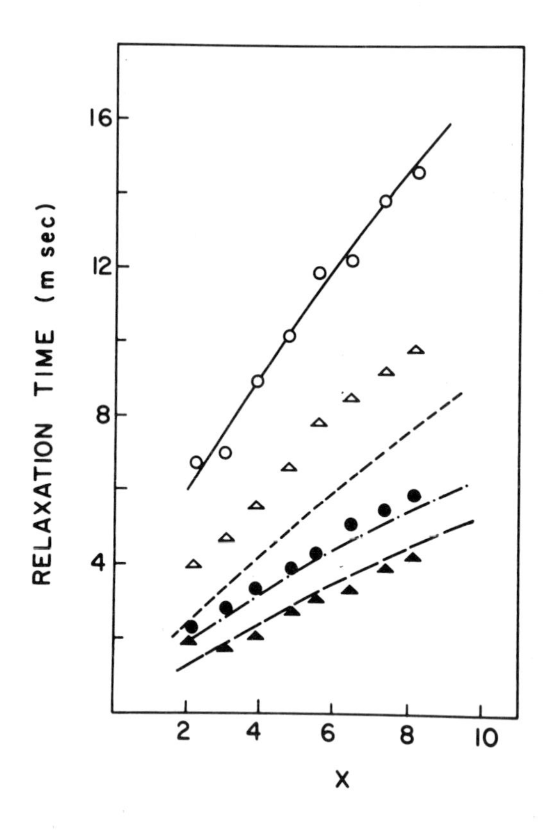

Figure 5. Calculated and experimental relaxation times to fit the proposed model.

——— O T_1 at 15.8 MHz; ------ Δ T_1 at 8.09 MHz;

-·-·-·- ● T_2 at 15.8 MHz; — — — — ▲ T_2 at 8.09 MHz.

References

1. Shporer M. and Civan M.M., in Current Topics in Membrane and Transport, Bronner F. and Kleinveller ed. (1977).
2. Cope F.W., Proc. Natl. Acad. Sci. U.S. (1977) 54, 225.
3. Jardetzky O. and Wertz J.E., J. Am. Chem. Soc. (1960) 82, 318.
4. Shporer M. and Civan M.M., Biophys. J. (1972) 12, 114.
5. Berendsen H.J.C. and Edzes H.T., Ann. N.Y. Acad. Sci. (1973) 204, 459.
6. Monoi H., Biochem. Biophys. Acta (1976) 451, 604.
7. Lyon R.C., Wekesser J., and Magnuson J.A., ACS Symposium Series (1976) 34, 360, Magnetic Resonance in Colloid and Interface Science, H.A. Resing and C.G. Wade Ed.
8. Baram A., Luz Z. and Alexander S., J. Chem. Phys. (1973) 58, 4558.
9. Hubbard P.S., J. Chem. Phys. (1970) 53, 985.
10. Abragam A., The Principles of Nuclear Magnetism, Oxford at the Clarendon Press (1961) p. 443.
11. McLachlan A.D., Proc. Roy. Soc. (1964) A280, 271.
12. Rubinstein M., Baram A. and Luz Z., Mol. Phys. (1971) 20, 67.
13. Bull. T.E., J. Mag. Res. (1972) 8, 344.
14. Rafaeli-Eshkol D., Biochem. J. (1968) 109, 679.
15. Goldberg M. and Gilboa H., Biochem. Biophys. Acta (1978) 538, 268.

NMR SPECTROSCOPY OF TRICYCLIC NON-ALTERNANT SYSTEMS CONTAINING SEVEN-MEMBERED RINGS, DIBENZ[*b,f*]OXEPINE AND DIBENZ[*b,f*]AZEPINE: PARENTS OF PHARMACOLOGICALLY ACTIVE SYSTEMS.

J.A.G. DRAKE and D.W. JONES
School of Chemistry, University of Bradford, Bradford, West Yorkshire, BD7 1DP, U.K.

1. INTRODUCTION

Tricyclic dibenzo derivatives of the seven-membered oxepine, azepine and cycloheptatriene rings are the parent systems of several clinically-active psychotropic drugs [1,2]. In six- and seven-membered-ring derivatives with central-nervous-system psychopharmacological activity [3,4,5], ability to induce such action may be associated with a degree of non-planarity and low aromaticity of the central ring. In the crystalline state, dibenz[*b,f*] oxepine (I) [6] and the isoelectronic dibenz[*b,f*]azepine (II) (studied independently [7], but found to have remarkably similar dimensions) take up an overall butterfly or saddle shape with the heterocyclic rings in boat conformation. The dihedral angle between the almost planar benzene rings in the crystal structure of I [6] is 134° and 145° in II [7]. Photo-electron [8] and electronic [9] spectra provide some evidence for the non-planarity of I and II in the gas phase and in methanol solution.

In this paper, detailed analyses are described of the 220 MHz ^{1}H nuclear magnetic resonance (NMR) spectra of solutions of I and II (aided by analyses of related compounds [10]) so that structural

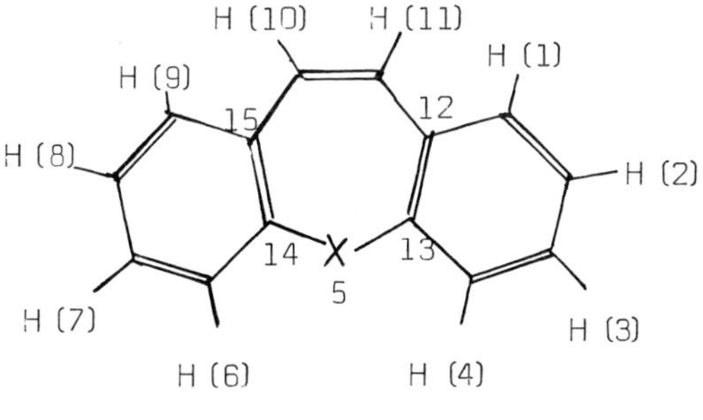

H (10) H (11) H (1) H (9) 15 12 H (8) H (2) 14 X 13 5 H (7) H (3) H (6) H (4)

I X = O

II X = NH

III X = C = O

IV X = CHOH

B. Pullman (ed.), *Nuclear Magnetic Resonance Spectroscopy in Molecular Biology*, 493-507.

inferences can be drawn from the chemical shifts and coupling constants derived. A preliminary examination is also reported of the ^{1}H-decoupled ^{13}C NMR spectrum of I in chloroform solution.

2. ANALYSIS OF NMR SPECTRA

2.1. Experimental Methods

High-resolution ^{1}H NMR spectra of solutions in carbon disulphide (Analar, B.D.H., Ltd.)and chloroform $-d_1$ (Prochem., B.O.C. Ltd.) (Figure 1(a)) of I (colourless tablets, m.pt. 383 K, from Dr. P.M.G. Bavin [11]) and II (orange platelets, m.pt. 478 K, R.N. Emanuel) were recorded at 220 MHz on a Varian HR220 field-sweep spectrometer (P.C.M.U., A.E.R.E., Harwell); 220 MHz proton-decoupling experiments were made by Dr. R.A. Spragg on a Perkin-Elmer R34 frequency-sweep spectrometer. ^{13}C high-resolution F.T. spectra were recorded at P.C.M.U. on a Brüker HX90E spectrometer. Internal reference for all solutions was 1% tetramethylsilane (TMS)(B.D.H., Ltd).

The Bradford University I.C.L. 1904A computer and Calcomp graph plotter were used for GPTF (a U.E.A. spectrum-plotting program) simulation of ^{1}H spectra, which had been solved for chemical shifts and coupling constants by LAOCOON III, and for ORTEP (crystallographic molecular-drawing program from the X-RAY system) plotting of substrate geometry when solvent-shift data for I were given a lanthanide-induced shift conformational analysis (LISCA program) treatment(U.E.A. program).

2.2. ^{1}H Spectra of Dibenz[*b,f*]oxepine.

For I [12] and II (as with III and IV [10]), molecular symmetry will be expected to result in one olefinic ^{1}H chemical shift and four (from the four magnetically non-equivalent pairs) close aromatic ^{1}H shifts. The 220 MHz four-proton aromatic sub-spectrum of I in carbon disulphide somewhat resembles the 60 MHz ^{1}H spectrum of 1-methylbenzotriazole in chloroform-d_1 [13], with a single-proton multiplet about 0.15 p.p.m. to low field of a relatively narrow overlapped three-proton multiplet around 7.0 p.p.m. In III, H(4) resonates 8.2 p.p.m. downfield from TMS, i.e. is appreciably deshielded by the carbonyl group by comparison with H(1-3), which resonate at around 7.5 p.p.m. However, analogous tentative assignment of the lowfield proton (7.15 p.p.m.) in I to H(4) did not enable the spectrum to be simulated; reassignment of the lowfield chemical shift to a proton, H(2) or H(3), which had two *ortho* coupling constants eventually led (with difficulty, owing to the close proximity ∿0.03 p.p.m.) to a solution (with coupling constants similar to those in III) for the other three aromatic proton chemical shifts (Table I). The remaining chemical shift, a singlet at 6.53 p.p.m., was assigned to H(11), by comparison with the spectrum of III. Ring-current considerations favour H(3) downfield of H(2). Also, between I spectra in chloroform-d_1

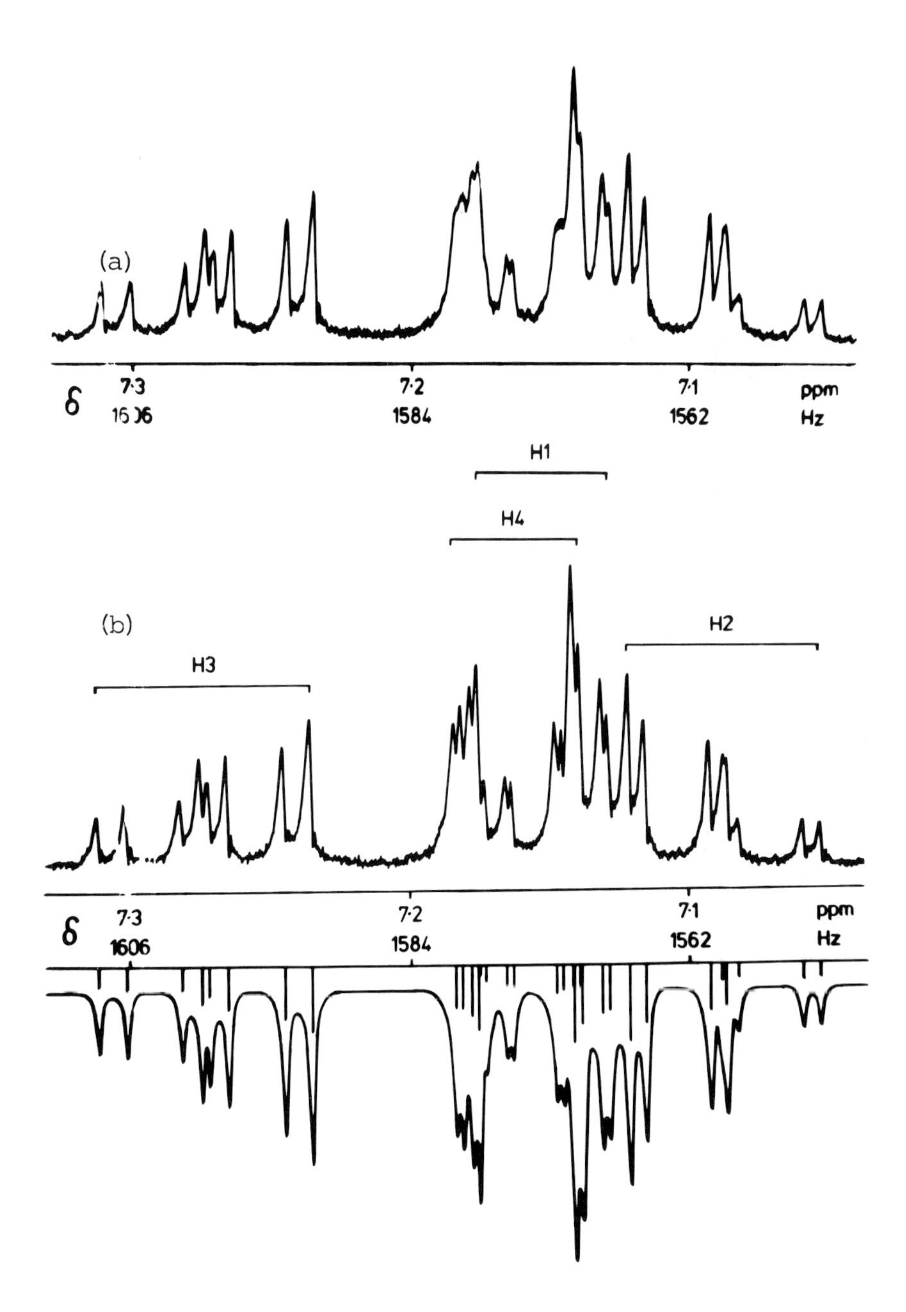

Figure 1. Dibenz[*b,f*]oxepine: 220 MHz ^{1}H NMR spectra of aromatic protons H(1-4) in 0.13 mol dm^{-3} solution in chloroform-d_1 at 293 K, (a) experimental undecoupled spectrum, and (b) experimental H(11)-decoupled spectrum (upper) and simulated spectrum with energy transitions represented as sticks at half the intensity scale (lower).

and carbon disulphide, the shift of one proton (either H (1) or H (4)) with a single *ortho* coupling is much greater than the solvent shifts of the other protons; this was taken to be H (4), which would be the most influenced by any clustering of polar solvent molecules. With H (11) decoupled from the aromatic protons (Figure 1 (b)), the chemical shift now assigned to H (4) showed a coupling (0.3 Hz) comparable to $J_{4,11}$ in III; this analogy favours the shift sequence in I of H (3) ≫ H (1) ≫ H (4) ≫ H (2) (rather than H (2) ≫ H (4) ≫ H (1) ≫ H (3)). Iterative analysis with LAOCOON of the H (11)-decoupled and -coupled spectra (Figure 1) gave the results in Table I.

2.3. ^{1}H Spectra of Dibenz[*b,f*]azepine.

At room temperature, dibenz[*b,f*]azepine (II) decomposes within several hours in chloroform-d_1 solution; it is only sparingly soluble in carbon disulphide. The spectrum (Figure 2 (a)) of the aromatic protons in a saturated carbon-disulphide solution (0.07 mol dm^{-3}) at 293 K bears a distinct resemblance to that of I in the same solvent with regard to the relative position of the lowfield single-proton multiplet to the two-proton mid-field multiplet (the three-proton multiplet in I); however the doublet (H (1) or H (4)) resonates at higher field in II. This similarity aided assignment: H (4), the aromatic proton closest to the heteroatom and so most affected by its replacement, can be assigned to the high-field doublet; LAOCOON simulation then confirms that H (3) is the low-field multiplet, with H (2) and H (1) forming the mid-field multiplet. Of the protons in the central seven-membered ring, the broad singlet of ~15 Hz half-height width at 4.62 p.p.m. can be assigned to the N*H* proton from the integration (half the intensity of other proton absorptions) and broadening by the nitrogen, while the narrow singlet, identical in appearance with that in I (and also with those in III and IV), is assigned to H (11).

Selective decoupling of the two singlets revealed noticeable coupling of the N*H* to H (1) (Figure 2 (c)), but none was detected to the other protons. Although no coupling was observed from H (11) to any proton, for H (4), which is only 0.23 p.p.m. downfield from H (11), decoupler-frequency ringing interference might obscure detection of any coupling (Figure 2 (b)). However, the shape of the H (4) doublet, decoupled from N*H*, could be reproduced in simulated spectra only by the introduction of some coupling between H (11) and H (4). Results of the LAOCOON iterative analyses of these spectra are given in Table II.

The spin system might be described as $[\mathrm{ABEHM}]_2\mathrm{X}$, where A = H (1), B = H (2), E = H (3), H = H (4), M = H (11), and X = H (5); the brackets and subscript represent the mirror symmetry (each half of the molecule coupled equally to X). The LAOCOON probable errors are larger in δ_1, δ_2, $J_{1,3}$ and $J_{2,3}$ because $(\delta_2 - \delta_1) < J_{1,2}$ so that only four, instead of eight, energy transitions can be assigned to δ_1 and six, rather than eight, transitions to δ_2.

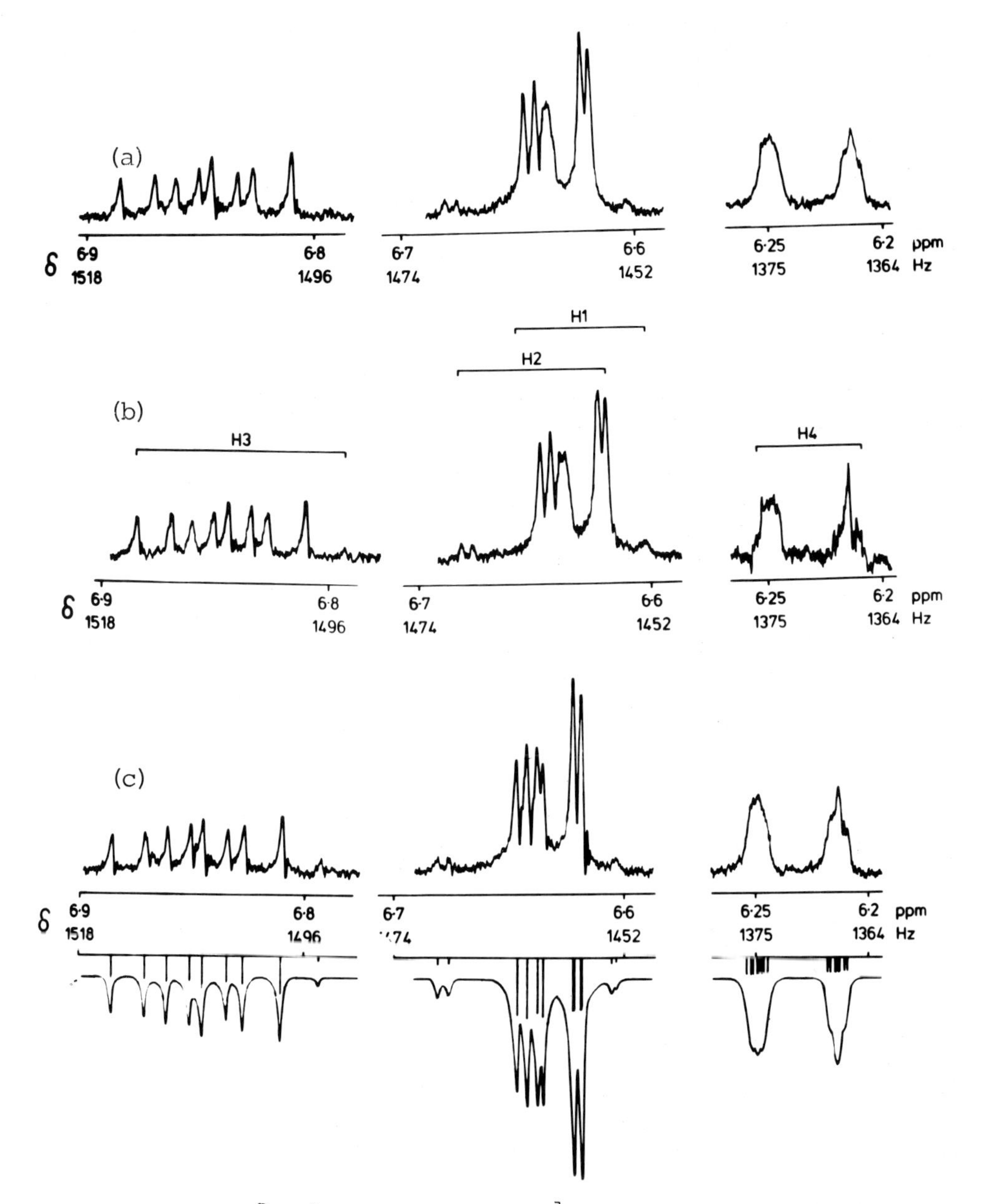

Figure 2. Dibenz[*b,f*]azepine: 220 MHz ^{1}H NMR spectra of aromatic protons H(1-4) in 0.07 mol dm^{-3} solution in carbon disulphide at 293 K, (a) experimental undecoupled spectrum, (b) experimental H(11)-decoupled spectrum (the H(4) doublet is distorted by decoupler-frequency ringing interference) and (c) experimental N-*H*-decoupled spectrum (upper) and simulated spectrum with energy transitions represented as sticks at half the intensity scale (lower).

TABLE I

^{1}H spectral parameters of dibenz[b,f]oxepine (I) from LAOCOON iterative analyses: chemical shifts (δ/p.p.m.) downfield from TMS from 220 MHz spectra in (i) 0.26 mol dm^{-3} CS_2 solution at 293.5 K, (ii) 0.26 mol dm^{-3} $CDCl_3$ solution at 293.5 K, (iii) H(11)-decoupled, 0.13 mol dm^{-3} $CDCl_3$ solution at 293 K; (iv) coupling constants J/Hz, under same conditions as (iii).

	(i)	(ii)	(iii)		(iv)[a,b]
δ_1	7.003	7.145	7.148	$J_{1,2}$	7.49
δ_2	6.976	7.092	7.094	$J_{1,3}$	1.68
δ_3	7.115	7.269	7.269	$J_{1,4}$	0.32
δ_4	6.988	7.169	7.165	$J_{2,3}$	7.30
δ_{11}	6.527	6.690	6.691	$J_{2,4}$	1.16
				$J_{3,4}$	8.01

[a] r.m.s. error 0.03 Hz; [b] estimates of $J_{1,11}$ = 0.1 and $J_{4,11}$ = 0.3 Hz from line broadening by superimposition of simulated spectra over experimental spectrum.

2.4 ^{13}C Spectra of Dibenz[b,f]oxepine.

Of the seven chemical shifts in the ^{13}C high-resolution proton-decoupled spectrum of a 0.26 mol dm^{-3} solution of I in chloroform-d_1 at 310 K, the two smaller absorptions are easily distinguished as quaternary carbons; their assignment, the lower field one to C(13) the other to C(12), is compatible with the electrogenativity of the oxygen atom adjacent to C(13). Some doubt is cast on the value from SCF calculations (with the equation of Pople and Beveridge [14]) of carbon electron densities (Table III), based on co-ordinates of I found in the crystal structure [6], in predicting the chemical-shift sequence C(13) > C(2) ∿ C(3) > C(11) ∿ C(12) > C(1) > C(4), since experimentally the C(12) and C(13) quaternary carbons have the biggest shifts. On the other hand, the additivity rules [15] give C(13) > C(3) ∿ C(12) ∿ C(1) > C(2) > C(4) (Table III). Since the experimental and calculated shifts of C(12) and C(13) are so close, similarity between the calculated C(4) and C(2) chemical shifts to two experimental shifts can be used for fairly confident assignment (Table III); C(4) had also been predicted as at highest field from the electron-density sequence.

TABLE II

^{1}H chemical shifts, δ/p.p.m., and coupling constants, J/Hz, from LAOCOON iterative analyses of H (11)-decoupled 220 MHz spectra of dibenz[b,f] azepine (II) in 0.07 mol dm^{-3} CS_2 solution at 293 K.

	δ/p.p.m.		J/Hz[b]
H (1)	6.624	H (1)-H (2)	7.62
H (2)	6.636	H (1)-H (3)	1.57
H (3)	6.845	H (1)-H (4)	0.31
H (4)	6.231	H (2)-H (3)	7.38
H (5)	4.618	H (2)-H (4)	1.19
H (11)	6.013	H (3)-H (4)	7.80

[a] Overall (shifts and couplings) r.m.s. probable errors from LAOCOON 0.03 Hz; [b] estimates of 0.30 Hz for $J_{1,5}$ and $J_{4,11}$ from line broadening by superimposition of simulated spectra over experimental spectrum.

Assignment of the three experimental chemical shifts close to 130 p.p.m. to C (1), C (3) or C (11) on the basis of either carbon electron densities or additivity rules would be uncertain since they do not necessarily correspond to the environment of I in chloroform solution. While selective proton decoupling of H (11) might enable C (11) to be assigned, this technique would be difficult for C (1) and C (3) because of the closeness of their proton chemical shifts. For two benzo[b] oxepines, V and VI, ^{13}C NMR shifts at about 121 p.p.m., not assigned by Günther and Jikeli [16], can probably be assigned to C (4) from analogy with the ^{13}C data for I.

3. SOLVENT, CONCENTRATION AND TEMPERATURE EFFECTS: RESULTS AND DISCUSSION

3.1. Solvent and Concentration Shifts in Dibenz[b,f]oxepine.

With previously derived chemical shifts and coupling constants as starting parameters in LAOCOON, 220 MHz ^{1}H spectra of I were solved at several temperatures and concentrations in chloroform-d_1 and carbon disulphide. In the iterative processes, H (11) chemical shifts and inter-ring coupling constants to H (11) were excluded; the four aromatic chemical shifts and six coupling constants were allowed to vary independently to r.m.s. errors generally between 0.03 and 0.04 Hz. Typically,

TABLE III

Experimental ^{13}C shifts (δ/p.p.m. downfield from T.M.S.) and relative peak areas, A, of dibenz[*b,f*]oxepine (I) and probable assignments, with additivity-rule shift predictions and SCF-calculated carbon electron densities; assignments of experimental shifts in benzo[*b*]oxepine (V) and 10-propylbenzo[*b*]oxepine (VI) [16].

Experimental δ_I	A	Probable[a] Assignment		Additivity-rule predicted δ	Shifts in related oxepines[b] δ_V	δ_{VI}
157.5	55	C (13)	Q	155.5	156.3	156.0
130.6	99	C (12)	Q	129.0	131.3	131.1
121.4	712	C (4)	P	121.5	120.9[c]	120.5[c]
124.8	869	C (2)	P	125.5	124.2	124.1
129.3	796	C (1)	P	128.5	126.5	125.6
129.8	826	C (3)	P	130.0	128.8	128.8
130.1	1000	C (11)	P		129.7	128.9
		C (10)			132.5	145.2
		C (14)			146.7	146.1
		C (15)			114.6	115.3

[a] P,Q denote Protonated, Quaternary; [b] Shifts from reference [16]; [c] assignments by comparison with shifts in I.

30 to 35 transitions with intensities above 0.05 could be assigned in each spectrum. Table IV shows the concentration gradients of ^1H shifts in I for (a) chloroform-d_1 and (b) carbon disulphide, together with shifts at infinite dilution. When the solvent shifts from carbon disulphide to chloroform, which are in the sequence H (4) > H (11) > H (1) > H (2) > H (3), are treated in LISCA as if they emanated from a point dipole, there is a suggestion of a weak dipolar complex or cluster of chloroform

TABLE IV

Dibenz[*b*,*f*]oxepine (I): ^{1}H shifts (from 220 MHz spectra), δ/p.p.m., extrapolated to infinite dilution, with corresponding solvent shifts.

Proton	Solvent $CDCl_3$ at 299 K		Solvent CS_2 at 300 K		Solvent shift	
	δ/p.p.m.	Dilution shift p.p.m. $mole^{-1}$	δ/p.p.m.	Dilution shift p.p.m. $mole^{-1}$%	$\Delta = \delta_{CDCl_3} - \delta_{CS_2}$	Ratio $\Delta\delta/\Delta\delta_3$
H(1)	7.165	-0.0098	7.014	-0.0092	0.150	1.25
H(2)	7.109	-0.0091	6.988	-0.0096	0.122	1.02
H(3)	7.287	-0.0090	7.167	-0.0091	0.120	1.00
H(4)	7.177	-0.0049	6.982	-0.0017	0.195	1.62
H(11)	6.708	-0.0096	6.550	-0.0091	0.158	1.32
$CHCl_3$	7.264	-0.0184				

molecules between 3 and 4 Å from the oxygen in an "axial" direction. Increased polarisation induced in the oxygen bonds of chloroform (compared with carbon disulphide) will increase the electric field associated with the polar oxygen; consequent modification of the electron distribution in the conjugated bonds may further deshield the protons and so give rise to solvent shifts. Although III [10] complexes axially and equatorially with Eu(fod)$_3$, I does not appear to form such a complex, presumably because access to the ring oxygen is hindered by H(4) and H(6).

The dilution shift of H(4) is much smaller than those of the other hydrogens in I (Table IV), as is also the case for III. The negative dilution shift of -0.0184 p.p.m. mole^{-1}% for chloroform protons in a chloroform solution of I (the corresponding values for chloroform solutions of III and IV are 0.0060 and -0.0247 p.p.m. mole^{-1}%, respectively) presumably results from the balance between shielding by a π-electron complex and deshielding by solvent-solute intermolecular hydrogen-bond formation. If there is solute-solute association in I, close similarity of dilution shifts for H(1), H(2), H(3) and H(11) suggests rather random interconversion of conformations; any solute-solute induced dipolar deshielding will be greatest on H(4), close to the oxygen.

3.2. Effect of Temperature on Dibenz[*b*,*f*]oxepine shifts.

Negative temperature coefficients for shifts of protons of I in chloroform-d_1 (Table V) suggest that the deshielding arising from changes in dielectric constant of the solvent (6.76 and 4.81 at 213 and 293 K, respectively, for chloroform) overrides the increased shielding due to closer solute association that is likely to occur with increasing temperature. The sequence of (dielectric) solvent shifts from carbon disulphide to chloroform-d_1, H(4)>H(11)∿H(1)>H(2)∿H(3), is closer to that of the temperature shifts, H(11)∿H(4)>H(1)∿H(2)∿H(3), than is the dilution-shift sequence, H(1)∿H(11)∿H(2)∿H(3)>>H(4). Evidently, with decreasing temperatures the hydrogen bond from chloroform to the oxygen of I may strengthen as the dielectric constant increases (the negative temperature coefficient in Table V contrasts with the small positive shift of 0.002 p.p.m. K^{-1} that chloroform displays as a pure solvent), but solute-solute association shows little change.

An alternative interpretation of the differential effects among aromatic hydrogens in I has been put forward by Mr. C.W. Haigh in terms of paramagnetic currents in the central ring. While such currents in markedly non-planar neutral compounds I and II would be much smaller than the (extremely large) paramagnetic ring currents in isoelectronic planar negative anions such as the nitranion of 5H-dibenz[*b*,*f*]azepine [17], for which the chemical shifts are at much higher fields than those in I and II, they might be sufficient to bring H(1) and H(4) upfield of H(2), leaving the position of H(3) to be accounted for by electronic effects. Moreover, the enhanced upfield shifts of H(1), H(4) and H(5) when the solvent is changed from chloroform to

TABLE V

Dibenz[*b*,*f*]oxepine: ^{1}H chemical shifts δ/p.p.m., from 0.12 mol dm^{-3} $CDCl_3$ solution extrapolated to 273 K, and temperature coefficients, θ, from measurements in the range 224-322 K.

Proton	δ/p.p.m.	$\theta/10^{3}$ p.p.m. K^{-1}
H (1)	7.167	-0.79
H (2)	7.113	-0.78
H (3)	7.290	-0.75
H (4)	7.191	-0.89
H (11)	6.716	-0.97
$C\mathit{H}Cl_3$	7.237	-0.32

carbon disulphide could have the same origin. However, C-C bond lengths of the central rings in I [6] and II [7] vary appreciably.

3.3. ^{1}H-^{1}H Coupling Constants.

No influence of solvent or concentration on H...H coupling constants, *J*, of I and II (Table VI) (and related compounds III and IV) was detected; values of *J* for I in Table VI are determined from 16 independent measurements. The mean LAOCOON error was appreciably smaller than the experimental standard deviations, which were much the same for all coupling constants in all four compounds.

Overall, there is a strong similarity between the H...H coupling constants in I, II and IV, with means $J_{1,2}$ ∿7.60, $J_{2,3}$ ∿7.39, $J_{2,4}$ ∿1.19, $J_{1,4}$ ∿0.34, $J_{4,11}$ ∿0.3$_0$ and $J_{1,11}$ ∿0.1$_0$ Hz, while $J_{3,4}$ is the same (7.8 Hz) for II and III. Mean *ortho* coupling constants for I-IV follow the sequence $J_{3,4}$ = 7.92 > $J_{1,2}$ = 7.65 > $J_{2,3}$ = 7.37 Hz, which is not unexpected from the evidence of coupling constants and π-bond-order calculations in polynuclear hydrocarbons [18], although the parallel between *J* values and bond lengths is not close (Table VI). Evidently, the bonds in the aromatic rings are not fully equivalent, as is shown by the bond lengths in the crystal structures of I [6] and II [7]: C (1) -C (2) and C (3) -C (4) are slightly shorter than in benzene, while C (2) -C (3) is longer, so that some π-electron density may be donated to the central ring 1.33Å. Apart from C (10) -C (11) of 1.33 Å, lengths of C-C bonds in the central ring in the crystal structure of I are dissimilar:

TABLE VI

Ortho and other $^{1}H...^{1}H$ coupling constants (J/Hz) (with experimental standard deviations[a] in parentheses) in I and II, together with some C-C bond lengths (/Å) observed in crystal structure (with apparent e.s.d.s in parentheses) and calculated [18].

	I			II		
H...H	J[b]	C-C bond lengths calculated	C-C bond lengths observed [6]	J	C-C bond lengths calculated	C-C bond lengths observed [7]
1,2	7.60(6)	1.385	1.369(8)	7.62(2)	1.385	1.372(8)
2,3	7.39(6)	1.391	1.392(8)	7.38(9)	1.391	1.374(8)
3,4	8.08(7)	1.373	1.385(7)	7.80(2)	1.380	1.387(7)
1,3	1.70(4)	-	-	1.57(9)	-	-
1,4	0.33(6)	-	-	0.31(4)	-	-
1,5	-	-	-	0.30[c]	-	-
1,11	0.10[c]	-	-	not detected	-	-
2,4	1.20(6)	-	-	1.19(3)	-	-
4,11	0.30[c]	-	-	0.30[c]	-	-

[a] LAOCOON probable errors for I are 0.02 Hz.
[b] No couplings were detected for $J_{2,11}$ or for $J_{3,11}$ or (in II) for $J_{2,5}$, $J_{3,5}$, $J_{5,4}$, or $J_{5,11}$.
[c] Couplings from line broadening; errors estimated at 0.05 Hz.

C (14-15) = 1.39 Å and C (15)-C (10) = 1.47 Å; corresponding lengths are almost identical in II. Also, in the crystal structures of I and II, H (1) deviates on the opposite side from H (2-4) from the mean plane through the benzene-ring carbon atoms; with H (1)-C (1)-C (3)-H (3) dihedral angles 24° and 9°, respectively, $J_{1,3}$ is 1.70 in I and 1.57 Hz in II, whereas $J_{2,4}$ is 1.20 and 1.19 Hz in I and II when the dihedral angles H (2)-C (2)-C (4)-H (4) are zero.

Consistently with the non-planarity of the four tricyclic compounds in solution, the *epi* long-range couplings $^5J_{4,11}$ in I-IV and $^5J_{1,5}$ in II are all shorter at 0.3 Hz than the 0.4-0.8 Hz usual for *epi* ^5J in planar polynuclears [19,20]. Non-planar pathways may also be responsible for the smaller *peri* $J_{1,11}$ of 0.1 Hz in I-IV than in substituted naphthalenes (0.2 - 0.5 Hz) and anthracenes (0.4 - 0.6 Hz) [20].

4. SUMMARY

The ^{1}H 220 MHz high-resolution NMR spectra of dibenz[*b,f*]oxepine (I) and dibenz[*b,f*]azepine (II) have been solved by means of the LAOCOON III program. NMR analyses in carbon-disulphide and, for I, in chloroform-d_1 solution at temperatures 224-322 K appear to be consistent with the shape found in the crystal structure. Chemical shifts are such that the central ring may carry a small paramagnetic ring current, while shift changes with solvent suggest weak clustering of solvent $CDCl_3$ near the oxygen of I. *Ortho*coupling constants $J_{1,2}$, $J_{2,3}$, and $J_{3,4}$ are, respectively, close to 7.6, 7.4 and 7.9 Hz in I, II and related compounds.

ACKNOWLEDGEMENTS

The authors thank Dr. P.M.G. Bavin (SKF LABORATORIES, Ltd) for a sample of dibenz[*b,f*]oxepine, Mr. C.W. Haigh for comments on ring currents, Professor C. Maclean and Dr. A. Geiren for communication of results in theses, and Dr. R.A. Spragg (Perkin-Elmer, Ltd) for 220 MHz ^{1}H decoupling experiments. They are also grateful to the Science Research Council (CASE) and B.S.C. (Chemicals) Ltd. for financial support and to the S.R.C. for access to ^{13}C and 220 MHz ^{1}H NMR facilities at P.C.M.U., Harwell.

REFERENCES

1. Zirkle, C.L., and Kaiser, C.: in M. Gordon (ed) Psychopharmacological Agents, Academic, New York, Vol 3, 1974, p.39.

2. Kricka, L.J., and Ledwith, A.: Chem. Rev. 74, 101 (1974).

3. Seidlova, V., Pelz, K., Adlerová, Jirkowský, I., Metysová, J., and Protiva, M.: Collect. Czech. Chem. Commun., 34, 2258 (1969).

4. Bergmann, E.D., and Aizenshtat, Z.: in E.D. Bergmann and B. Pullman (eds), Quantum Aspects of Heterocyclic Compounds in Chemistry (Jerusalem Symposia on Quantum Chem. and Biochem., II), Israel Academy of Sciences and Humanities, 1970, p.349.

5. Coscia, L., Causa, P., and Giuliani, E.: Arzneim.-Forsch., 25, 1261 (1975).

6. Drake, J.A.G., and Jones, D.W.: J. Pharm. Pharmac., 29, 303 (1977).

7. Geiren, A., Hoppe, W., Schäffer, J.B., and Wimmer, O.: Unpublished measurements.

8. Güsten, H., Klasinc, L., Tóth, T., and Knop, J.V.: J. Electron Spectros., 8, 417 (1976).

9. Tóth, T., and Klasinc, L.: Z. Naturforschung, 29A, 1371 (1974).

10. Drake, J.A.G., and Jones, D.W.: Unpublished measurements.

11. Anet, F.A.L., and Bavin, P.M.G.: Canad. J. Chem., 35, 1084 (1957).

12. Bavin, P.M.G., Bartle, K.D., and Jones, D.W.: J. Heterocyclic Chem., 5, 327(1968).

13. Rondeau, R.E., Rosenberg, H.M., and Dunbar, D.J.: J. Mol. Spectrosc., 29, 305 (1969).

14. Pople, J.A., and Beveridge, D.L.: Approximate Molecular Orbital Theory, McGraw-Hill, New York, 1970, Appendix A, p.163.

15. Wehrli, F.W., and Wirthlin, T.: Interpretation of Carbon-13 NMR Spectra, Heyden, London, 1976, p.47.

16. Günther, H., and Jikeli, G.: Chem. Ber., 106, 1863 (1973).

17. Vos, H.W., Bakker, Y.W., Maclean, C., and Velthorst, N.H.: Chem. Phys. Letters, 25, 80 (1974).

18. Bartle, K.D., Jones, D.W., and Matthews, R.S.: J. Mol. Structure, 4, 445 (1969).

19. Landis, P.S., and Givens, E.N.: Bull. N. J. Acad. Sci., 7, 15 (1972).

20. Bartle, K.D., Jones, D.W., and Matthews, R.S.: Rev. Pure Appl. Chem., 19, 191 (1969).

DISCUSSION

Mallion: You were very properly hesitant about describing the magnetic properties you were discussing in these non-planar molecules as being due to "ring-current" effects since the "ring-current" concept relies on the ability to talk of electrons as being 'α' - or 'π'-type. However, if non-planarity between adjacent bonds is mild, so that the distinctions 'σ' and 'π' do have an approximate local validity, it is still legitimate to attribute some magnetic effects to inter-atomic motion of those electrons which, in a planar molecule, would rigorously be described as "π"- even though the non-planarity between well-separated bonds may in fact be very large (See Haigh, C.W. & Mallion, R.B. : 1971, Molec. Phys., 22, 955).

Could you therefore give any indication of the maximum dihedral angle between the "$2p_z$" - orbitals on adjacent carbon-atoms in these molecules ? Is it greater or less than that in the helicene hydrocarbons ?

D.W. Jones: I think that the angular deviations in the benzene rings of dibenz [b,f] oxepin are smaller than in the helicenes. In the crystal structure of dibenz [b,f] oxepin, the biggest dihedral angle between C-H bonds attached to adjacent carbon atoms in the benzene ring is 18° across C(1)-C(2). The corresponding angles across bonds C(1)-C(12)-C(13)-C(4) close to the seven-membered ring are 10-12°, while the dihedral angles across C(2)-C(3) and C(3)-C(4) are both very small.

APPENDIX. DISCUSSION DOCUMENT ON NOMENCLATURE

1. CONFORMATIONS OF BIOPOLYMERS

2. NOTATION FOR DESCRIPTIONS OF CONFORMATIONS

3. NOTATIONS FOR NMR PARAMETERS USED IN ANALYSIS OF NUCLEOTIDES

David B. Davies,
Department of Chemistry, Birkbeck College,
Malet Street, London WC1E 7HX, England

1. NOMENCLATURE FOR CONFORMATIONS OF BIOPOLYMERS

Abbreviations and Symbols used to describe conformations of biopolymers are necessary so that different conformations may be specified unequivocally and in a convenient manner. The IUPAC Recommendations (1971) for description of polypeptide (PP) conformations are published whereas rules for polynucleotides (PN) and polysaccharides (PS) are under active discussion. There is scope for confusion if the present systems are adopted. For example, the descriptions of conformations for PP, PN and PS involve different atoms used to define torsion angles (ie, main chain for PP and PN whereas PS involves *H* atoms), different magnitudes of angles (ie, 0 - 360 for PN and 0 ± 180 for PP and PS) and different ways of defining the torsion angle (PP and PS are similar but PN is different). A further complication results from the same symbols (ϕ, ψ, ω) being used for both PP and PS nomenclature whereas the most widely used nomenclature for PN (Sundaralingam *et al*, 1973) involves not only ϕ, ψ and ω but also ϕ', ψ' and ω' in the same nucleotidyl unit. This document is presented for discussion so that the opportunity for standardising the nomenclature for these three classes of biopolymers is not missed. A scheme is proposed which would involve changes in PN and PS nomenclature but leave PP nomenclature unchanged.

Nomenclature for polypeptides is well established (IUPAC, 1971) and good reasons are needed if present systems used for PN (Sundaralingam *et al*, 1973) and PS (Marchessault *et al*, 1975) are to be changed. A sufficiently compelling reason for change is the rationalisation of both PN and PS with PP nomenclature (as outlined below) whilst retaining those features unique to each biomolecular system.

(i) An unambiguous nomenclature for PN, PS and PP is needed which can be extended to glycoproteins, nucleoproteins, polysaccharides with phosphate linkages, nucleotide-peptide interactions, etc. It is suggested that different Greek letters be used for PP (ϕ, ψ, ω),

B. Pullman (ed.), Nuclear Magnetic Resonance Spectroscopy in Molecular Biology, 509-516.

Chain direction →

(i) Polypeptides

				Angles
	-CO-NH — CH (R, χ) — CO — NH — CH (R) — CH-CO-			0 ± 180
a	φ	ψ	ω	

(ii) Polynucleotides

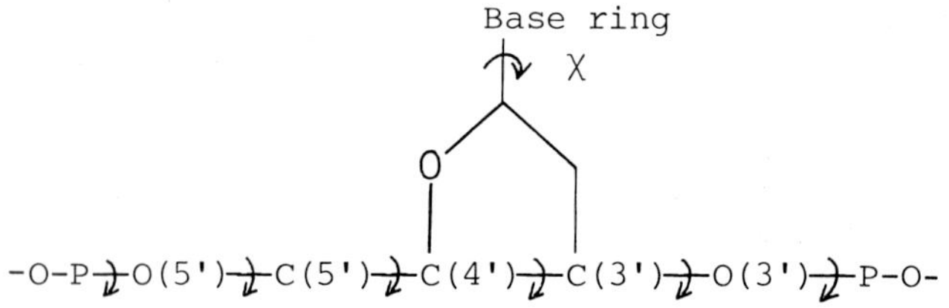

Base ring
χ
O
-O-P-O(5')-C(5')-C(4')-C(3')-O(3')-P-O-

0 - 360

b	ω	φ	ψ	ψ'	φ'	ω'
c	γ	δ	ε	ζ	α	β
d	β	γ	δ	ε	ζ	α
e	ψ	θ	ε	σ	ω	φ
This work (and *c*)	α	β	γ	δ	ε	ζ

(iii) Polysaccharides

eg,

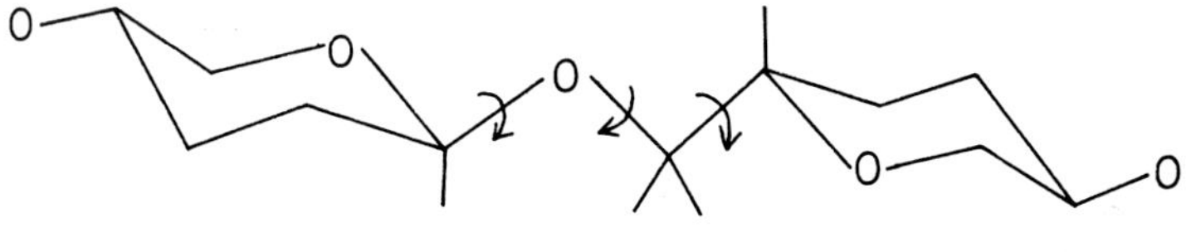

f	φ	ψ	ω	(*H* atoms)	0 ± 180
This work	κ	λ	μ	(Main chain)	0 - 360

Figure 1 Nomenclature for descriptions of Conformations of Biopolymers

a IUPAC, 1971; *b* Sundaralingam *et al*, 1973; *c* Seeman *et al*, 1976; *d* Kim, 1976; *e* Arnott, 1970; *f* Marchessault, 1975.

PN (α - ζ, see note (iv) below) and PS (κ, λ, μ or other symbols) as shown in Figure 1.

(ii) A consistent method of determining torsion angles should be used. The one established for polypeptides (IUPAC, 1971) is recommended, ie,

Main chain atoms define the bonds,
Zero angle defined by atoms in synplanar arrangements,
Magnitude of angle determined by rotation of front (or back) bond being rotated clockwise (+) or anticlockwise (-) with respect to the back (or front) bond in order that it may eclipse the bond to the back (or front),
Angles 0 - 360 unless 0 ± 180 is chosen for symmetry reasons (eg, PP).

(iii) Retention of as much common ground as possible between related nomenclature systems is desirable, eg,

Bond angles τ
General angle θ
Endocyclic torsion angles of sugar rings denoted by ν (PN and PS)
Exocyclic group torsion angle denoted by χ.

(iv) Polynucleotides. Different systems of nomenclature for polynucleotides (Arnott, 1970; Kim, 1976; Seeman *et al*, 1976; Sundaralingam *et al*, 1973) are summarised in Figure 1(ii). The scheme suggested in this work favours the α - ζ nomenclature in order to avoid confusion with PP nomenclature though choice of starting position of the nucleotidyl unit varies (Seeman *et al*, 1976; Kim, 1976). As the phosphate group is a unique feature of polynucleotide chains, it is suggested that the α - ζ nomenclature starts at the phosphate group and proceeds along the chain whose direction can be defined as 3' $\longrightarrow$ 5' *through the phosphate* as well as 5' $\longrightarrow$ 3' *through the sugar ring*. This scheme which is shown in Figure 1(ii) was provided as one of the alternatives by Seeman *et al* (1976).

2. NOTATION FOR STAGGERED CONFORMATIONS

The conformational properties of many single bonds in molecules are determined by analysis of NMR parameters in terms of the time-average of contributions from the three staggered conformations for rotation about the bond, eg, side chains of PP, backbone O(5')-C(5') and C(5')-C(4') bonds of PN and main chain bonds of PS. The terms *gauche* and *trans* are not suitable for descriptions of conformations and have lead to confusion in the designation of particular conformations for polynucleotides, eg, staggered conformations for O(5')-C(5') bonds may be designated g^+, t, g^- (X-ray terminology), *g'g'*, *g't'*, *t'g'* (using ^1H NMR measurements) and *g* and *t* (using ^{13}C NMR measurements).

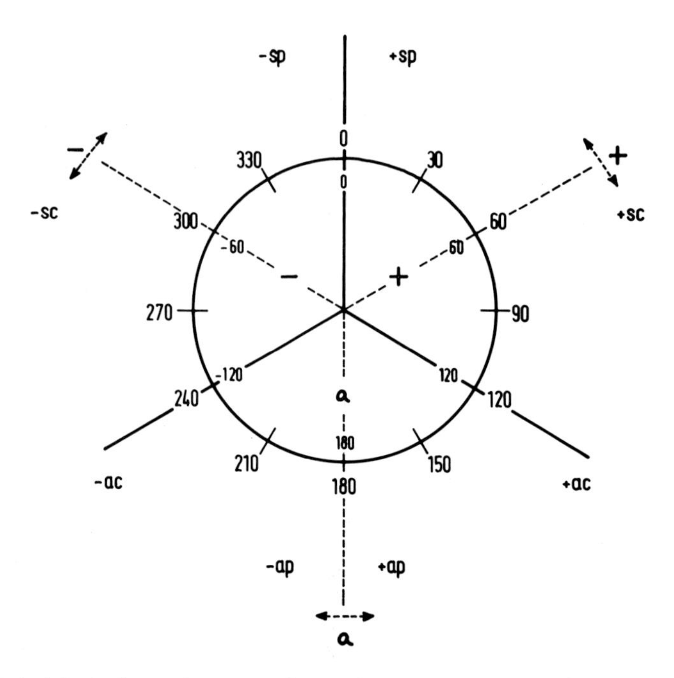

Figure 2 Notation for staggered conformations +(60°, +sc) a(180°, ap) and -(300°, -sc) based on the Klyne-Prelog (1960) description of conformation for single bonds.
NB : sp (syn-periplanar), sc (syn-clinal), ac (anti-clinal) and ap (anti-periplanar)

A conformational notation is needed to provide an unequivocal description of the particular conformational region of interest and to provide a convenient notation for NMR parameters. The following conformational notation based on the Klyne-Prelog (1960) description of conformation shown in Figure 2 has been suggested (Davies, 1978).

ie, Bonds are designated by the appropriate Greek symbol, eg, θ

Specific conformations are denoted by the magnitude of the angle, eg, $\theta(79)$ or θ_{79}.

Staggered conformations are denoted by +(60°, +sc), a (180°, ap) and -(300°, -sc) conformational regions.

NMR parameters corresponding to these conformations may be designated by the same notation, eg, J_+, J_a, J_- and δ_+, δ_a, δ_-.

This conformational notation may be used for any single bond. The terms *gauche* and *trans* may still be used to denote spin-coupling relationships ie, J_g and J_t.

Table 1 Staggered conformations for nucleotide backbone bonds

Bond Conformation	O(5')-C(5') 60°	180°	300°	C(5')-C(4') 60°	180°	300°
^{1}H NMR	g't'	g'g'	t'g'	gg	gt	tg
^{13}C NMR	g	t	g			
X-ray	g^+	t	g^-	g^+	t	g^-
Davies (1978)	ϕ_+	ϕ_a	ϕ_-	ψ_+	ψ_a	ψ_-
This work (α - ζ)	β_+	β_a	β_-	γ_+	γ_a	γ_-

Examples are shown for polynucleotides :

(i) Staggered conformations for backbone bonds are listed in Table 1 together with various notations used to describe different rotamers.

The conformational description suggested in this work is succinct and unequivocal. The use of the notation for analysis of NMR parameters for each bond (eg, J_+, J_a, J_-) has been developed in a recent review (Davies, 1978). The three staggered conformations for O(5')-C(5') and C(5')-C(4') bonds are shown diagrammatically in Figures 3 and 4 respectively.

(ii) Other combinations of conformations may be specified uniquely eg, preferred conformations for a nucleotidyl unit involved in stacked conformations is χ_a,^{3}E, gg, g'g' (Ezra *et al*, 1977). This conformational unit may be specified unequivocally and succinctly as χ_a, N, ϕ_a, ψ_+ (using the nomenclature of Sundaralingam *et al*, 1973) or χ_a, N, β_a, γ_+ (using the nomenclature suggested in this work). In the latter example the pseudorotational description of sugar ring conforma-

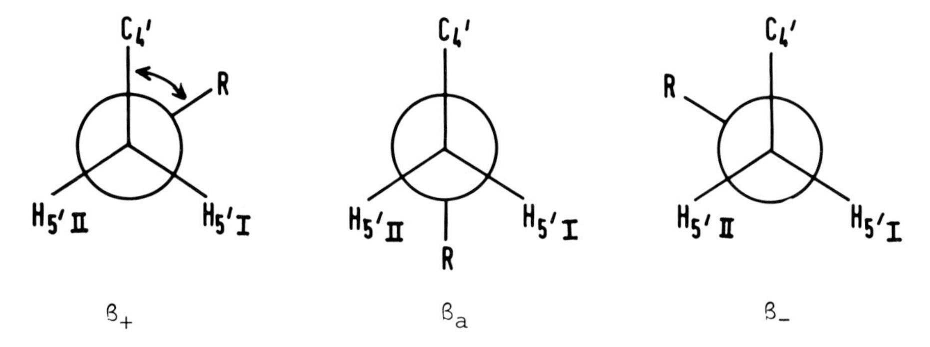

Figure 3 Notation for classical staggered rotamers for O(5')-C(5') bonds, β (this work), including designation of non-equivalent methylene protons (this work).

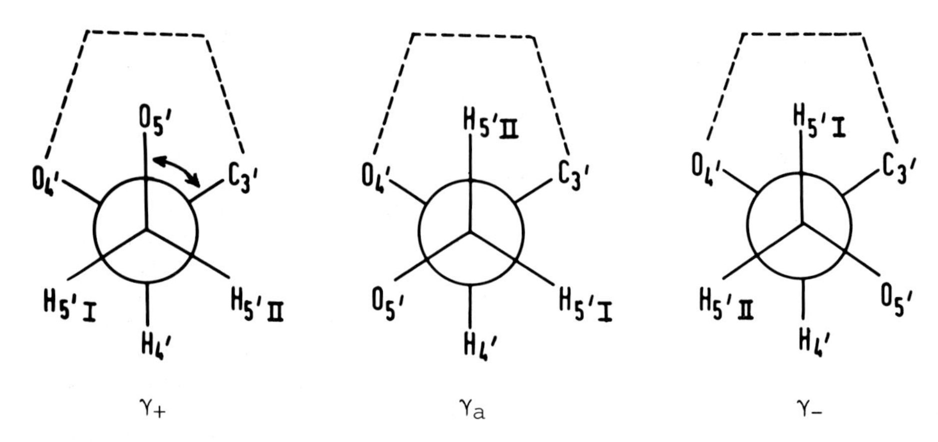

Figure 4 Notation for classical staggered rotamers for C(5')-C(4') bonds, γ (this work) including designation of non-equivalent methylene protons (this work).

tions (Altona and Sundaralingam, 1972; 1973) is preferred over the envelope (E) and twist (T) conformational description for ribofuranosyl rings in solution.

(iii) Stacked conformations have been designated by conformations about O(3')-P and P-O(5') bonds, eg, g^-g^- (RH) and g^+g^+ (LH) stacks (Kim *et al*, 1973). In this terminology the sequence is implied. Use of the Greek symbol to denote a particular bond allows the stacked conformation to be specified unequivocally (Davies, 1978). The α - ζ notation used in this work stresses that the bonds are from adjacent nucleotidyl units.

Table 2 Possible designation of RH and LH stacked conformations, etc.

	RH stack	LH stack
Kim *et al* (1973)	g^-g^-	g^+g^+
Davies (1978)	$\omega'_-\omega_-$	$\omega'_+\omega_+$
This work	$\zeta_-\alpha_-$	$\zeta_+\alpha_+$

3. NOTATION FOR NUCLEOTIDES

3.1 Sugar ring conformation

A number of descriptions of the conformations of β-D-Ribo- and β-D-2'-Deoxyribofuranosyl rings in nucleic acid derivatives have been defined

(Sundaralingam *et al*, 1973). Initial descriptions in terms of the *endo* and *exo* terminology have their corresponding descriptions in terms of the envelope (E) and twist (T) terminology. The pseudo-rotational analysis of five-membered rings (Altona and Sundaralingam, 1972; 1973) provides an unequivocal description of the sugar ring conformations in terms of two parameters, P (the angle of pucker) and τ_m (the degree of pucker). It is found that sugar rings of β-D-Ribo- and β-D-2'-Deoxyribo-nucleoside and -nucleotide derivatives exhibit two narrow conformational regions in the complete pseudorotational cycle (N-type, P=0±90; S-type, P=180±90) which are characterised by ^{N}P *ca* 10 and ^{S}P *ca* 160 (Altona and Sundaralingam, 1972). It is suggested that special terms such as $\bar{N}$ and $\bar{S}$ be used to designate these 'normal' conformational regions. Use of the terms $\bar{N}$ and $\bar{S}$ implies a conformational flexibility for the sugar ring for molecules in solution. For those authors who prefer the approximate analysis of sugar ring conformations in terms of C(2')-*endo* or C(3')-*endo* (ie, ^{2}E or ^{3}E) conformations, it is suggested that the nomenclature is changed to the twist (T) notation to be more in line with results from X-ray crystallography.

3.2 Non-equivalent methylene protons

It is necessary to distinguish between observed signals (usually designated n' for downfield and n" for upfield) and their unequivocal assignment to particular methylene protons, eg, C(5') protons in ribo- and deoxyribo- derivatives and C(2') protons in 2'-deoxyribo derivatives. A notation which was based on the original designation of these protons (Sundaralingam *et al*, 1973) was developed (Davies, 1978) in order to overcome possible confusion in the literature where both observed and assigned C(5') proton signals {and C(2') for deoxyribose rings} have been labelled H(5') and H(5") {and H(2'), H(2")}; for example, the C(5') methylene proton signals were designated $H^{2}_{5'}$ and $H^{1}_{5'}$ and their corresponding spin-coupling constants, $J^{1}_{4'5'}$ and $J^{2}_{4'5'}$. (An example of the use of this numbering scheme is shown in Figures 2 and 3 in the author's contribution to this Symposium.)

An alternative description of these methylene protons has been recommended (Cramer *et al*, 1975). The designation for C(5') methylene protons is H(5'I) and H(5'II) and for C(2') methylene protons is H(2'I) and H(2'II) as shown in Figure 5. The corresponding vicinal proton

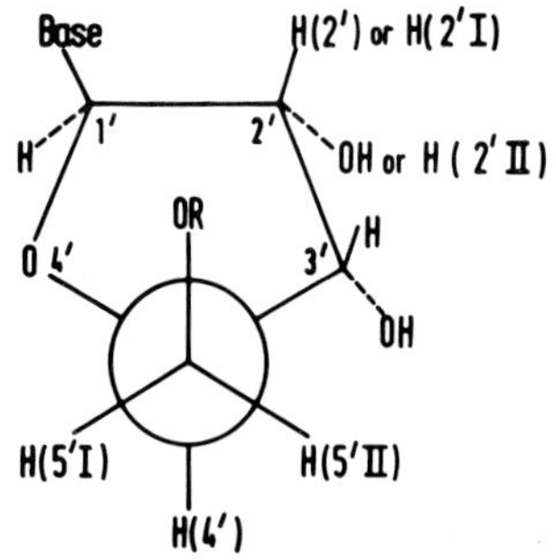

Base
H(2') or H(2'I)
H
1'
2'
OH or H(2'II)
OR
O 4'
3'
H
OH
H(5'I)
H(5'II)
H(4')

Figure 5 Designation of non-equivalent methylene protons in β-D-Ribo(C2'-OH) and β-D-2'-Deoxyribo-Nucleoside derivatives: when viewed along the O(5')-C(5')-C(4') bond, H_i(5'I) is to the right and H_i(5'II) the left; when viewed along the C(1')-C(2')-C(3') bond, H_i(2'I) is to the right and H_i(2'II) is to the left (Cramer *et al*, 1975)

spin coupling constants may be designated J(4'5'I), J(4'5'II) and J(1'2'I), J(1'2'II) when the assignment is unequivocal. Otherwise the terminology based on the observed signals should be used, eg, J(4'5'), J(4'5") and J(1'2'), J(1'2"). The designation of the C(5') methylene protons for each of the three classical staggered conformers for O(5')-C(5') and C(5')-C(4') bonds is shown in Figures 3 and 4 respectively. Notations based on the numbers I and II make the designations of appropriate NMR parameters more clear than those based on the numbers 1 and 2.

NB : A further reason for such a change may be that modern printing technology makes it expensive (a hand operation) to print a superscript and subscript in a single vertical space.

3.3 Oligonucleotides

Extension of NMR notations to oligonucleotides is necessary. Similar parameters for different nucleotidyl units can be differentiated for $n \leqslant 3$, eg, for ApApA the three units can be differentiated by Ap (or A-) pAp (or -A-) and pA (-A) and the corresponding $J_{1'2'}$ magnitudes by $J_{1'2'}$(A-), $J_{1'2'}$(-A-) and $J_{1'2'}$(-A). For $n \geqslant 4$ and separate units of large molecules such as tRNA it is necessary to specify each unit, eg, $J_{1'2'}$ magnitudes in $(A)_n$ might be denoted by $J_{1'2'}(A_i)$ where A_i represents the *i*th residue in a series.

References

Altona, C. and Sundaralingam, M., 1972, *J.Amer.Chem.Soc.*, 94, 8205
Altona, C. and Sundaralingam, M., 1973, *J.Amer.Chem.Soc.*, 95, 2333
Arnott, S., 1970, in *Progress in Biophysics and Molecular Biology* (Butler, J.A.V. and Noble,D., eds) Pergamon Press, Oxford, Vol.21.
Cramer, F., Gauss, D. and Saenger, W., 1975, document submitted to IUPAC-IUB nomenclature commission
Davies, D. B., 1978, *Progress in NMR spectroscopy* (Emsley, J.W., Feeney, J. and Sutcliffe, L.H., eds) Pergamon Press, Oxford, in Press
Ezra, F. S., Lee, C-H., Kondo, N. S., Danyluk, S. S. and Sarma, R. H., 1977, *Biochemistry*, 16, 1977
IUPAC, 1971, Information Bulletin, Appendices on Tentative Nomenclature, Symbols, Units and Standards, No. 10
eg : *Biochemistry*, 9 (1970) 3471
Kim, S-H., Berman, H. M., Seeman, N. C. and Newton, M. D., 1973, *Acta Crystallogr. Section B*, 29, 703-710
Kim, S-H., 1976, document submitted to IUPAC-IUB nomenclature commission
Klyne, W. and Prelog, V., 1960, *Experentia*, 16, 521
Marchessault, R. H., 1975, document SCBN/PS/7 submitted as IUPAC-IUB Subcommission on Polysaccharide Nomenclature
Seeman, N. C., Rosenberg, J. M., Suddath, F. L., Kim, J. J. P. and Rich, A., 1976, *J.Mol.Biol.*, 104, 142
Sundaralingam, M., Pullman, B., Saenger, W., Sasisekharen, V. and Wilson, H. R., 1973, in *Conformations of Biological Molecules and Polymers* (Bergman, E.D. and Pullman, B., eds) Academic Press, New York pp. 815-820

INDEX OF SUBJECTS

INDEX OF NAMES